W9-CEV-925

ION CHANNELS

Volume 4

ION CHANNELS

Editorial Board:

Phillippe Ascher
Robert L. Barchi
Arthur Brown
Bertil Hille
Lily Y. Jan
Rodolfo R. Llinas
Ricardo Miledi
Erwin Neher
Ole Peterson
Charles F. Stevens
W. Trautwein
Richard W. Tsien
P. N. R. Usherwood

A Continuation Order Plan is available for this series. A continuation order will bring delivery of each new volume immediately upon publication. Volumes are billed only upon actual shipment. For further information please contact the publisher.

ION CHANNELS
Volume 4

Edited by
Toshio Narahashi
Northwestern University Medical School
Chicago, Illinois

PLENUM PRESS • NEW YORK AND LONDON

Library of Congress Catalog Card Number 88-647703

ISBN 0-306-45224-3

© 1996 Plenum Press, New York
A Division of Plenum Publishing Corporation
233 Spring Street, New York, N. Y. 10013

10 9 8 7 6 5 4 3 2 1

All rights reserved

No part of this book may be reproduced, stored in a retrieval system, or transmitted in
any form or by any means, electronic, mechanical, photocopying, microfilming,
recording, or otherwise, without written permission from the Publisher

Printed in the United States of America

CONTRIBUTORS

ROGELIO O. ARELLANO • Laboratory of Cellular and Molecular Neurobiology, Department of Psychobiology, University of California, Irvine, Irvine, California 92717

PASCAL BARBRY • Institute of Molecular and Cellular Pharmacology, CNRS, 06560 Valbonne, France

D. BENKE • Institute of Pharmacology, Federal Institute of Technology (ETH), and University of Zurich, CH-8057 Zurich, Switzerland

J. BENSON • Institute of Pharmacology, Federal Institute of Technology (ETH), and University of Zurich, CH-8057 Zurich, Switzerland

ELIZABETH BLACHLY-DYSON • Vollum Institute, Oregon Health Sciences University, Portland, Oregon 97201

KEVIN P. CAMPBELL • Howard Hughes Medical Institute, Department of Physiology and Biophysics, University of Iowa College of Medicine, Iowa City, Iowa 52242

DAVID O. CARPENTER • Wadsworth Center for Laboratories & Research, New York State Department of Health, Albany, New York 12201, and School of Public Health, State University of New York at Albany, Albany, New York 12203-3727

MARCO COLOMBINI • Department of Zoology, University of Maryland, College Park, Maryland 20742

MICHEL DE WAARD • Howard Hughes Medical Institute, Department of Physiology and Biophysics, University of Iowa College of Medicine, Iowa City, Iowa 52242

JIU PING DING • Department of Anesthesiology, Washington University School of Medicine, St. Louis, Missouri 63110

STEWART R. DURELL • Laboratory of Mathematical Biology, National Cancer Institute, National Institutes of Health, Bethesda, Maryland 20892-5677

MICHAEL FEJTL • Wadsworth Center for Laboratories & Research, New York State Department of Health, Albany, New York 12201, and Institute of Neurophysiology, University of Vienna, A-1090 Vienna, Austria

MICHAEL FORTE • Vollum Institute, Oregon Health Sciences University, Portland, Oregon 97201

J. M. FRITSCHY • Institute of Pharmacology, Federal Institute of Technology (ETH), and University of Zurich, CH-8057 Zurich, Switzerland

CHRISTINA A. GURNETT • Howard Hughes Medical Institute, Department of Physiology and Biophysics, University of Iowa College of Medicine, Iowa City, Iowa 52242

H. ROBERT GUY • Laboratory of Mathematical Biology, National Cancer Institute, National Institutes of Health, Bethesda, Maryland 20892-5677

TORU IDE • Department of Biophysical Engineering, Faculty of Engineering Science, Osaka University, Toyonaka, Osaka 560, Japan

MICHIKI KASAI • Department of Biophysical Engineering, Faculty of Engineering Science, Osaka University, Toyonaka, Osaka 560, Japan

MICHEL LAZDUNSKI • Institute of Molecular and Cellular Pharmacology, CNRS, 06560 Valbonne, France

JON LINDSTROM • Department of Neuroscience, Medical School of the University of Pennsylvania, Philadelphia, Pennsylvania 19104-6074

CHRISTOPHER J. LINGLE • Department of Anesthesiology, Washington University School of Medicine, St. Louis, Missouri 63110

B. LÜSCHER • Institute of Pharmacology, Federal Institute of Technology (ETH), and University of Zurich, CH-8057 Zurich, Switzerland

RICARDO MILEDI • Laboratory of Cellular and Molecular Neurobiology, Department of Psychobiology, University of California, Irvine, Irvine, California 92717

H. MOHLER • Institute of Pharmacology, Federal Institute of Technology (ETH), and University of Zurich, CH-8057 Zurich, Switzerland

MURALI PRAKRIYA • Department of Anesthesiology, Washington University School of Medicine, St. Louis, Missouri 63110

U. RUDOLPH • Institute of Pharmacology, Federal Institute of Technology (ETH), and University of Zurich, CH-8057 Zurich, Switzerland

CHRISTOPHER R. SOLARO • Department of Anesthesiology, Washington University School of Medicine, St. Louis, Missouri 63110

RICHARD M. WOODWARD • Laboratory of Cellular and Molecular Neurobiology, Department of Psychobiology, University of California, Irvine, Irvine, California 92717; *present address*: Acea Pharmaceuticals, Irvine, California 97218

PREFACE

The importance of ion channels in biomedical sciences is increasingly recognized. Technological developments in recent years, especially in the fields of biophysics, electrophysiology, and molecular biology, have made it possible to advance our knowledge of ion channels. Ion channels are now being studied in relation not only to their roles in normal physiological function of excitable cells but also to pharmacology, toxicology, and pathophysiology of cells. It should be noted that the study of ion channels is no longer limited to excitable cells such as nerve and muscle. Cells in any tissues are endowed with ion channels, and study of the role of ion channels in inexcitable cells is actively being pursued.

This volume, the fourth in the series, deals with a variety of ion channels. H. R. Guy and S. R. Durell have used molecular modeling to assimilate the available functional and structural data into three-dimensional conceptions of ion channel proteins. Among the various types of voltage-gated Ca^{2+} channels, there are important structural and functional similarities and differences in subunit-subunit and protein-Ca^{2+} channel interactions. M. De Waard, C. A. Gurnett and K. P. Campbell describe several avenues of research that should provide significant clues about the structural features involved in the biophysical and functional diversity of voltage-gated Ca^{2+} channels. $GABA_A$ receptors are prime examples of receptor heterogeneity in the CNS, comprising at least 15 subunits. H. Mohler, J. M. Fritschy, B. Lüscher, U. Rudolph, J. Benson, and D. Benke give an overview of the structure, cellular localization, pharmacology, and regulation of the major $GABA_A$ receptors in the CNS.

Amiloride-sensitive Na^+ channels are the site of sodium reabsorption by the epithelial lining. P. Barbry and M. Lazdunski describe the molecular properties of the highly Na^+-selective and highly amiloride-sensitive Na^+ channel in relation to its electrophysiological properties. The voltage-dependent anion channel (VDAC) of the mitochondrial outer membrane produces one of the largest aqueous pathways from a single 30-kDa protein. M. Colombini, E. Blachly-Dyson, and M. Forte give the properties of the channel as assayed following reconstitution into planar phospholipid membranes. *Xenopus* oocytes have been used extensively as a system for expressing exogenous mRNA or cRNA and studying the transcribed proteins. However, it is important to recognize that the membrane of the follicular cells contains receptors and ion channels. R. O. Arellano, R. M. Woodward, and R. Miledi address this issue.

Many cells express K^+-selective channels whose activation requires the elevation of the submembrane Ca^{2+} concentration. Thus Ca^{2+}-dependent K^+ channels permit the status of the internal biochemical milieu to influence the electrical activity of a cell. C. J. Lingle, C. R. Solaro, M. Prakriya, and J. P. Ding summarize their work on this type of K^+ channel in adrenal chromaffin cells. Calcium release from the sarcoplasmic reticulum plays a pivotal role in excitation-contraction coupling in muscle cells. M. Kasai and T. Ide give the results obtained by using a variety of methods, including light scattering, permeation of ions and neutral molecules trace, fluorescence quenching, and lipid bilayer studies. Molluscan neurons have been used extensively in the study of ion channels. M. Fejtl and D. O. Carpenter summarize their analyses on voltage- and ligand-gated ion channels at the single-channel level. J. Lindstrom describes recent developments in the molecular structural and functional aspects of muscle and neuronal nicotinic acetyl-choline receptors. The significance of the acetylcholine receptors, particularly their pathophysiology and pharmacology, has been increasingly recognized in recent years.

In conjunction with Volumes 1, 2, and 3 of this series, I hope that this book will serve as a useful reference not only for experts in the field, but also for those who wish to obtain updated knowledge in the area of ion channels.

Toshio Narahashi

Chicago, Illinois

CONTENTS

CHAPTER 2

STRUCTURAL AND FUNCTIONAL DIVERSITY OF VOLTAGE-
ACTIVATED CALCIUM CHANNELS

MICHEL DE WAARD, CHRISTINA A. GURNETT,
and KEVIN P. CAMPBELL

CHAPTER 5

VDAC, A CHANNEL IN THE OUTER MITOCHONDRIAL MEMBRANE

MARCO COLOMBINI, ELIZABETH BLACHLY-DYSON,
and MICHAEL FORTE

CHAPTER 6

ION CHANNELS AND MEMBRANE RECEPTORS IN FOLLICLE-
ENCLOSED *XENOPUS* OOCYTES

ROGELIO O. ARELLANO, RICHARD M. WOODWARD,
and RICARDO MILEDI

CHAPTER 7

CALCIUM-ACTIVATED POTASSIUM CHANNELS IN ADRENAL
CHROMAFFIN CELLS

CHRISTOPHER J. LINGLE, CHRISTOPHER R. SOLARO,
MURALI PRAKRIYA, and JIU PING DING

CHAPTER 8

REGULATION OF CALCIUM RELEASE CHANNEL IN
SARCOPLASMIC RETICULUM

MICHIKI KASAI and TORU IDE

CHAPTER 9

SINGLE-CHANNEL STUDIES IN MOLLUSCAN NEURONS

MICHAEL FEJTL and DAVID O. CARPENTER

CHAPTER 10

NEURONAL NICOTINIC ACETYLCHOLINE RECEPTORS

JON LINDSTROM

CHAPTER 1

DEVELOPING THREE-DIMENSIONAL MODELS OF ION CHANNEL PROTEINS

H. ROBERT GUY and STEWART R. DURELL

1. INTRODUCTION

Ion channel–forming proteins are an exceptionally important class of membrane proteins that are involved in many complex biological processes. Within this class there is a wide variety of structures and functions. For example, the types of proteins reviewed here include antibiotics, toxins, molecular sieves, a shark repellent, and the voltage- and ligand-gated types of channels responsible for electrical conduction in the nervous system. The sizes of these proteins range from about 20 amino acid residues, for some of the antibiotics and toxins, to about 2000 residues for the pore-forming subunit of voltage-gated sodium and calcium channels. While the transmembrane region of many membrane proteins appears to be composed primarily of α-helices, some channel proteins, such as those in the outer membranes of bacteria and mitochondria, are primarily β-sheet structures. Despite some advances, much remains to be learned about the full three-dimensional structures of membrane proteins. Unfortunately, efforts in this direction have been severely hampered by difficulties in isolation, purification, and crystallization. To date, only a few integral membrane proteins have yielded to structure determination in native-like environments. On the other hand, the explosive developments in cell biology, molecular biology, and genetic engineering have led to the discovery of many ion channel proteins and the identification of their genes in the genome. By means of mutation, cloning, and expression in oocyte membranes, many groups have been able to investigate the relationship between amino acid sequence and function of these channels. Although more ambiguous than the data obtained from X-ray crystallography or

H. ROBERT GUY and STEWART R. DURELL • Laboratory of Mathematical Biology, National Cancer Institute, National Institutes of Health, Bethesda, Maryland 20892-5677.

Ion Channels, Volume 4, edited by Toshio Narahashi, Plenum Press, New York, 1996.

cryomicroscopyy, these functional studies have provided a wealth of information from which structural information can be deduced.

The purpose of this review is to describe how molecular modeling has been used to assimilate the available functional and scant structural data into three-dimensional conceptions of ion channel proteins. In turn, these models greatly aid in designing new functional-type experiments to further test and refine our understanding of the real protein structures. For example, atomic-scale models provide a picture of the interrelationships of the amino acid residues in space, which is extremely useful for deciding which of the potentially thousands of residues would be worth mutating or deleting.

In this review we also describe the particular strategy that has developed in our lab for modeling large integral membrane proteins, such as voltage-gated ion channels and their homologues. Specifically, we use an iterative, hierarchical approach that is analogous to increasing the resolution, or precision, of the model. For example, the initial phase involves predicting which segments span the membrane and which segments are involved in functional mechanisms such as ion selectivity, gating, and the binding of drugs, toxins, or substrates. If these predictions are tested by experiment then the next phase is to construct a low-resolution three-dimensional model which predicts the relative orientations of the segments and their positions relative to the membrane plane. Similarly, the results of experiments designed to test these models are used to refine the three-dimensional structures by building in the actual atoms of the residues. Ideally, this cycle of modeling and experimentation continues until an accurate representation of the structure emerges.

This approach, and the more detailed theoretical methods described below, has proven very successful in deducing the structure of the voltage-gated ion channels. For example, almost all the first phase predictions for the voltage-gated Na^+ channel have been verified by experiment. This includes the transmembrane topology and the segments responsible for the activation and inactivation gating, ion selectivity, and toxin binding sites. Of particular importance was the prediction that the ion-selective part of the channel is formed by four short hairpin loops that dip into the pore from the extracellular side of the membrane. This was contrary to the prevailing thinking at the time that all intermembrane protein segments are in the form of fully transversal α-helices. Incidently, this assumption was also shattered by the finding that the bacterial porin channels are large β-barrel structures rather than groupings of transmembrane α-helices.

Our current efforts with voltage-gated ion channels are directed toward refining the atomic structure of the ion-selective region of the pore. Although the accuracy of these models is unlikely to equal that obtainable by X-ray crystallography, they may still be of considerable importance to the design of new drugs. In particular, one method involves predicting mutations in the sequence that would affect the binding affinities of known pore-binding drugs and natural toxins. Beyond this, theoretical modeling has much to contribute to the understanding of ion channel proteins in general. For example, even if a structure is available from crystallographic methods, theoretical modeling is often needed for understanding the many dynamic mechanisms by which the ion channel proteins function. These would include conformational changes responsible for opening and closing of the channel, how drugs and toxins alter these mechanisms, and how some natural mutations have led to diseases. Likewise, if structural information is available for a particular channel, then theoretical modeling methods are very useful for building working structures of other homologous sequences in the superfamily. By this means, much can be learned from correlating the changes in sequence, structure, and function. Theoretical principles and modeling techniques can also be used to refine molecular structures from low-resolution physical data. For example, these methods have been very useful for building in the residues of models developed from electron diffraction studies of two-dimensional crystals, which typically only identify the relative locations of the transmembrane α-helices.

Finally, theoretical models can also aid in determining the structure of membrane proteins by more direct physical methods. For example, a three-dimensional model can be used to engineer disulfide bonds into the structure, which might enable the protein to withstand the harsh treatments required for membrane extraction and crystallization. Another approach would be to synthesize a water-soluble analog of the protein by substituting polar residues for the nonpolar residues predicted to be in contact with the hydrophobic core of the membrane. In this case, conventional crystallographic and/or spectroscopic methods could be used to determine the structure of the protein.

2. PHYSICAL FACTORS INFLUENCING MEMBRANE PROTEIN STRUCTURE AND MODELING

While considerable progress has been made in the pursuit of predicting the structure of a protein from its amino acid sequence, it is safe to say that considerable work remains to be done on the "protein folding problem." While automated, energy-based methods have been developed to predict the folding pattern of small water-soluble proteins, larger proteins generally require the existence of a homologous protein with a known structure to produce a model with confidence. In this sense, the prediction problem is greater for integral membrane proteins, for which only a few crystal structures have been determined. And unfortunately, automated methods have only begun to be developed for integral membrane proteins of any size (Taylor *et al.*, 1994).

There are a number of reasons why the modeling of membrane proteins is more difficult and less advanced than the work for globular-type, water-soluble proteins. Perhaps the biggest problem again is the lack of a database of known structures from which statistically based "rules" can be developed to guide the structure prediction process. This is in contrast to the aqueous proteins, for which the rapidly increasing database of structures has been used to construct many types of mean-field potential energy functions and parameters for modeling proteins on various levels of resolution and organization. Another confounding problem is the complex nature of the membrane environment. Unlike the aqueous proteins, which basically have just a nonpolar environment on the inside and a polar environment on the outside, the integral membrane protein channels can have as many as five distinct environments to consider. These include the very hydrophobic membrane core, the somewhat less nonpolar interior of the protein, the polar inside of the channel containing motionally restricted waters and polar residues, the polar head group layers of the membrane lipids, and the hydrophilic aqueous environments of the intracellular and extracellular spaces. This variety of environments complicates the process of deciding where to position residues of different polarity types in the model. For example, the most energetically favorable location for amphipathic side chains (such as arginine, lysine, tyrosine, and tryptophan) may actually be at an interface between polar and nonpolar phases rather than in a strictly hydrophilic environment. In such a situation, the residues would likely be oriented in such a way as to distribute the polar and nonpolar atoms into the appropriate environments (Guy, 1985). As an example of the problem this causes, based on the sequence alone we would find it impossible to predict that the P segments of voltage-gated Na^+, Ca^{2+}, and K^+ channels (see Sections 7 and 8) and the internal loop of porins (see Section 3.3.2) actually comprise the lining of the pore rather than just extend into the hydrophilic extracellular or intracellular space.

Conversely, the unusual nature of the membrane environment also provides some advantages for modeling the embedded proteins. As seen in the crystal structures of the photosynthetic reaction center and bacteriorhodopsin, the hydrophobic environment of the transmembrane region appears to impose a simpler order on the structural elements than does the solvent for the typical water-soluble protein. For example, the photosynthetic reaction center consists of an extracellular subunit, two homologous subunits that each have five transmembrane α-helices, and a fourth subunit with a large intracellular domain and one transmembrane α-helix. The portions of the reaction center outside of the membrane are similar to the average water-soluble protein; that is, the majority of residues are not part of either an α-helix or β-sheet, the lengths of the helices and the ways they come in contact vary greatly, and the particular secondary structure elements are difficult to predict from the sequence alone. Conversely, for the portions of the protein in the transmembrane region every residue is uncharged and part of an α-helix (assuming the neutral state for the histidines), and all the adjacent helices pack in either a parallel or antiparallel manner; actually they cross at the small angles predicted by the "knobs-into-holes" (Crick, 1953) and "3-4 ridges into grooves" (Chothia et al., 1981) theories (see Section 3.3.3). The fact that hydrophobic portions of all of the transmembrane helices begin and end at the two membrane surfaces makes them easy to identify by simply searching for sufficiently long hydrophobic stretches in the sequence. The reason for the overwhelming preponderance of regularized α-helix or β-barrel secondary structure observed in the transmembrane regions is that the nonpolar membrane core doesn't provide any other opportunity for the amide groups to form hydrogen bonds. This also simplifies the modeling because it avoids the ambiguities associated with the random coil structure commonly observed in water-soluble proteins.

Another advantage of working on membrane proteins is that there are a number of unique ways to determine aspects of the structure and test the models which are not available for soluble proteins. For example, numerous methods have been developed that exploit the physical barrier of the membrane to determine the transmembrane topology of the protein. In particular, the protein can be exposed on just a single side of the membrane to a wide variety of labeling or proteolytic reagents, after which the identity of the affected segments can be determined experimentally. Antibodies and high-recognition segments engineered into the sequence have also been used to facilitate this process. In addition, sites on the protein that are naturally phosphorylated in the cytoplasm or glycosylated on the extracellular side can also be identified. For the channels, reagents thought to enter and block the pore can be covalently bound to the protein, and the groups to which they bind can be identified (Giraudat *et al.*, 1986; Hucho *et al.*, 1986). In addition, many natural toxins and drugs act from only one side of the membrane, and the residues involved in binding these molecules can be identified by a variety of methods involving antibodies or mutagenesis experiments. Mutagenesis has also been used to identify the residues involved in the gating and conductance properties of channels. But perhaps the most significant advantage to modeling ion channels is the ability to measure the functional properties of a single protein. Single-channel patch-clamp methods have made it possible to not only record the conformational transitions of a single protein associated with gating, but also measure the current flowing through a single-channel molecule. Combined with mutagenesis experiments, this provides an unprecedented way to identify specific residues involved in different functions of the protein. It is also enables determination of specific modes of action such as whether an observed reduction in current is caused by altering the activation or inactivation gating, the conductance properties of the pore, or the amount of proteins that is expressed.

3. MODELING ION CHANNEL PROTEIN STRUCTURES

3.1. Overview

The hierarchical approach we have developed to model ion channels can be thought of as separated into three broad phases. These are:

Phase I: Predict the transmembrane topology, secondary structure, and segments involved in the active sites and functional mechanisms.

Phase II: Predict the general three-dimensional pattern, such as the relative positions and orientations of the transmembrane segments. This also includes predicting contacts of specific residues with other residues, the lipids, and/or the solvent.

Phase III: Build in the actual atoms using computer graphics and simple energy calculations.

Initially, the modeling process proceeds sequentially, starting with the first phase. However, as the resolution, or rather precision, of the model structure increases with the phases, the probability of errors also increases. For example, given a Phase I model of the transmembrane topology there are generally many possible Phase II models of how the structural elements are arranged in three-dimensional space. Likewise, for any given Phase II folding pattern there tend to be many Phase III models delineating specific arrangements of contacts between the atoms. However, despite the greater ambiguity of the Phase III models, they have the advantage of being amenable to analysis by more sophisticated, atom-based computational methods, and by enabling the design of specific residue mutations to test them. For this reason, the modeling becomes a long-range, iterative process, where the results of tests based on the Phase III models might be used to update the Phase I and/or Phase II models.

Different methods are used to develop the models of each phase; likewise, different sets of criteria and experiments are used to judge their viability. The methods and criteria that we commonly employ are described below.

3.2. Phase I: Transmembrane Topology, Functional Mechanisms, and Ligand Binding Sites

The first step in modeling the transmembrane topology of a protein is to predict which segments of the sequence cross the membrane as well as is their secondary structure.

3.2.1. Hydrophobic and Amphipathic Transmembrane α-Helices

Perhaps the most common method used to identify transmembrane helices is to create a hydropathy plot. In its most basic incarnation, the strategy is to identify hydrophobic regions in a sequence approximately 20 residues long, which corresponds to the length of an α-helix that spans the average membrane core. A number of parameter sets have been developed for making the plots, which define the relative hydrophobicity of each residue and its secondary structure "preferences" (Cornette et al., 1987). Unfortunately, however, a parameter set has yet to be developed that can confidently pick out the transmembrane helices of membrane proteins that are not predominantly hydrophobic. This is an especially serious problem for ion channel proteins, which are often modeled as bundles of transmembrane helices that form the water-filled pore on the inside and are in contact with the hydrophobic region of the membrane on the outside (see sections on δ-hemolysin and colicin). The basic hydrophobic algorithm may work for proteins that have only a single transmembrane segment or proteins for which each helix inserts individually into the membrane and then aggregates to form the final structure. Indeed, the latter case has been indicated for an α-helical hairpin of bacteriorhodopsin (Popot et al., 1987). However, for the partially polar or amphipathic transmembrane helices of some ion channels one can envision mechanisms whereby the helices are inserted into the membrane in a preaggregated state that buries and shields the polar residues from making contact with the hydrophobic core (see Section 4). Other limitations of the basic hydropathy analysis are due to the increasingly discredited assumptions that all transmembrane helices completely span the membrane and are roughly orthogonal to the plane.

To extend the analysis of membrane proteins, we and other groups have developed methods to predict the presence of amphipathic helices from the sequences. These are generally based on searching for a periodic pattern of polar and nonpolar residues with the characteristic α-helical pitch of 3.6 residues/turn. However, the problem of determining whether the identified helices span the membrane, lie on the surface, or are fully part of the aqueous domain often still remains. For example, we and other groups initially postulated that a particular amphipathic α-helix that just precedes the last hydrophobic segment of the nicotinic acetylcholine receptor forms the lining of the channel (Guy, 1984; Finer-Moore and Stroud, 1984; Kosower, 1983, 1987). However, subsequent experiments demonstrated instead that this segment resides on the intracellular side of the membrane (Guy and Hucho, 1987).

3.2.2.   Transmembrane β-Strands

Given all the historical emphasis on hydropathy analysis and α-helices, it is ironic that the first (and only, to date) known structure of an intact channel is that of the porins, which are in the outer membrane of gram-negative bacteria. They have 16 (Cowan *et al.*, 1992; Weiss *et al.*, 1991; Kreusch and Schulz, 1994), or for maltoporin (Schirmer *et al.*, 1995), 18 transmembrane β-strands that form a large β-barrel. Also, one or more loop segments dip into the β-barrel and span part of the transmembrane region. Much of the interior of the barrel is filled with water, and there are many hydrophilic side chains which extend into the pore. Methods have been developed to predict the transmembrane topology of the β-barrel (Fischbarg *et al.*, 1993; Forte *et al.*, 1987; Jeanteur *et al.*, 1991). The simplest of these assumes that all segments ten residues long that have hydrophobic residues at every other position are transmembrane strands and that all residues on the exterior of the β-barrel are hydrophobic. Tyrosine and tryptophan are counted as hydrophobic if they occur at the first or last position in the strand. These amphiphilic aromatic side chains are frequently observed to occur in the lipid head group region. It should be noted that assuming that all residues inside the barrel are hydrophilic is not as successful a predictor, due to the fact that the ones covered by the loop or loops tend to be hydrophobic. The more complicated methods also include the use of secondary structure prediction algorithms that are based on the statistical preferences of soluble protein structures (Fischbarg *et al.*, 1993). Unfortunately, the ability of these potentials to distinguish correctly between α-helices and β-strands in membrane proteins has not been established; however, they do appear to correctly predict most of the loop regions between the strands. It should be considered, however, that these methods may work only on those proteins in which all transmembrane segments are β-strands. Although predictions for other types of proteins have not been reported, we suspect that these methods will incorrectly predict the presence of numerous transmembrane β-stands for proteins that only have transmembrane α-helices and for water-soluble proteins with a β-sheet structure. One of the methods has been used to predict that facilitative glucose transporters form a β-barrel structure. It will be interesting to determine whether this, or the 12-transmembrane helix model predicted by hydropathy analysis, is correct.

3.2.3.   Transmembrane Segments without Lipid Contact

The most difficult transmembrane segments to predict are those that, like the loop inside the β-barrel of porin, do not interact with the lipids but rather only come in contact with the other protein segments and/or the water and ions inside the pore. This is because the environment of these segments is similar to what is experienced on the intracellular and extracellular sides of the membrane and is thus free from the unique structural influences imposed by the hydrophobic core. This can be an unfortunate situation, because it is these types of segments which are often involved in the interesting, functional aspects of the protein such as ion selectivity, gating, and ligand binding (see the sections on voltage-gated channels). As an example, because of the presence of hydrophilic residues, regular hydropathy analysis predicts that the P segments of voltage-gated Na^+ channels are in an aqueous region rather than in the transmembrane domain forming the ion-selective filter of the pore. Likewise, hydropathy analysis of the more hydrophobic P segments of K^+ channels incorrectly predicts them to be trans-membrane helices. The correct prediction was actually made from the functional data that indicated that the selectivity filter is negatively charged and near the extracellular entrance (Guy and Seetharamulu, 1985; see Section 6). However, the situation is not so clear-cut. In our current Na^+ and Ca^{2+} channel models many of the polar residues of the P segments are exposed to the solvent in the extracellular vestibules of the channels.

The initial prediction for the P segment of Na^+ channels (Guy and Seetharamulu, 1985) would have been much easier and more convincing had we had the relative plethora of ion channel sequences now available. To correctly identify the ion-selective region one simply needs to identify the segment that is conserved best among channels that select for the same ion but at the same time is not well conserved among channels that select for different ions. Thus, comparisons of homologous proteins can be a very important tool in predicting the transmembrane topologies of proteins with unorthodox folding patterns and in identifying functionally important segments.

3.2.4. Sequence Properties that May Help Determine Topology

One observation that can help predict the transmembrane topology of membrane proteins is that the extracellular side tends to be glycosylated, and the intracellular side tends to be phosphorylated. Of course, models that place most potential glycosylation sites on the outside of the cell and most potential phosphorylation sites on the inside of the cell would be favored.

Another fact that may be useful in predicting the topology is that some proteins have a hydrophobic leader sequence at their N-terminus that aids in transport across the endoplasmic reticulum membrane before being cleaved off. Therefore, identification of this type of leader sequence in the gene would suggest that the N-terminus of the protein is on the extracellular side of the plasma membrane.

3.3. Phase II: General Position and Orientation of Transmembrane Segments

The second stage of modeling involves predicting a low-resolution picture of the positions and relative orientations of the transmembrane segments without specifying the exact locations of all the atoms. The methods and criteria used to develop these models are described below.

3.3.1. Residue–Solvent and Residue–Residue Contacts Based on Residue Type

The primary constraint on positioning the segments is that most hydrophilic side chain groups should be at least partially exposed to water, and most hydrophobic side chains should be either buried within the protein or exposed to lipid. Another requirement for most ion channels is that the protein forms a continuous barrier between the lipid and the water-filled pore. This is generally accomplished by positioning the postulated lipid-exposed segments so that they completely span the membrane. Once the global fold has been designed to satisfy the different environment requirements of the membrane and solvent, smaller adjustments are made to the segments to optimize the number of stabilizing disulfide bonds, salt bridges, hydrogen bonds, contacts among aromatics, and hydrophobic interactions. Parameters based on statistical analyses of soluble proteins can be used to evaluate these interactions (Miyazawa and Jernigan, 1985).

3.3.2. Use of Homologous Sequences

A tremendous amount of useful information can be obtained by comparing members of a protein family or superfamily. The first step is generally to align all the sequences and determine the patterns of residue conservation and variability. As described below, there are a number of ways in which this information can be used to help predict the structure of membrane proteins.

1. Homologous proteins have similar backbone folding patterns. Probably the most basic modeling assumption is that all segments that can be aligned unambiguously from homologous proteins have similar backbone folding patterns. Of course, this is especially useful if the structure of at least one member of the homologous group is known.

2. Assemblies of homologous subunits or domains are usually arranged symmetrically. Many membrane proteins with multiple subunits and/or homologous domains are known to have real or pseudo symmetry about an axis orthogonal to the membrane surface. For example, some voltage-gated K^+ channels have fourfold symmetry, some transmitter-activated receptors have fivefold symmetry, and gap junction channels have sixfold symmetry. Other transmitter- and voltage-activated channels are composed of homologous subunits or repetitions within a polypeptide chain. When this occurs, the homologous regions are believed to have similar backbone structures and be arranged in a pseudo-symmetrical manner. For example, two transmembrane subunits of the photosynthetic reaction center have similar backbone structures and are oriented with approximate twofold symmetry even though most of the residues are not identical. If appropriate, then the inclusion of a symmetrical design greatly facilitates the modeling by constraining how the adjacent subunits or homologous domains can pack next to each other. For this reason we have also included symmetry into all our models of channels formed by peptides.

3. Nonfunctional residues on the surface are poorly conserved among closely related proteins. Statistical studies show that the surface residues of soluble proteins that are not directly involved in forming an active site or binding subunits are much more variable among homologous family members than are the residues buried in the core (Go and Miyazawa, 1980). The reason for this is probably that the geometry and interactions of the buried residues are important to maintain a specific packing arrangement and global fold for the specific function of the protein. In contrast, the polar residues on the surface that simply extend into the solvent can generally be altered without substantially affecting the protein's overall structure and function. Fortunately, a similar "surface variability" effect seems to occur for the residues of the transmembrane seg-

ments of membrane proteins. In particular, many putative transmembrane α-helices are found to have a remarkably clear-cut pattern of "unilateral conservation," in which the residues on one face of the helix are conserved to a considerably greater degree than residues of the other (Baldwin, 1993; Guy, 1986, 1988, 1990; Rees *et al.*, 1989). But unlike the soluble proteins, the variable face is typically more hydrophobic than the conserved one. As confirmed by the photosynthetic reaction center (Rees *et al.*, 1989) and bacteriorhodopsin structures (Durell and Guy, 1994; Guy and Durell, 1994), these patterns have been interpreted to mean that the helix is part of a transmembrane bundle, where the conserved face is part of the intraprotein core and the variable, hydrophobic face is exposed to the lipids on the surface. Thus for modeling, finding these combined periodic patterns of residue polarity and conservation greatly aid in identifying the transmembrane α-helices in the sequence and predicting their relative orientations in the protein structure.

4. Conserved residues among a superfamily of proteins tend to be clustered together. Another important phenomenon that has been observed in soluble protein structures is that the "functional" residues, e.g., those forming a subunit or substrate binding site, are generally found to be the best conserved among a family of homologous sequences. This is easily understood as due to the interplay of structure, function, and evolutionary selection pressures. For example, in the globin superfamily it is generally observed that the most highly conserved residues are those directly involved in binding the heme prosthetic group. In addition, it is observed that the variability of the residues gradually increases as a function of distance from these conserved sites in the protein, with surface residues being less well conserved than residues buried within the protein or between subunits. The same phenomenon also occurs in the superfamily of membrane proteins that includes bacteriorhodopsin, where the conservation of the residues is found to decrease as a function of distance from the retinal group and proximity to the surface (Guy and Durell, 1994). Thus, knowledge of the relative conservation of the residues in the sequence helps to identify the functionally important ones, and attempting to reproduce the clustering pattern aids in predicting the global fold and relative positions of the residues.

5. Conservation of sequence correlates with conservation of function. A subtle corollary to the clustering phenomenon described above is that conservation of the residues between homologous proteins correlates with conservation of the functional properties. Perhaps the best example of this phenomenon in membrane proteins is the conservation pattern of the ion-selective P segment of voltage-gated and some ligand-gated ion channels. For the K^+-, Na^+- and Ca^{2+}-selective channels the P segments are the most highly conserved regions of the sequence among a broad range of proteins that select for the same ion, but differ greatly between homologous proteins that select for different ions. Thus, determining correlations between sequence conservation and protein function can help identify the residues responsible for those functions.

3.3.3. Helix Packing Theory

There are a number of features of protein helices that influence how they can pack together in energetically favorable ways. From analyzing α-helical coiled coils in fibrous proteins Crick (1953) first noted that if the axes cross at either −20° or 50° then the residue side chains can pack in between each other in a manner described as "knobs-into-holes." Subsequently, Chothia *et al.* (1981) expanded on this by noting that in an α-helical conformation every fourth side chain of the sequence forms a right-handed spiral ridge, while every third side chain forms a left-handed spiral ridge. Similar to Crick's studies, they concluded that the two most common packing arrangements are "4-4 ridges-into-grooves," in which the right-handed ridges are interlaced and the axes cross at ca. 50°, and "3-4 ridges-into-grooves," in which the left-handed ridges of one helix interlace with the right-handed ridges of the other helix and the axes cross at ca. −20°. The exact crossing angles for these arrangements depend upon the radii of the helices, the number of residues per turn, and the pitch. It must be pointed out, however, that there are many exceptions to these rules, and helices are observed to cross at a wide variety of angles. For example, some helices pack with the ridges crossing each other rather than being interlaced. In this case the point of contact is usually occupied by a small residue, such as glycine, to permit close contact. While unfortunately the theory is not sufficient to predict unambiguously the relative orientations of helices from their sequences, these packing motifs are still useful for designing a series of initial models.

3.3.4. Sequentially Adjacent Helices with Short Linkers Tend To Be Structurally Adjacent

For modeling a membrane protein with multiple transmembrane helices that thread back and forth across the membrane it is helpful to know that each helix must be in an antiparallel orientation relative to its nearest neighbors in the sequence. In addition, if the segment of residues linking two transmembrane helices is short then the helices are likely to be packed together in an antiparallel, hairpin configuration. This motif is typified by the transmembrane helices of the bacteriorhodopsin structure. It is important to note that in known protein structures helices are more often found packed in an antiparallel rather than parallel orientation, probably influenced by the favorable interactions of the electrostatic helix dipoles (Hol *et al.*, 1981). In contrast, if there is a long linking segment between two helices then they may not actually be packed together in the membrane, but rather packed with other helices not adjacent in the sequence. This can be seen in the structure of the bacterial photosynthetic reaction center and results in a mixture of parallel and antiparallel arrangements.

3.3.5. Established Structural Motifs Should Be Favored

There are many examples of convergent evolution in soluble proteins, where similar conformation motifs have developed for nonhomologous proteins. This likely is an indication that these motifs are exceptionally energetically stable. Assuming that a similar phenomenon occurs in membrane proteins, the modeling is facilitated if a known structural motif can be identified from the amino acid sequence. One such example is the bundle of four antiparallel α-helices, which is suggested for the transmembrane portion of each subunit of the transmitter-activated channels (Guy, 1990). Another is the eight-stranded α/β, barrel motif, which is similar to what we postulate for the pardaxin channel (see Section 4.5.).

3.3.6. Experimental Data

In some cases one may have experimental data that is insufficient to determine the structure directly but still very helpful in constraining the models. Generally, there are three types of data: structural, functional, and that obtained by mutagenesis. Potential types of structural data could include the dimensions of the protein from electron microscopy or diffraction; transmembrane topology from many types of labeling, binding, and/or digestion experiments; secondary structure from circular dichroism, IR absorption, Raman scattering, or X-ray diffraction; and distances from disulfide bridge formation or spin-spin interactions between labeled groups. While information of this sort can easily be incorporated by simply excluding inconsistent models, one must not rely too heavily on limited results. For example, it is well known that secondary structure measurements can vary widely depending on the environmental conditions and how perturbed the sample is from the native state. For ion channels, functional types of data could include the permeability characteristics of the pore, the kinetics of gating processes, and the ways that various drugs and toxins modify these processes. Indeed, this type of information was extremely important in the models of the voltage-activated Na^+ channel developed by us and other groups (see Section 6). Finally, mutagenesis experiments, which assess the effects of modifying specific residue targets provide a unique way to extend studies of both the structural and functional properties of the protein. Although this approach must be used with caution, especially when the three-dimensional structure is not known, it is proving to be an extremely powerful tool for studying ion channel proteins.

3.4. Phase III: Approximation of Atomic Positions

The models can be made more precise by approximating the positions of all atoms in at least some parts of the protein. Thus far we have had to use a "manual," subjective approach to accomplish this, due to the lack of proven, comprehensive, automated methods. In general, we use one or two of the commercially available molecular mechanics software packages to build the models on the computer. In the case of a helical protein, idealized α-helices are initially constructed in which each residue has the conformation that occurs most frequently in the α-helices of known structures. Then, the helices are docked and the side chain conformations are adjusted to eliminate atomic overlap and enhance energetically favorable interactions. For proteins predicted to contain a specific β-structural element (e.g., a β-barrel of α/β TIM barrel), an idealized model is constructed based on the geometry observed in known protein structures. From this initial model we begin an iterative procedure of automated energy minimization [using a prepackaged program such as CHARMM (Brooks *et al.*, 1983)] and "manual" manipulation to optimize the packing and refine the structure. In the process the structure is directed to conform to the following energy-based theoretical criteria:

1. Large cavities should not exist inside the hydrophobic portion of the protein.
2. For most ion channels, the transmembrane portion of the protein should form a continuous barrier between the lipid on the outside and the water inside the pore.
3. The hydrophobic surface at the transmembrane region should approximate the thickness of the average alkyl-chain portion of a membrane.
4. All hydrophobic and hydrophilic residues should be in the appropriate environments: i.e., hydrophobic residues should either be exposed to the lipids or buried in the protein, and the hydrophilic residues should either be exposed to the solvent and/or in contact with other polar residues.
5. All hydrogen bond donors and acceptors should form hydrogen bonds or be exposed to water.
6. All charged residues should form salt bridges with other residues or be exposed to the solvent.
7. All residues should be in an energetically favorable conformation.

We hope that with advances in computing power and algorithms a more automated approach will become available for developing the Phase III models. Such methods will require evaluating accurately the relative free energies of trial protein models in their native environments, which would include influences from the water, ions, lipids, and membrane voltage. Despite the uncertainties in our more subjective Phase III models, they are still very useful for the following reasons:

1. They test the steric feasibility of the Phase II models, including being able to position the residues in energetically stable environments as described above.
2. They suggest experimental tests of the models (especially residue targets for mutagenesis) that are not apparent from the Phase II models alone.
3. They provide a convenient way of illustrating the models and sharing them with other groups.
4. They provide a starting point for more extensive computational analyses.

4. CHANNELS FORMED BY SMALL α-HELICAL PROTEINS

4.1. General Concepts

If one wishes to model transmembrane regions of many membrane proteins it is important to be able to accurately predict the conformations of the α-helical segments, their side chain conformations, and how they are packed together. In developing methods to achieve this we began with small α-helical peptides, e.g., δ-lysin, magainin, cecropin, and pardaxin. These models postulated how monomers could assemble on one surface of the membrane, insert into the transmembrane region, and form the channels. While the relatively small size of the peptides makes Phase III modeling easier than for larger membrane proteins, they also have unique disadvantages. One major problem is polymorphism. The peptide's secondary structure often changes when it moves from water into a membrane. Once associated with a membrane they may cause detergent-like lipid perturbation effects that are difficult to separate from real channel formation. Even when this can be resolved, a variety of different types of channels may continuously form and break down. This makes experimental determination and testing of a unique channel structure very difficult.

Before describing the models and how they were developed, let us first consider how amphipathic α-helical peptides could interact with the membrane and form ion channels. One can envision several general mechanisms:

1. Peptide monomers insert individually across the membrane and then aggregate to form channels. This mechanism has been postulated for the alamethicin channel and for how the helices of larger membrane proteins, such as bacterio-rhodopsin, become incorporated into the membrane (Popot *et al.*, 1987). For peptides only long enough to span the membrane once this mechanism would lead to channels formed by parallel rather than antiparallel aggregates of helices. For example, if the peptides are inserted into the membrane under the influence of a transmembrane voltage then the electric dipole would cause each helix to have the same orientation.

2. Amphipathic α-helices may aggregate on the surface of the membrane to form "rafts" that are hydrophobic on one side and hydrophilic on the opposite side. A peptide monolayer of this type has been postulated to be formed by mellitin at the water-air interface (Terwilliger *et al.*, 1982). These peptide "rafts" would displace the lipid monolayer. If the monomers of the "rafts" form a linear array then they could fold through the membrane when two interact. We have proposed this mechanism for channels formed by δ-lysin and other peptides (Guy and Raghunathan, 1989; Raghunathan *et al.*, 1990). This mechanism favors the formation of channels in which the monomers are antiparallel because of the favorable electrostatic dipole interactions. However, if the "rafts" form with radial symmetry then channels with parallel monomers could also result. Pardaxin channels may be formed this way. "Raft" insertion mechanisms might be favored if the polar face of the helix is too hydrophilic to allow the peptide to span the membrane as a monomer. In this case the aggregate would shield the polar residues from contact with the hydrophobic lipid alkyl chains.

3. Interactions of the peptides with lipids on the membrane surfaces could cause some of the lipids to become oriented parallel to the membrane so that their polar head groups form the channel lining. This mechanism works best if the peptides are on both sides of the membrane. These types of channels may be formed by magainin (see Fig. 1; Cruciani *et al.*, 1992).

4. When the peptides fold into the membrane some lipids may accompany them, creating a channel in which the lining is formed by both the hydrophilic face of the peptide helix and by lipid head groups. These types of channels are also plausible for magainin and cecropin (see Figs. 1 and 2).

5. One portion of a peptide may lie within the lipid monolayer on one side of the membrane parallel to the membrane while another part spans the second mono-

layer to form a channel. We postulate that cecropin, sarcotoxin, and pardaxin may form these types of channels (see Figs. 2 and 3).

The general approach described above for Phase III modeling was used to generate and pack the α-helical peptides next to each other. To simplify the analysis we usually assume that all monomers in an aggregate have identical conformations and that each monomer has identical interactions with surrounding monomers, solvent, and lipids. These assumptions reduce the number of conceivable packing arrangements greatly. The assembly can be generated by a series of translations and/or rotations of the original monomer so that the energies of only one monomer and surrounding atoms within a certain distance of the monomer need be calculated.

4.2. δ-Lysin

δ-Lysin, a lytic peptide from staphylococcus, forms ion channels with several conductances. It has been crystallized (Thomas *et al.*, 1986) and an α-helical backbone structure has been determined by NMR (Tappin *et al.*, 1988). It can form an almost ideal amphipathic α-helix in which all residues on one face are hydrophobic and residues on the opposite face are hydrophilic. Both positively and negatively charged residues are distributed along the hydrophilic face. We have developed both channel and "raft" models of δ-lysin assemblies (Raghunathan *et al.*, 1990). In both models adjacent helices are antiparallel. With the exception of a tryptophan side chain, all residues on one surface of the raft are hydrophilic and all residues on the opposite surface are hydrophobic. Transition from the membrane raft to a channel structure may be energetically favorable. The Phase III calculations suggest that channels formed by six and eight monomers are about energetically equivalent and the energy increases as the size of the channel increases, being least favorable for the raft. Of course, these energy values alone cannot be used to discriminate between alternative models because the calculations do not include effects of water, lipid, ions, membrane voltage, or entropy.

The methods we used to develop Phase III models contain subjective elements that can be rigorously tested only on known structures. To test these methods we have attempted to predict the crystal structure of δ-lysin using α-helices and the known unit-cell dimensions of the crystal. After developing initial raft models of δ-lysin we noted that the distance between adjacent dimers was close to the length of one crystal unit-cell dimension (23.2Å), and the perpendicular "height" of the raft was close to the length of a perpendicular unit-cell dimension (35 Å). We thus developed models of crystal structures in which the rafts form monolayers and the hydrophobic surfaces of the monolayers pack together to form bilayers. Unfortunately, the atomic coordinates of the crystal structure have not yet been determined, so the validity of the modeling process cannot be evaluated.

4.3. Magainin

Magainin is a potent antimicrobial agent secreted from *Xenopus* granular glands (Zasloff *et al.*, 1988). It has little regular secondary structure in water but becomes highly helical when associated with membranes (Jackson *et al.*, 1992; Duclohier, 1994). Magainin can form amphipathic α-helices in which positively charged groups are distributed all along the hydrophilic face. Magainin induces channels in artificial membranes (Cruciani *et al.*, 1992) with conductances that range from 1 to 300 picosiemen. The ion selectivity of the channels depends upon the lipids used to make the membranes. For example, if the membrane is composed of phosphatidylcholine and phosphatidylethanolamine, which have no net charge, then the channels are more permeable to anions than cations (Duclohier *et al.*, 1989); however, if half of the lipids are phosphatidylserine, which has a net charge of -1, then the channels are more permeable to cations (Cruciani *et al.*, 1992). This is best explained by models in which lipid head groups form some or all of the channel lining. Figure 1 illustrates how such channels could form in a way that does not expose the lipid alkyl chains to water or require that any of the polar peptide groups be buried in the alkyl phase. In these models magainin helices are postulated to first assemble on the surface of the membrane. When the surface concentration is sufficiently high the peptides are postulated to form aggregates that thin the bilayer and subsequently insert into the membrane to form channels. In the largest putative pores, peptides are present on both membrane surfaces to cover the alkyl chains of the channel-forming lipids that are oriented parallel to the membrane surface. These types of models require that magainin monomers move across the membrane by forming transient pores composed of several monomers. This prediction has been confirmed experimentally (Matsuzaki *et al.*, 1995). The modeling suggests that magainin may form antiparallel dimers. The primary assumption of these models is that the hydrophobic portions of the magainin helices are exposed to lipid alkyl chains and that the positive charges along the hydrophilic face of the helices bind strongly to the lipid head groups. This latter interaction may prevent formation of larger peptide rafts or channels as proposed for δ-lysin and may cause lipid to accompany the peptides as they move into and/or through the membrane.

4.4.   Cecropin and Sarcotoxin

Cecropins (Boman and Hultmark, 1987), bactericidins (Dickinson *et al.*, 1988), and sarcotoxins (Okada and Natori, 1985) are homologous antimicrobial peptides made by insects. The cecropins form channels in lipid bilayers that are more permeable to anions than cations (Christensen *et al.*, 1988). They resemble magainin in that their N-terminal region can form a positively charged amphipathic α-helic of about 20 residues; however, they contain an additional C-terminal segment that can form a second, more hydrophobic, amphipathic helix of about 12 residues. The junction between these segments is formed by glycine and (in cecropin and bactericidin) proline residues that are likely to disrupt the helical pattern between the segments. This helix-bend-helix secondary structure has been confirmed by NMR analysis of cecropin in apolar solvent (Holak *et al.*, 1988). Comparison of N-terminal segments from different homologues reveals a strong pattern of "unilateral conservation," i.e., the residues at one interface region between the hydrophobic and hydrophilic faces are well conserved, whereas the residues at the opposite interface are poorly conserved. Modeling has indicated that the well-conserved interfaces of two helices can pack together in an energetically favorable way

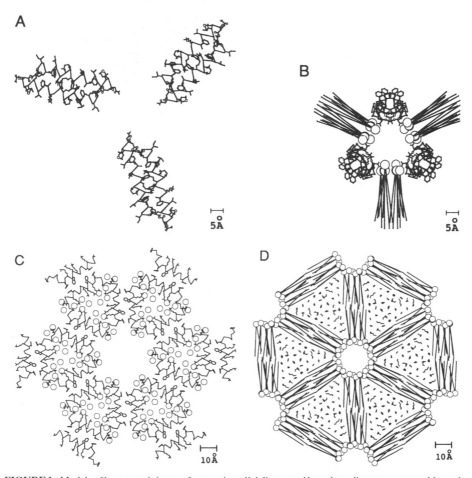

FIGURE 1. Models of how magainin may form antiparallel dimers and how these dimers may assemble on the surface of the membrane (A and C) and insert into the membrane (B) to form pores that are partially or completely lined with lipid head groups. In the model of A and B, some dimers (in A) are postulated to lie on the membrane surfaces where they cover alkyl chains of pore-lining lipids (straight lines in B). Other dimers (in B) span the membrane and form a pore lined by the polar faces of magainin dimers and lipid head groups (circles). In the model of C and D the magainin dimers lie only on the two membrane surfaces and the pore is lined entirely by lipid head groups. Note that in both models alkyl chains (straight lines) of lipids postulated to line the pores are covered on the surfaces by the hydrophobic faces of the magainin helices. (From Cruciani *et al.*, 1992.)

to form an antiparallel dimer that is hydrophobic on one side and hydrophilic on the other side. In support of this model interactions between cecropin monomers in membranes have been demonstrated experimentally (Mchaourab *et al.*, 1994). Several types of channel structures can be imagined (Durell *et al.*, 1992): 1) the C-terminal helices span the membrane and aggregate to form the pore (Fig. 2A, D); 2) the aggregated C-terminal helices move to the *trans* side of the membrane, causing the N-terminal helices to tilt into

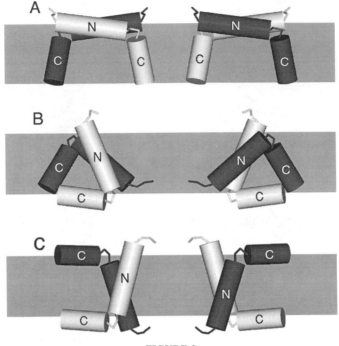

FIGURE 2.

the membrane and form a large funnel-shaped pore (Fig. 2B); and 3) the other C-terminal helices move to the *cis* side of the membrane, causing the N-terminal helices to form a cylindrical pore (Fig. 2C). This conformational transition could pull negatively charged lipid head groups which are bound to the sides of the N-terminal helices into the lining of the pore. Figure 2D illustrates how these dimers could assemble into a hexagonal array on the membrane surface so that channels are formed in the regions of sixfold symmetry when the relatively hydrophobic C-terminal helices span the second lipid monolayer (Durell *et al.*, 1992). This model suggests that it may be possible to grow two-dimensional crystals of these peptide–lipid complexes and to determine the structures experimentally.

4.5.   Pardaxin

Pardaxin is a potent shark repellent made by the Red Sea Moses sole fish. Unlike most of the other peptides we studied, pardaxin forms membrane channels with only one single-channel conductance (Lazarovici *et al.*, 1990, 1992). The channel is permeable to both anions and cations but will not allow tris base, which has a diameter of about 7 Å, to pass. The dose-response curve for channel formation is very cooperative, suggesting that eight to twelve monomers form the channel. The pardaxin sequence can be divided into three segments: residues 2–11 are primarily hydrophobic but also contain a proline and lysine; residues 13–26 can form an amphipathic α-helix that begins with a proline

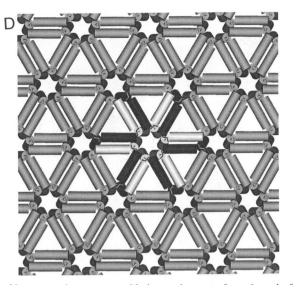

FIGURE 2. Models of how cecropins may assemble in membranes to form channels. The monomers are postulated to have two α-helices, N and C, that are connected by a bend. N-terminal helices of two cecropins are postulated to form an amphipathic antiparallel dimer in which the apolar surface is exposed to lipid alkyl chains and the polar surface is exposed to water and lipid head groups. The more hydrophobic C-terminal helices may penetrate the monolayer in A and D to form small pores. Larger pores may form when the N-terminal helices tilt into (B) or span (C) the bilayer. (D) Representation of how cecropins may form a hexagonal lattice in a membrane. (From Durell et al., 1992.)

and in which the hydrophilic face has one lysine, one threonine, and one serine residue; and residues 27–33 have the polar sequence Ser-Ser-Ser-Gly-Gln-Glu-COOH. Circular dichroism studies in solution indicate that pardaxin has both α-helical and β-strand segments, and NMR studies indicate that it is the first two segments that form the α-helices. We propose that the channel is composed of eight pardaxin monomers that form a structure similar to the α/β TIM barrel structure of numerous soluble proteins (Chou et al., 1990). In these structures an eight-stranded β-barrel is surrounded by eight α-helices. In the pardaxin model the C-terminal segments form the β-barrel, which is surrounded by the amphipathic α-helices of the middle segments. The N-terminal segment is postulated to be primarily helical and oriented about parallel to the membrane surface. Alternate models have been developed in which the monomers are either parallel (see Fig. 3) or antiparallel to each other. We tentatively favor the parallel models because they better account for the voltage dependence of channel formation and because the parallel structure is more like the barrels observed in many soluble proteins. In both structures the two negatively charged groups at the C-terminus can form salt bridges with the two lysine side chains that are near the N-terminus. The pore through the channel is large throughout the membrane except in the region formed by the glutamine side chains. Channels with large lumens and short and narrow selectivity filters have been postulated for numerous ion channels including voltage-gated channels of nerves and muscles.

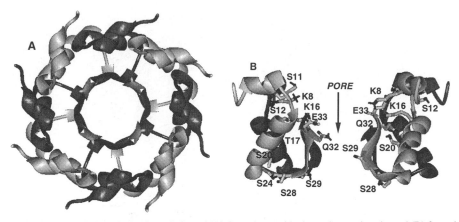

FIGURE 3. Model of pardaxin channel viewed (A) from the outside down the pore's axis, and (B) from the side showing half the subunits. The large outer spirals represent α-helices and the inner ribbons represent β-strands. The structure is postulated to be an eight-stranded αβ-barrel. The N-terminal segments are postulated to form short α-helices that are parallel to and on the membrane's outer surface. Polar side chains are illustrated and labeled for one subunit on each side of B. Polar groups of the side chains are shaded black. The narrowest part of the pore is formed by the Gln32 side chains.

4.6.   Amyloid β-Protein

Amyloid β-protein is suspected of being at least part of the cause of Alzheimer's disease. It is a ca. 40-residue peptide that forms ion channels in lipid bilayers (Arispe *et al.*, 1993). Features of its sequence resemble some aspects of both cecropin and pardaxin. Like cecropin it has a relatively short hydrophobic C-terminal segment, which is preceded by a segment that may form a small α-helix that is hydrophobic in the middle but contains charged residues at the ends. The linking segment between these putative helices has a high propensity for a coil or turn conformation. The helices are preceded by a segment which may form an amphipathic β-hairpin that is hydrophobic on one side and highly charged on the other. We (Durell *et al.*, 1994) have developed ion channel models for amyloid β-protein similar to that of cecropin, in which the hydrophobic C-terminal segments of several monomers penetrate the *trans* bilayer and the other segments form a larger, more hydrophilic pore in the *cis* bilayer, which may also include lipid head groups. We have also developed alternative ion channel models, more like that for pardaxin, in which the N-terminal segments form a large transmembrane β-barrel pore that is surrounded by the more hydrophobic C-terminal α-helices. Unfortunately, neither of these very tentative models has been tested experimentally.

5.   WATER-SOLUBLE PROTEINS THAT FORM ION CHANNELS

Colicins are antimicrobial proteins that form large pores in microbe membranes and synthetic bilayers. Colicins exist both in solution and interact with membranes to form channels. The crystal structure has been determined for the water-soluble form of the

FIGURE 4. A dimer model of a colicin channel. Helices 7 and 8 are hydrophobic and do not form part of the pore's lining. Helices 1, 2, 4, and 5 are amphipathic with their polar residues lining the pore. (From Guy, 1983.)

portion of colicin A that forms the channel. A long hydrophobic portion forms two α-helices near the C-terminus, while much of the remaining structure forms amphipathic α-helices. Dimer models of colicin E1 and A had been proposed previously (Guy, 1982). In this model, two hydrophobic α helices were on the exterior of the transmembrane segments, and a large water-filled pore was formed by amphipathic helices that precede the hydrophobic ones. The secondary structure of these models was very similar to that of the crystal structure: except for the fact that the transition between helices 2 and 3 occurred sooner in the model than in the crystal, all the helices were in the same approximate places in the sequence. The major difficulty with this model is that several experiments suggest that only one colicin molecule forms the channel. Recent experiments indicate that when the channel opens under the influence of voltage, large polar portions of the protein cross the membrane (Slatin *et al.*, 1994). If, after this transition, some amphipathic helices are present on both membrane surfaces, then the pore could be at least partially lined by lipids as we envision for magainin pores. Also, colicin channels have numerous conformations with differing conductances (see Cramer *et al.*, 1995, for review), which suggests that there is no single correct model of the colicin channel.

6. PHASE I MODELS OF VOLTAGE-GATED CHANNELS

6.1. Models and Rationale

Potassium channels are composed of four identical or homologous subunits, whereas the pore-forming portions of Na^+ and Ca^{2+} channels are formed by a single polypeptide chain that contains four homologous repeats, each of which has been postulated to have a transmembrane topology similar to that illustrated in Fig. 5 (Guy and Seetharamulu, 1985). Each subunit was postulated to have six transmembrane α-helices (S1–S6) and one hairpin (P) that spans only the outer part of the transmembrane region and that forms the ion-selective portion of the channels. Although this model was entirely hypothetical when proposed, many of its features have been confirmed experimentally. There is now evidence that each of the putative extracellular loops of K^+ channels are indeed exposed on the extracellular surface (Chua *et al.*, 1992; MacKinnon and Yellen, 1990; Shen *et al.*, 1993), that the N- and C-terminal ends are on the cytoplasmic side of the membrane (Liman *et al.*, 1992; Zagotta *et al.*, 1990), and that the central part of the K^+ channel P segment is accessible to tetraethylammonium (TEA)

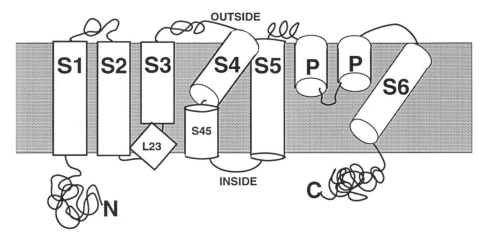

FIGURE 5. Postulated transmembrane topology of a voltage-grated channel subunit (K^+ channel) or homologous repeat (Na^+ and Ca^{2+} channels). This topology was first proposed by Guy and Seetharamulu (1985). Cylinders represent α-helices.

which enters and blocks the pore from the inside (Choi *et al.*, 1993; Hartmann *et al.*, 1991; Yellen *et al.*, 1991).

A schematic representation of our predictions (Guy, 1988) about which segments of Na^+ channel proteins form functional sites or mechanisms previously identified by membrane biophysieists is illustrated in Fig. 6. Activation gating had been postulated to be triggered by the movement of several charged "voltage sensors" through the membrane's electric field (Armstrong, 1981). The S4s were postulated to be the voltage sensors because they have an unusual pattern of positively charged residues at every third position separated by hydrophobic residues. The S4s were placed in the transmembrane region so they could respond to voltage changes. The activation gate had been postulated to occlude the cytoplasmic entrance of the pore when the channel closed and then to move out of the way when it opened (Armstrong, 1981; Hille, 1977). The segments linking S4s to S5s were postulated to be the activation gate because they are linked to S4 and because they are located near the cytoplasmic entrance of the channel in the model. The channel's ion-selective region, or "selectivity filter," had been postulated to be a narrow negatively charged region near the extracellular entrance of the pore (Armstrong, 1981; Hille, 1975). The P segments were postulated to form the selectivity filter because their negatively charged residues could balance some of the positively charged residues of S4 and because they could form a narrow region near the pore's extracellular entrance. Tetrodotoxin (TTX) and saxitoxin (STX) had been postulated to block the extracellular entrance of Na^+ channels by binding to negatively charged carboxyl groups (Hille, 1975). The C-terminal halves of the P segments were postulated to form these toxin binding sites because they have the required negatively charged residues and appropriate location in the model. The inactivation gate had been postulated to be a positively charged protein moiety that enters and blocks the intracellular entrance to the pore when

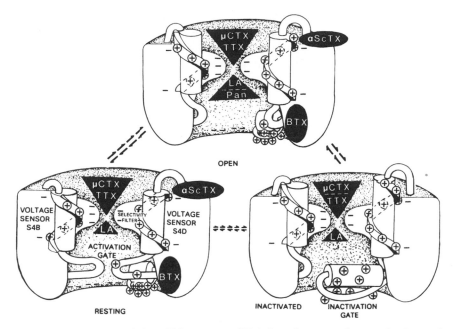

FIGURE 6. Early model postulating which segments of Na$^+$ channel sequences form previously postulated mechanisms or binding sites. Postulated binding sites for tetrodotoxin (TTX), μ-conotoxin (μCTX), local anesthetics (LA), pancuronium (Pan), batrachotoxin (BTX), and α-scorpion toxin (αScTX) are shown in black. Predictions that have been confirmed or supported experimentally are that the voltage sensors are formed by S4's, that the inactivation gate is within the segment linking repeat III to repeat IV, that inactivation is coupled to a transmembrane movement of S4 in repeat IV, that the selectivity filter is formed by P's, and the TTX binding site by charged residues is near the ends of the P's. (From Guy, 1988.)

the channel is in an open or activated conformation (Armstrong and Bezanilla, 1977). In Na$^+$ channels the cytoplasmic segment linking homologous repeats III to IV was postulated to form the inactivation gate because it is positively charged, well conserved between eel and rat brain species, and can be located near the cytoplasmic entrance to the channel in the models. Inactivation was postulated to be linked most closely with transmembrane movement of S4 of repeat IV because it has a longer pattern of positive charges than the other charges, because the segments linking it to S3 and S5 are better conserved than in the other repeats, and because extracellular α scorpion toxin alters inactivation (Catterall, 1984).

6.2. Experimental Verification of Models

Almost all of these predictions have been verified or supported by experiments. The best verified predictions are that the P segments form the ion-selective portion of the pore and binding sites for toxins. For example, it has been shown that mutations of the charged residues near the C-terminal ends of the Na$^+$ channel's P segments alter channel conductance, selectivity, and blockade by TTX and STX as anticipated by the model (Terlau *et al.*, 1991). In addition, mutations near the beginning and end of the P segment of K$^+$ channels affect blockade of the pore from the outside by TEA and charybdotoxin (MacKinnon and Yellen, 1990), mutations near the middle affect blockade of the pore from the inside by TEA (Choi *et al.*, 1993; Hartmann *et al.*, 1991; Yellen *et al.*, 1991), and mutations of glycines in the latter part of P eliminate the selectivity of the pore (Heginbotham *et al.*, 1994).

A strategy that has proven effective in analyzing P segments is to identify homologous proteins with known sequences that have different pore properties and then demonstrate that those properties can be transferred by mutating residues of the P segment of one protein to aligned residues of the other protein. For example, one position (p14 in Fig. 7) of the alignment of P segments is occupied by glutamic acid in all four homolgous repeats of the Ca^{2+} channel α-subunit, whereas in the Na$^+$ channel it is occupied by glutamic acid in repeat I, by aspartic acid in repeat II, by lysine in repeat III, and by alanine in repeat IV. Heinemann *et al.* (1992) showed that mutating the lysine and alanine to glutamic acid in Na$^+$ channels created a channel with pore properties very

```
              1p   5p   10p        15p   20p
   Na  I      FDTFswAFLaLFRlMT----qDfWEnLyqlT
   Na  II     mhdFfhSFLiVFRVLc----GEwiETmwDcM
   Na  III    FDNVGlgyLSLLQVAT----FkGWmDIMYAA
   Na  IV     FeTFGnSMIcLFQitT----saGWdgLLapi
   Ca  I      FDNFGfSMLtVyQCiT----mEGWtDVLYwv
   Ca  II     FDNFPQALISVFQVLT----GEdWnsVMYng
   Ca  III    FDNVlSAMmSLFTVsT----FEGWPqLLYrA
   CA  IV     FqTFPQAVLLLFRCAT----GEaWqEILLAc
   cGMP       ARkYVySLYWStlTLTTi--GEtpPpVrDse
   K  AKT1    wmrYVTSMYWSITTLTTVGYGDlhPVnTkEM
   K  EAG     ksmYVTALYFtMTCMTsVGfGnVaAETdnEK
   K  Shak    FkSiPdAFwWavVTMTTVGYGDmtPVgfwGK
   K  mSlo    AlTYwecVYLLMVTMSTVGYGDVyAkTTLGR
   K  Eco     pRSlmTAFYFSIETMSTVGYGDivPVsesAR
   K  ROMK1   ingmtSAFLFSLETqvTIGYGfrfvTeqcAT
              1p   5p   10p   15p   20p   25p
```

FIGURE 7. Sequences of P segments from a Na$^+$ rat brain channel (Noda *et al.*, 1986), a skeletal muscle Ca^{2+} channel (Tanabe *et al.*, 1987), a cyclic nucleotide–gated channel (cGMP) (Kaupp *et al.*, 1989), KAT1 *Arabidose* K$^+$ channel (Anderson *et al.*, 1992), EAG K$^+$ channel (Warmke *et al.*, 1991), *Shaker* voltage-gated K$^+$ channel (Tempel *et al.*, 1987), mSlo Ca^{2+}-activated K$^+$ channel (Butler *et al.*, 1993), a putative K$^+$ channel from *E. coli* (Milkman and McKane-Bridges, 1993), and an inward rectifying K$^+$ channel, ROMK1 (Ho *et al.*, 1993). Sodium and Ca^{2+} channel segments are numbered beginning with the first proline in repeats II and IV of Ca^{2+} channels (top) and K$^+$ channel segments beginning with the first proline in *Shaker* P segment (bottom). Residues that occur in three or more sequences in this alignment are bold, those that occur in two sequences are in normal upper case, and those that occur in only one sequence are in lower case. Alignment of Na$^+$ and Ca^{2+} with K$^+$ channels is difficult and ambiguous. (From Guy and Durell, 1995.)

similar to those of Ca^{2+} channels. As another example, the first part of the P segment of nonselective cyclic nucleotide–gated channels is quite similar to the analogous sequence of K^+ channels. However, the latter part of the P segment, which is highly conserved among K^+ channels, is very different in the two families. In fact, a deletion occurs in the latter portion of the cyclic nucleotide–gated channel P segment (Guy *et al.*, 1991). Heginbotham *et al.* (1992) deleted a glycine and tyrosine in the *Shaker* K^+ channel at the position of the apparent natural deletion and found that the properties of the mutated pore resembled those of the cyclic nucleotide channel; i.e., it was permeable to both Na^+ and K^+ and was blocked by relatively low concentrations of Ca^{2+}.

Our predictions concerning the activation and inactivation gating mechanisms have also been supported by mutagenesis experiments. Almost all mutations S4's of Na^+ and K^+ channels alter activation gating, and most of those that reduce the magnitude of the positive charge tend to reduce the voltage dependence of the gating (Papazian *et al.*, 1991; Shao and Papazian, 1993; Stühmer *et al.*, 1989). Unfortunately, it is difficult to interpret the effects of mutations on the voltage dependence of activation gating. Some mutations that alter the charge on S4 have little effect, and some mutations that do not alter charge on S4 or that are in other segments do alter the voltage dependence. Mutations, phosphorylation, and antibody binding to the segment linking repeats III and IV of Na^+ channels alter inactivation (Stühmer *et al.*, 1989; Vassilev *et al.*, 1988), which is consistent with the models. Mutagenesis experiments have also identified an inactivation segment at the N-terminal of some K^+ channels and have shown that free peptides with the same or similar sequence can mimic inactivation by plugging the intracellular entrance of the pore (Zagotta *et al.*, 1990). Remarkably, a peptide with the sequence of the putative Na^+ inactivation gate can also mimic the effect of the K^+ inactivation particle; i.e., it can restore inactivation to K^+ channels that do not inactivate (Patton *et al.*, 1993), even though the K^+ and Na^+ inactivation segments exhibit no apparent homology. Data supporting our hypothesis that inactivation is coupled to transmembrane movement of S4 of repeat IV comes from studies of patients with the genetic disease paramyotonia congenita. Sodium channel inactivation is altered in these patients. In some, mutations occur in the putative extracellular link between S3 and S4 of repeat IV (Chahine *et al.*, 1994).

7. POTASSIUM CHANNELS

7.1. Phase III Models of Ion-Selective P Segments

Potassium channels are an ancient superfamily of proteins that form many types of channels in both prokaryotic and eukaryotic cells. Voltage-gated K^+ channels are composed of four similar or identical subunits. This is probably true for the other families as well. When sequences of distantly related families of K^+ channel are compared, the only regions that are conserved well enough to align unambiguously are the P (also called H5) regions that determine the selectivity of the pore. Our current strategy is to develop Phase III models of four symmetrical P segments and then model the surrounding segments around them.

The most crucial questions about the P segments are how they determine the pore's selectivity for K^+ ions and how they bind ligands and toxins that block the pores. In developing our models we have assumed that the atomic structure of the region that determines the pore's selectivity is almost identical for all homologous K^+ channels. A crucial issue is whether the selectivity is determined by the residue side chain or backbone atoms. Two glycines at positions 13p and 15p (see Fig. 7) are the only residues that are identical in all homologous K^+ channels. This has two implications that are unfortunate for developing and testing these models: 1) Glycines have more conformational freedom than any other amino acid residue, and all secondary structure algorithms would predict them to be in a random coil or turn conformation. Furthermore, studies of sequence conservation in known soluble protein structures indicate that highly conserved glycine residues tend to be buried in the protein and have backbone conformations that are energetically unfavorable for other amino acid residue types (Overington *et al.*, 1992). Thus, these crucial residues are unlikely to have an α-helical or β-strand conformation. This greatly increases the number of plausible protein structures and thus makes the segment more difficult to model. 2) Since glycine residues lack side chains and the other P segment residues are not absolutely conserved, one is left with the somewhat surprising hypothesis that the pore's ion selectivity is determined primarily by the backbone atoms. This possibility makes the structure more difficult to test, since the backbone atoms cannot be altered without drastically affecting the structure with insertions or deletions. Nonetheless, mutagenesis experiments have supported these conclusions. Heginbotham *et al.* (1994) mutated all the residues throughout the most conserved portion of the P segment in *Shaker* voltage-gated K^+ channels. With the exception of the two conserved glycines, all the residues could be mutated without measurably altering the selectivity of the pore for K^+ over Na^+; however, any mutation in either glycine made the pore nonselective among monovalent cations. Since all K^+ binding sites in known protein structures are formed by oxygens, Heginbotham *et al.* (1994) suggested that it is the amide oxygens of the backbone which form the K^+ binding sites in the pore. If this is correct then the pore must not have a β-barrel structure as previously proposed (Durell and Guy, 1992), because in that case the narrowest regions of the pore would have to be formed by residue side chains. Assuming that it is backbone atoms which line the pore, it is important to determine which amide groups actually form the K^+ binding site or sites. Barium is known to block K^+ channels with high affinity. Analyses of the effects of external and internal K^+ concentrations on Ba^{2+} blockade of Ca^{2+}-activated K^+ channels strongly suggest that the pores have several high-affinity K^+ binding sites (Neyton and Miller, 1988). Studies of ion flux through K^+ channels also indicate that the channel is simultaneously occupied by at least three ions (Stampe and Begenisich, 1995). The most efficient way to form a series of K^+ binding sites from the G15p and G17p residues is to place this portion of the four P segments parallel to the axis of the pore in a relatively extended conformation, with all the backbone amide group oxygens pointing toward the center as illustrated in Fig. 8 (Guy and Durell, 1994). Although this structure is rather unorthodox, it is not structurally unreasonable. For example, the L14p and Y16p residues have backbone torsion angles similar to those of a right-handed α-helix, and the G15p and G17p residues have conformations similar to that of a left-handed α-helix. Since this is a commonly observed conformation for glycine residues in known protein structures, it explains why these uniquely flexible

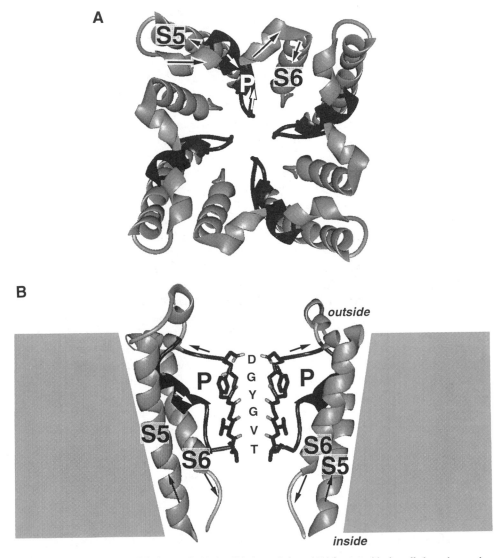

FIGURE 8. Structural model of pore of a *Shaker* K$^+$ channel viewed (A) from outside the cell along the pore's axis, and (B) from the side with only two of the four subunits shown. The P segments are shaded black. Thick ribbons represent α-helices. In B all heavy atoms are shown for residues postulated to form the ion-selective portion of the P segments and one letter code for these residues is indicated on the axis of the pore; oxygens of this segment that are postulated to bind K$^+$ ions are colored gray.

residues are conserved in all the K^+ channel sequences. Having K^+ ions binding to the amide oxygens should act to stabilize this conformation of the pore. However, when K^+ ions are not in the pore the conformation may change, possibly switching to a standard β-strand conformation. This type of conformational transition could occur simply by rotation of the backbone amide groups, without significant movement of the strands or side chains (Guy and Durell, 1995).

Additional constraints on the structure of the P segments can be deduced from the effects of mutations on the blocking efficiency of TEA at both the internal and external binding sites. Mutations at residues 2p, 9p, 16p, 18p, 20p, and 22p alter TEA binding at the extracellular site (DeBiasi *et al.*, 1993; Hartmann *et al.*, 1991; MacKinnon and Yellen, 1990; Pascual *et al.*, 1995; see Fig. 9). The 20p residue appears to affect the binding of extracellular TEA in a direct steric manner (Heginbotham and MacKinnon, 1992), whereas mutations at positions 2p and 22p appear to act by distant electrostatic effects. In our models the D18p residues are the only other residues that form direct contact with TEA at the extracellular site. The effects of the mutations at 9p and 16p may be due to small conformational changes. Near the middle of the P segment mutations at residues 11p, 12p, and 14p alter binding of TEA from the inside (Choi *et al.*, 1993; Hartmann *et al.*, 1991; Yellen *et al.*, 1991; Aiyar *et al.*, 1994). Our models have been constrained so that these are some of the innermost residues of the P region. In addition, blockade by intracellular TEA has a small voltage dependence, in which the molecule is calculated to transverse about 20% of the transmembrane electric field to reach the internal binding site and blockade by external TEA has little or no voltage dependence (Newland *et al.*, 1992). Therefore, if most of the voltage drop across the membrane occurs between the internal and external TEA binding sites, then the sites are probably relatively far apart in the protein. The location of the intracellular TEA binding site relatively near the intracellular entrance is supported also by effects of mutations of S5 and S6 residues on intracellular blockade by TEA and 4-aminopyridine (4-AP) as discussed below. This provides some of the rationale for modeling the segment from 13p to 20p as a relatively extended structure that transverses much of the membrane. However, less extended structures cannot be excluded since some experiments have suggested that the two TEA sites are close enough to interact electrostatically (Newland *et al.*, 1992). A relative proximity of the two sites is also suggested by the observations that mutations at 9p and 16p affect extracellular TEA binding, whereas mutations only two residues away at 11p and 14p affect intracellular TEA binding.

There are several reasons to suspect that the first part of the P segment forms an α-helix as illustrated in Figs. 8 and 9. Circular dichroism experiments of a peptide with the *Shaker* P segment sequence indicate that in inorganic solvents the N-terminal is α-helical (Peled and Shai, 1993). Mutagenesis experiments also favor a helical model. If residues 2p and 9p are near the extracellular TEA binding site, it is more likely that the segment connecting them is an α-helix rather than a much longer β-strand. Also, in most β-strand models 9p should be closer to the intracellular entrance than the extracellular entrance. In addition, Lü and Miller (1995) have mutated each residue of the P segment one at a time to cysteine and observed the effects of extracellular silver ions in blocking the K^+ channel. Silver is the only small monovalent cation that binds covalently to sulfhydryl groups. They found that in the N-terminal region the residues that are more

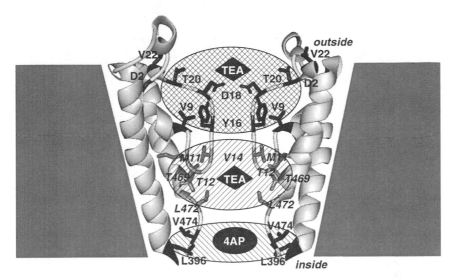

FIGURE 9. Structural model of a *Shaker* pore illustrating binding sites for TEA from the outside and inside and 4-aminopyridine from the inside. Residues where mutations have been shown to alter binding of TEA from the outside (DeBiasi *et al.*, 1993; Hartmann *et al.*, 1991; MacKinnon and Yellen, 1990; Pascual *et al.*, 1995) and 4-aminopyridine from the inside (Shieh and Kirsch, 1994) are shaded black; those where mutations alter TEA binding from the inside are colored gray (Aiyar *et al.*, 1994; Choi *et al.*, 1993; Lopez *et al.*, 1994; Hartmann *et al.*, 1991; Yellen *et al.*, 1991). Note clustering of residues around putative binding sites.

exposed to the outer vestibule of the pore in our helical model (i.e., 1p, 2p, 5p, 6p, and 9p) are indeed the ones accessible from the outside to the silver ions (see Fig. 10). In contrast, the residues predicted to be more buried (4p, 7p, 8p, and 10p) were inaccessible. Residue 3p is labeled slowly, suggesting that it is accessible only part of the time. With the exception of G17p, residues predicted to be in the pore lining and the coiled portion of the outer vestibule (14p, 16p, and 18p–21p) were also accessible to the silver ions. Residues in the putative intracellular portion (11p–13p) and 17p were not labeled. It is unclear whether these residues line part of the open pore since the channels were closed when exposed to silver. The only P segments residues that, when replaced with cysteine, have been shown to react with large extracellular sulfhydryl reagents are at positions 1p, 2p, 18p, 19p, 20p, and 22p (Kürz *et al.*, 1995; Pascul, 1995). In our models these residues are in the outer vestibule where they should be readily accessible to large molecules.

The structure of the outer vestibule is also being analyzed by studying effects of mutations on channel blockade by scorpion toxins. The outer entrances of many voltage-gated and Ca^{2+}-activated channels are blocked by a series of homologous scorpion toxins. The NMR structures of charybdotoxin and some of its homologues have been determined (Bontems *et al.*, 1992). These relatively small (37-residue) toxins are stabilized by three disulfide bridges in the peptide's core. Most of the side chains are on the peptide's surface where substitutions by mutagenesis have little effect on the backbone structure. Experiments have measured the effects of charybdotoxin mutations on the blockade of both modified voltage-gated *Shaker* channels (Goldstein *et al.*, 1994) and a

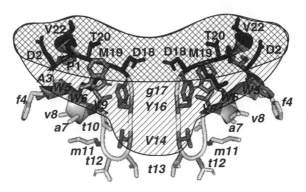

FIGURE 10. Illustration of effects of cysteine-scanning mutation mutagenesis on voltage-gated K$^+$ channels. Residues where side chains are accessible to large extracellular sulfhydryl reagents when mutated to cysteine are shaded black and are exposed in the crossed region (Pascual *et al.*, 1995; Kürz *et al.*, 1995). Residues where cysteine side chains can react with extracellular silver but not with large sulfhydryl reagents (Lü and Miller, 1995) have italicized letters and are in the lower shaded region. Residues where cysteine side chains cannot react with extracellular silver have lower case letters and, with the exception of G17p, are buried in the unshaded region.

Ca^{2+}-activated K$^+$ channel (Stampe *et al.*, 1994). In both cases the crucial residues for activity occur on the positively charged surface of the toxin. Studies strongly suggest that one of these positively charged residues, Lys27, extends into the pore where it competes with intracellular K$^+$ at one of the K$^+$ binding sites (Park and Miller, 1992). Using these guidelines, we docked an NMR-determined structure of charybdotoxin into our models of the pore's outer vestibule (Guy and Durell, 1995; see Fig. 11). To our surprise, the fit was remarkably good in the models we favor most. Lys27 extends into the pore where it interacts with D18p carboxyl groups and backbone amide groups; flanking Arg25 and Arg34 residues interact with D18p groups on each side of the pore's entrance; two additional lysines, Lys11 and Lys31, interact with negatively charged Glu or Asp residues at position 2p; and other crucial residues on the toxins interact with complementary groups on the channel. The fit is quite tight and all four subunits are involved extensively in toxin binding. Recent studies have identified some specific residue-residue interactions between the toxins and channels (Aiyar *et al.*, 1995; Goldstein *et al.*, 1994; Hidalgo and MacKinnon, 1995; Naranjo and Miller, 1995). These too have been taken into account in our current models.

7.2.   Segments Surrounding K$^+$ Channel P Segments

Models of the P segments alone do not consider the structure influences of the transmembrane segments that surround them. Most K$^+$ channels have complicated structures with numerous transmembrane segments per subunit. We have modeled S5 and S6 as transmembrane α-helices: the putative S6 helix is postulated to be broken near the intracellular entrance where two prolines occur in voltage-gated K$^+$ channels, with its final C-terminal portion on and parallel to the intracellular membrane surface. The

FIGURE 11. Model of how charybdotoxin binds in the outer vestibule of a *Shaker* K$^+$ channel. The channel is shaded gray. CTX residues where mutations have a large effect on the dissociation rate of the toxin are labeled and colored black; those where mutations have a moderate effect are colored gray. The remaining white residues either are buried within the core of the protein and were not mutated or are exposed on the upper surface where mutations have little effect. Data are from Goldstein *et al.* (1994). The amine group of K27 is postulated to extend into the pore where it competes with K$^+$ that enters the pore from inside the cell (Park and Miller, 1992).

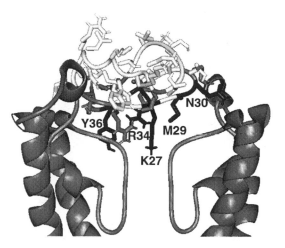

additional S1–S4 segments have not yet been included in these models. The most compelling direct evidence that S5 and S6 surround the P segments in voltage-gated K$^+$ channels comes from effects of mutations on channel blockade by intracellular TEA and 4-AP. Mutations of S6 residues at *Shaker* positions T469 and L472 alter blockade by intracellular TEA and/or their analogs (Aiyar *et al.*, 1994; Choi *et al.*, 1993; Lopez *et al.*, 1994; Shieh and Kirsch, 1994). In our models these residues interact with the P segment residues 11p and 12p, which we postulate to form at least part of the intracellular TEA binding site. Mutations of residues at position 396, which is the first hydrophobic residue of S5, and position 474, which occurs between the two prolines in S6, alter blockade by intracellular 4-AP (Shieh and Kirsch, 1994). In our model these two hydrophobic residues pack next to each other in the intracellular entrance of the pore. The effects of these mutations cannot be attributed to distant electrostatic effects since they involve only hydrophobic residues. These results have several implications. Segments S5 and S6 probably surround the P segments and form the intracellular portion of the pore that is in series with the P portion of the pore. If the relatively small TEA and 4-AP molecules bind to S5 and S6 near the pore's axis, then S5 and S6 must be tilted relative to the pore as illustrated in Fig. 9. The intracellular 4-AP binding site is probably at the pore's intracellular entrance since the first hydrophobic residue of S5 may form part of the 4-AP binding site. The intracellular TEA binding site is probably near the intracellular entrance since TEA and 4-AP are competitive antagonists and since S6 residues that alter the binding of TEA and 4-AP are only two residues apart.

Another reason to suspect that S5 and S6 surround the P segments is that some K$^+$ channels have no segments homologous to S1–S4. The family of K$^+$ channels that contains the inward rectifiers (IRK's) appear to have only two transmembrane segments in addition to the P segments, i.e., a hydrophobic M1 segment preceding P and a hydrophobic M2 segment following P (Ho *et al.*, 1993). Some bacterial K$^+$ channels may also have this simplified topology (Joiner *et al.*, 1995). M1 and M2 are probably homologous to the S5 and S6 segments in voltage-gated channels. Although the inward rectifying channel models were based primarily on the *Shaker* voltage-gated channel models, they satisfy our modeling criteria described in Section 3 extremely well. This is remarkable because only a few residues are identical in these two sequences. The break between the two putative helical parts of M2 occurs at a position occupied by aspartate in strongly rectifying channels and by asparagine in weakly rectifying channels. Mutagenesis experiments implicate this residue in the blockade of the pore from the inside by Mg^{2+} (Lu and MacKinnon, 1994; Wible *et al.*, 1994) and polyamines (Lopatin *et al.*, 1994). These blockades account for much or all of the voltage-dependent gating of the channel.

8.   MODELS OF Na^+ AND Ca^{2+} CHANNEL P SEGMENTS

We have recently developed a new generation of models of Na^+ and Ca^{2+} channel P segments (Guy and Durell, 1995). Like our original models (Guy and Seetharamulu, 1985) the P segments are postulated to form α-helical hairpins; however, now the ion-selective region and tetrodotoxin (TTX) and saxitoxin (STX) binding site is postulated to be formed by the central residues that link the two helices and/or by the residues that initiate the second helix. In addition, the helices are no longer postulated to be oriented parallel to the pore's axis, but rather tilted so that they form a cone-shaped outer vestibule (see Fig. 12). Although the initial P segment helix in these models is positioned similar to the helix in the K^+ channel model, the models differ in that the segment linking the putative helices is much longer in the K^+ channel models, and side chain rather than backbone atoms form the ion selectivity filter in the Na^+ and Ca^{2+} channel models.

To simplify the comparison of homologous sequences, the numbers in Fig. 7 have been adjusted to match the aligned prolines of the K^+ channels and of repeats II and IV of the Ca^{2+} channels. Mutagenesis experiments have identified the ion selective residues at position 14p as discussed earlier. The exact effect of mutating the E14p's of Ca^{2+} channels to lysine or glutamine depends upon which repeat is altered (Kim *et al.*, 1993; Mikala *et al.*, 1994; Yang *et al.*, 1993). Thus, in spite of the sequence similarity of the four repeats the pore is functionally asymmetric. Calcium channels appear to have two Ca^{2+} binding sites near the extracellular surface (Almers and McCleskey, 1984; Hess and Tsien, 1984). To explain these results, Yang *et al.* (1993) suggested that some of the 14p glutamic acids form one Ca^{2+} binding site near the extracellular entrance of the pore and others form an additional site farther down inside the pore.

In our models the four P segment hairpin structures are assembled with approximate fourfold symmetry about the axis of pore to form a cone-shaped outer vestibule. Residues exposed to water in the vestibule are primarily hydrophilic, and the residues buried between the helices and the other transmembrane segments are hydrophobic. The negatively charged E14p carboxylate groups form the selectivity filter at the narrowest portions of the pore. The E14p of repeats II and IV form a Ca^{2+} binding site near the extracellular entrance of the pore, and the E14p of repeats I and III form a second site farther down in the pore. This staggering of the E14p residues is made possible by modeling the segments connecting the two helices in repeats I and III differently from those in repeats II and IV (see Guy and Durell, 1995, for rationale). In addition to

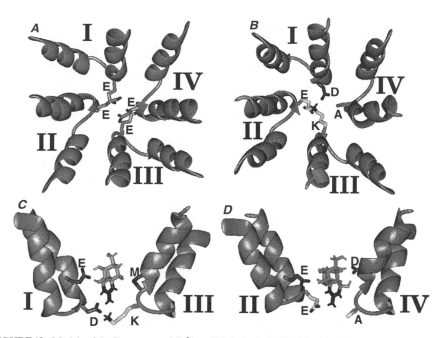

FIGURE 12. Models of the P segments of Ca^{2+} and Na^{+} channels. The 14p side chains crucial for selectivity are shown and labeled. Ribbons are α-helices and smaller tubes are loops connecting the helices. (A) Model of the four Ca^{2+} channel P segments as assembled in the pore viewed from outside the cell. E14p side chains in repeats II and IV are nearer the extracellular entrance than those in repeats I and III. (B) Model of the four P segments of Na^{+} channel P segments viewed from outside the cell. (C) Side view of Na^{+} channel P segments for repeats I and III with TTX bound in the outer entrance of the pore. Side chains crucial for TTX binding at positions 14p and 17p are shown and labeled. Positively charged guanidinium group of TTX is shaded black. The extracellular surface is at the top. (D) Same as C but with the channel and TTX rotated 90° about the pore's axis so that P segments of repeats II and IV are shown.

differences in the backbone structure, side-chain differences introduce additional asymmetry; e.g., in this model Ca^{2+} may bind off the pore's axis to D17p of repeat I and E17p of repeat IV in the outer entrance of the selectivity filter and to D15p of repeat II in the inner entrance.

The P segments of the Na^{+} channel are more difficult to model because their sequences differ more among the four repeats than they do for the Ca^{2+} channels. Thus, backbone conformations of the P segment from different repeats are likely to differ even more than in Ca^{2+} channels. Fortunately, however, more experimental data are available for Na^{+} channels than for Ca^{2+} channels. Not only have mutagenesis experiments indicated that the 14p residues of Na^{+} channels are crucial for ion selectivity, but also the binding of TTX and STX depends strongly upon the identity of the 14p and 17p residues in all four repeats (Terlau *et al.*, 1991). Conversely, mutation of the residues at positions 9p, 13p, 16p, and 18p appear to have little effect on the binding of these toxins. Figure 12 displays one of several helical hairpin models we developed for the binding of TTX in Na^{+} channels. In accordance with the data described above, each model satisfies the criterion of having almost all of the polar atoms of the toxin forming salt bridges or hydrogen bonds with the 14p and 17p residues of the P segments. Unfortunately, the positions of the helices, conformations of the connecting loops and side chains, and even the order of the repeats remain ambiguous in the Na^+ channel model.

9. FUTURE OF MEMBRANE CHANNEL PROTEIN MODELING

The need to develop reliable methods to model membrane channel protein structures and functional mechanisms should be obvious given the present rate at which membrane channel sequences are being determined, the prospect that this rate will increase sharply in the next few years as efforts to clone the human and other genomes increases, the rapidly increasing number of groups using site-directed mutagenesis and patch-clamp methods to study membrane channel proteins, the difficulty in determining membrane channel structure experimentally, and the importance of these proteins, especially in pharmacology. Present efforts to model membrane proteins are still in their infancy. Although existing methods have produced some promising models, most of these models have not been proven. Methods we use to develop the models usually have subjective elements and are often very time-consuming. These methods need to be verified, quantified, and automated to make the modeling objective and rapid at each phase of the hierarchy. Unfortunately, it appears that this may be difficult to accomplish for even Phase I models of many membrane channel proteins. The shortage of known protein structures limits our ability to quantify and evaluate predictive methods. Methods that scan the sequences for predominantly hydrophobic segments long enough to span the membrane as α-helices may work for many membrane proteins; however, it appears that they do not work for most of the channel proteins or peptides described here. If other channel proteins are as complex as our models of voltage-gated channels, the problem of structural prediction from sequence alone will indeed be very difficult.

The creation of our models was influenced greatly by our attempt to be consistent with the general picture of voltage-gated channels that has been developed over the last four decades of experimental work, including the more recent results using site-directed mutagenesis. Unfortunately, predictions that can be made based on the models are not usually quantitative. For example, we would like to be able to quantify the molecular basis for the relative ion permeabilities and conductances of a channel, why it opens and closes at a particular rate, why drugs and toxins bind with a particular affinity, and how modifying the protein sequence or drug structure affects these properties. If we could do this then mutagenesis experiments would be much better tests of models. There are several reasons to be optimistic that at least some of these goals can be accomplished: 1) Massive efforts are currently under way to understand how proteins fold and to quantify energies involved in these processes. Many methods developed for proteins and macromolecules in general should be applicable to membrane proteins. Several groups (Chiu et al., 1993; Roux and Karplus, 1994) are attempting to develop methods to calculate the energy an ion encounters as it passes through a complex channel. 2) Computer technology is improving rapidly. Computer graphics systems coupled to powerful computers to perform extensive energy calculations are now available at a reasonable price. The power of computation will continue to increase while the cost will continue to decrease. 3) Experimental determination of the photosynthetic reaction center structure (Deisenhofer *et al.*, 1985), bacteriorhodopsin (Henderson *et al.*, 1990), and porins (Cowan *et al.*, 1992; Schirmer *et al.*, 1995; Weiss *et al.*, 1992) has stimulated much interest in determining the structures of other membrane proteins. As more structures are determined it should be possible to test and improve modeling methods.

It is important to remember that experimental determination and theoretical prediction of protein structures are complementary rather than competing methodologies. For example, structures determined by methods such as X-ray crystallography are usually static representations, whereas functional processes such as the insertion of proteins into membranes, conformational changes, binding of substrate molecules, and the movement of ions through channels are dynamic processes. Understanding how these processes work will almost certainly require theoretical analyses. In addition to knowing what the structure is, it is also important to know how and why it forms. In summary, the future of structural determination of membrane channel proteins remains uncertain, but substantial progress has already been made and we have every reason to be optimistic that great progress is on the horizon.

10. REFERENCES

Aiyar, J., Ngyen, A. N., Chandy, K. G., and Grissmer, S. 1994, The P-region and S6 of Kv3.1 contribute to the formation of the ion conduction pathway, *Biophys. J.* **67**:2261.

Aiyar, J., Rizzi, J., Withka, J., Lee, C.-L., Andrews, G., Dethleffs, B., Simon, M., Gutman, G. A., and Chandy, K. G., 1995, Topological map of the receptor site for immunosuppressive peptides on a lymphocyte K$^+$ channel, Kv1.3, *Biophys. J.* **68**:A23.

Almers, W., and McCleskey, E. W., 1984, Non-selective conductance in calcium channels of frog muscle: Calcium selectivity in a single-file pore, *J. Physiol. (London)* **353**:585–608.

Anderson, J. A., Huprikar, S. S., Kochian, L. V., Lucas, W. J., and Gaber, R. F., 1992, Functional expression of a probable *Arabidopsis thaliana* potassium channel in *Saccharomyces cerevisiae*, *Proc. Natl. Acad. Sci. USA* **89**:3736–3740.

Arispe, N., Pollard, H. B., and Rojas, E., 1993, Giant multilevel cation channels formed by Alzheimer disease amyloic β-protein[Aβp-(1-40)] in bilayer membranes, *Proc. Natl. Acad. Sci. USA* **90**:10573–10577.

Armstrong, C. M., 1981, Sodium channels and gating currents, *Physiol. Rev.* **61**:644–683.

Armstrong, C. M., and Bezanilla, F., 1977, Inactivation of the sodium channel II: Gating current experiments, *J. Gen. Physiol.* **70**:567–590.

Baldwin, J. M., 1993, The probable arrangement of the helices in G protein-coupled receptors, *EMBO J.* **12**:1693–1703.

Boman, H. G., and Hultmark, D., 1987, Cell-free immunity in insects, *Annu. Rev. Microbiol.* **41**:103–126.

Bontems, F., Gilquin, G., Roumestand, C., Menez, A., and Toma, F., 1992. Analysis of side-chain organization on a refined model of charybdotoxin: Structural and functional implications, *Biochemistry* **31**:7756–7764.

Brooks, B. R., Bruccoleri, R. E., Olafson, B. D., States, D. J., Swaminathan, S., and Karplus, M., 1983, CHARMM: A program for macromolecular energy, minimization, and dynamics calculations, *J. Comput. Chem.* **4**:187–217.

Butler, A., Tsunoda, S., McCobb, D. P., Wei, A., and Salkoff, L., 1993, mSlo, a complex mouse gene encoding 'maxi' calcium-activated potassium channels, *Science* **261**:221–224.

Catterall, W. A., 1984, The molecular basis of neuronal excitability, *Science* **223**:653–661.

Chahine, M., George, A. L., Zhou, M., Ji, S., Sun, W., Barchi, R. L., and Horn, R., 1994, Sodium channel mutations in paramyotonia congenita uncouple inactivation from activation, *Neuron* **12**:281–294.

Chiu, S. W., Novotny, J. A., and Jakobsson, E., 1993, The nature of ion and water barrier crossings in a simulated ion channel, *Biophys. J.* **64:**98–108.

Choi, K. L., Mossman, C., Aubie, J., and Yellen, G., 1993, The internal quaternary ammonium receptor site of *Shaker* potassium channels, *Neuron* **10:**533–541.

Chothia, C., Levitt, M., and Richardson, D., 1981, Helix to helix packing in proteins, *J. Mol. Biol.* **145:**215.

Chou, H.-C., Heckel, A., Némethy, G., Rumsey, S., Carlacci, L., and Scheraga, H., 1990, Energetics of the structure and chain tilting of antiparallel β-barrels in proteins, *Proteins: Struct. Funct. Genet.* **8:**14–22.

Christensen, B., Fink, J., Merrifield, R. B., and Mauzerall, D., 1988, Channel-forming properties of cecropins and related model compounds incorporated into planar lipid membranes, *Proc. Natl. Acad. Sci. USA* **85:**5072–5076.

Chua, K., Tytgat, J., Liman, E., and Hess, P., 1992, Membrane topology of RCK1 K-channels, *Biophys. J.* **61:**A289.

Cornett, J. L., Cease, K. B., Margalit, H., Berzofsky, J. A., and Celisi, C., 1987, Hydrophobicity scales and computational techniques for detecting amphipathic structures in proteins, *J. Mol. Biol.* **195:**659–685.

Cowan, S. W., Schirmer, T., Rummel, G., Steiert, M., Ghosh, R., Pauptit, R. A., Jansonius, J. N., and Rosenbusch, J. P., 1992, Crystal structures explain functional properties of two *E. coli* porins, *Nature* **358:**727–733.

Cramer, W. A., Heymann, J. B., Schendel, S. L., Deriy, B. N., Cohen, F. S., and Elkins, P. A., and Stauffacher, 1994, C. V., Structure-function of the channel-forming colicins, *Annu. Rev. Biophys. Biomol. Struct.* **24:**611–641.

Crick, F. H. C., 1953. The packing of α-helices: Simple coiled-coils, *Acta Crystallogr.* **6:**689–697.

Cruciani, R. A., Barker, J. L., Durell, S. R., Raghunathan, G., Guy, H. R., Zasloff, M., and Stanley, E. F., 1992, Magainin 2: A natural antibiotic from frog skin forms ion channels in lipid bilayer membranes, *Eur. J. Pharmacol.* **226:**287–296.

DeBiasi, M., Kirsch, G. E., Drewe, J. A., Hartmann, H. A., and Brown, A. M., 1993, Cesium selectivity conferred by histidine substitution in the pore of the potassium channel Kv 2.1, *Biophys. J.* **64:**A341.

Deisenhofer, J., Epp, O., Mike, K., Huber, R., and Michel, H., 1985, Structure of the protein subunits in the photosynthetic reaction center of *Rhodopseudomonas viridis* at 3 Å resolution, *Nature (London)* **318:**618–623.

Dickinson, L., Russell, V., and Dunn, P. E., 1988, A family of bacteria-regulated, cecropin D–like peptides from *Manduca sexta*, *J. Biol. Chem.* 263:19424–19429.

Duclohier, H., 1994, Anion pores from magainin and related defensive peptides, *Toxicology* **87:**175–188.

Duclohier, H., Molle, G., and Spach, G., 1989, Antimicrobial peptide magainin I from *Xenopus* skin forms anion-permeable channels in planar lipid bilayers, *Biophys. J.* **56:**1017–1021.

Durell, S. R., and Guy, H. R., 1992, Atomic scale structure and functional models of voltage-gated potassium channels, *Biophys. J.* **62:**238–250.

Durell, S. R., and Guy, H. R., 1994, Modelling transmembrane helices in membrane proteins, *Biophys. J.* **66:**A63.

Durell, S. R., Raghunathan, G., and Guy, H. R., 1992, Modeling the ion channel structure of cecropin, *Biophys. J.* **63:**1623–1631.

Durell, S. R., Guy, H. R., Arispe, N., Rojas, E., and Pollard, H. B., 1994, Theoretical models of the ion channel structure of amyloid β-protein, *Biophys. J.* **67:**2137–2145.

Finer-Moore, J., and Stroud, R. M., 1984, Amphipathic analysis and possible formation of the ion channel in an acetylcholine receptor, *Proc. Natl. Acad. Sci. USA* **81:**155–159.

Fischbarg, J., Cheung, M., Czegledy, F., Li, J., Iserovich, P., Kuang, K., Hubbard, J., Garner, M., Rosen, O. M., Golde, D. W., and Vera, J. C., 1993, Evidence that facilitative glucose transporters may fold as β-barrels, *Proc. Natl. Acad. Sci. USA* **90:**11658–11662.

Forte, M., Guy, H. R., and Mannella, C. A., 1987, Molecular genetics of the VDAC ion channel: Structural model and sequence analysis, *J. Bioenerg. Biomembr.* **19:**341–350.

Giraudat, J., Dennis, M., Heidman, T., Chang, J. Y., and Changeux, J. P., 1986, Structure of the high-affinity binding site for noncompetitive blocker of the acetylcholine receptor: Serine-262 of the δ subunit is labeled by chlorpromazine, *Proc. Natl. Acad. Sci. USA* **83:**2719.

Go, M., and Miyazawa, S., 1980, Relationship between mutability, polarity and exteriority of amino acid residues in protein evolution, *Int. J. Pept. Protein. Res.* **15:**211–224.

Goldstein, S. A. N., Pheasant, D. J., and Miller, C., 1994, The charybdotoxin receptor of a Shaker K^+ channel: Peptide and channel residues mediating molecular recognition, *Neuron* **12:**1377–1388.

Guy, H. R., 1983, A model of colicin E1 membrane channel protein structure, *Biophys. J.* **41:**363a.

Guy, H. R., 1984, A structural model of the acetylcholine receptor channel based on partition energy and helix packing calculations, *Biophys. J.* **45:**249–261.

Guy, H. R., 1985, Amino acid side-chain partition energies and distribution of residues in soluble proteins, *Biophys. J.* **47:**61–70.

Guy, H. R., 1986, Review and revision of structural models for the transmembrane portion of the acetylcholine receptor channel, in: *Nicotinic Acetylcholine Receptor Structure and Function: NATO ASI Series*, Vol. H3 (A. Maelicke, ed.), Springer-Verlag, Berlin, Heidelberg, pp. 447–463.

Guy, H. R., 1988, A model relating the sodium channel's structure to its function, in: *Molecular Biology of Ionic Channels: Current Topics in Membrane Transport*, Vol. 33 (W. S. Agnew, T. Claudio, and F. J. Sigworth, eds.), Academic Press, San Diego, pp. 289–308.

Guy, H. R., 1990, Models of voltage- and transmitter-activated channels based on their amino acid sequences, in: *Monovalent Cations in Biological Systems* (C. A. Pasternak, ed.), CRC Press, Boca Raton, pp. 31–58.

Guy, H. R., and Durell, S. R., 1994, Using homology in modeling the structure of voltage-gated ion channels, In: *Molecular Evolution of Physiological Processes* (D. Fambrough, ed.), The Rockefeller University Press, New York, pp. 197–212.

Guy, H. R., and Durell, S. R., 1995, Structural model of Na^+, Ca^{2+}, and K^+ channels, in: *Ion Channels and Genetic Diseases* (D. Dawson, ed.), The Rockefeller University Press, New York, pp. 1–16.

Guy, H. R., and Hucho, F., 1987, The ion channel of the nicotinic acetylcholine receptor, *Trends Neurosci.* **10:**318–321.

Guy, H. R., and Raghunathan, G., 1989, Structural models of membrane insertion and channel formation by antiparallel α-helical membrane peptides, in: *Transport Through Membranes: Carriers, Channels and Pumps* (A. Pullman, J. Jortner, and B. Pullman, eds.), Kluwer Academic Publishers, Dordrecht, Boston, London, pp. 369–379.

Guy, H. R., and Seetharamulu, P., 1985, Molecular model of the action potential sodium channel, *Proc. Natl. Acad. Sci. USA* **83:**508–512.

Guy, H. R., Durell, S. R., Warmke, J., Drysdale, R., and Ganetzky, B., 1991, Similarities in amino acid sequences of *Drosophila eag* and cyclic nucleotide gated channels, *Science* **254:**730.

Hartmann, H. A., Kirsch, G. E., Drewe, J. A., Taglialatela, M., Joho, R. H., and Brown, A. M., 1991, Exchange of conduction pathways between two related K^+ channels, *Science* **251:**942–944.

Heginbotham, L., and MacKinnon, R., 1992, The aromatic binding site for tetraethylammonium ion on potassium channels, *Neuron* **8:**483–491.

Heginbotham, L., Abramson, T., and MacKinnon, R., 1992. A functional connection between the pores of distantly related ion channels as revealed by mutant K^+ channels, *Science* **258:**1152–1155.

Heginbotham, L., Lu, Z., Abramson, T., and MacKinnon, R., 1994, Mutations in the K^+ channel signature sequence, *Biophys. J.* **66:**1061–1067.

Heinemann, S. H., Terlau, H., Stühmer, W., Imoto, K., and Numa, S., 1992, Calcium channel characteristics conferred on the sodium channel by single mutations, *Nature* **356:**441–443.

Henderson, R., Baldwin, J. M., Ceska, T. A., Zemlin, F., Beckmann, E., and Downing, K. H., 1990, Model for the structure of bacteriorhodopsin based on high-resolution electron cryomicroscopy, *J. Mol. Biol.* **213:**899–929.

Hess, P., and Tsien, R. W., 1984, Mechanism of ion permeation through calcium channels, *Nature* **309:**453–456.

Hidalgo, P., and MacKinnon, R., 1995, Revealing the architecture of a K^+ channel pore through mutant cycles with a peptide inhibitor, *Science* **268:**307–310.

Hille, B., 1975, The receptor for tetrodotoxin and saxitoxin: A structural hypothesis, *Biophys. J.* **15:**615–619.

Hille, B., 1977, Local anesthetics: Hydrophilic and hydrophobic pathways for the drug-receptor reaction, *J. Gen. Physiol.* **69:**497–515.

Ho, K., Nichols, C. G., Lederer, W. J., Lytton, J., Vassilev, P. M., Kanazirska, M. V., and Hebert, S. C., 1993, Cloning and expression of an inwardly rectifying ATP-regulated potassium channel, *Nature* **262:**31–38.

Hol, W. G. J., Halie, L. M., and Sander, C., 1981, Dipoles of the α-helix and β-sheet: Their role in protein folding, *Nature* **294:**532–536.

Holak, T. A., Engström, Å., Kraulis, P. J., Lindeberg, G., Bennich, H., Jones, T. A., Gronenborn, A. M., and

Clore, G. M., 1988, The solution conformation of the antibacterial peptide cecropin A: A nuclear magnetic resonance and dynamical simulated annealing study, *Biochemistry* **27**:7620–7629.

Hucho, F., Oberthür, W., and Lottspeich, F., 1986, The ion channel of the nicotinic acetylcholine receptor is formed by the homologous helices M II of the receptor subunits, *FEBS Lett.* **205**:137.

Jackson, M., Mantsch, H. H., and Spencer, J. H., 1992, Conformation of magainin-2 and related peptides in aqueous solution and membrane environments probed by fourier transform infrared spectroscopy, *Biochemistry* **31**:7289–7293.

Jeanteur, D., Lakey, J. H., and Pattus, F., 1991, The bacterial porin superfamily: Sequence alignment and structural prediction, *Mol. Microbiol.* **5**:2153–2164.

Joiner, W., Yang, Y., Sigworth, F., and Kaczmarek, L., 1995, LCTB is a K-selective prokaryotic relative of eukaryotic ion channels, *Biophys. J.* **68**:A148.

Kaupp, U. B., Niidome, T., Tanabe, T., Terada, S., Bonigk, W., Stühmer, W., Cook, N. J., Kangawa, K., Matsuo, H., Hirose, T., Miyata, T., and Numa, S., 1989, Primary structure and functional expression from complementary DNA of the rod photoreceptor cyclic GMP–gated channel, *Nature* **342**:762–766.

Kim, M.-K., Morii, T., Sun, L.-X., Imoto, K., and Mori, Y., 1993, Structural determinants of ion selectivity in brain calcium channels, *FEBS Lett.* **318**:145–148.

Kosower, E. M., 1983, Partial tertiary structure assignment for the acetylcholine receptor on the basis of the hydrophobicity of amino acid sequences and channel location using single group rotation theory, *Biochem. Biophys. Res. Commun.* **111**:1022.

Kosower, E. M., 1987, A structural and dynamic model for the nicotinic acetylcholine receptor, *Eur. J. Biochem.* **168**:431.

Kreusch, A., and Schulz, G. E., 1994, Refined structure of the porin from *Rhodobacter blastica, J. Mol. Biol.* **243**:891–905.

Kürz, L. L., Zühlke, R. D., Zhang, H-J., and Joho, R. H., 1995, Side-chain accessibilities in the pore of a K⁺ channel probed by sulfhydryl-specific reagents after cysteine-scanning mutagenesis, *Biophys. J.* **68**:900–905.

Lazarovici, P., Primor, N., Gennaro, J., Fox, J., Shai, Y., Lelkes, P. I., Caratsch, C. G., Raghunathan, G., Guy, H. R., Shih, Y. L., and Edwards, C., 1990, Origin, chemistry and mechanisms of action of pardaxin; a repellent, presynaptic excitatory, ionophore polypeptide, in: *Marine Toxins: Origin, Structures, and Pharmacology* (S. Hall, ed.), American Chemical Society, pp. 348–364.

Lazarovici, P., Edwards, C., Raghunathan, G., and Guy, H. R., 1992, Secondary structure, permeability, and molecular modeling of paradaxin pores, *J. Nat. Toxins* **1**:1–15.

Lopatin, A. N., Makhina, E. N., and Nichols, C. G., 1994, Potassium channel block by cytoplasmic polyamines as the mechanism of intrinsic rectification, *Nature* **372**:366–369.

Lopez, G. A., Jan, Y. N., and Jan, L. Y., 1994, Evidence that the S6 segment of the *Shaker* voltage-gated channel comprises part of the pore, *Nature* **367**:179–182.

Lü, Q., and Miller, C., 1995, Silver as a probe of pore-forming residues in a potassium channel, *Science* **268**:304–307.

Lü, Z., and MacKinnon, R., 1994, Electrostatic tuning of Mg²⁺ affinity in an inward-rectifier K⁺ channel, *Nature* **371**:243–246.

MacKinnon, R., and Yellen, G., 1990, Mutations affecting TEA blockade and ion permeation in voltage-activated K⁺ channels, *Science* **250**:276–279.

Matsuzaki, K., Murase, O., Fujii, N., and Miyajima, H., 1995, Translocation of an antimicrobial peptide, magainin 2, across lipid bilayers through forming a multimeric pore, *Biophys. J.* **68**:A126.

Mchaourab, H. S., Hyde, J. S., and Feix, J. B., 1994, Binding and state of aggregation of spin-labeled cecropin AD in phospholipid bilayers: Effects of surface charge and fatty acyl chain length, *Biochemistry* **33**:6691–6699.

Mikala, G., Bahinski, A., Yatani, A., Tang, S., and Schwartz, A., 1994, Differential contribution by conserved glutamate residues to an ion-selectivity site in the L-type Ca²⁺ channel pore, *FEBS Lett.* **335**:265–269.

Milkman, R., and McKane-Bridges, M., 1993, An *E. coli* homologue of eukaryotic potassium channels, Genebank entry ECOKCH.

Miyazawa, S., and Jernigan, R. A., 1985, Estimation of effective inter-residue contact energies from protein crystal structures: Quasi-chemical approximation, *Macromolecules* **18**:534–552.

Naranjo, D., and Miller, C., 1995, Counting points of interaction between a *Shaker* K⁺ channel vestibule and a peptide toxin, *Biophys. J.* **68**:A24.

Newland, C. F., Adelman, J. P., Tempel, B. L., and Almers, W., 1992, Repulsion between tetraethylammonium ions in cloned voltage-gated potassium channels, *Neuron* **8**:978–982.

Neyton, J., and Miller, C., 1988, Discrete Ba^{2+} block as a probe of ion occupancy and pore structure in the high-conductance Ca^{2+}-activated K$^+$ channel, *J. Gen. Physiol.* **92**:569–586.

Noda, M., Ikeda, T., Kayano, T., Suzuki, H., Takeshima, H., Takahashi, T., Kuno, M., and Numa, S., 1986, Existence of distinct sodium channel messenger RNAs in rat brain, *Nature* **320**:188–192.

Okada, M., and Natori, S., 1985, Primary structure of sarcotoxin I, an antibacterial protein induced in the hemolymph of *Sarcophaga peregrina* (flesh fly) larvae, *J. Biol. Chem.* **260**:7174–7177.

Overington, J., Donnelly, D., Johnson, M. S., Sali, A., and Blundell, T. L., 1992, Environment-specific amino acid substitution tables: Tertiary templates and prediction of protein folds, *Protein Sci.* **1**:216–226.

Papazian, D. M., Timpe, L. C., Jan, Y. N., and Jan, L. Y., 1991, Alteration of voltage-dependence of *Shaker* potassium channel by mutations in the S4 sequence, *Nature* **349**:305–310.

Park, C. S., and Miller, C., 1992, Interaction of charybdotoxin with permeant ions inside the pore of a K$^+$ channel, *Neuron* **9**:307–313.

Pascual, J. M., Shieh, C. C., and Brown, A. M., 1995, Multiple charged and aromatic amino acids contribute to block by tetraethylammonium in potassium channels, *Biophys. J.* **68**:A31.

Patton, D. E., West, J. W., Catterall, W. A., and Goldin, A. L., 1993, A region critical for sodium channel inactivation functions as an inactivation gate in a potassium channel, 11th International Biophysics Congress, p. 133.

Peled, H., and Shai, Y., 1993, Membrane interaction and self-assembly within phospholipid membranes of synthetic segments corresponding to the H-5 region of the *Shaker* K$^+$ channel, *Biochemistry* **32**:7879.

Popot, J.-L., Gerchman, S.-E., and Engelman, D. M., 1987, Refolding of bacteriorhodopsin in lipid bilayers: A thermodynamically controlled two stage process, *J. Mol. Biol.* **198**:655.

Raghunathan, G., Seetharamulu, P., Brooks, B. R., and Guy, H. R., 1990, Models of δ-hemolysin membrane channels and crystal structures, *Proteins: Struct. Funct. Genet.* **8**:213–225.

Rees, D. C., DeAntonio, L., and Eisenberg, D., 1989, Hydrophobic organization of membrane proteins, *Science* **245**:510–513.

Roux, B., and Karplus, M., 1994, Molecular dynamics simulations of the gramicidin channel, *Annu. Rev. Biophys. Biomol. Struct.* **23**:731–761.

Schirmer, T., Keller, T. A., Wang, Y. F., and Rosenbusch, J. P., 1995, Structural basis for sugar translocation through maltoporin channels at 3.1 Å resolution, *Science* **267**:512–514.

Shen, N. V., Chen, X., Boyer, M. M., and Pfaffinger, P. J., 1993, Deletion analysis of K$^+$ channel assembly, *Neuron* **11**:67–76.

Shieh, H.-C., and Kirsch, G. E., 1994, Mutational analysis of ion conduction and drug binding sites in the inner mouth of voltage-gated K$^+$ channels, *Biophys. J.* **67**:2316–2325.

Shao, X. M., and Papazian, D. M., 1993, S4 mutations alter the single-channel gating kinetics of *Shaker* K$^+$ channels, *Neuron* **11**:343–352.

Slatin, S. L., Qiu, X-Q., Jakes, K. S., and Finkelstein, A., 1994, Identification of a translocated protein segment in a voltage-dependent channel, *Nature* **371**:158–161.

Stampe, P., and Begenisich, T., 1995, The pore of *Shaker* K$^+$ channels can simultaneously be occupied by at least three ions, *Biophys. J.* **68**:A129.

Stampe, P., Kolmakova-Partensky, L., and Miller, C., 1994, Intimations of K$^+$ channel structure from a complete functional map of the molecular surface of charybdotoxin, *Biochemistry* **33**:443–450.

Stühmer, W., Conti, F., Suzuki, H., Wang, X., Noda, M., Yahagi, N., Kubo, H., and Numa, S., 1989, Structural parts involved in activation and inactivation of the sodium channel, *Nature* **339**:597–603.

Tanabe, T., Takeshima, H., Mikami, A., Flockerzi, V., Takahashi, H., Kangawa, K., Kojima, M., Matsuo, H., Hirose, T., and Numa, S., 1987, Primary structure of the receptor for calcium channel blockers from skeletal muscle, *Nature* **328**:313–318.

Tappin, M. J., Pastore, A., Norton, R. S., Freer, J. H., and Campbell, I. D., 1988, High-resolution NMR study of the solution structure of δ-hemolysin, *Biochemistry* **27**:1643–1647.

Taylor, W. R., Jones, D. T., and Green, N. M., 1994, A method for α-helical integral membrane protein fold prediction, *Proteins: Struct. Funct. Genet.* **18**:281–294.

Tempel, B. L., Papazian, D. M., Schwarz, T. L., Jan, Y. N., and Jan, L. Y., 1987, Sequence of a probable potassium channel component encoded at *Shaker* locus of *Drosophila*, *Science* **237**:770–775.

Terlau, H., Heinemann, S. H., Stühmer, W., Pusch, M., Conti, F., Imoto, H., and Numa, S., 1991, Mapping the site of tetrodotoxin and saxitoxin of sodium channel II, *FEBS Lett.* **293:**93–96.

Terwilliger, T. C., Weissman, L., and Eisenberg, D., 1982, The structure of melittin in the form I crystals and its implication for melittin's lytic and surface activities, *Biophys. J.* **37:**353–361.

Thomas, D. H., Rice, D. W., and Fitton, J. E., 1986, Crystallization of the delta toxin of *Staphylococcus aures, J. Mol. Biol.* **192:**675.

Vassilev, P. M., Scheuer, T., and Catterall, W. A., 1988, Identification of an intracellular peptide segment involved in sodium channel inactivation, *Science* **241:**1658–1664.

Warmke, J., Drysdale, R., and Ganetzky, B., 1991, A distinct potassium channel polypeptide encoded by the *Drosophila eag* locus, *Science* **252:**1560–1562.

Weiss, M. S., Abele, U., Weckesser, J., Welte, W., Schiltz, E., and Schulz, G. E., 1991, Molecular architecture and electrostatic properties of a bacterial porin, *Science* **254:**1627–1630.

Wible, B. A., Taglialatela, M., Ficker, E., and Brown, A. M., 1994, Gating of inwardly rectifying K^+ channels localized to a single negatively charged residue, *Nature* **371:**246–349.

Yang, J., Ellinor, P. T., Sather, W. A., Zhang, J.-F., and Tsien, R. W., 1993, Molecular determinants of Ca^{2+} selectivity and ion permeation in L-type Ca^{2+} channels, *Nature* **366:**158–161.

Yellen, G., Jurman, M. E., Abramson, T., and MacKinnon, R., 1991, Mutations affecting internal TEA blockade identify the probable pore-forming region of a K^+ channel, *Science* **251:**939–942.

Zasloff, M., Martin, B., and Chen, H-C., 1988, Antimicrobial activity of synthetic magainin peptides and several analogues, *Proc. Natl. Acad. Sci. USA* **85:**910–913.

Zagotta, W. N., Hoshi, T., Aldrich, R. W., 1990, Restoration of inactivation in mutants of *Shaker* potassium channels by a peptide derived from ShB, *Science* **250:**568–571.

CHAPTER 2

STRUCTURAL AND FUNCTIONAL DIVERSITY OF VOLTAGE-ACTIVATED CALCIUM CHANNELS

MICHEL DE WAARD, CHRISTINA A. GURNETT, and KEVIN P. CAMPBELL

1. INTRODUCTION

1.1. Voltage-Dependent Ca^{2+} Channels as Pathways for Intracellular Ca^{2+} Elevation

Action potentials and neurotransmitters are the vehicle of information transfer in the nervous system. The diversity of this information is in part encoded by the quantitative contribution and properties of the ion channels underlying both the action potential and the events that follow. As such, voltage-dependent Ca^{2+} channels represent one class of these ion channels that have both a remarkable ubiquity and importance in a variety of cell types. At rest, there is normally a 10,000-fold concentration difference between intracellular and extracellular Ca^{2+} concentrations. Intracellular Ca^{2+} concentrations are normally in the 10–100 nM range, whereas outside cells, the Ca^{2+} concentration is about 1–2 mM. Elevations of the intracellular Ca^{2+} concentrations can occur not only as a result of the increase in activity of voltage-sensitive Ca^{2+} channels, but also in response to the activation of ligand-gated ion channels (NMDA-sensitive glutamatergic receptors or nicotinic ACh receptors) or through the release of Ca^{2+} from intracellular pools (IP_3- and/or ryanodine-sensitive). During increases in voltage-dependent Ca^{2+} channel activity, the cytosolic Ca^{2+} can reach concentrations up to 100 μM immediately beneath the

MICHEL DE WAARD, CHRISTINA A. GURNETT, and KEVIN P. CAMPBELL • Howard Hughes Medical Institute, Department of Physiology and Biophysics, University of Iowa College of Medicine, Iowa City, Iowa 52242.

Ion Channels, Volume 4, edited by Toshio Narahashi, Plenum Press, New York, 1996.

plasma membrane. This rise in intracellular Ca^{2+} is generally not homogeneous but instead seems to be highly organized in space and time. The buffering activity of numerous cellular proteins, which bind Ca^{2+} tightly and rapidly, limits the spatial diffusion of transient cytosolic Ca^{2+} increases and keeps the overall Ca^{2+} concentration to a low free level. Several other mechanisms also ensure an effective buffering of free cytoplasmic Ca^{2+} in the cell. In the plasma membrane, the Ca^{2+}-ATPase and the electrogenic Na^+/Ca^{2+} exchanger can effectively extrude Ca^{2+} from the cell. Several intracellular Ca^{2+} stores (endoplasmic reticulum and mitochondria) will also accumulate Ca^{2+} by activating a Ca^{2+}-ATPase or a $Ca^{2+}/2H^+$ exchanger.

Physiologically, Ca^{2+} acts as an intracellular second messenger by initiating or regulating numerous biochemical and electrical events of the cell. Calcium ions are implicated in the regulation of several enzymes and in the control of gene expression. They are essential for contraction in all types of muscle and for the control of the activity of several other ion channels (i.e., Ca^{2+}-activated K^+ and Cl^- channels). They also control many neuronal events such as neurotransmitter release (Katz, 1969; Augustine *et al.*, 1987), synaptogenesis, and neurite outgrowth.

Thus it is increasingly evident that a diverse number of voltage-dependent Ca^{2+} channels are required to account for the complex spatiotemporal characteristics of Ca^{2+} entry and for the numerous cellular functions of Ca^{2+} ions. In fact, several types of voltage-dependent Ca^{2+} channels have now been identified that differ by their functional properties. In this chapter, we will review the structural features that are the basis of the biophysical, pharmacological, and functional diversity of voltage-dependent Ca^{2+} chan-nels. In particular, we will emphasize the structural determinants that control the subunit composition of Ca^{2+} channels and the role of ancillary subunits in Ca^{2+} channel diversity and function.

1.2. Biophysical Diversity of Voltage-Dependent Ca^{2+} Channels

Five classes of voltage-dependent Ca^{2+} channels have clearly been defined so far based on their biophysical and pharmacological properties (Table I). The brain has the most complex and diverse expression of Ca^{2+} channels as it expresses at least one member of each class. The nomenclature adopted is the one proposed by Nowycky *et al.* (1985) and Fox *et al.* (1987a,b) based on recordings of the peripheral nervous system. This nomenclature was subsequently expanded to include Ca^{2+} current recordings from cells of the central nervous system (Mintz *et al.*, 1992a,b; Usowicz *et al.*, 1992; Ellinor *et al.*, 1993). To date, two main categories of Ca^{2+} channels have been identified in excitable cells: a low voltage–activating (LVA) channel (T-type), which is also rapidly inactivating, and several high voltage–activating (HVA) channels (L-, N-, P-, and R-type). Other than their activation threshold, numerous functional criteria have been used to discriminate among voltage-dependent Ca^{2+} channels. The identification of these channels relies primarily on their conductance, the time- and voltage-dependence of inactivation, the relative permeabilities to several divalent cations, the distribution of open- and close-time durations, and the identification of several gating models. The following sections illustrate the properties of these LVA and HVA channels.

T-type Ca^{2+} channels have been identified in several tissues, including neuronal, smooth muscle, cardiac, and skeletal muscle cells. They are activated by relatively small

TABLE I
Biophysical Classification of LVA and HVA Ca^{2+} Channels[a]

Property	T	L	N	P	R
Conductance	7 to 10 pS	11 to 25 pS	10 to 22 pS	9 to 19 pS	14 pS
Activation threshold	>−70 mV LVA	>−30 mV HVA	>−30 mV HVA	>−40 mV HVA	>−40 mV HVA
Inactivation range	−110 to −50 mV	−60 to −10 mV	−120 to −30 mV	?	−100 to −40 mV
Inactivation rate	14 to 80 msec	>500 msec	50 to >500 msec	>500 msec	20 to 40 msec
Relative conductance	Ba^{2+} = Ca^{2+}	Ba^{2+} > Ca^{2+}	Ba^{2+} > Ca^{2+}	?	Ba^{2+} > Ca^{2+}
Open-time duration	0.5 to 2 msec	0.5 to 10 msec	0.7 to 1.5 msec	<1 msec	?
Gating modes	?	2	3	?	?

[a]The conductance values are estimated from recordings using 90–110 mM Ba^{2+} as the charge carrier. The two gating modes of L-type channels are based on switching of channel activity from a short-lived open-time duration to a long-lived one. The N-type channel has three gating modes that are due to the switching of the channel between low-, medium-, and high-open state probabilities (Delcour and Tsien, 1993).

depolarizations compared to other Ca^{2+} channel types (> −70 mV) and reach peak current amplitude at about −30 mV. Remarkably, these LVA channels inactivate very rapidly with time constants on the order of 10–80 msec and at very negative membrane potentials ($V_{1/2}$ < −50 mV). This set of properties explains the importance of LVA channels in pacemaker activity (Hagiwara et al., 1988). T-type channels also have the lowest conductance of all voltage-dependent Ca^{2+} channels, with values that range between 7 and 10 pS with 100 mM Ba^{2+} as the charge carrier. Unlike HVA channels, LVA channels are equally permeable to Ca^{2+} and Ba^{2+}, suggesting differences in the ion channel structure responsible for ionic selectivity (Nilius et al., 1985).

L-type Ca^{2+} channels probably represent the most heterogeneous group of voltage-dependent Ca^{2+} channels, as several subtypes have been identified in a wide range of tissues from muscle to brain. These HVA channels typically have a conductance between 11 and 25 pS and are long-lasting with little time- or voltage-dependent inactivation. The functional properties of L-type Ca^{2+} channels are most similar in cardiac, secretory, and neuronal cells. However, the skeletal muscle L-type channel is significantly different in that it activates one order of magnitude more slowly than cardiac or smooth muscle L-type channels. This channel is also characterized by a smaller conductance (~11 pS) than most other L-type Ca^{2+} channels. The time course of inactivation can also help in discriminating between the various L-type Ca^{2+} channels. While the kinetics of inactivation of the L-type channels are very slow in neurons and secretory cells, they are more rapid in cardiac and vascular muscle (Kass and Sanguinetti, 1984; Fedulova et al., 1985). There is evidence that functionally different L-type Ca^{2+} channels can coexist in the same cell type. Forti and Pietrobon (1993) have described the presence of no less than three different dihydropyridine (DHP)-sensitive Ca^{2+} channels in rat cerebellar granule neurons. Two of the channels have cardiac-like properties, and another channel was characterized by an unusual increase in gating activity upon membrane repolarization.

N-type Ca^{2+} channels are probably the best characterized of the neuronal HVA channels. They are characterized by a greater sensitivity to holding potential and complex gating behaviors (Delcour and Tsien, 1993; Plummer and Hess, 1991). Their conductances are on average smaller than L-type channels and range between 12 and 20 pS (Fox et al., 1987b; Plummer et al., 1989; Hirning et al., 1988). These channels are completely inactivated at -30 mV, a membrane potential where L-type channels remain largely active. The inactivation time course of N-type Ca^{2+} channels can be both fast (time constants in the tens of milliseconds) or slow (time constants in the hundreds of milliseconds). In fact, individual N-type Ca^{2+} channels can carry two kinetically distinct components, and the switching between both components seems to be under the control of an unidentified cellular factor (Plummer and Hess, 1991).

The existence of a distinct P-type Ca^{2+} channel was recently proposed on the basis of Ca^{2+} current recordings in Purkinje neurons (Mintz et al., 1992a). P-type channels activate for potentials above -40 mV and show no or very little time-dependent inactivation during depolarization (time constant in the range of a second) (Regan et al., 1991; Mintz et al., 1992a,b). The unitary conductance of P-type channels remains a complex issue. Usowicz et al. (1992) has recently demonstrated the coexistence of no less than three different conductances (9, 13, and 19 pS) for this channel in Purkinje neurons. It has not yet been resolved whether these conductances are due to the activity of a single Ca^{2+} channel type or of several subtypes of the P channel. Overall, these channels are probably best identified by a different pharmacological sensitivity than N- and L-type Ca^{2+} channels.

In contrast to T-, L- and N-type Ca^{2+} channels, there is more controversy concerning the identity of R-type Ca^{2+} channels that have recently emerged in the biophysical classification (Randall et al., 1993). R-type Ca^{2+} channels are found in cerebellar granule cells and represent another class of HVA channels. Like N- and T-type Ca^{2+} channels, the R-type channel inactivates at fairly hyperpolarized potentials ($V_{1/2} = -70$ mV). Interestingly, R-type channels also have very fast time-dependent inactivation (time constant between 20 and 40 msec), further enhancing their resemblance to T-type channels.

It is worth mentioning that voltage-sensitive Ca^{2+} channels can be distinguished by their properties of cell regulation, tissue-specific expression, and cell compartmentalization. For instance, LVA Ca^{2+} channels have been successfully isolated after selective run-down of HVA currents (Carbone and Lux, 1987; Nilius et al., 1985). Also, L-type Ca^{2+} channels in smooth or cardiac muscles can be identified by their process of Ca^{2+}-dependent inactivation (Kalman et al., 1988). However, due to overlapping biophysical properties of the various Ca^{2+} channel types, it is now believed that the best discriminating criteria of these channels are their pharmacological susceptibilities to blockage by specific drugs and toxins.

1.3. Pharmacological Diversity of Voltage-Dependent Ca^{2+} Channels

Table II summarizes the pharmacological properties of the various Ca^{2+} channel types. Although the list of compounds that target voltage-dependent Ca^{2+} channels seems to increase exponentially, currently only three classes of drugs efficiently discriminate between functionally different Ca^{2+} channels. First, 1,4-dihydropyridines are synthetic organic compounds that have been useful in identifying a class of Ca^{2+}

TABLE II
Pharmacological Sensitivity of LVA and HVA Ca^{2+} Channels[a]

Drug	T	L	N	P	R
Dihydropyridines	− (10 μM)	+ Kd < 37 nM	− (10 μM)	− (10 μM)	− (10 μM)
Phenylalkylamines	− (50 μM)	+ Kd < 10 μM	− (10 μM)	?	?
Benzothiazepines	− (10 μM)	+ Kd < 1 μM	− (10 μM)	?	?
ω-CgTx GVIA	− (5 μM)	−/+ (R) (1 μM)	+ (I) Kd = 0.7–30 nM	− (5 μM)	− (5 μM)
ω-CgTx MVIIC	− (5 μM)	− (10 μM)	+ Kd = 1–10 μM	+ Kd < 1 μM	− (5 μM)
ω-Aga IVA	− (200 nM)	− (200 nM)	− (800 nM)	+ Kd = 2–10 nM	− (100 nM)
ω-Aga IIIA	− (100 nM)	+ Kd = 1 nM	+ Kd = 1 nM	+ Kd = 0.5 nM	?
Cd^{2+}	+ Kd > 40 μM	+ Kd = 1.5 μM	+ Kd = 1 μM	+ Kd ?	+ Kd = 1 μM
Ni^{2+}	+ Kd < 40 μM	+ Kd = 230 μM	+ Kd = 270 μM	+ Kd ?	+ Kd = 65 μM

[a]The sensitivity of voltage-dependent Ca^{2+} channels to organic (classical Ca^{2+} channel antagonists and various toxins) and inorganic molecules (Ni^{2+} and Cd^{2+}) is shown. DHPs, ω-CgTx GVIA, and ω-Aga IVA are the most specific channel ligands and thus discriminate effectively between L-, N-, and P-type channels, respectively. The highest ineffective drug concentration is shown in parentheses. The effective concentrations of the various drugs are given for Ca^{2+} current recording conditions. When the effect of the molecule depends on the extracellular Ca^{2+} concentration, the effective drug concentrations are given for an extracellular permeant ion concentration of 5–10 mM.

channels found in muscle and several neurons and neuroendocrine cells. For instance, L-type Ca^{2+} currents in rat insulinoma cells are blocked by nimodipine with an apparent K_d less than 37 nM (Pollo et al., 1993). Although L-type Ca^{2+} channels are all sensitive to DHPs, several subtypes of L channels exist that are characterized by small variations in their pharmacological and biophysical properties. For instance, skeletal and cardiac muscle L-type channels have distinctive biophysical properties despite closely related structures and functions. The dissociation constant of DHP for skeletal muscle L-type channel is greater than that for brain and cardiac muscle (Glossmann and Striessnig, 1988). Also, higher concentrations of DHP are required to block HVA channels in neurons than in smooth muscle cells (Triggle and Janis, 1987). Skeletal muscle L-type channels are not only sensitive to DHPs, but also to phenylalkylamines, benzothiazepines, and piperazines, thus defining four pharmacological binding sites that interact allosterically among each other (Catterall and Striessnig, 1992). It is still not well established whether all L-type Ca^{2+} channels, particularly those in brain, are also sensitive to these additional classes of drugs. Also, DHP antagonists are more selective and potent blockers for smooth muscle than for cardiac muscle L-type Ca^{2+} channels despite a minimum of 95% amino acid identity between the α_1 subunits of both channels (Welling *et al.*, 1993). Subtle pharmacological differences could be very useful to distinguish among subtypes of the L-type family.

Second, ω-conotoxin GVIA (ω-CgTx GVIA), a Ca^{2+} channel antagonist from the venom of the piscivorous marine mollusk *Conus geographus*, specifically blocks N-type channels. In early studies of the N-type Ca^{2+} channels in PC12 cells, ω-CgTx GVIA was shown to irreversibly block a component of the high threshold–activated, DHP-insensitive, Ca^{2+} current (Plummer *et al.*, 1989). Also, the toxin blocks N-type Ca^{2+} channels from sympathetic neurons with high affinity ($K_d = 0.7$ nM) (Boland *et al.*, 1994). DHP-sensitive Ca^{2+} channels were not affected by the application of ω-CgTx GVIA, demonstrating the specificity of the block. In addition, $[^{125}I]$-ω-CgTx GVIA binding in the brain is not inhibited by DHPs. However, the toxin specificity is probably not absolute because in rare cases some L-type Ca^{2+} channels can be reversibly blocked by high concentrations of ω-CgTx GVIA (Kasai and Neher, 1992). Thus, irreversible and high-affinity blocking by ω-CgTx GVIA is adopted as the defining characteristic of N-type Ca^{2+} channels.

Third, ω-agatoxin IVA (ω-Aga IVA) is a 48–amino acid peptide antagonist isolated from the spider venom of *Agelenopsis aperta* that is selective for the P-type Ca^{2+} current (Mintz *et al.*, 1992a,b). The venom of this spider contains a second toxin ω-Aga IVB that has 71% amino acid identity with ω-Aga IVA and shares its specificity and potency ($K_d = 3$ nM) for P-type channels (Adams *et al.*, 1993). Interestingly, the blocking of P-type currents by both ω-Aga IVA and ω-Aga IVB can quickly be reversed by large depolarizations (Mintz *et al.*, 1992a), demonstrating that the toxin binding site is dependent on the conformation of the channel in the same manner that the DHP binding site depends on the L-type channel conformation. Although P-type Ca^{2+} channels were initially characterized by their lack of sensitivity to DHP antagonists and to ω-CgTx GVIA, they are potently inhibited by ω-CgTx MVIIC (or SNX-230). This result was somewhat surprising because ω-CgTx MVIIC, a 26–amino acid peptide toxin from the marine snail *Conus magus*, is structurally related to ω-CgTx GVIA (Hillyard *et al.*, 1992). This toxin seems, however, to lack the specificity required for the unambiguous identification of P-type channels since it also blocks N-type Ca^{2+} channels, albeit with a lower affinity.

In contrast, R-type and T-type Ca^{2+} channels, also present in a variety of neurons, are mainly characterized by their insensitivity to DHPs, ω-CgTx GVIA, and ω-Aga IVA. The lack of selective antagonists has made the identification of these channels more complicated at the unitary level. However, by minimizing the amplitude of membrane depolarizations, T-type channels are easier to isolate than R-type channels. Also, LVA channels are blocked by Ni^{2+} concentrations lower than those required for Cd^{2+}, an order of potency opposite to that displayed by HVA channels. Again, this is consistent with the proposal of a difference in selectivity between these two groups of channels (Nilius *et al.*, 1985; Hagiwara *et al.*, 1988). R-type channels also have a slightly higher sensitivity to Ni^{2+} than other HVA channels, which could aid in their identification. Overall, it is believed that R-type channels may be a heterogeneous population of voltage-sensitive channels mainly characterized by their lack of sensitivity to all three specific Ca^{2+} channel antagonists.

Although many Ca^{2+} channel types are routinely distinguished by their pharmacological properties, not much is known about the structural requirements underlying their pharmacological differences. Clearly, some structural diversity is expected from channels that have differential sensitivities to structurally unrelated drugs. In contrast, it is generally assumed that Ca^{2+} channels, sensitive to one or several related drugs, may differ only by minor structural features. However, several examples seem to contradict these two widely accepted assumptions. For instance, it was found that two peptide toxins from the *Agelenopsis aperta* spider venom (ω-Aga I and ω-Aga IIIA) are potent blockers of both L and N channels despite extensive structural differences between these two channel types (Mintz *et al.*, 1991; Cohen *et al.*, 1992). Thus, these results suggest that very different channels can share some degree of homology restricted to their toxin binding sites, as recently suggested by Adams and Olivera (1994). Also, it seems that only slight differences in channel structure are required to account for marked differences in DHP sensitivities between cardiac or smooth muscle L-type Ca^{2+} channels (Welling *et al.*, 1993) or even between splice variants of the cardiac channel (Soldatov *et al.*, 1995). This suggests that an extensive pharmacological subdivision of voltage-dependent Ca^{2+} channels on the basis of their sensitivity to structurally related drugs may not necessarily translate into a greater functional and biophysical diversity of Ca^{2+} channels.

1.4. Correlation between Ca^{2+} Channel Type and Function in Excitable and Nonexcitable Cells

L-type Ca^{2+} channels play important roles in excitation-contraction (E-C) coupling in skeletal, cardiac, and smooth muscle by triggering Ca^{2+} release from the sarcoplasmic reticulum (SR) through the ryanodine receptor. In cardiac muscle, Ca^{2+} release from the SR is induced by the entry of Ca^{2+} through the L-type channel. In contrast, in skeletal muscle, conformational changes in the L-type channel appear sufficient to trigger Ca^{2+} release from intracellular stores in the absence of any extracellular Ca^{2+} permeation. This has led to the proposal that a direct or indirect molecular coupling exists between the L-type channel in the plasma membrane and the ryanodine receptor in the SR membrane. Finally, neuronal L-type Ca^{2+} channels may play an important role in basal cellular activity and a lesser role in neurotransmission, a proposal that is consistent with their localization in cell bodies and proximal dendrites (Westenbroek *et al.*, 1990).

N-, P-, and R-type Ca^{2+} channels are predominantly neuronal (Tsien *et al.*, 1988). However, N-type Ca^{2+} channels have also been found in endocrine cells such as human small-cell lung carcinoma (Sher *et al.*, 1990a) and rat insulinoma (Sher *et al.*, 1990b) cell lines. Immunocytochemical results demonstrate that both N- and P-type Ca^{2+} channels are located along dendrites as well as at synapses (Westenbroek *et al.*, 1992).

There is a general consensus that DHP-resistant (P-, N-, and R-type) Ca^{2+} channels reside in nerve terminals and are involved in exocytosis in mammalian central neurons (Hirning *et al.*, 1988; Takahashi and Momiyama, 1993; Weeler *et al.*, 1994). For instance, the release of glutamate and GABA from hippocampal neurons can be inhibited by ω-CgTx GVIA, illustrating the role of N-type Ca^{2+} channels in transmitter release (Burke *et al.*, 1993). Also, ω-Aga IVA blocks the GABAergic transmission in cerebellar and spinal cord slices with potencies close to that required for Ca^{2+} current blockade of the P-type current from Purkinje cells (Mintz et al., 1992a; Takahashi and Momiyama, 1993). Both N- and P-type Ca^{2+} channels have also been found at the neuromuscular junction (Robitaille et al., 1990; Sugiura et al., 1995). In frog neuromuscular junction, the transmitter release is entirely blocked by ω-CgTx GVIA demonstrating the predominant role of N-type Ca^{2+} channels in presynaptic activity (Kerr and Yoshikami, 1984). In contrast, in mammalian neuromuscular junction, synaptic transmission is insensitive to the N-type channel blocker, implicating a role for other voltage-dependent Ca^{2+} channels. Double labeling of mouse neuromuscular junction with biotinylated SNX-260, a structural analog of ω-CgTx MVIIC, and α-bungarotoxin indicates a presynaptic localization of P-type Ca^{2+} channels at the active zone and suggests a role of these channels in mammalian peripheral transmitter release (Sugiura et al., 1995). In many preparations, however, neurotransmission is not completely abolished by the combination of both saturating concentrations of ω-CgTx GVIA and ω-Aga IVA, providing evidence for the involvement of other voltage-dependent Ca^{2+} channels, likely R-type. Although neurosecretion in brain does not appear to be directly dependent on DHP-sensitive L-type Ca^{2+} channels, they have been shown to be intimately involved in secretion in adrenal chromaffin cells. While both P- and N-type Ca^{2+} channels were able to trigger some secretion, activation of L-type Ca^{2+} channels appeared to be even more efficiently coupled to secretion in these cells (Artalejo et al., 1994).

The activity of both N- and P-type Ca^{2+} channels is heavily regulated by several receptor-activated pathways. The N-type channel can be inhibited by norepinephrine in rat sympathetic neurons (Bean, 1989), by GABA in rat hippocampal neurons (Scholz and Miller, 1991), and by glutamate in rat CA3 pyramidal neurons (Swartz and Bean, 1992). P-type Ca^{2+} channels in cerebellar Purkinje neurons are also inhibited by the $GABA_B$ agonist baclofen (Mintz and Bean, 1993). Inhibition of Ca^{2+} channel activity is a reasonable mechanism for regulation of transmitter release by these neuromodulators. Several questions are currently under analysis, in particular which Ca^{2+} channel subtypes control neurosecretion and whether these channels are differently modulated by receptor-activated pathways.

Most nonexcitable cells, including lymphocytes and mast cells, do not contain significant components of voltage-dependent Ca^{2+} channels. While many intracellular signaling events, such as T-cell activation and histamine secretion from mast cells, rely on extracellular Ca^{2+} influx, the Ca^{2+} current is clearly distinguished from voltage-dependent Ca^{2+} current based on its gating, unitary conductance, and ionic selectivity (Tsien et al., 1987).

2.   SUBUNIT STRUCTURE OF VOLTAGE-DEPENDENT Ca^{2+} CHANNELS

2.1.   Purification of the Skeletal Muscle L-Type Ca^{2+} Channel

The skeletal muscle transverse tubule system represents the richest and most homogeneous source of voltage-dependent L-type Ca^{2+} channels in the body. This channel can also be specifically labeled by a tritiated derivative of PN200-110, a 1,4-DHP antagonist. This set of properties has helped the first purification of a voltage-dependent Ca^{2+} channel (Flockerzi *et al.*, 1986; Leung *et al.*, 1987; Takahashi *et al.*, 1987). The skeletal muscle L-type Ca^{2+} channel is a complex of four subunits (α_1, $\alpha_2\delta$, β, and γ) which copurify in a stoichiometric ratio (1:1:1:1) with a total molecular mass of about 400 kDa (Fig. 1).

The purified α_1 subunit has a molecular mass of about 170 kDa and is responsible for the process of ionic permeation and the mechanical coupling to the sarcoplasmic reticulum Ca^{2+} release channel. The 170-kDa α_1 subunit appears to be a proteolytic product of a larger and minor 212-kDa protein in which approximately 300 C-terminal amino acids have been posttranslationally removed (De Jongh *et al.*, 1989). The truncated skeletal muscle α_1 subunit, like its full-length counterpart, can function both for Ca^{2+} permeation and E-C coupling (Beam *et al.*, 1992). Although deletions in the C-terminal sequence of the cardiac α_1 subunit resulted in increases in the open probability of the channel (Wei *et al.*, 1994), similar C-terminal deletions in the skeletal α_1 subunit had no effect on Ca^{2+} channel activity. Thus, the physiological relevance of this

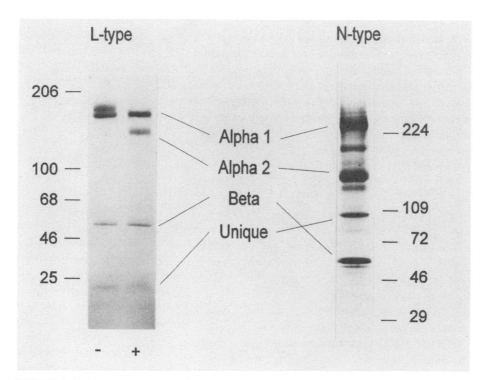

FIGURE 1. Purified L- and N-type Ca^{2+} channel subunits. (A) The four subunits of the voltage-dependent L-type Ca^{2+} channel are shown under reducing (+) and nonreducing conditions (−) (Leung *et al.*, 1987). On reduction, the $\alpha_2\delta$ subunit undergoes a characteristic mobility shift to 150 kDa. The γ subunit is unique to the L-type Ca^{2+} channel complex. (B) The purified subunits of the N-type Ca^{2+} channel complex are shown with the unique 95-kDa subunit (Witcher *et al.*, 1993a).

posttranslational modification is not yet understood. The α_1 subunit also contains the binding sites for DHP (Striessnig *et al.*, 1991), for phenylalkylamines (Striessnig *et al.*, 1990), and for benzothiazepines (Catterall and Striessnig, 1992). These sites have been well characterized by the use of photoaffinity ligands (Striessnig *et al.*, 1990; Nakayama *et al.*, 1991; Regulla *et al.*, 1991; Tang *et al.*, 1993). The phenylalkylamines bind to a cytoplasmic sequence located downstream from the S6 helix of the fourth homologous repeat (refer to the membrane topology of the α_1 subunit in Section 3.3). Instead, the DHP binding site is essentially extracellular and is formed by two extracellular sequences just upstream from the S6 helices of repeats III and IV and the loop between helices S5 and S6 of the third repeat.

The skeletal muscle β subunit has a molecular mass of about 52 kDa. Since it is a nonglycosylated, hydrophilic, and peripheral membrane protein, its localization is completely cytoplasmic (Takahashi *et al.*, 1987). Consistent with this proposal, this subunit is also phosphorylated by cAMP-dependent protein kinase, protein kinase C, and cGMP-dependent protein kinase (Jahn *et al.*, 1988). The $\alpha_2\delta$ subunit is a 170-kDa complex that is cleaved to an α_2 (140 kDa) and a δ (19–32 kDa) peptide under reducing conditions, demonstrating their linkage by disulfide bonds. This subunit is heavily glycosylated, both on α_2 and δ, with 30% of its molecular mass composed of carbohydrates.

The γ subunit has a molecular mass of 32 kDa and, like the $\alpha_2\delta$ subunit, is also a glycosylated transmembrane protein. This subunit is very hydrophobic and is proposed to contain four transmembrane α-helices. Northern blot analyses have demonstrated that this subunit is unique to skeletal muscle (Jay *et al.*, 1990). The specific association of all these subunits within a multisubunit complex was proven, not only by the copurification of $\alpha_2\delta$, β, and γ subunits with the α_1 DHP receptor, but also by coimmunoprecipitation experiments of the complex with antibodies that selectively recognize each subunit (Lazdunski *et al.*, 1987; Takahashi *et al.*, 1987; Leung *et al.*, 1988; Sharp *et al.*, 1988; Sharp and Campbell, 1989).

2.2.   Purification of the N-Type Ca^{2+} Channel Complex

There are several structural homologies between L- and N-type Ca^{2+} channels, as demonstrated by the usefulness of several cross-reacting antibodies used in purifying Ca^{2+} channels from brain (Sakamoto and Campbell, 1991; Witcher *et al.*, 1993a). The use of [125I]-ω-CgTx GVIA as a specific marker of N-type channels was critical to its successful purification. Toxin-labeled N-type channels have been purified to homogeneity by successive enrichment steps that include the use of heparin-agarose chromatography, immunoaffinity chromatography with a β-specific monoclonal antibody, and a sucrose density gradient centrifugation (Witcher *et al.*, 1993a). Four proteins from brain of 230, 140, 95, and 57 kDa molecular weight copurify with the toxin binding activity in approximately equivalent stoichiometric ratios (Witcher *et al.*, 1993a) (Fig. 1). These proteins also comigrate on the sucrose gradient with the peak of [125I]-ω-CgTx GVIA activity demonstrating that they all represent tightly interacting subunits of the same Ca^{2+} channel complex.

The N-type α_1 subunit from rabbit brain has a molecular mass of 230 kDa and binds ω-CgTx GVIA, but not DHPs or phenylalkylamines. Two size forms of the α_1 subunit

(235 and 210 kDa) appear in purification of the N-type Ca^{2+} channel (Westenbroek *et al.*, 1992) much as is found in L-type channels of skeletal muscle (De Jongh *et al.*, 1991) or neurons (Hell *et al.*, 1993). Structurally, the lower molecular mass subunit has been shown to result from a C-terminal truncation of the full-length protein. The two size forms have been shown to be differentially phosphorylated by calcium- and calmodulin-dependent protein kinase II (CaM kinase II) *in vitro* (Hell *et al.*, 1994). Differential phosphorylation of the two size forms *in vitro* may indicate the possible relevance of the C-terminal truncations seen *in vivo*. The 140-kDa $\alpha_2\delta$ subunit is biochemically quite similar to the skelctal muscle $\alpha_2\delta$ subunit. It is extensively glycosylated and has a characteristic mobility shift on reducing SDS-PAGE with the appearance of the disulfide-linked δ peptides. The β-subunit, while of another type, shares several immunogenic epitopes with the skeletal muscle β subunit. A unique 95-kDa protein was shown to copurify with the channel complex. Although this protein is less well characterized than the other proteins that had skeletal muscle homologues, it remains possible that the 95-kDa protein is an isoform of the γ subunit. Like the γ subunit, this protein is lightly glycosylated and hydrophobic as suggested by 3-(trifluoromethyl)-3-(*m*-iodophenyl)diazirine labeling.

    In conclusion, the purification of a muscle L-type and a brain N-type Ca^{2+} channel strongly suggests that the minimum subunit composition of most native HVA Ca^{2+} channels is $\alpha_1\alpha_2\delta\beta$ and that some structural variability can be introduced by the association of a fourth subunit whose expression is tissue-specific.

2.3.   Characterization of Other Voltage-Dependent Ca^{2+} Channels

    There is much biochemical evidence demonstrating that other voltage-dependent Ca^{2+} channels are also multisubunit complexes composed of at least an α_1, an $\alpha_2\delta$, and a β subunit. Two monoclonal antibodies directed against the skeletal muscle $\alpha_2\delta$ subunit immunoprecipitate a very significant fraction of [^{3}H]-PN200-110 from brain homogenates (Ahlijanian *et al.*, 1990), demonstrating that neuronal L-type Ca^{2+} channels also contain the $\alpha_2\delta$ subunit. Moreover, biochemical characteristics of the $\alpha_2\delta$ subunit, such as the shift in molecular mass upon disulfide reduction, were similar to that found in skeletal muscle. In fact, the $\alpha_2\delta$ subunit has been immunologically detected in a variety of tissues suggesting a wide distribution and an important sequence conservation (Morton and Froehner, 1989).

    DHP-sensitive Ca^{2+} channels are 30 to 80 times less abundant in cardiac than in skeletal muscle (Perez-Reyes *et al.*, 1989), and the difficulty in obtaining sufficient starting material has made large-scale purification difficult. Nonetheless, several researchers have shown the presence of a 190–200-kDa α_1 subunit and an associated $\alpha_2\delta$ subunit, which is heavily glycosylated and has the characteristic mobility shift with disulfide reduction due to the dissociation of the α_2 and δ peptides (Cooper *et al.*, 1987; Yoshida *et al.*, 190; Haase *et al.*, 1991; Tokumaru *et al.*, 1992). No β subunit has been identified in these cardiac channel purifications, although some candidate proteins of suggestive molecular weight copurify (Kuniyasu *et al.*, 1992). In addition, there is molecular evidence for the expression of several β subunits in cardiac muscles (Perez-Reyes *et al.*, 1992; Hullin *et al.*, 1992).

Characterization of vertebrate Ca^{2+} channels of evolutionary distance from mammals, such as those of *Cyprinus carpio* (carp) has demonstrated striking structural conservation among DHP-sensitive Ca^{2+} channel subunits (Grabner *et al.*, 1991). For instance, the carp DHP-binding α_1 subunit has molecular masses of 211 and 190 kDa, which may indicate a C-terminal truncation. In addition, highly glycosylated $\alpha_2\delta$ subunits copurify with the carp α_1 DHP receptor, although these have higher molecular masses than their skeletal muscle counterparts, and appear as two distinct bands prior to disulfide reduction.

Considering the association of Lambert-Eaton Myasthenic Syndrome (LEMS) with small-cell lung carcinoma, there has been much interest in identifying the Ca^{2+} channel composition of both small carcinoma cells and peripheral nerves. Recent biochemical characterization of Ca^{2+} channel components from small-cell lung carcinoma and neuroblastoma cell lines have revealed shared epitopes in both L-type Ca^{2+} channel α_1 and $\alpha_2\delta$ subunits, suggesting that similarities between the Ca^{2+} channels in these two cells may result in the presynaptic autoimmune disease (Morton *et al.*, 1994). Many other investigations have focused on the ability of LEMS sera to immunoprecipitate ω-CgTx GVIA binding from the small-cell lung carcinoma cell lines (Sher *et al.*, 1990a) and the biophysical and molecular characterization of Ca^{2+} channels in both cell types (Carbone *et al.*, 1990; Oguro-Okano *et al.*, 1992; Condignola *et al.*, 1993).

Finally, it is noteworthy that, in the absence of good ligands, the subunit composition of the T-type Ca^{2+} channel remains totally elusive. The complete lack of structural data on this important Ca^{2+} channel has further hampered its characterization. The channel properties suggest that it may have important differences in sequence that have so far precluded the cloning of its permeable subunit. Alternative cloning strategies (i.e., expression cloning) might be required to finally resolve the identity of LVA channels.

3. MOLECULAR DIVERSITY OF VOLTAGE-DEPENDENT Ca^{2+} CHANNELS

3.1. Cloning of Ca^{2+} Channel Subunits

The skeletal muscle L-type Ca^{2+} channel was the first channel described in which all the subunits (α_{1S}, $\alpha_2\delta_a$, β_{1a}, and γ) were identified and cloned (Tanabe *et al.*, 1987; Ruth *et al.* 1989; Ellis *et al.*, 1988; Jay *et al.*, 1991). Cloning of these subunits followed their purification through peptide sequencing or the development of antibodies to each protein. The full-length clone of the α_{1S} subunit encodes 1873 amino acids with a predicted mass of 212 kDa, consistent with the molecular mass of the purified protein. The proteolytic cleavage of this protein to a lower molecular mass of 170 kDa is believed to occur between residues 1685 and 1699 (De Jongh *et al.*, 1991). Despite the fact that the expression of the α_{1S} subunit failed to produce currents when injected into *Xenopus* oocytes, sequence similarity between the cloned Ca^{2+} channel α_1 subunit (Tanabe *et al.*, 1987) and the previously cloned voltage-dependent Na^+ channel (Noda *et al.*, 1984) suggested that the protein indeed encoded a voltage-dependent channel. Later, experiments in myotubes from muscular dysgenic mice that lack DHP-sensitive currents and E-C coupling proved that the cloned α_{1S} sequence was responsible for these skeletal muscle–specific channel functions (Tanabe et al., 1988). The β subunit cDNA of the DHP receptor encodes a 524–amino acid protein with a predicted mass of about 58 kDa, whereas the $\alpha_2\delta$ subunit cDNA encodes a protein of 1106 amino acids with a predicted size of 125 kDa.

Northern blot analysis carried out with cDNA probes from this L-type Ca^{2+} channel provided the first evidence that similar subunits exist in other tissues (Biel et al., 1991; Ruth et al., 1989; Ellis et al., 1988). For instance, a β_{1a} subunit cDNA probe identified a 1.6- and a 1.9-kb transcript in skeletal muscle and a 3-kb transcript in brain (Ruth et al., 1989). An $\alpha_2\delta$ cDNA probe identified an 8-kb message in skeletal muscle and also in heart, aorta, and brain (Ellis et al., 1988). Degenerate oligonucleotide probes were subsequently used to clone several other Ca^{2+} channel α_1 subunits. The following sections summarize what is known about the structural diversity of Ca^{2+} channel subunits.

3.1.1. Molecular Classification of α_1 Subunits

Mammalian α_1 subunits are encoded by at least six different genes, defined as S, A, B, C, D, and E, of which five (A to E) are expressed in brain (Snutch et al., 1990; Soong et al., 1993). In contrast, the α_{1S} subunit appears to be expressed predominantly, if not exclusively, in skeletal muscle (Ellis et al., 1988; Morton and Froehner, 1989). The different classes of α_1 subunits define two major groups based on sequence homologies (Fig. 2). First, classes S, C, and D are most closely related and different from the other group of α_1 subunits. Sequence analysis of these clones show that the class C and class D gene products share approximately 66–71% amino acid identity with the α_{1S} subunit. Second, classes A, B, and E are highly homologous to each other and form a second separate group of α_1 subunits. Classes A, B, and E, which share as much as 54–64% identity, are more distantly related to the other classes with 33–43%, 34%, and 41–42% identity to the α_{1S} subunit, respectively. It is likely that these Ca^{2+} channel groups have evolved from a single ancestral Ca^{2+} channel subunit and that gene duplication and subsequent genomic divergence form the basis of this structural diversity. In addition to these six classes, alternative splicing of a primary transcript (described by a lower case letter) can also produce distinct Ca^{2+} channels by increasing the structural, functional, and, possibly, pharmacological diversity of this subunit (Snutch et al., 1991; Welling et al., 1993; Hui et al., 1991; Diebold et al., 1992; Williams et al., 1992a). Thus, at present, about 20 structurally distinct α_1 subunits are known. For instance, the cardiac DHP-sensitive α_{1C} subunit encodes a 243-kDa protein of 2171 amino acids that has only 66% homology with the α_{1S} subunit (Mikami et al., 1989). Alternatively spliced forms of the cardiac α_{1C} subunit clone were isolated from rabbit lung (Biel et al., 1990), rat aorta (Koch et al., 1990), and rat brain (Snutch et al., 1991), demonstrating a wide tissue distribution for this channel. Both the cardiac and smooth muscle α_{1C} subunit differ only in the N-terminus, the hydrophobic segments IS6 and IVS3, and by an insertion in the I-II cytoplasmic linker for the smooth muscle α_{1C} subunit (Mikami et al., 1989; Biel et al., 1990; Koch et al., 1990).

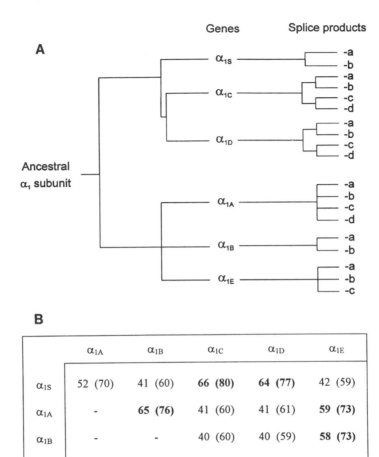

FIGURE 2. Gene family tree and amino acid sequence homologies between α_1 subunits. (A) The nomenclature is from Birnbaumer *et al.* (1994). (B) The overall percentage identity or similarity (bracketed values) are shown for each α_1 subunit pair. Identity values that are the highest are in bold to emphasize the structural relatedness of each α_1 subunit.

3.1.2. Molecular Classification of β Subunits

Mammalian β subunits are encoded by four different genes, all expressed in brain, as shown by Northern blot analysis (Birnbaumer *et al.*, 1994). Additional molecular diversity arises from alternative splicing of the transcripts of these genes (Powers *et al.*, 1992). These β subunits have been assigned numerals (1, 2, 3, and 4) based on the order of gene discovery followed by a letter identifying the splice variant (Fig. 3A). The β_1 gene can be alternatively spliced within two defined regions to give four different β_1 subunits (β_{1a}, β_{1b}, β_{1c}, and β_{1d}). First, there is an internal region of β subunits that is highly variable. One of the exons encodes a 52–amino acid sequence in the skeletal

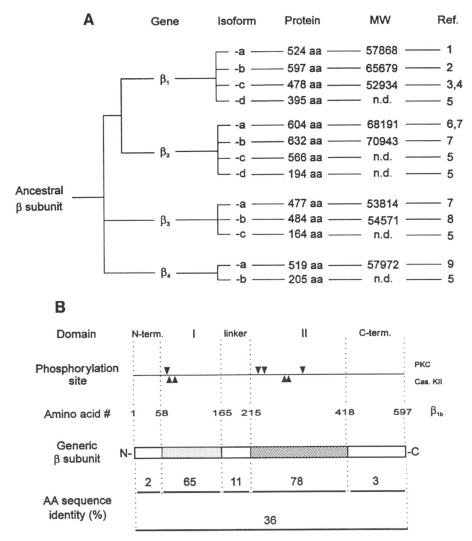

FIGURE 3. Classification of β subunits and amino acid sequence homologies. (A) Gene family tree and protein amino acid length and predicted molecular weight. References: 1, Ruth *et al.* (1989); 2, Pragnell *et al.* (1991); 3, Williams *et al.* (1992a); 4, Collin *et al.* (1993); 5, Castellano and Perez-Reyes (1994); 6, Perez-Reyes *et al.* (1992); 7, Hullin *et al.* (1992); 8, Castellano *et al.* (1993a); 9, Castellano *et al.* (1993b). (B) Schematic representation of β subunits and division in five structural domains. Conserved consensus protein kinase C and casein kinase II phosphorylation sites are shown as arrowheads.

muscle β_{1a} subunit (Ruth *et al.*, 1989), whereas another exon encodes a 7–amino acid sequence for the brain β_{1b} isoform (Pragnell *et al.*, 1991). Second, the β_{1b} subunit has a different and longer C-terminus than the β_{1a} subunit, suggesting that one or several alternative exons are used to construct the 3′ end of the β_{1b} mRNA. Thus overall, the neuronal β_{1b} subunit encodes a protein of predicted molecular mass of 66 kDa which is 8 kDa more than the skeletal muscle β_{1a} subunit. The β_{1c} subunit is another brain splice variant that expresses the short internal exon of β_{1b} and the exons coding the small C-terminal region of β_{1a} (Williams et al., 1992a). Four splice variants have also been identified for the β_2 gene (β_{2a}, β_{2b}, β_{2c}, and β_{2d}). The β_{2a} (Perez-Reyes et al., 1992), β_{2b} (Hullin et al., 1992), β_{2c}, and β_{2d} subunits (Castellano and Perez-Reyes, 1994) were alternatively spliced at the internal exon ($\beta_{2a} = \beta_{2b} \neq \beta_{2c} \neq \beta_{2d}$), whereas β_{2a} and β_{2b} differed by alternative splicing of the first exon resulting in two different N-terminal sequences. The β_2 gene is expressed abundantly in heart and to a lesser extent in aorta, trachea, and lung (Hullin et al., 1992). Finally, three alternative splice variants were found for β_3, β_{3a} (Hullin et al., 1992), β_{3b} (Castellano et al., 1993b), and β_{3c} (Castellano and Perez-Reyes, 1994), and two for β_4, β_{4a} (Castellano et al., 1993b), and β_{4b} (Castellano and Perez-Reyes, 1994). The β_3 gene is expressed in brain and smooth muscle–containing tissues such as aorta, trachea, and lung (Hullin et al., 1992). Instead, the β_4 gene is almost exclusively expressed in brain and in high levels in cerebellum. The β_{1d}, β_{2d}, β_{3b}, and β_{4b} subunits all contain a reading frame shift resulting in the expression of truncated β subunits that lack the entire second conserved domain and the more variable C-terminus. The physiological significance of these large truncations is not known.

A comparison of the amino acid sequence between all four β subunit gene products and splice variants has defined two high- (conserved domains I and II) and three low-homology domains (the N- and C-terminal domains and the linker between domains I and II). For instance, domains I and II extend from amino acid 58 to 165 and 215 to 418, respectively, in the β_{1b} subunit (Fig. 3B). Overall, these β subunits vary in molecular mass from ~50 to 85 kDa. Figure 3B also locates several conserved consensus phosphorylation sites on β subunits. Four potential phosphorylation sites for protein kinase C (PKC), but not protein kinase A (PKA), are present in all the β subunits. One such PKC site is located at residue 64 in domain I and three others in domain II at residues 228, 238, and 348 in the β_{1b} subunit. The skeletal muscle β_{1a} subunit can indeed be phosphorylated in vitro by various kinases (PKC, cAMP- and cGMP-dependent PK, calmodulin-dependent kinase II, and casein kinase II) (Nastainczyk et al., 1987; Jahn et al., 1988). Phosphorylation by cAMP-dependent kinase was shown to occur on threonine 205 and serine 182, although the latter is not a consensus site (Ruth et al., 1989; De Jongh et al., 1989). The functional importance of these phosphorylations are not yet known but such a high convergence of intracellular signaling suggests an important role of β subunits in the regulation of Ca^{2+} channel activity or function.

3.1.3. Primary Structure of the $\alpha_2\delta$ Subunit

Northern blot analysis and antibody cross-reactivity have indicated that $\alpha_2\delta$ is a well-conserved subunit in different tissues (Ellis et al., 1988; Morton and Foehner, 1989). So far, only a single gene has been identified that encodes both the α_2 and δ proteins (Ellis et al., 1988). This gene is expressed in multiple tissues as a glycosylated protein of 175 kDa that represents the disulfide-linked α_2 and δ peptides. It also encodes a 26–amino acid signal sequence that is cleaved after insertion into the plasma membrane. Despite the lack of a recognizable α_2-δ proteolytic cleavage site in the primary sequence, most biochemical analyses have shown that the α_2 protein is formed from the N-terminal sequence and the δ peptide is derived from the remaining C-terminus (De Jongh et al., 1990). There are three hydrophobic potential transmembrane domains, equally spaced in the protein sequence, although the third lies within five amino acids of the carboxyl terminus. Analysis of the primary sequence indicates that there are 18 potential gly-cosylation sites and numerous cysteine residues, which may make the elucidation of the extract structure difficult (Ellis *et al.*, 1988).

Five alternative splice products of the $\alpha_2\delta$ gene are now known (Williams *et al.*, 1992a; Kim *et al.*, 1992). The resulting differences in sequence are quite limited since they all occur between the first and second hydrophobic domains and are restricted to short amino acid segments. Although these splice variants are differentially expressed in brain and skeletal muscle, their significance remains unknown.

3.2. Subunit Identification of Native L- and N-Type Ca^{2+} Channels

Correlation of the cloned subunits with those found in native Ca^{2+} channel com-plexes relies on the use of toxin sensitivities, sequence-specific antibodies, and Northern blot analysis. The α_1 subunit associated with the L-type Ca^{2+} channel in skeletal muscle is molecularly classified as α_{1S}, which was originally cloned from a skeletal muscle library (Tanabe *et al.*, 1987) and appears to be uniquely expressed in skeletal muscle by Northern blot analysis (Ellis *et al.*, 1988). Expression of the α_{1s} subunit in L cells resulted in DHP-sensitive, voltage-gated Ca^{2+} current (Perez-Reyes *et al.*, 1989). The first β subunit cloned was also from a skeletal muscle library (Ruth *et al.*, 1989). Although this cDNA cross-hybridized with a larger mRNA from brain, the smaller hybridizing species in skeletal muscle corresponds to the β_{1a} clone, the small molecular weight β subunit of skeletal muscle, and the larger species is a splice variant of this gene found in brain (Pragnell *et al.*, 1991).

The α_1 subunit associated with [^{125}I]-ω-CgTx GVIA binding activity in purified N-type Ca^{2+} channels in brain has been identified as the α_{1B} subtype, based on reactivity of the 230-kDa protein with antibodies produced to a fusion protein containing the α_{1B} II-III loop (Witcher *et al.*, 1993a). Likewise, when the α_{1B} protein was expressed in HEK 293 cells, ω-CgTx GVIA sensitive currents were produced (Williams *et al.*, 1992b). The β subunit associated with the N-type Ca^{2+} channel has been shown to be the β_3 subtype, based on recognition of the 57-kDa protein by β_3 subunit–specific antibodies (Witcher *et al.*, 1993a). Four peptide sequences from the N-type $\alpha_2\delta$ subunit identified it as the neuronal $\alpha_2\delta_b$ splice variant of the $\alpha_2\delta$ gene (Williams *et al.*, 1992b). It differs from the skeletal muscle $\alpha_2\delta_a$ by a 19–amino acid deletion (residues 507–527 of $\alpha_2\delta_a$) and a 7–amino acid insertion (residues 602–608 of $\alpha_2\delta_b$).

3.3. Membrane Topology of Voltage-Dependent Ca^{2+} Channel Subunits

The three main subunits of voltage-dependent Ca^{2+} channels (α_1, $\alpha_2\delta$, and β) are proposed to be arranged in the membrane as illustrated in Fig. 4. The α_1 subunit bears strong similarities to other voltage-gated channels. This subunit possesses four succes-sive motifs (I to IV) of very similar structure. Each repeat contains six transmembrane segments (S1–S6). There are thus a total of 24 transmembrane segments orderly arranged to define a central ionic pore. The transmembrane segments are highly con-served sequences in Ca^{2+} channel α_1 subunits and also in Na^+ and K^+ channels (Noda *et al.*, 1984). The sequence variability is mostly located in the intracellular loops

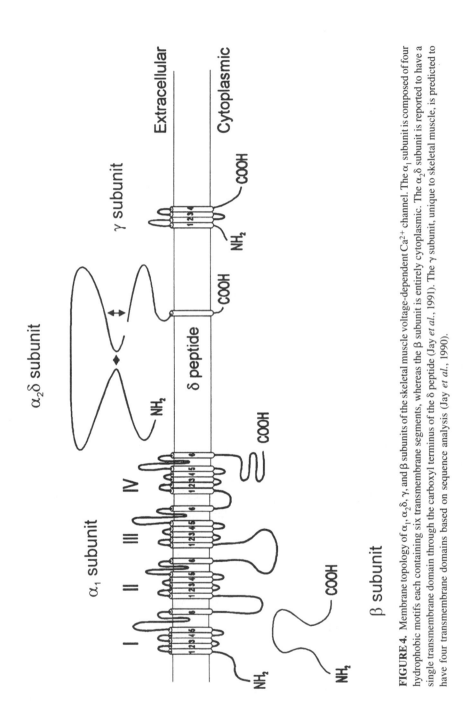

FIGURE 4. Membrane topology of α_1, $\alpha_2\delta$, γ, and β subunits of the skeletal muscle voltage-dependent Ca^{2+} channel. The α_1 subunit is composed of four hydrophobic motifs each containing six transmembrane segments, whereas the β subunit is entirely cytoplasmic. The $\alpha_2\delta$ subunit is reported to have a single transmembrane domain through the carboxyl terminus of the δ peptide (Jay *et al.*, 1991). The γ subunit, unique to skeletal muscle, is predicted to have four transmembrane domains based on sequence analysis (Jay *et al.*, 1990).

connecting the segments and the motifs and in the carboxyl terminus. The fourth segment (S4) in each repeat is 20 amino acids in length and contains repeated motifs of one positively charged amino acid at every third or fourth position followed by several hydrophobic amino acids. Single-point mutations of the positively charged amino acids of the S4 segments suggest that these segments represent portions of the voltage sensors that may initiate channel opening during activation (Stuhmer *et al.*, 1989). In support of this notion, changes in conformation of the S4 segments are held responsible for part of the gating currents that are triggered in response to membrane depolarization (Guy and Conti, 1990). However, it is still not well understood how these changes in S4 conformation, and subsequently elsewhere in the channel, are transduced into modifications of the gating process. It appears that this transduction can be regulated since the association of the β subunit to α_1 does not affect voltage sensing itself but increases the efficiency of the coupling between voltage sensing and pore opening (Olcese *et al.*, 1994). Two short segments, SS1 and SS2 located between S5 and S6 in each of the four repeated motifs, and the α-helical segments S5 and S6 themselves, are proposed to form the permeation pore (Guy and Conti, 1990). Single amino acid mutations within the pore region can alter the ion selectivity of the pore, effectively converting a Ca^{2+} channel into a Na^+ channel (Heinemann *et al.*, 1992). The connecting loops between S5 and S6 of the third repeat and the transmembrane segment S6 of the fourth repeat were identified as binding sites for both tetrodotoxin and DHP, two drugs that block the entry of Na^+ and Ca^{2+} via their respective channels (Catterall and Striessnig, 1992).

In contrast to the complex membrane topology of the α_1 subunit, the β subunit does not contain any putative transmembrane domain. However, it contains seven α-helical domains and four heptad repeats in which most of the first and fourth residues of every seven are hydrophobic. These heptad repeats are thought to be involved in interactions with cytoskeletal proteins (Fuchs and Hanukoglu, 1983).

The membrane topology of the $\alpha_2\delta$ subunit remains a significant controversial issue, as the topology may suggest or refute a direct interaction with the entirely cytoplasmic β subunit. If all three hydrophobic regions are transmembrane domains then a significant fraction of the protein between the first and second transmembrane domains would be intracellular and could possibly interact with β subunits. However, experimental data demonstrate that only a single transmembrane domain is present in the C-terminus of the δ subunit (Jay *et al.*, 1991). Alkaline extraction of skeletal muscle membranes in the presence of reducing agents is able to extract α_2 but not the transmembrane δ peptide. In the single transmembrane model, only five amino acids in the C-terminus would be cytoplasmic.

4. FUNCTIONAL RECONSTITUTION OF PURIFIED VOLTAGE-DEPENDENT Ca^{2+} CHANNELS

4.1. Functional Properties of Purified Ca^{2+} Channels in Bilayers

The functional activity of both L- and N-type Ca^{2+} channels can be reconstituted by insertion of the purified proteins in artificial lipid bilayers (Flockerzi *et al.*, 1986; De

Waard *et al.*, 1994a). Reconstitution of the α_{1S} subunit proves that this protein is sufficient to form a functional channel in the absence of other associated subunits (Pelzer *et al.*, 1989). This simplified channel also retains its sensitivity to numerous drugs (the DHP Bay K 8644 and the phenylalkylamine D600) and to phosphorylation by protein kinase A, unequivocally demonstrating that it is a major target for drug binding and cytoplasmic regulation. Similarly, the activity of the purified N-type channel is blocked by the addition of micromolar concentrations of ω-CgTx GVIA in the presence of 10 mM divalent Ba^{2+} as the charge carrier (Witcher *et al.*, 1993a). However, the channel is insensitive to L-type channel ligands such as the DHP and phenylalkylamine classes of drugs (Witcher *et al.*, 1993a; De Waard *et al.*, 1994a). Figure 5 demonstrates, for instance, that the mean current amplitude and open-time duration of the N-type Ca^{2+} channel in a low-P_0 mode is not affected by the application of 1 μM Bay K 8644. These pharmacological characterizations constitute properties essential for identification of the Ca^{2+} channel type. Once the pharmacological sensitivity of the purified channel is firmly established, further characterization of the biophysical properties of the channel could be undertaken.

Like its native counterpart, the conductance of the purified N-type Ca^{2+} channel is somewhat variable and ranges from 7 to 27 pS (De Waard *et al.*, 1994a). This variability may seem surprising for what appears to be a homogeneous population of channels. These results may suggest the existence of several ω-CgTx GVIA–sensitive α_{1B} isoforms in brain that differ by their unitary conductance. Such a hypothesis is, however, more difficult to sustain in face of the conductance variability of the purified L-type channel from skeletal muscle (Hymel *et al.*, 1988; Ma and Coronado, 1988). This tissue is regarded as the least heterogeneous source of DHP-sensitive channels, and very little sequence variability has been reported for the α_{1S} subunit. Alternatively, it is also possible that differential posttranslational processing (limited proteolysis or specific phosphorylations) is responsible for an alteration of the conductance values. In any case, these results demonstrate that the unitary conductance of a channel cannot be used as a reliable parameter of channel identification until the determinants of conductance variability have been identified. Also similar to native N-type Ca^{2+} channels, purified N-type channels have several gating modes that differ by their probability of opening (Delcour and Tsien, 1993; De Waard *et al.*, 1994a). Purified N-type channels have at least two gating patterns, one characterized by a low opening probability and a second defined by a much higher opening probability. In both cases, however, the distribution of open-time durations reveals short mean values between 1.5 and 8 msec.

Both L- and N-type Ca^{2+} channels are permeable to a large number of inorganic cations including Ba^{2+} and Sr^{2+} in addition to Ca^{2+} ions. However, HVA channels are not equally permeable to these cations. For instance, Ba^{2+} and Sr^{2+} currents are larger than Ca^{2+} currents in native HVA Ca^{2+} channels as demonstrated in rabbit myocytes (Droogmans *et al.*, 1987). Likewise, purified L- and N-type Ca^{2+} channels have a higher permeability to Ba^{2+} than to Ca^{2+} (Fig. 6). For instance, N-type Ca^{2+} channels have a 2.8-fold higher permeability when Ba^{2+} is the charge carrier. These results indicate that the selectivity process is well conserved during the purification and reconstitution procedures (De Waard *et al.* 1994a).

The amplitude of Ca^{2+} currents through voltage-dependent Ca^{2+} channels is a

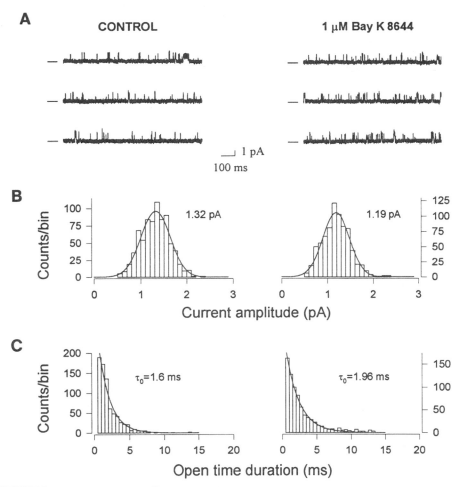

FIGURE 5. The purified N-type Ca^{2+} channel activity is not affected by DHP agonists. (A) Trace examples of purified N-type Ca^{2+} channel activity reconstituted by the tip-dip technique (De Waard *et al.*, 1994a). The holding potential is 100 mV and the permeant cation is 500 nM Ba^{2+}. The channel activity is shown before (left) and after (right) 2 min application of 1 μM Bay K 8644 in the extrapipette medium. Filter frequency is 1.5 kHz and sampling frequency is 10 kHz. (B) Amplitude histograms. Mean ± SD current amplitude is $I_{Ba} = 1.32 \pm 0.32$ pA (control) and $I_{Ba} = 1.19 \pm 0.29$ pA (Bay K 8644). Bin width is 0.1 pA/div. (C) Open-time duration histograms. The mean open-time duration is 1.6 msec (control) and 1.96 msec (Bay K 8644). Bin width is 0.5 msec div.

function of the extracellular Ca^{2+} concentration. In rat skeletal muscle, a twofold increase in Ca^{2+} current occurs in response to a fivefold increase in extracellular Ca^{2+} concentration from 2 to 10 mM (Donaldson and Beam, 1983). Similarly, the mean unitary conductance of purified N-type Ca^{2+} channels increases with elevating symmetrical Ba^{2+} concentrations. Plateau conductances are reached at 60 mM Ba^{2+} and half-maximal conductance occurs at 19 mM Ba^{2+} (De Waard *et al.*, 1994a).

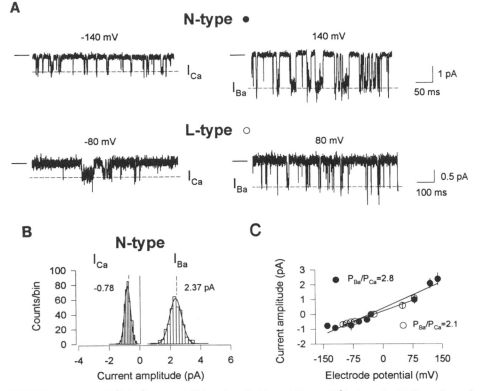

FIGURE 6. Relative Ca^{2+}/Ba^{2+} permeabilities of purified L- and N-type Ca^{2+} channels. (A) Recordings of purified N- and L-type Ca^{2+} channels in asymmetrical ionic conditions (100 mM Ba^{2+} in the electrode and 100 mM Ca^{2+} in the extrapipette solution). At negative driving forces, Ca^{2+} permeability is favored over Ba^{2+} permeability and results in smaller current amplitudes than at positive driving forces. (B) Amplitude histograms of N-type channel at pipette potentials of −140 and 140 mV. Mean current amplitudes are −0.78 ± 0.2 pA (−140 mV) and 2.37 ± 0.41 pA (140 mV). Bin widths are 0.1 and 0.2 pA/div. (C) Unitary conductances of both N- and L-type channels in asymmetrical ionic conditions. The conductances are 12.7 pS (N-type) and 10 pS (L-type) and the reversal potentials are −30 mV (N-type) and −22 mV (L-type). The estimated permeability ratio P_{Ba}/P_{Ca} are 2.8 (N-type) and 2.1 (L-type), assuming a Cl^- permeability of zero. The data for the N-type Ca^{2+} channel are from De Waard *et al.* (1994a).

4.2. Similarities between Purified L- and N-Type Ca^{2+} Channels

Purified channels from both skeletal muscle and brain retain several of the functional and pharmacological characteristics of native Ca^{2+} channels in planar lipid bilayers. They are selective to Ca^{2+} and exhibit open-time durations, conductances, and pharmacological sensitivities to antagonists that are consistent with native channels. Although there may be some cellular regulatory factors that modulate these elementary channel properties, these characteristics seem to be intrinsic to the Ca^{2+} channel multisubunit complexes themselves. The similarities in properties between purified and native voltage-dependent Ca^{2+} channels present strong evidence that the pharmacological identity, ionic selectivity, permeability, and gating activity of the channel are intrinsic qualities of the protein complex itself. Unfortunately, several important properties of native Ca^{2+} channels are either lost or difficult to observe after purification and reconstitution. For instance, the purified L-type and N-type Ca^{2+} channels do not inactivate or "run down" and both channels are surprisingly voltage-independent (Flockerzi et al., 1986; De Waard et al., 1994a). These results could be interpreted in either of two ways: first, the consecutive processes of purification and reconstitution could lead to abnormal channel activity due to partial denaturation or aberrant membrane insertion, or, second, the altered properties may instead reflect an absolute requirement for cellular factors that do not biochemically copurify with the channel. The first hypothesis of a conformational abnormality to explain the lack of voltage dependence of the purified channel in lipid bilayers is difficult to reconcile with the observation that purified channels behave normally with regard to complex mechanisms such as the binding of channel ligands and the gating process that should result as a consequence of voltage sensing. Obviously, the alternative hypothesis may prevail suggesting that important cell regulatory factors are required for 1) Ca^{2+}- and voltage-dependent inactivation, 2) run down, and 3) the transduction between voltage sensing per se and changes in the gating behavior.

In spite of the technical limitations of Ca^{2+} channel purification and reconstitution and the difficulties in interpreting the data, there is obviously more information to gain from this type of experiment. First, reconstitution of purified channels allows the analysis of the intrinsic properties of voltage-dependent Ca^{2+} channels in the absence of associated cellular factors, which cannot easily be achieved with expression experiments. A renewed interest in channel reconstitution is therefore to be expected for a rigorous analysis of the direct modulatory effects of phosphorylations, second messengers or intracellular proteins such as G proteins, or components of the excitation-contraction or excitation-secretion machinery. Also, valuable data on the mechanisms of action of various toxins (i.e., conotoxins or agatoxins) are to be gained from the analysis of the changes in elementary channel properties. An important controversy remains regarding the mechanism of blockade of ω-CgTx GVIA at the single-channel level, despite its extensive use in functional studies of the cellular role of N-type Ca^{2+} channels. It is not clear whether the toxin blocks Ca^{2+} permeability by simple occlusion of the pore (Ellinor et al., 1994) or by a reduction in opening probability (Witcher et al., 1993b). Likewise, it will be interesting to analyze the mechanisms of P-type current block by ω-Aga IVA when the subunits composing this channel have been identified and reconstituted into lipid bilayers.

5. EXPRESSION OF Ca^{2+} CHANNELS

5.1. Functional Classification of Cloned α_1 Subunits

It is now firmly established that α_1 subunits can form minimum voltage-dependent Ca^{2+} channels with clearly defined functional and pharmacological properties (Lacerda et al., 1991; Mori et al., 1991; Stea et al., 1993; Mikami et al., 1989; Soong et al., 1993). These subunits contain the basic functional elements of voltage-dependent Ca^{2+} channels such as the voltage sensor and the ionic pore and binding sites for most drugs. These properties were useful to tentatively assign the α_1 subunits to the various classified LVA and HVA Ca^{2+} channels. Table III summarizes the pharmacological and functional properties of the expressed α_1 subunits. On pharmacological grounds, these channels can be divided into two groups based on their sensitivity to DHPs: the L-type (S, C, and D) and the non-L-type (A, B, and E). Surprisingly, this subdivision matches well the division that was made on the basis of molecular α_1 homologies.

Stable expression of the α_{1S} subunit in mouse L cells results in the appearance of DHP-sensitive Ca^{2+} channels (Lacerda et al., 1991). The kinetics of activation of this subunit were excessively slow, suggesting that additional components were required to fully restore the properties of the native channel. The role of the α_{1S} subunit in excitation-contraction coupling was confirmed by expressing its cDNA in mouse dysgenic myotubes (mdg). This manipulation restores both the mechanical contraction and the release of Ca^{2+} from SR, two processes previously lacking in mdg because of a nucleotide deletion in the fourth transmembrane repeat of the endogenous α_1 subunit (Tanabe et al., 1988). Two other α_1 subunits are proposed to be components of L-type Ca^{2+} channels based on their homology to α_{1s} and their sensitivity to DHPs. The α_{1C} subunit is expressed in brain and in cardiac and smooth muscle, where it functions in the cardiac type of E-C coupling (Tanabe et al., 1990b). Similar to the cardiac-type channel that it encodes, the DHP-sensitive currents carried by the α_{1C} subunit are prone to Ca^{2+}-dependent inactivation (Neely et al., 1994) and facilitation by depolarizing prepulses (Kleppisch et al., 1994). No functional differences have yet been observed between the various alternative spliced forms of the α_{1C} transcripts (Itagaki et al., 1992). The α_{1D} subunit constitutes the third kind of L-type Ca^{2+} channel based on its sensitivity to low concentrations of DHP agonists and antagonists (Williams et al., 1992a). Curiously, this channel seems also to be sensitive to micromolar concentrations of the peptide ω-CgTx GVIA, although the affinity of the toxin for the channel is low and the inhibition of the current is reversible. These properties are consistent with previous observations that some native L-type Ca^{2+} channels could be reversibly blocked by ω-CgTx GVIA (Kasai and Neher, 1992). The low affinity and reversibility of α_{1D} block by ω-CgTx GVIA distinguishes this channel from a true N-type Ca^{2+} channel.

Expression of a cDNA that encodes the α_{1A} Ca^{2+} channel produces a current that is insensitive to both DHP antagonists and ω-CgTx GVIA (Mori et al., 1991; Sather et al., 1993). The channel can also be blocked by 200 nM ω-Aga IVA and even more potently by ω-CgTx MVIIC ($IC_{50} < 150$ nM). Figure 7A illustrates, for instance, the rate of current amplitude decay by blockade of the channel activity by an extracellular application of 2 μM ω-CgTx MVIIC. These pharmacological properties are apparently related to those reported for the native P-type channel and seem consistent with the high level of expression of the α_{1A} subunit in cerebellum that parallels well the localization of P-type channels (Mori et al., 1991). There are, however, a number of substantial differences that may indicate that the α_{1A} subunit is not a constituent of the P-type channel: 1) Ca^{2+} currents carried by α_{1A} lack the ~1 nM IC_{50} sensitivity for ω-Aga IVA that is normally displayed by native P-type channels in Purkinje neurons (Tsien et al., 1991; Sather et al., 1993), 2) α_{1A} currents activate in a range of potentials (threshold -30 mV) that are more depolarized than P-type channels (threshold -50 mV), and 3) the time-dependent

TABLE III

Properties of Cloned α_1 Subunits[a]

Property	Gene					
Classification	S	A	B	C	D	E
Splice variants	2	4	2	4	4	3
Primary tissue location	Skeletal	Neuronal Cardiac	Neuronal	Neuronal Aortic-Cardiac	Neuronal Endocrine	Neuronal
Homology	100%	33–44%	34%	66%	71%	41–42%
Pharmacology	DHPs Antagonists/agonists	FTX ω-Aga IV ω-Cx MVIIC	ω-CTx GVIA (irreversible)	DHPs Antagonists/agonists	DHPs ω-CTx GVIA (reversible)	Resistant to all
Component of	Skeletal muscle DHP receptor	?	Brain ω-Cx receptor	?	?	?
Functional type	HVA L	HVA P or Q ?	HVA N	HVA L	HVA L	HVA R
Cell function	E-C coupling	E-S coupling	E-S coupling	Other	Other	E-S coupling

[a]cDNA sequence of the various α_1 subunits are compared to the sequence of the α_{1S} subunit. The α_{1S} and α_{1B} subunits are identified components of the skeletal muscle L-type and brain N-type Ca^{2+} channels, respectively. The pharmacological sensitivity of the other α_1 subunits has aided their identification as L- or non-L-type Ca^{2+} channels. There is still some controversy about the functional correlate of the α_{1A} subunit (P- or Q-type).

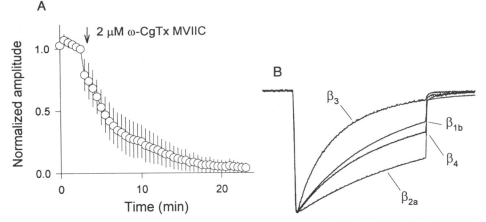

FIGURE 7. Toxin sensitivity and kinetic regulation by β subunits of the HVA α_{1A} Ca^{2+} channel. (A) The currents carried by α_{1A} channels can be totally blocked by 2 μM ω-CgTx MVIIC. The external Ba^{2+} concentration is 2 mM to minimize the antagonism that occurs between divalent cations and the toxin binding. (B) Different β subunits induce variable rates of inactivation with β_3 inducing the fastest rate and β_{2a} the slowest. The holding potential is −90 mV and the test pulses were delivered at 30 mV. The data are from De Waard and Campbell (1995).

inactivation is more rapid (time constant of 150 to 700 msec depending on the β subunit being coexpressed) than P-type currents (little inactivation). This has recently led to the proposal that the α_{1A} channel represents a novel channel type defined as Q-type. However, the precise structural bases for these distinctions remain to be established with more certainty. There is a precedent case for such a discrepancy between cloned and native Ca^{2+} channels. The α_{1C} channel is reportedly 100-fold less sensitive to ω-Aga IIIA than L-type channels in myocytes (Mintz et al., 1991) but is unquestionably a component of these cardiac channels. Clearly, some caution is needed when comparing the properties of native and expressed channels. The subunit composition of the P-type channel is not yet known and the inactivation kinetics of the α_{1A} channel are largely controlled by the association of the ancillary subunits (Fig. 7B).

The α_{1B} subunit has also been characterized pharmacologically, as expression of its cDNA gives rise to [^{125}I]-ω-CgTx GVIA binding and to ω-CgTx GVIA–sensitive currents in HEK 293 cells (Williams et al., 1992b; Fujita et al., 1993; Stea et al., 1993). The high-affinity binding of this toxin to the α_{1B} expressed protein is irreversible, making it a component of N-type Ca^{2+} channels (Williams et al., 1992a,b). These properties distinguish the α_{1B} Ca^{2+} channel from the α_{1D} Ca^{2+} channel that is less specifically blocked by ω-CgTx GVIA. Recently, several α_{1E} Ca^{2+} channels have been cloned (human α_{1E}, rat rbE, rabbit BII, and doe-1 from *Discopyge ommata*) and expressed in *Xenopus* (Soong et al., 1993; Wakamori et al., 1994; Ellinor et al., 1993) or HEK 293 cells (Williams et al., 1994). Initially, this channel was proposed to represent a member of the low voltage–activated Ca^{2+} channel class, possibly a T-type channel (Soong et al., 1993). In low divalent cation concentration, the class E channel activates at negative membrane potentials (threshold at −50 mV in 4 mM [Ba^{2+}]), has a rapid current decay during inactivation ($\tau \sim 100$ msec), and is blocked by low concentrations of Ni^{2+} (IC_{50} of 28 μM). The conductance of the expressed channel is 12 pS (Wakamori *et al.*, 1994). In fact, it is now clear that the voltage dependence of the α_{1E} channel does not differ from that displayed by the α_{1A} channel (De Waard and Campbell, 1995) and, as all HVA channels, is blocked by lower concentrations of Cd^{2+} (IC_{50} of 1 μM) than Ni^{2+} (Williams *et al.*, 1994). Such a cation sensitivity is thus still opposite to that displayed by T-type channels. The α_{1E} channel is also characterized by its lack of sensitivity to all the discriminating drugs (DHP, ω-CgTx GVIA, and ω-Aga IVA) that block L-, N- and P-type channels. Based on these results, it is therefore proposed that the α_{1E} subunit represents a member of the R-type Ca^{2+} channel family recently identified in cerebellar granule cells (Ellinor *et al.*, 1993).

5.2. Role of Ancillary Subunits

The gating mechanisms of ion channels are often under the control of several biochemical signals including protein–protein interactions and covalent modifications of the channel such as phosphorylation (Kaczmarek and Levitan, 1987; Levitan, 1988). Subunit-subunit interactions have also recently been recognized as a mechanism of gating control by altering the conformation of α_1 subunits. We will review the evidence demonstrating that two of the ancillary subunits (β and $\alpha_2\delta$) control the biophysical and pharmacological properties of the α_1 pore-forming protein.

5.2.1. Effects of β Subunits on Channel Properties

The proliferation of cloned Ca^{2+} channel α_1 and β subunits in the absence of any systematic biochemical characterization has led to the coexpression of numerous α_1-β combinations. Expression of most α_1 subunits alone gives rise to very small current densities, and since β subunits substantially increase the level of current amplitude, there has been a natural tendency to coexpress both subunits in various cells. In fact, current amplitude stimulation by β subunits has been observed for almost any α_1 subunit (Williams *et al.*, 1992a,b; Wei *et al.*, 1991; Mori *et al.*, 1991; Ellinor *et al.*, 1993; Stea *et al.*, 1993). The potency of current amplitude stimulation is somewhat variable and depends on the class of α_1 and β subunits being coexpressed, the cDNA constructs, the cDNA/cRNA transfection/injection technique, and the cell type used to assay the protein expression level. Estimates of current amplitude stimulation by β subunits have been most reliably obtained from injection of cRNAs into *Xenopus laevis* oocytes. In this system, coexpression of β_{1b} with either α_{1A} subunit (De Waard *et al.*, 1994b), α_{1B} (Stea *et al.*, 1993), or α_{1E} (Wakamori *et al.*, 1994) produces, respectively, an 18-fold, a fourfold, or a threefold stimulation in peak current amplitude. Also, β subunits of each of the four genes can produce between a twofold to 19-fold stimulation in current amplitude of the α_{1C} channel (Wei *et al.*, 1991; Hullin *et al.*, 1992; Perez-Reyes *et al.*, 1992; Castellano *et al.*, 1993a,b) or of the α_{1A} channel (Mori *et al.*, 1991; De Waard *et al.*, 1994b). These observations strongly suggest that different β subunits can interact with the same α_1 subunit and that the mechanisms responsible for current stimulation are well conserved among different voltage-dependent Ca^{2+} channels.

Several hypotheses may explain how expression of the β subunit causes an increase in measured current amplitude. First, the direct interaction of the α_1 subunit with the β subunit may result in an increase in the conductance or in the opening probability of the channel by inducing a conformational change in the pore subunit. Conformational effects of the β subunit are likely, as the β subunit has both large effects on DHP binding affinity (Mitterdorfer et al., 1994) and on channel kinetics and voltage dependence. However, estimates of β subunit effects at the unitary level demonstrate that they do not modify the conductance of the expressed Ca^{2+} channels (Wakamori et al., 1994). Second, β subunits may cause a recruitment of new channels from intracellular stores or stabilization of channel complexes at the cell surface. Reports that have evaluated changes in the level of α_1 protein expression in the plasma membrane are contradictory. Perez-Reyes et al. (1992) report a fourfold increase in the number of [^{3}H]-PN200-110 binding sites upon coexpression of the α_{1C} with the β_2 subunit, while Nishimura et al. (1993) report a similar increase in DHP binding without an increase in the amount of α_1 protein. Mitterdorfer et al. (1994) failed to see an increase in DHP binding sites. In any case, DHP binding to L-type Ca^{2+} channels has always been difficult to interpret, as the affinity of these ligands to the channel is very state-dependent in native channels. For instance, it is known that the affinity of DHP antagonists for L-type channels increases upon inactivation, a state that is attained by depolarization. It is not known how the association of a β subunit affects the voltage-dependent behavior of DHP binding. This information would, however, be useful in the interpretation of β subunit effects on DHP binding in heterologous Ca^{2+} channels.

Analysis of the voltage dependence of the recombinant Ca^{2+} channels has revealed that β subunits also have systematic effects on other biophysical parameters encoded by the α_1 channels (De Waard and Campbell, 1995). In all cases studied, it was found that the stimulation of current amplitude by β subunits is greater at more depolarized potentials. This increase in stimulation efficiency by β subunits is due to a hyperpolarizing shift of the threshold and peak of the Ca^{2+} current. In several cases, β subunits also modify the voltage dependence of inactivation by shifting the midpoint of channel inactivation toward hyperpolarized values (De Waard and Campbell, 1995; Soong et al., 1993; Stea et al., 1993). This shift was, however, not consistently observed since β subunits had no effect on the voltage dependence of the α_{1C} subunit (Tomlinson et al., 1993). Finally, kinetic modifications by β subunits have been reported as well (De Waard et al., 1994b; Lacerda et al., 1991; Singer et al., 1991; Varadi et al., 1991). In mouse L cells, β subunits normalize the α_{1S} currents by accelerating the activation kinetics by a factor of 10 (Lacerda et al., 1991; Varadi et al., 1991). These accelerations in activation kinetics are not systematically observed for every α_1 subunit. For instance, β subunits have no effect on the activation kinetics of the α_{1A} subunit (Mori et al., 1991; De Waard and Campbell, 1995) and even slow the activation kinetics of the α_{1E} subunit (Wakamori et al., 1994). In contrast to activation kinetics, β subunits modify the inactivation kinetics of all α_1 subunits (Wakamori et al., 1994; Varadi et al., 1991; Castellano et al., 1993a,b; Lacerda et al., 1991; Sather et al., 1993; Ellinor et al., 1993). Qualitatively, the similarity of β subunits in regulating the inactivation kinetics is not as great as for the current stimulation and the changes in voltage-dependence and kinetics of activation. The β_3 subunit induces the fastest inactivation kinetics, whereas β_2 induces the slowest decay (De Waard and Campbell, 1995; Ellinor et al., 1993; Sather et al., 1993).

In conclusion, it appears that β subunits can potently modulate a fixed set of α_1 channel properties. This extensive cross-reactivity suggests that the various α_1 and β subunits share common interactions. However, the relative differences in extent of current stimulation and modulation of kinetics also suggest that these interactions can be regulated by the structural divergences in the β subunits. Finally it should be mentioned that the importance of β subunit association to the α_1 channel is not limited to the regulation of the biophysical and pharmacological properties of the channel since β subunits also modify the regulation of α_1 subunits by phosphorylation events (Klockner et al., 1992).

5.2.2. Effects of $\alpha_2\delta$ Subunit on Channel Properties

The effects of $\alpha_2\delta$ are not quite as dramatic as those of β subunits and, in most cases, are observed only on current amplitude and inactivation kinetics. Effects of the $\alpha_2\delta$ subunit on current amplitude range from a 1.5-fold increase when expressed with α_{1D} and β_2 in Xenopus oocytes (Williams et al., 1992a) to more than a 17-fold increase when expressed with α_{1C} and β_{1a} in the same cells (Singer et al., 1991). The large variability of effects seen may be due to differences in subunit compositions (correct $\alpha_1\alpha_2\delta\beta$ combination) and differences in glycosylation in the various expression systems.

Coexpression of $\alpha_2\delta$ subunit alters the time course of activation and inactivation of the cardiac α_{1C} channel expressed in Xenopus oocytes (Singer et al., 1991). However, these results are variable since no effect of $\alpha_2\delta$ was reported on the brain α_{1C} channel, a closely related isoform (Tomlinson et al., 1993). Likewise, no effect of $\alpha_2\delta$ subunit on activation and inactivation time courses was seen in transfected HEK 293 cells (Brust et al., 1993). In only a single case $\alpha_2\delta$ shifted the voltage dependence of inactivation of the α_{1E} channel toward more depolarized values (Wakamori et al., 1994). Once again, environmental factors, such as the appropriateness of subunit combinations and the expression system used, appear to play a large role in determining the effectiveness of the $\alpha_2\delta$ subunit expression on current kinetics.

Recent data demonstrate that the coexpression of the β subunit is required to observe the maximal regulation of the $\alpha_2\delta$ subunit (De Waard and Campbell, 1995). Expression of β and $\alpha_2\delta$ subunits with the α_{1A} subunit produces a current amplitude that is larger than if the effects of both subunits were merely additive (Mori et al., 1991; De Waard and Campbell, 1995). Also, changes in α_{1A} channel kinetics by β subunits are regulated by $\alpha_2\delta$, which increases the rate of inactivation of the current (De Waard and Campbell, 1995). Finally, the maximum number of [^{125}I]-ω-CgTx GVIA binding sites were more than additive when both $\alpha_2\delta$ and β subunits were coexpressed with the α_{1B} subunit in HEK 293 cells than when the subunits were expressed individually (Brust et al., 1993). Thus, there seems to exist a synergistic action between $\alpha_2\delta$ and β subunits. It is likely that α_1-β interactions strengthen the interactions existing between α_1 and $\alpha_2\delta$ and thereby reinforce the effects of $\alpha_2\delta$ on current amplitude and kinetics. Conversely, it is possible that α_1-$\alpha_2\delta$ interactions modulate the conformation of the α_1 subunit and increase the affinity of α_1 for β subunits, which could explain the potentiation in current amplitude increase by β subunits. Coexpression of the $\alpha_2\delta$ subunit has been shown to increase the affinity of recombinant channel complexes 1.5-fold for ω-CgTx GVIA in HEK 293 cells (Brust et al., 1993). This is probably the strongest proof that the association of the $\alpha_2\delta$ subunit alters the conformation of the α_1 subunit and thereby the toxin binding site.

Overall, expression experiments provide additional evidence that the minimal subunit composition of voltage-dependent Ca^{2+} channels can be described as $\alpha_1\alpha_2\delta\beta$. The data obtained so far suggest that both $\alpha_2\delta$ and β subunits directly interact with the α_1 subunit. These interactions regulate the biophysical properties of the channel, its pharmacological sensitivity, and its regulation by cytoplasmic factors. Overall, these data indicate the complexity that is involved in fully and reliably characterizing native voltage-dependent Ca^{2+} channels without a good understanding of their subunit composition and the regulatory environment provided by the cell. Future investigations should therefore be aimed at identifying the ancillary subunits that are associated with each α_1 subunit in native voltage-dependent Ca^{2+} channels.

6. SUBUNIT INTERACTION SITES IN VOLTAGE-DEPENDENT Ca^{2+} CHANNELS

6.1. Identification of an Alpha 1 Interaction Domain (AID) that Binds β Subunits

There is overwhelming evidence that most voltage-dependent, and probably all HVA channels, contain a cytoplasmic β subunit as an element of the protein channel complex. This ancillary subunit is a constitutive component of both the purified N- and L-type channels and is a strong regulatory component of all cloned α_1 subunits. Moreover, there seems to be an important cross-reactivity among various α_1 and β subunits isoforms, suggesting that the interactions between both α_1 and β subunits are well conserved. For instance, both the α_{1C} and the α_{1A} subunits can be regulated by various isoforms of the β_1, β_2, β_3, and β_4 subunits (Wei *et al.*, 1991; Hullin *et al.*, 1992; Perez-Reyes *et al.*, 1992; Castellano *et al.*, 1993; De Waard and Campbell, 1995). In turn, β_{1b} is reportedly known to regulate several isoforms of α_1 subunits, α_{1A}, α_{1B}, α_{1C}, and α_{1E} (De Waard *et al.*, 1994b; Stea *et al.*, 1993; Wakamori *et al.*, 1994). As demonstrated by several biochemical reports, the β subunit is entirely cytoplasmic, which confines the α_1-β interaction sites to intracellular loci. Such a localization should avoid the potential structural complexity associated with membrane-spanning segments. It was therefore expected that at least one of the cytoplasmic loops of the α_1 subunit was interacting with conserved sequences on β subunits. *In vitro* translated and [35S] methionine-labeled β subunits were shown to interact with the purified α_{1S} and α_{1B} subunits of the skeletal muscle DHP receptor and the brain N-type Ca^{2+} channel (Pragnell *et al.*, 1994). The same probe could also interact with proteolytic fragments of these α_1 subunits, thereby strongly indicating that the α_1 interaction domain (AID) responsible for the binding of β subunits is poorly conformation-dependent. This unique property of the AID helped the experimental design of the locus determination of the β binding site on the α_1 subunit. The AID site was determined on several α_1 subunits (α_{1S}, α_{1A}, α_{1B}, α_{1C}) by screening epitope libraries of these proteins with [35S]-labeled β subunits. All the epitopes that were obtained using this strategy mapped to the same locus on the α_1 subunit, namely the cytoplasmic loop separating the first from the second hydrophobic repeats according to the putative transmembrane topology of this subunit. Figure 8 shows a cDNA sequence alignment of representatives of all six α_1 subunit genes that should contain the β binding site AID. The overall sequence identity of this cytoplasmic loop is only 19%, which greatly facilitates the identification of the AID as a minimal conserved sequence, QQ-E-L-GY--WI--E. Thus the AID has a length of 18 amino acids, of which nine are well conserved and expected to participate in the interaction with β subunits. Although there are identical amino acids upstream from the 18–amino acid epitope, these were not required for binding, as demonstrated by the clone detected by overlap method (underlined in Fig. 8). The importance of the AID was demonstrated by single-point mutations that either disrupted or altered the binding of β subunits. These point mutations also altered the β subunit regulation of the α_1 current properties (Fig. 9). Mutations that did not affect the association of β subunits to the α_1 subunit modified only the efficiencies of β current amplitude stimulation, suggesting that the conformation of the AID is essential to this β regulation. Conversely, mutations that affected the ability of α_1 subunits to bind β subunits completely abolished all the regulation by β subunits, demonstrating that the AID is not only a primary attachment site for the β subunit but is also required for all its effects on current properties.

6.2. Identification of a Beta Interaction Domain (BID) that Interacts with AID

The identification of a β subunit sequence involved in an interaction with an α_1 subunit stemmed from our interest in the mechanism of β-induced current stimulation. A

α_1	**Interaction sequences**	**Ref.**
S	GEFTKEREKAKSRGTFQKLREK**QQLEEDLRGYMSWIT**QGEVMDVEDLREG-------KLSL	1
A	GEFAKERERVENRRAFLKLRRQ**QQIERELNGYMEWIS**KAEEVILAEDETDVEQRHPFDGAL	2
B	GEFAKERERVENRRAFLKLRRQ**QQIERELNGYLEWIF**KAEEVMLAEEDRNAEEKSPLD-VL	3
C	GEFSKEREKAKARGDFQKLREK**QQLEEDLKGYLDWIT**QAEDIDPENEDEGMDEEKPRNMSM	4
D	GEFSKEREKAKARGDFQKLREK**QQLEEDLKGYLDWIT**QAEDIDPRENEEEGGEEGKRNTSM	5
E	GEFAKERERVENRRAFMKLRRQ**QQIERELNGYRAWID**KAEEVMLAEENKNSGTSALEVLRR	6

$\textbf{AID}_x$ |QQ-E--L-GY--WI---E|

FIGURE 8. AID interaction domain is formed by nine highly conserved amino acids in the I-II cytoplasmic linker of all α_1 subunits. Ca^{2+} channel class A through class E, as well as class S, have been shown to interact with the BID site of the β subunit through this well-defined AID_x epitope (X = S, A, B, C, D, or E). Although there are several conserved amino acids 5' to the AID_x epitope, these are by definition not required for interaction with the β subunit, as demonstrated by the underlined sequences shown to bind β subunits *in vitro* (Pragnell *et al.*, 1994). References: 1, Tanabe *et al.* (1987); 2, Mori *et al.* (1991); 3, Williams *et al.* (1992b); 4, Mikami *et al.* (1989); 5, Williams *et al.* (1992a); 6, Niidome *et al.* (1992).

A

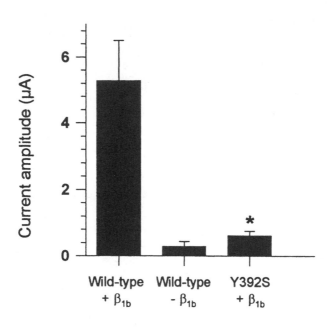

B

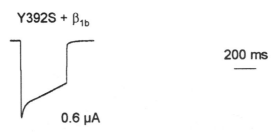

FIGURE 9. The Y392S mutation in AID of the α_{1A} subunit disrupts all the β_{1b} subunit regulation. The data presented in (A), (B), and (C) are from Pragnell *et al.* (1994). The current properties of mutant α_{1A} subunit coexpressed with β_{1b} (filled symbol) were compared to the properties of wild-type α_{1A} subunit in the absence

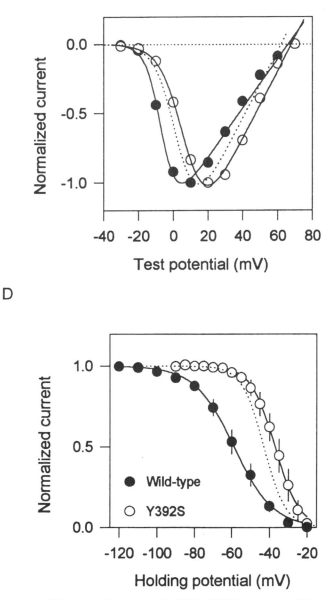

(dashed line) and presence of β_{1b} subunit (open symbol). (D) The Y392S mutation in AID_A also abolishes the hyperpolarizing shift induced by the β_{1b} subunit.

β interaction domain (BID) was initially identified as the minimum β subunit sequence capable of triggering a stimulation in the current amplitude. It was expected that since all β subunits are almost equally potent with respect to amplitude stimulation, BID probably represented a conserved sequence. As previously demonstrated in Fig. 3B, the β subunit can be divided into five structural regions depending on the level of amino acid homologies among various β subunits. Several cDNA constructs were made that encoded various structural domains of the β_{1b} subunit. The truncated forms of this subunit were then tested for their efficiencies of α_{1A} current stimulation into *Xenopus* oocytes (De Waard *et al.*, 1994b). The results demonstrated that a sequence of 30 amino acids, located at the N-terminus of the second most conserved domain of β subunits, was responsible for a significant fraction of current stimulation. The structural complexity of the BID is demonstrated by the presence of five proline residues that could introduce several β-turns in the interaction site. The BID may potentially be a regulatory target by a protein kinase with the presence of two consensus sites for protein kinase C. Direct interaction of this sequence with the α_1 subunit was proven, not only by current amplitude stimulation, but also by nonspecific changes in the activation and inactivation kinetics of the α_{1A} current (De Waard *et al.*, 1994b). For instance, the rather drastic changes in inactivation kinetics that occur by truncating the β subunit suggest that the BID is strongly influenced by changes in conformation of adjacent sequences in the β subunit. The presence of a splicing region, thus variable for each β subunit isoform, immediately upstream from the BID might be of relevance to this observation. Interestingly, the truncated β_{1b} subunit that encodes the entire second conserved domain (BID + 173 amino acids downstream from BID) was shown to interact directly with the AID on the α_1 subunit in an overlay assay. These data strongly suggested either that there are two independent interaction sites in the second conserved domain of β subunits (one for β stimulation called BID and another for the overlay interaction) or that both the stimulation and the overlay interaction with the AID occurred via the BID. To resolve this dilemma, single-point mutations were performed in BID on the full-length β_{1b} subunit. The abilities of the mutant β subunits to interact with α_1 subunits or AID itself and to regulate α_1 currents were compared to those of the wild-type β subunit. The data obtained were strikingly similar to the results obtained with single-point mutations of the AID site. Again, two categories of mutants could be identified. First, some of the mutations affected the ability of β subunits to bind to the AID and to the full-length α_1 subunit, thereby demonstrating that the AID and BID are directly interacting with each other. These β mutants were unable to regulate the properties of α_1 currents (amplitude, kinetics, and voltage dependence), confirming that the AID-BID association is absolutely required for channel regulation by β subunits. Second, mutations that did not affect the association of β subunits to α_1 subunits only modified the efficiencies of current amplitude stimulation by β subunits, while the properties of kinetic and voltage-dependence modulations were left intact. These results again suggest that the mechanisms that are at the basis of current stimulation are different from those responsible for the changes in current kinetics and voltage dependence. The differences in extent of stimulation by the interacting β subunit mutants support the hypothesis that conformational changes contributed by the β subunit can significantly alter channel activity. This would then be achieved by either facilitating or hindering the coupling between movement of the voltage sensors and opening of the pore (Neely *et al.*, 1993).

6.3.   Perspectives on the Primary Subunit Interaction Sites

Due to substantial molecular diversities of α_1 and β subunits, there is an impressive potential for combinatorial heterogeneity in voltage-sensitive Ca^{2+} channels at the structural level and consequently at the functional level. This possibility is particularly true in brain where five out of six α_1 genes and all four β genes are expressed and is further strengthened by the identification of two conserved structural elements required for the interaction of α_1 and β subunits. It is not known whether such a potential for Ca^{2+} channel diversity is exploited by cells. Preliminary evidence from two purified channels (skeletal muscle L-type and brain N-type Ca^{2+} channel) might indicate that this is not the case. The data gathered so far show that in skeletal muscle, the α_{1S} subunit specifically assembles with the β_{1a} subunit. Also, the α_{1B} subunit associates with the β_3 subunit in brain despite the coexpression of both β_1 and β_4 genes in the same tissue. There may thus exist unidentified mechanisms of specificity in subunit recognition and assembly in native voltage-dependent Ca^{2+} channels. Two hypotheses may explain this selectivity in subunit assembly: 1) It is possible that the affinity between individual α_1 and β subunits governs their interactions. This would then represent a true specificity in subunit assembly. Recent biochemical evidence demonstrates that different β subunits do indeed have different affinities for each α_1 subunit (De Waard *et al.*, 1995). In this respect, it will be important to analyze the importance of nonconserved residues interspersed among the conserved residues of the AID in determining the affinity of various β subunits for each α_1 subunit. 2) The level and timing of expression of each Ca^{2+} channel subunit might determine the composition of the native channels. Recently, Tarroni *et al.* (1994) have suggested that the α_{1B} subunit of IMR 32 human neuroblastoma cell line is not associated with the β_3 subunit, but instead with a β_2 subunit. This is confirmed by expression experiments that demonstrate the functional interchangeability of β subunits in their association to α_1 subunits. Thus, it is possible that the α_{1S} subunit of skeletal muscle L-type channel is associated with a β_{1a} subunit, because these two subunits are predominantly, if not exclusively, expressed in this tissue. Such a mechanism would represent an apparent specificity of subunit assembly. It should be emphasized that both hypotheses of true and apparent specificity in subunit assembly are not mutually exclusive. Future investigations should be aimed at resolving the subunit composition of voltage-dependent Ca^{2+} channels in several other homogeneous cell populations and the effectiveness of β subunit interchangeability in native voltage-dependent Ca^{2+} channels.

7.   SUMMARY AND CONCLUSIONS

Data gathered from the expression of cDNAs that encode the subunits of voltage-dependent Ca^{2+} channels have demonstrated important structural and functional similarities among these channels. Despite these convergences, there are also significant differences in the nature and functional importance of subunit-subunit and protein–Ca^{2+} channel interactions. There is evidence demonstrating that the functional differences between Ca^{2+} channel subtypes is due to several factors, including the expression of distinct α_1 subunit proteins, the selective association of structural subunits and modulatory proteins, and differences in posttranslational processing and cell regulation. We summarize several avenues of research that should provide significant clues about the structural features involved in the biophysical and functional diversity of voltage-dependent Ca^{2+} channels.

7.1. Secondary Subunit Interaction Sites

We have previously demonstrated that the AID and BID are primary interaction sites required for anchoring the β subunit to α_1 in voltage-dependent Ca^{2+} channels. Single-point mutations in either one of these sites proved conclusively that this attachment serves an obligatory role in the regulation of Ca^{2+} channel properties by β subunits. Several research directions remain at this stage. First, it is possible that all β regulation (current amplitude stimulation and changes in kinetics and voltage dependence of channel activity) are transduced via this primary α_1-β attachment site. Two independent observations would be consistent with this possibility. Screening of an α_1 epitope library with [^{35}S]-labeled β subunit failed to identify α_1 sequences other than those containing the AID (Pragnell *et al.*, 1994). Different β subunits induce qualitatively similar regulation in spite of important sequence divergences in the N- and C-terminal regions. Nonconserved β subunit sequences may define their protein conformation and in turn influence the regulatory input of β subunits at the primary interaction site, thereby explaining the quantitative differences seen with different β subunits. Second, there may be functionally important secondary interaction sites operational only upon association of β subunits to the AID in the α_1 subunit. Several observations are in favor of this interpretation: 1) important stretches of conserved amino acid sequence exist in addition to BID itself, 2) truncated β subunits capable of interacting with AID regulate the α_1 inactivation kinetics in an anomalous manner, and 3) mutations in BID on the full-length β subunit that preserve the α_1-β interaction alter the β stimulation efficiency but leave intact the other regulation. This latter observation would not be expected if all β subunit regulation were transduced via the BID and AID sites. Recent experimental evidence suggests that the different β regulation may in fact physically occur at different loci on the β subunit. Olcese *et al.* (1994) have presented evidence that the nonconserved N-termini of β subunits set the rate of channel inactivation. This regulation in inactivation kinetics occurs independently of the effects of the β subunit on activation, suggesting that activation and inactivation gates on the α_1 subunit can be regulated by different parts of β subunits. Future investigations should be aimed at mapping additional interaction sites between α_1 and β subunits.

7.2. Interactions of Ca^{2+} Channels with Other Cellular Proteins

The cytoplasmic loop between repeats II and III of various α_1 subunits has been implicated in several protein-channel interactions. Chimeras in which regions of the α_{1S}

subunit were replaced with equivalent domains of the α_{1C} subunit have demonstrated that the II-III cytoplasmic linker determines the type of E-C coupling. This α_{1S} loop is therefore believed to interact directly, or indirectly via unidentified proteins, with the ryanodine receptor to produce the extracellular Ca^{2+}-independent type of E-C found in skeletal muscle (Catterall, 1991; Tanabe et al., 1990a). Recently, syntaxin, a protein required for transmitter release at the synapse, was shown to interact with the α_{1B} subunit of the N-type channel (Leveque et al., 1994; Sheng et al., 1994). This interaction was mapped to an 87–amino acid sequence also located on the cytoplasmic loop between hydrophobic repeats II and III. This finding demonstrates at a molecular level the colocalization of a Ca^{2+} source (the Ca^{2+} channel) and a Ca^{2+} effector (a protein of the exocytosis complex). This interaction has potential importance for the mechanism of excitation-secretion (E-S) coupling because syntaxin also interacts with αSNAP (Soluble NSF Attachment Protein), SNAP-25, and synaptobrevin by forming a tight exocytotic complex (Sollner et al., 1993). The interaction of this complex with the N-type Ca^{2+} channel is essential not only for the docking of synaptic vesicles at the active zone, but also for an effective coupling between Ca^{2+} entry and transmitter release via mechanisms that are the subject of intense investigations. In both cases, however, it is not yet understood how conformational changes in the α_1 subunit could be transduced to trigger Ca^{2+} release in skeletal muscle or secretion at neuronal synapses. Also, it is not known how these interactions may affect the biophysical or pharmacological properties of the channels. However, the data available clearly indicate that sequence differences in the II-III loop can define the nature of protein–Ca^{2+} channel interactions and thereby the function of each Ca^{2+} channel subtype. Future investigations may prove the existence of additional types of interaction between the II-III loop of other α_1 subunits (A, C, D, and E) and yet unidentified proteins. This may also ultimately lead to a better understanding of the functional significance of alternative splicing in the II-III loop domain of various α_1 subunits (Snutch et al., 1991).

Voltage-dependent Ca^{2+} channels, and N-type channels in particular, are targets of various neuromodulators (Diversé-Pierluissi et al., 1995). Frequently, these modulations involve G proteins acting directly on the channel by a fast and membrane-delimited pathway, although channel activity can also be regulated by indirect mechanisms such as phosphorylation (Hille, 1992). There is also evidence that biochemical modifications of the N-type channel by one pathway (protein kinase C phosphorylation, for example) may alter the susceptibility of the channel to alternative regulations, including G protein inhibition (Swartz, 1993; Diversé-Pierluissi et al., 1995). So far, it is well documented that G proteins can modulate the activity of several types of Ca^{2+} channels by directly interacting with one or more subunits of the protein complex. Direct G protein interaction with N-type Ca^{2+} channels is believed to decrease the current by altering the prevalence of a high-P_o gating mode of the channel activity (Delcour and Tsien, 1993). Purified G_S protein stimulates the activity of the skeletal muscle L-type Ca^{2+} channel (Yatani et al., 1988). Also, the endogenous $G_{o\alpha}$ subunit partially copurifies with the N-type Ca^{2+} channel, indicating the existence of a high-affinity site between both protein complexes (McEnery et al., 1994). Although the subjects of intense investigation, the site of G protein interaction with voltage-dependent Ca^{2+} channels and the precise mechanisms of regulation have not yet been elucidated. It is also not known how the subunit composition and/or posttranslational modifications of the Ca^{2+} channel complex may affect G protein modulation of the currents.

It is believed that answers to several of these questions may also provide some clues regarding the structural determinants involved in the spatial distribution of calcium channels at the cellular level. Differential distribution of voltage-gated Ca^{2+} channels in cells is required for the compartmentalized activation of Ca^{2+}-dependent processes. For instance, it is not known what factors define the localization and the clustering of L-type Ca^{2+} channels at the branching points of dendrites in hippocampal pyramidal cells (Westenbroek *et al.*, 1990) or similarly the clustering of N-type Ca^{2+} channels at the active zones of the synapse.

7.3. Functional Map of Ca^{2+} Channels

To appreciate the real diversity of Ca^{2+} channels, we must first identify the structural elements that control the pattern of activity of any given channel and then understand the molecular mechanisms whereby channel gating responds to biochemical changes at their site of regulation. The mapping of several functionally important elements within various α_1 subunits has paved much of the way (Fig. 10). Chimeras between the α_{1S} and α_{1C} subunits have shown that the IS3 segment and the extracellular IS3-S4 linker are critical sequences implicated in the control of the current activation kinetics of both subunits (Nakai *et al.*, 1994). The different inactivation kinetics of α_{1A}, α_{1C}, and α_{1E} seem to be under the control of specific sequences within the IS6

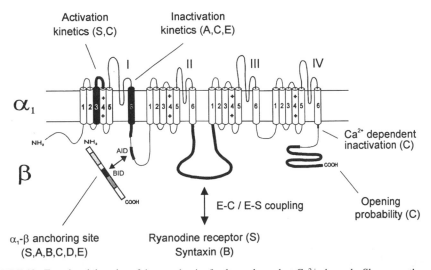

FIGURE 10. Functional domains of the α_1 subunit of voltage-dependent Ca^{2+} channels. Shown are the well-characterized β anchoring site in the I-II cytoplasmic linker, the regions involved in activation kinetics, and inactivation kinetics. The highly variable II-III cytoplasmic linker is important in excitation-contraction secretion in the α_{1S} subtype and in syntaxin binding in the α_{1B} subtype. Carboxyl terminal to IVS6 is a region with slight EF-hand homology that appears to be involved in Ca^{2+}-dependent inactivation. Deletions of the C-terminus result in increased channel open probability, suggesting its importance in channel gating.

segment and of a proximal sequence in the I-II cytoplasmic loop downstream from this IS6 segment (Zhang *et al.*, 1994). Finally, the C-terminal region of α_1 subunits seems to contain important structural elements for the regulation of Ca^{2+} channel activity. The extreme C-terminal portion of the α_{1C} subunit is involved in tonic inhibition of channel opening probability (Wei *et al.*, 1994), whereas a cytoplasmic EF-hand Ca^{2+} binding sequence immediately downstream from the IVS6 segment is essential for Ca^{2+}-dependent inactivation of the α_{1C} subunit (de Leon *et al.*, 1995).

The mapping of subunit-subunit and protein-channel interaction sites has added additional structural elements to the functional complexity of α_1 subunits. We have demonstrated that the primary α_1-β interaction site occurs within the cytoplasmic linker separating the first and second hydrophobic repeats (Pragnell *et al.*, 1994; De Waard *et al.*, 1994b). Also, the intracellular loop between motifs II and III of the α_1 subunit is either involved in skeletal muscle E-C coupling (α_{1S}) or E-S coupling (α_{1B}) via a direct interaction with syntaxin (Sheng *et al.*, 1994). It will be important to determine how conformational changes introduced by β subunits in the α_1 channel may regulate the contribution of the various structural elements of the α_1 subunit to Ca^{2+}-dependent inactivation, to contraction and secretion, or to the control of activation and inactivation kinetics. It is interesting that the implication of the IS6 segment in Ca^{2+} channel inactivation is not consistent with either the N-terminal localization of the "ball-and-chain" inactivation of K^+ channels (Hoshi *et al.*, 1990) or the fast inactivation by the III-IV linker in voltage-dependent Na^+ channels (Catterall, 1993). These differences in the localization of the structural elements may also underlie a different mechanism of inactivation in voltage-dependent Ca^{2+} channels. In this respect, the proximity of AID to the IS6 segment is conspicuous. It will therefore be important to analyze whether the interactions between AID and the β subunit can allosterically modulate the conformation of the IS6 segment and its contribution to the inactivation process.

Another prevalent cellular mechanism for the modulation of ion channel function is phosphorylation. The phosphorylation state of the channel subunits or the proteins associated with the channel influences the amplitude (Yang and Tsien, 1993), voltage dependence (Sculptoreanu *et al.*, 1993), or time course of the current (Werz *et al.*, 1993). Phosphorylation may also affect the effectiveness of several channel regulations (Swartz, 1993). The underlying biophysical and molecular mechanisms of these modulatory processes remain to be analyzed. In particular, it will be necessary to analyze how the interactions between AID and the β subunit may influence the effects of these phosphorylations on channel activity.

ACKNOWLEDGMENTS. Kevin P. Campbell is an Investigator of the Howard Hughes Medical Institute. We thank Drs. Rachelle Crosbie and Daniel Jung for critical reading of the manuscript.

8. REFERENCES

Adams, M. E., and Olivera, B. M., 1994, Neurotoxins: Overview of an emerging research technology, *Trends Neurosci.* **17:**151–155.

Adams, M. E., Mintz, I. M., Reily, M. D., Thanabal, V., and Bean, B. P., 1993, Structure and properties of ω-agatoxin IVB, a new antagonist of P-type calcium channels, *Mol. Pharmacol.* **44:**681–688.

Ahlijanian, M. K., Westenbroek, R. E., and Catterall, W. A., 1990, Subunit structure and localization of dihydropyridine-sensitive calcium channels in mammalian brain, spinal cord, and retina, *Neuron* **4:** 819–832.

Artalejo, C. R., Adams, M. E., and Fox, A. P., 1994, Three types of Ca^{2+} channel trigger secretion with different efficacies in chromaffin cells, *Nature* **367:**72–74.

Augustine, G. J., Charlton, M. P., and Smith, S. J., 1987, Calcium action in synaptic transmitter release, *Annu. Rev. Neurosci.* **10:**633–693.

Beam, K. G., Adams, B. A., Niidome, T., Numa, S., and Tanabe, T., 1992, Function of a truncated dihydro-pyridine receptor as both voltage sensor and calcium channel, *Nature* **360:**169–171.

Bean, B. P., 1989, Multiple types of calcium channels in heart muscle and neurons: Modulation by drugs and neurotransmitters, *Ann. N.Y. Acad. Sci.* **560:**334–345.

Biel, M., Ruth, P., Bosse, E., Hullin, R., Stuhmer, W., Flockerzi, V., and Hofmann, F., 1990, Primary structure and functional expression of a high voltage activated calcium channel from rabbit lung, *FEBS Lett.* **269:**409–412.

Biel, M., Hullin, R., Freudner, S., Singer, D., Dascal, N., Flockerzi, V., and Hofmann, F., 1991, Tissue-specific expression of high-voltage-activated dihydropyridine-sensitive L-type calcium channels, *Eur. J. Biochem.* **200:**81–88.

Birnbaumer, L., Campbell, K. P., Catterall, W. A., Harpold, M. M., Hofmann, F., Horne, W. A., Mori, Y., Schwartz, A., Snutch, T. P., Tanabe, T., and Tsien, R. W., 1994, The naming of voltage-gated calcium channels, *Neuron* **13:**505–506.

Boland, L. M., Morrill, J. A., and Bean, B. P., 1994, ω-Conotoxin block of N-type calcium channels in frog and rat sympathetic neurons, *J. Neurosci.* **14:**5011–5027.

Brust, P. F., Simerson, S., McCue, A. F., Deal, C. R., Schoonmaker, S., Williams, M., Velicelebi, G., Johnson, E. C., Harpold, M. M., and Ellis, S. B., 1993, Human neuronal voltage-dependent calcium channels: Studies on subunit structure and role in channel assembly, *Neuropharmacology* **32:**1089–1102.

Burke, S. P., Adams, M. E., and Taylor, C. P., 1993, Inhibition of endogenous glutamate release from hippocampal tissue by Ca^{2+} channel toxins, *Eur. J. Pharmacol.* **238:**383–386.

Carbone, E., and Lux, H. D., 1987, Kinetics and selectivity of a low-voltage-activated calcium current in chick and rat sensory neurones, *J. Physiol.* **386:**547–570.

Carbone, E., Sher, E., and Clementi, F., 1990, Ca currents in human neuroblastoma IMR32 cells: Kinetics, permeability, and pharmacology, *Pflügers Arch.* **416:**170–179.

Castellano, A., and Perez-Reyes, E., 1994, Molecular diversity of Ca^{2+} channel β subunits, *Biochem. Soc. Trans.* **22:**483–488.

Castellano, A., Wei, X., Birnbaumer, L., and Perez-Reyes, E., 1993a, Cloning and expression of a third calcium channel β subunit, *J. Biol. Chem.* **268:**3450–3455.

Castellano, A., Wei, X., Birnbaumer, L., and Perez-Reyes, E., 1993b, Cloning and expression of a neuronal calcium channel β subunit, *J. Biol. Chem.* **268:**12359–12366.

Catterall, W. A., 1991, Excitation-contraction coupling in vertebrate skeletal muscle: A tale of two calcium channels, *Cell* **64:**871–874.

Catterall, W. A., 1993, Structure and function of voltage-gated ion channels, *Trends Neurosci.* **16:**500–506.

Catterall, W. A., and Striessnig, J., 1992, Receptor sites for Ca^{2+} channel antagonists, *Trends Pharmacol. Sci.* **13:**256–262.

Cohen, C. J., Ertel, E. A., Smith, M. M., Venema, V. J., Adams, M. E., and Leibowitz, M. D., 1992, High affinity block of myocardial L-type calcium channels by the spider toxin ω-Aga-toxin IIIA: Advantages over 1,4-dihydropyridines, *Mol. Pharmacol.* **42:**947–951.

Collin, T., Wang, J.-J., Nargeot, J., and Schwartz, A., 1993, Molecular cloning of three isoforms of the L-type voltage-dependent calcium channel β subunit from normal human heart, *Circ. Res.* **72:**1337–1344.

Condignola, A., Tarroni, P., Clementi, F., Pollo, A., Lovallo, M., Carbone, E., Sher, E., 1993, Calcium channel subtypes controlling serotonin release from human small cell lung carcinoma cell lines, *J. Biol. Chem.* **268:**26240–26247.

Cooper, C. L., Vandaele, S., Barhanin, J., Fosset, M., Lazdunski, M., and Hosey, M. M., 1987, Purification and characterization of the dihydropyridine-sensitive voltage-dependent calcium channel from cardiac tissue, *J. Biol. Chem.* **262:**509–512.

De Jongh, K. S., Merrick, D. K., and Catterall, W. A., 1989, Subunits of purified calcium channels: A 212-kDa form of α_1 and partial amino-acid sequence of a phosphorylation site of an independent β subunit, *Proc. Natl. Acad. Sci. USA* **86:**8585–8589.

De Jongh, K. S., Warner, C., and Catterall, W. A., 1990, Subunits of purified calcium channels, *J. Biol. Chem.* **265:**14738–14741.

De Jongh, K. S., Warner, C., Colvin, A. A., and Catterall, W. A., 1991, Characterization of two size forms of the α_1 subunit of skeletal muscle L-type calcium channels, *Proc. Natl. Acad. Sci. USA* **88:**10778–10782.

Delcour, A. H., and Tsien, R. W., 1993, Altered prevalence of gating modes in neurotransmitter inhibition of N-type calcium channels, *Science* **259:**980–984.

de Leon, M., Jones, L., Perez-Reyes, E., Wei, X., Soong, T. W., Snutch, T. P., and Yue, D. T., 1995, An essential structural domain for Ca-sensitive inactivation of L-type Ca^{2+} channels, *Biophys. J.* **68:**A13.

De Waard, M., and Campbell, K. P., 1995, Subunit regulation of the neuronal α_{1A} Ca^{2+} channel, *J. Physiol.* **485:**619–634.

De Waard, M., Witcher, D. R., and Campbell, K. P., 1994a, Functional properties of the purified N-type Ca^{2+} channel from rabbit brain, *J. Biol. Chem.* **269:**6716–6724.

De Waard, M., Pragnell, M., and Campbell, K. P., 1994b, Ca^{2+} channel regulation by a conserved β subunit domain, *Neuron* **13:**495–503.

De Waard, M., Witcher, D. R., Pragnell, M., Liu, H., and Campbell, K. P., 1995, Properties of the α_1-β anchoring site in voltage-dependent Ca^{2+} channels, *J. Biol. Chem.* **270:**12056–12064.

Diebold, R. J., Koch, W. J., Ellinor, P. T., Wang, J., Muthuchamy, M., Wieczorek, D. F., and Schwartz, A., 1992, Mutually exclusive exon splicing of the cardiac calcium channel α_1 subunit gene generates developmentally regulated isoforms in the heart, *Proc. Natl. Acad. Sci. USA* **89:**1497–1501.

Diversé-Pierluissi, M., Goldsmith, P. K., and Dunlap, K., 1995, Transmitter-mediated inhibition of N-type calcium channels in sensory neurons involves multiple GTP-binding proteins and subunits, *Neuron* **14:**191–200.

Donaldson, P. L., and Beam, K. G., 1983, Calcium currents in a fast-twitch skeletal muscle of the rat, *J. Gen. Physiol.* **82:**449–468.

Droogmans, G., Declerck, I., and Casteels, R., 1987, Effect of adrenergic agonists on Ca^{2+}-channel currents in single vascular smooth muscle cells, *Pflügers Arch.* **409:**7–12.

Ellinor, P. T., Zhang, J.-F., Randall, A. D., Zhou, M., Schwarz, T. L., Tsien, R. W., and Horne, W. A., 1993, Functional expression of a rapidly inactivating neuronal calcium channel, *Nature* **363:**455–458.

Ellinor, P. T., Zhang, J.-F., Horne, W. A., and Tsien, R. W., 1994, Structural determinants of the blockade of N-type calcium channels by a peptide neurotoxin, *Nature* **372:**272–275.

Ellis, S. B., Williams, M. E., Ways, N. R., Brenner, R., Sharp, A. H., Leung, A. T., Campbell, K. P., McKenna, E., Koch, W. J., Hui, A., Schwartz, A., and Harpold, M. M., 1988, Sequence and expression of mRNAs encoding the α_1 and α_2 subunits of a DHP-sensitive calcium channel, *Science* **241:**1661–1664.

Fedulova, S. A., Kostyuk, P. K., and Vesolovski, N. S., 1985, Two types of calcium channels in the somatic membrane of new-born rat dorsal root ganglion neurones, *J. Physiol.* **359:**431–446.

Flockerzi, V., Oeken, H.-J., Hofmann, F., Pelzer, D., Cavalie, A., and Trautwein, W., 1986, Purified dihydropyridine-binding site from skeletal muscle t-tubules is a functional calcium channel, *Nature* **323:**66–68.

Forti, L., and Pietrobon, D., 1993, Functional diversity of L-type calcium channels in rat cerebellar neurons, *Neuron* **10:**437–450.

Fox, A. P., Nowycky, M. C., and Tsien, R. W., 1987a, Kinetic and pharmacological properties distinguishing three types of calcium currents in chick sensory neurones, *J. Physiol.* **394:**149–172.

Fox, A. P., Nowycky, M. C., and Tsien, R. W., 1987b, Single channel recordings of three types of calcium channels in chick sensory neurones, *J. Physiol.* **394:**173–200.

Fuchs, E., and Hanukoglu, I., 1983, Unraveling the structure of intermediate filaments, *Cell* **34:**332–334.

Fujita, Y., Mynlieff, M., Dirksen, R. T., Kim, M.-S., Niidome, T., Nakai, J., Friedrich, F., Iwabe, N., Miyata, T., Furuichi, T., Furutama, D., Mikoshiba, K., Mori, Y., and Beam, K. G., 1993, Primary structure and functional expression of the ω-conotoxin sensitive N-type Ca^{2+} channel from rabbit brain, *Neuron* **10:**585–598.

Glossmann, H., and Striessnig, J., 1988, Structure and pharmacology of voltage-dependent calcium channels, *ISI Atlas Sci. Pharmacol.* **2:**202–210.

Grabner, M., Friedrich, K., Knaus, H.-G., Streissnig, J., Scheffauer, F., Staudinger, R., Koch, W. J., Schwartz, A., and Glossmann, H., 1991, Calcium channels from *Cyprinus carpio* skeletal muscle, *Proc. Natl. Acad. Sci. USA* **88:**727–731.

Guy, H. R., and Conti, F., 1990, Pursuing the structure and function of voltage-gated channels, *Trends Neurosci.* **13:**201–206.

Haase, H., Streissnig, J., Holtzhauser, M., Vetter, R., and Glossman, H., 1991, A rapid procedure for the purification of cardiac 1,4-dihydropyridine receptors from porcine heart, *Eur. J. Pharmacol.* **207:**51–55.

Hagiwara, N., Trisawa, H., and Kameyama, M., 1988, Contribution of two types of calcium currents to the pacemaker potentials of rabbit sino-atrial node cells, *J. Physiol.* **395:**233–253.

Heinemann, S. H., Terlau, H., Stuhmer, W., Imoto, K., and Numa, S., 1992, Calcium channel characteristics conferred on the sodium channel by single mutations, *Nature* **356:**441–443.

Hell, J. H., Yokoyama, C. T., Wong, S. T., Warner, C., Snutch, T. P., and Catterall, W. A., 1993, Differential phosphorylation of two size forms of the neuronal class C L-type calcium channel alpha 1 subunit, *J. Biol. Chem.* **268:**19451–19457.

Hell, J. H., Appleyard, S. M., Tokoyama, C. T., Warner, C., and Catterall, W. A., 1994, Differential phosphorylation of two size forms of the N-type calcium channel α_1 subunit which have different COOH termini, *J. Biol. Chem.* **269:**7390–7396.

Hille, B., 1992, G. protein coupled mechanisms and nervous signaling, *Neuron* **9:**187–195.

Hillyard, D. R., Monje, V. D., Mintz, I. M., Bean, B. P., Nadasdi, L., Ramachandran, J., Miljanich, G., Azimi-Zoonooz, A., McIntosh, J. M., Cruz, L. J., Imperial, J. S., and Olivera, B. M., 1992, A new conus peptide ligand for mammalian presynaptic Ca^{2+} channels, *Neuron* **9:**69–77.

Hirning, L. D., Fox, A. P., McCleskey, E. W., Olivera, B. M., Thayer, S. A., Miller, R. J., and Tsien, R. W., 1988, Dominant role of N-type Ca^{2+} channels in evoked release of norepinephrine from sympathetic neurons, *Science* **239:**57–60.

Hoshi, T., Zagotta, W. N., and Aldrich, R. W., 1990, Biophysical and molecular mechanisms of shaker potassium channel inactivation, *Science* **250:**533–538.

Hui, A., Ellinor, P. T., Krizanova, O., Wang, J.-J., Diebold, R. J., and Schwartz, A., 1991, Molecular cloning of multiple subtypes of a novel rat brain isoform of the α_1 subunit of the voltage-dependent calcium channel, *Neuron* **7:**35–44.

Hullin, R., Singer-Lahat, D., Freichel, M., Biel, M., Dascal, N., Hofmann, F., and Flockerzi, V., 1992, Calcium channel β subunit heterogeneity: Functional expression of cloned cDNA from heart, aorta and brain, *EMBO J.* **11:**885–890.

Hymel, L., Striessnig, J., Glossmann, H., and Schindler, H., 1988, Purified skeletal muscle 1,4-dihydropyridine receptors form phosphorylation-dependent oligomeric calcium channels in planar bilayers, *Proc. Natl. Acad. Sci. USA* **85:**4290–4294.

Itagaki, K., Koch, W. J., Bodi, I., Klockner, U., Slish, D. F., and Schwartz, A., 1992, Native-type DHP-sensitive calcium channel currents are produced by cloned rat aortic smooth muscle and cardiac α_1 subunits expressed in *Xenopus laevis* oocytes and are regulated by α_2- and β-subunits, *FEBS Lett.* **297:**221–225.

Jahn, H., Nastainczyk, W., Rohrkasten, A., Schneider, T., and Hofmann, F., 1988, Site specific phosphorylation of the purified receptor for calcium channel blockers by cAMP- and cGMP-dependent protein kinases, protein kinase C, calmodulin-dependent protein kinase II and casein kinase II, *Eur. J. Biochem.* **178:**535–542.

Jay, S. D., Ellis, S. B., McCue, A. F., Williams, M. E., Vedvick, T. S., Harpold, M. M., and Campbell, K. P., 1990, Primary structure of the γ subunit of the DHP-sensitive calcium channel from skeletal muscle, *Science* **248:**490–492.

Jay, S. D., Sharp, A. H., Kahl, S. D., Vedvick, T. S., Harpold, M. M., and Campbell, K. P., 1991, Structural characterization of the dihydropyridine-sensitive calcium channel α_2-subunit and the associated δ peptides, *J. Biol. Chem.* **266:**3287–3293.

Kaczmarek, L. K., and Levitan, I. B., 1987, *Neuromodulation: The Biochemical Control of Neuronal Excitability*, Oxford University Press, New York.

Kalman, D., O'Lague, P. H., Erxleben, C., and Armstrong, D. L., 1988, Calcium-dependent inactivation of the DHP-sensitive calcium channel in GH3 cells, *J. Gen. Physiol.* **92:**531–548.

Kasai, H., and Neher, E., 1992, Dihydropyridine-sensitive and ω-conotoxin-sensitive calcium channels in a mammalian neuroblastoma-glioma cell line, *J. Physiol.* **448:**161–188.

Kass, R. S., and Sanguinetti, M. C., 1984, Inactivation of calcium channel current in the calf cardiac Purkinje fiber. Evidence for voltage- and calcium-mediated mechanisms, *J. Gen. Physiol.* **84:**705–726.

Katz, B., 1969, *The Release of Neural Transmitter Substances*, Liverpool University Press, Liverpool.

Kerr, L. M., and Yoshikami, D., 1984, A venom peptide with a novel presynpatic blocking action, *Nature* **308:**282–284.

Kim, H.-L., Kim, H., Lee, P., King, R., and Chin, H., 1992, Rat brain expresses an alternatively spliced form of the dihydropyridine-sensitive L-type calcium channel α_2 subunit, *Proc. Natl. Acad. Sci. USA* **89:**3251–3255.

Kleppisch, T., Pedersen, K., Strubing, C., Bosse-Doenecke, E., Flockerzi, V., Hofmann, F., and Hescheler, J., 1994, Double-pulse facilitation of smooth muscle α_1-subunit Ca^{2+} channels expressed in CHO cells, *EMBO J.* **13:**2502–2507.

Klockner, U., Itagaki, K., Bodi, I., and Schwartz, A., 1992, Beta-subunit expression is required for cAMP-dependent increase of cloned cardiac and vascular calcium channel currents, *Eur. J. Physiol.* **420:**413–415.

Koch, W. J., Ellinor, P. T., and Schwartz, A., 1990, cDNA cloning of a dihydropyridine-sensitive calcium channel from rat aorta, *J. Biol. Chem.* **265:**17786–17791.

Kuniyasu, A., Oka, K., Ide-Yamada, T., Hatanaka, Y., Abe, T., Nakayama, H., and Kanaoka, Y., 1992, Structural characterization of the dihydropyridine receptor–linked calcium channel from porcine heart, *J. Biochem.* **112:**235–242.

Lacerda, A. E., Kim, H. S., Ruth, P., Perez-Reyes, E., Flockerzi, V., Hofmann, F., Birnbaumer, L., and Brown, A. M., 1991, Normalization of current kinetics by interaction between the α_1 and β subunits of the skeletal muscle dihydropyridine-sensitive Ca^{2+} channel, *Nature* **352:**527–530.

Lazdunski, M., Schmid, A., Romey, G., Renaud, J. F., Galizzi, J. P., Fosset, M., Borsotto, M., and Barhanin, J., 1987, Dihydropyridine-sensitive Ca^{2+} channels: Molecular properties of interaction with Ca^{2+} channel blockers, purification, subunit structure, and differentiation, *J. Cardiovasc. Pharmacol.* **9:**S10–S15.

Leung, A. T., Imagawa, T., and Campbell, K. P., 1987, Structural characterization of the 1,4-dihydropyridine receptor of the voltage-dependent Ca^{2+} channel from rabbit skeletal muscle, *J. Biol. Chem.* **262:**7943–7946.

Leung, A. T., Imagawa, T., Block, B., Franzini-Armstrong, C., and Campbell, K. P., 1988, Biochemical and ultrastructural characterization of the 1,4-dihydropyridine receptor from rabbit skeletal muscle. Evidence for a 52,000 Da subunit, *J. Biol. Chem.* **263:**994–1001.

Leveque, C., El Far, O., Martin-Moutot, N., Sato, K., Kato, R., Takahashi, M., and Seagar, M. J., 1994, Purification of the N-type calcium channel associated with syntaxin and synaptotagmin. A complex implicated in synaptic vesicle exocytosis, *J. Biol. Chem.* **269:**6306–6312.

Levitan, I. B., 1988, Modulation of ion channels in neurons and other cells, *Annu. Rev. Neurosci.* **11:**119–136.

Ma, J., and Coronado, R., 1988, Heterogeneity of conductance states in calcium channels of skeletal muscle, *Biophys. J.* **53:**387–395.

McEnery, M. W., Snowman, A. M., and Snyder, S. H., 1994, The association of endogenous $G_{o\alpha}$ with the purified ω-contoxin GVIA receptor, *J. Biol. Chem.* **269:**5– 8.

Mikami, A., Imoto, K., Tanabe, T., Niidome, T., Mori, Y., Takeshima, H., Narumiya, S., and Numa, S., 1989, Primary structure and functional expression of the cardiac dihydropyridine-sensitive calcium channel, *Nature* **340:**230–233.

Mintz, I. B., and Bean, B. P., 1993, $GABA_B$ receptor inhibition of P-type Ca^{2+} channels in central neurons, *Neuron* **10:**889–898.

Mintz, I. M., Venema, V. J., Adams, M. E., and Bean, B. P., 1991, Inhibition of N- and L-type Ca^{2+} channels by the spider venom toxin ω-Aga IIIA, *Proc. Natl. Acad. Sci. USA* **88:**6628–6631.

Mintz, I. M., Venema, V. J., Swiderek, K., Lee, T., Bean, B. P., and Adams, M. E., 1992a, P-type Ca^{2+} channels blocked by the spider toxin ω-Aga-IVA, *Nature* **355:**827–829.

Mintz, I. M., Adams, M. E., and Bean, B. P., 1992b, P-type calcium channels in rat central and peripheral neurons, *Neuron* **9:**85–95.

Mitterdorfer, J., Froschmayr, M., Grabner, M., Striessnig, J., and Glossman, H., 1994, Calcium channels: The β subunit increases the affinity of dihydropyridine and Ca^{2+} binding sites of the α_1 subunit, *FEBS Lett.* **352**:141–145.

Mori, Y., Friedrich, T., Kim, M.-S., Mikami, A., Nakai, J., Ruth, P., Bosse, E., Hofmann, F., Flockerzi, V., Furuichi, T., Mikoshiba, K., Imoto, K., Tanabe, T., and Numa, S., 1991, Primary structure and functional expression from complementary DNA of a brain calcium channel, *Nature* **350**:398–402.

Morton, M. E., and Froehner, S. C., 1989, The α_1 and α_2 polypeptides of the dihydropyridine-sensitive calcium channel differ in developmental expression and tissue distribution, *Neuron* **2**:1499–1506.

Morton, M. E., Cassidy, T. N., Froehner, S. C., Gilmour, B. P., and Laurens, R. L., 1994, α_1 and α_2 Ca^{2+} channel subunit expression in neuronal and small cell carcinoma cells, *FASEB J.* **8**:884–888.

Nakai, J., Adams, B. A., Imoto, K., and Beam, K. G., 1994, Critical roles of the S3 segment and S3-S4 linker of repeat I in activation of L-type calcium channels, *Proc. Natl. Acad. Sci. USA* **91**:1014–1018.

Nakayama, H., Taki, M., Striessnig, J., Glossmann, H., Catterall, W. A., and Kanaoka, Y., 1991, Identification of 1,4-dihydropyridine binding regions within the α_1 subunit in skeletal muscle Ca^{2+} channels by photo-affinity labeling with diazepine, *Proc. Natl. Acad. Sci. USA* **88**:9203–9207.

Nastainczyk, W., Rohrkasten, A., Sieber, M., Rudolph, C., Schachtele, C., Marme, D., and Hofmann, F., 1987, Phosphorylation of the purified receptor for calcium channel blockers by cAMP kinase and protein kinase C, *Eur. J. Biochem.* **169**:137–142.

Neely, A., Wei, X., Olcese, R., Birnbaumer, L., and Stefani, E., 1993, Potentiation by the β-subunit of the ratio of the ionic current to the charge movement in the cardiac calcium channel, *Science* **262**:575–578.

Neely, A., Olcese, R., Wei, X., Birnbaumer, L., and Stefani, E., 1994, Ca^{2+}-dependent inactivation of a cloned cardiac Ca^{2+} channel α_1 subunit (α_{1C}) expressed in *Xenopus* oocytes, *Biophys. J.* **66**:1895–1903.

Niidome, T., Kim, M.-S., Friedrich, T., and Mori, Y., 1992, Molecular cloning and characterization of a novel calcium channel from rabbit brain, *FEBS Lett.* **308**:7–13.

Nilius, B., Hess, P., Lansman, J. B., and Tsien, R. W., 1985, A novel type of cardiac calcium channel in ventricular cells, *Nature* **316**:443–446.

Nishimura, S., Takeshima, H., Hofmann, F., Flockerzi, V., and Imoto, K., 1993, Requirement of the calcium channel β subunit for functional conformation, *FEBS Lett.* **324**:283–286.

Noda, M., Shimuzu, S., Tanabe, T., Takai, T., Kayano, T., Ikeda, T., Takahashi, H., Nakayama, H., Kanaoka, Y., Minamino, N., Kangawa, K., Matsu, H., Raftery, M., Hirose, T., Inayama, S., Hayashida, H., Miyata, T., and Numa, S., 1984, Primary structure of Electrophorus electricus sodium channel deduced from cDNA sequence, *Nature* **312**:121–127.

Nowycky, M. C., Fox, A. P., and Tsien, R. W., 1985, Three types of neuronal calcium channels with different calcium agonist sensitivity, *Nature* **316**:440–443.

Oguro-Okano, M., Griesmann, G. E., Wieben, E. D., Slaymaker, S. J., Snutch, T. P., and Lennon, V. A., 1992, Molecular diversity of neuronal-type calcium channels identified in small cell lung carcinoma, *Mayo Clin. Proc.* **67**:1150–1159.

Olcese, R., Qin, N., Schneider, T., Neely, A., Wei, X., Stefani, E., and Birnbaumer, L., 1994, The amino terminus of a calcium channel β subunit sets rates of channel inactivation independently of the subunit's effect on activation, *Neuron* **13**:1433–1438.

Pelzer, D., Grant, A. O., Cavalie, A., Pelzer, S., Sieber, M., Hofmann, F., and Trautwein, W., 1989, Calcium channels reconstituted from the skeletal muscle dihydropyridine receptor protein complex and its α_1 peptide subunit in lipid bilayers, *Ann. N.Y. Acad. Sci.* **560**:138–154.

Perez-Reyes, E., Kim, H. S., Lacerda, A. E., Horne, W., Wei, C., Rampe, D., Campbell, K. P., Brown, A. M., and Birnbaumer, L., 1989, Induction of calcium currents by the expression of the α_1-subunit of the dihydropyridine receptor from skeletal muscle, *Nature* **340**:233–236.

Perez-Reyes, E., Castellano, A., Kim, H. S., Bertrand, P., Baggstrom, E., Lacerda, A. E., Wei, X., and Birnbaumer, L., 1992, Cloning and expression of a cardiac/brain β subunit of the L-type calcium channel, *J. Biol. Chem.* **267**:1792–1797.

Plummer, M. R., and Hess, P., 1991, Reversible uncoupling of inactivation of N-type calcium channels, *Nature* **351**:657–659.

Plummer, M. R., Logothetis, D. E., and Hess, P., 1989, Elementary properties and pharmacological sensitivities of calcium channels in mammalian peripheral neurons, *Neuron* **2**:1453–1463.

Pollo, A., Lovallo, M., Biancardi, E., Sher, E., Socci, C., and Carbone, E., 1993, Sensitivity to dihydropyridines, ω-conotoxin and noradrenaline reveals multiple high-voltage-activated Ca^{2+} channels in rat insulinoma and human pancreatic β-cells, *Pflügers Arch.* **423:**462–471.

Powers, P. A., Liu, S., Hogan, K., and Gregg, R. G., 1992, Skeletal muscle and brain isoforms of a β-subunit of human voltage-dependent calcium channels are encoded by a single gene, *J. Biol. Chem.* **267:**22967–22972.

Pragnell, M., Sakamoto, J., Jay, S. D., and Campbell, K. P., 1991, Cloning an tissue-specific expression of the brain calcium channel β-subunit, *FEBS Lett.* **291:**253–258.

Pragnell, M., De Waard, M., Mori, Y., Tanabe, T., Snutch, T. P., and Campbell, K. P., 1994, Calcium channel β-subunit binds to a conserved motif in the I-II cytoplasmic linker of the $α_1$-subunit, *Nature* **368:** 67–70.

Randall, A. D., Wendland, B., Schweizer, F., Miljanich, G., Adams, M. E., and Tsien, R. W., 1993, Five pharmacologically distinct high voltage-activated Ca^{2+} channels in cerebellar granule cells, *Soc. Neurosci. Abstr.* **19:**1478.

Regan, L. J., Sah, D. W. Y., and Bean, B. P., 1991, Ca^{2+} channels in rat central and peripheral neurons: High-threshold current resistant to dihydropyridine blockers and ω-conotoxin, *Neuron* **6:**269–280.

Regulla, S., Schneider, T., Nastainczyk, W., Meyer, H. E., and Hofmann, F., 1991, Identification of the site of interaction of the dihydropyridine channel blockers nitrendipine and azidopine with calcium-channel $α_1$ subunit, *EMBO J.* **10:**45–49.

Robitaille, R., Adler, E. M., and Charlton, M. P., 1990, Strategic location of calcium channels at transmitter release sites of frog neuromuscular synapses, *Neuron* **5:**773–779.

Ruth, P., Rohrkasten, A., Biel, M., Bosse, E., Regulla, S., Meyer, H. E., Flockerzi, V., and Hofmann, F., 1989, Primary structure of the β subunit of the DHP-sensitive calcium channel from skeletal muscle, *Science* **245:**1115–1118.

Sakamoto, J., and Campbell, K. P., 1991, A monoclonal antibody to the B subunit of the skeletal muscle dihydropyridine receptor immunoprecipitates the brain ω-conotoxin GVIA receptor, *J. Biol. Chem.* **266:**18914–18919.

Sather, W. A., Tanabe, T., Zhang, J.-F., Mori, Y., Adams, M. E., and Tsien, R. W., 1993, Distinctive biophysical and pharmacological properties of class A (BI) calcium channel $α_1$ subunits, *Neuron* **11:**291–303.

Scholz, K. P., and Miller, R. J., 1991, $GABA_B$ receptor-mediated inhibition of Ca^{2+} currents and synaptic transmission in cultured rat hippocampal neurones, *J. Physiol.* **444:**669–686.

Sculptoreanu, A., Scheuer, T., and Catterall, W. A., 1993, Voltage-dependent potentiation of L-type Ca^{2+} channels due to phosphorylation by cAMP-dependent protein kinase, *Nature* **364:**240–243.

Sharp, A. H., and Campbell, K. P., 1989, Characterization of the 1,4-dihydropyridine receptor using subunit-specific polyclonal antibodies. Evidence for a 32,000 Dalton subunit, *J. Biol. Chem.* **264:**2816–2825.

Sharp, A. H., Gaver, M., Kahl, S. D., and Campbell, K. P., 1988, Structural characterization of the 32 kDa subunit of the skeletal muscle 1,4-dihydropyridine receptor, *Biophys. J.* **53:**231a.

Sheng, Z.-H., Rettig, J., Takahashi, M., and Catterall, W. A., 1994, Identification of a syntaxin-binding site on N-type calcium channels, *Neuron* **13:**1303–1313.

Sher, E., Pandiella, A., and Clementi, F., 1990a, Voltage-operated calcium channels in small cell lung carcinoma cell lines: Pharmacological, functional and immunological properties, *Cancer Res.* **50:**3892–3896.

Sher, E., Biancardi, E., Pollo, A., Carbone, E., and Clementi, F., 1990b, Neuronal-type, ω-conotoxin sensitive, calcium channels are expressed by an insulin-secreting cell line, *Cell Biol. Int. Rep. Suppl.* **14:**111.

Singer, D., Biel, M., Lotan, I., Flockerzi, V., Hofmann, F., and Dascal, N., 1991, The roles of the subunits in the function of the calcium channel, *Science* **253:**1553–1557.

Snutch, T. P., Leonard, J. P., Gilbert, M. M., Lester, H. A., and Davidson, N., 1990, Rat brain expresses a heterogeneous family of calcium channels, *Proc. Natl. Acad. Sci. USA* **87:**3391–3395.

Snutch, T. P., Tomlinson, W. J., Leonard, J. P., and Gilbert, M. M., 1991, Distinct calcium channels are generated by alternative splicing and are differentially expressed in the mammalian CNS, *Neuron* **7:**45–57.

Soldatov, N. M., Bouron, A., and Reuter, H., 1995, Different voltage-dependent inhibition by dihydropyridines of human Ca^{2+} channel splice variants, *J. Biol. Chem.* **270:**10540–10543.

Sollner, T., Whiteheart, S. W., Brunner, M., Erdjument-Bromage, H., Geromanos, S., Tempst, P., and Rothman, J. E., 1993, SNAP receptors implicated in vesicle targeting and fusion, *Nature* **362:**318–324.

Soong, T. W., Stea, A., Hodson, C. D., Dubel, S. J., Vincent, S. R., and Snutch, T. P., 1993, Structure and functional expression of a member of the low voltage-activated calcium channel family, *Science* **260:**1133–1136.

Stea, A., Dubel, S. J., Pragnell, M., Leonard, J. P., Campbell, K. P., and Snutch, T. P., 1993, A β-subunit normalizes the electrophysiological properties of a cloned N-type Ca^{2+} channel α_1-subunit, *Neuropharmacology* **32:**1103–1116.

Striessnig, J., Glossmann, H., and Catterall, W. A., 1990, Identification of a phenylalkylamine binding region within the α_1 subunit of skeletal muscle Ca^{2+} channels, *Proc. Natl. Acad. Sci. USA* **87:**9108–9112.

Striessnig, J., Murphy, B. J., and Catterall, W. A., 1991, Dihydropyridine receptor of L-type Ca^{2+} channels: Identification of binding domains for [^{3}H](+)-PN200-110 and [^{3}H]azidopine within the α_1 subunit, *Proc. Natl. Acad. Sci. USA* **88:**10769–10773.

Stuhmer, W., Conti, F., Suzuki, H., Wang, X., Noda, M., Yahagi, N., Kubo, H., and Numa, S., 1989, Structural parts involved in activation and inactivation of the sodium channel, *Nature* **339:**597–603.

Sugiura, Y., Woppmann, A., Miljanich, G. P., and Ko, C.-P., 1995, A novel ω-conopeptide for the presynaptic localization of calcium channels at the mammalian neuromuscular junction, *J. Neurocytol.* **24:**15–27.

Swartz, K. J., 1993, Modulation of Ca^{2+} channels by protein kinase C in rat central and peripheral neurons: Disruption of G protein-mediated inhibition, *Neuron* **11:**305–320.

Swartz, K. J., and Bean, B. P., 1992, Inhibition of calcium channels in rat CA3 pyramidal neurons by a metabotropic glutamate treceptor, *J. Neurosci.* **12:**4358–4371.

Takahashi, T., and Momiyama, A., 1993, Different types of calcium channels mediate central synaptic transmission, *Nature* **366:**156–168.

Takahashi, M., Seagar, M. J., Jones, J. F., Reber, B. F. X., and Catterall, W. A., 1987, Subunit structure of dihydropyridine-sensitive calcium channels from skeletal muscle, *Proc. Natl. Acad. Sci. USA* **84:**5478–5482.

Tanabe, T., Takeshima, H., Mikami, A., Flockerzi, V., Takahashi, H., Kangawa, K., Kojima, M., Matsuo, H., Hirose, T., and Numa, S., 1987, Primary structure of the receptor for calcium channel blockers from skeletal muscle, *Nature* **328:**313–318.

Tanabe, T., Beam, K. G., Powell, J. A., and Numa, S., 1988, Restoration of excitation-contraction coupling and slow calcium current in dysgenic muscle by dihydropyridine receptor complementary DNA, *Nature* **336:**134–139.

Tanabe, T., Beam, K. G., Adams, B. A., Niidome, T., and Numa, S., 1990a, Regions of the skeletal muscle dihydropyridine receptor critical for excitation-contraction coupling, *Nature* **346:**567–579.

Tanabe, T., Mikami, A., Numa, S., and Beam, K. G., 1990b, Cardiac-type excitation-contraction coupling in dysgenic skeletal muscle injected with cardiac dihydropyridine receptor cDNA, *Nature* **344:**451–453.

Tang, S., Yatani, A., Bahinski, A., Mori, Y., and Schwartz, A., 1993, Molecular localization of regions in the L-type calcium channel critical for dihydropyridine action, *Neuron* **11:**1013–1021.

Tarroni, P., Passafaro, M., Pollo, A., Popoli, M., Clementi, F., and Sher, E., 1994, Anti-β_2 subunit antisense oligonucleotides modulate the surface expression of the α_1 subunit N-type ω-CTX sensitive Ca^{2+} channels in IMR 32 human neuroblastoma cells, *Biochem. Biophys. Res. Commun.* **201:**180–185.

Tokumaru, H., Anzai, K., Abe, T., and Kirino, Y., 1992, Purification of the cardiac 1,4-dihydropyridine receptor using immunoaffinity chromatography with a monoclonal antibody against the $\alpha_2\delta$ subunit of the skeletal muscle dihydropyridine receptor, *Eur. J. Pharmacol.* **227:**363–370.

Tomlinson, W. J., Stea, A., Bourinet, E., Charnet, P., Nargeot, J., and Snutch, T. P., 1993, Functional properties of a neuronal class C L-type calcium channel, *Neuropharmacology* **32:**1117–1126.

Triggle, D. J., and Janis, R. A., 1987, Calcium channel ligands, *Annu. Rev. Pharmacol. Toxicol.* **27:**347–369.

Tsien, R. W., Hess, P., McCleskey, E. W., and Rosenberg, R. L., 1987, Calcium channels: Mechanisms of selectivity, permeation, and block, *Annu. Rev. Biophys. Chem.* **16:**265–290.

Tsien, R. W., Lipscombe, D., Madison, D. V., Bley, R. K., and Fox, A. P., 1988, Multiple types of neuronal calcium channels and their selective modulation, *Trends Neurosci.* **11:**431–438.

Tsien, R. W., Ellinor, P. T., and Horne, W. A., 1991, Molecular diversity of voltage-dependent Ca^{2+} channels, *Trends Pharmacol. Sci.* **12:**349–354.

Usowicz, M., Sugimori, M., Cherksey, B., and Llinas, R., 1992, P-type calcium channels in the somata and dendrites of adult cerebellar Purkinje cells, *Neuron* **9:**1185–1199.

Varadi, G., Lory, P., Schultz, D., Varadi, M., and Schwartz, A., 1991, Acceleration of activation and inactivation by the β subunit of the skeletal muscle calcium channel, *Nature* **352:**159–162.

Wakamori, M., Niidome, T., Furutama, D., Furuichi, T., Mikoshiba, K., Fujita, Y., Tanaka, I., Katayama, K., Yatani, A., Schwartz, A., and Mori, Y., 1994, Distinctive functional properties of the neuronal BII (class E) calcium channel, *Recept. Channels* **2:**303–314.

Weeler, D. B., Randall, A., and Tsien, R. W., 1994, Roles of N-type and Q-type Ca^{2+} channels in supporting hippocampal synaptic transmission, *Science* **264:**107–111.

Wei, X., Perez-Reyes, E., Lacerda, A. E., Schuster, G., Brown, A. M., and Birnbaumer, L., 1991, Heterologous regulation of the cardiac Ca^{2+} channel α_1 subunit by skeletal muscle β and γ subunits, *J. Biol. Chem.* **266:**21943–21947.

Wei, X., Neely, A., Lacerda, A. E., Olcese, R., Stefani, E., Perez-Reyes, E., and Birnbaumer, L., 1994, Modification of Ca^{2+} channel activity by deletions at the carboxyl terminus of the cardiac α_1 subunit, *J. Biol. Chem.* **269:**1635–1640.

Welling, A., Kwan, Y. W., Bosse, E., Flockerzi, V., Hofmann, F., and Kass, R. S., 1993, Subunit-dependent modulation of recombinant L-type calcium channels, *Circ. Res.* **73:**974–980.

Werz, M. A., Elmslie, K. S., and Jones, W. S., 1993, Phosphorylation enhances inactivation of N-type calcium channel current in bullfrog sympathetic neurons, *Eur. J. Physiol.* **424:**538–545.

Westenbroek, R. E., Ahlijanian, M. K., and Catterall, W. A., 1990, Clustering of L-type Ca^{2+} channels at the base of major dendrites in hippocampal pyramidal neurons, *Nature* **347:**281–284.

Westenbroek, R. E., Hell, J., Warner, C., Dubel, S., Snutch, T., and Catterall, W. A., 1992, Biochemical properties and subcellular distribution of an N-type calcium channel α_1 subunit, *Neuron* **9:**1099–1115.

Williams, M. E., Feldman, D. H., McCue, A. F., Brenner, R., Velicelebi, G., Ellis, S. B., and Harpold, M. M., 1992a, Structure and functional expression of α_1, α_2, and β subunits of a novel human neuronal calcium channel subtype, *Neuron* **8:**71–84.

Williams, M. E., Brust, P. F., Feldman, D. H., Patthi, S., Simerson, S., Maroufi, A., McCue, A. F., Velicelebi, G., Ellis, S. B., and Harpold, M. M., 1992b, Structure and functional expression of an ω-conotoxin-sensitive human N-type calcium channel, *Science* **257:**389–395.

Williams, M. E., Marubio, L. M., Deal, C. R. Hans, M., Brust, P. F., Philipson, L. H., Miller, R. J., Johnson, E. C., Harpold, M. M., and Ellis, S. B., 1994, Structure and functional characterization of neuronal α_{1E} calcium channel subtypes, *J. Biol. Chem.* **269:**22347–22357.

Witcher, D. R., De Waard, M., Sakamoto, J., Franzini-Armstrong, C., Pragnell, M., Kahl, S. D., and Campbell, K. P., 1993a, Subunit identification and reconstitution of the N-type Ca^{2+} channel complex purified from brain, *Science* **261:**486–489.

Witcher, D. R., De Waard, M., and Campbell, K. P., 1993b, Characterization of the purified N-type Ca^{2+} channel and cation sensitivity of ω-conotoxin GVIA binding, *Neuropharmacology* **32:**1127–1139.

Yang, J., and Tsien, R. W., 1993, Enhancement of N- and L-type calcium channel currents by protein kinase C in frog sympathetic neurons, *Neuron* **10:**127–136.

Yatani, A., Imoto, Y., Codina, J., Hamilton, S. L., Brown, A. M., and Birnbaumer, L., 1988, The stimulatory G protein of adenylyl cyclase, G_s, also stimulates dihydropyridine-sensitive Ca^{2+} channels, *J. Biol. Chem.* **263:**9887–9895.

Yoshida, A., Takahashi, M., Fujimoto, Y., Takisawa, H., and Nakamura, T., 1990, Molecular characterization of 1,4-dihydropyridine-sensitive calcium channels of chick heart and skeletal muscle, *J. Biochem.* **107:**608–612.

Zhang, J. F., Ellinor, P. T. Aldrich, R. W., and Tsien, R. W., 1994, Molecular determinants of voltage-dependent inactivation in calcium channels, *Nature* **372:**97–100.

CHAPTER 3

THE GABA$_A$ RECEPTORS
From Subunits to Diverse Functions

H. MOHLER, J. M. FRITSCHY, B. LÜSCHER,
U. RUDOLPH, J. BENSON, and D. BENKE

1. INTRODUCTION

Transmitter-gated ion channels are multisubunit membrane-spanning receptors that serve as rapid signal transduction devices regulating the flow of cations or anions through the cell membrane. Cell type–specific flexibility in neurotransmission is accomplished by a multiplicity of channel variants based on the combinatorial assembly of structurally related subunits. The heteromeric subunit structure also permits synaptic plasticity by changes in subunit composition and subunit-specific posttranslational modification. In addition, receptor heterogeneity offers the potential for cell type–specific drug targeting.

By gating the flow of chloride ions, GABA$_A$ receptors mediate the major inhibitory neurotransmission in the central nervous system (CNS) and, as apparent from their drug-induced modulation, contribute to the regulation of anxiety, vigilance, memory, epileptogenic activity, and muscle tension. Based on a repertoire of subunits encoded by at least 15 genes (α1–6, β1–3, γ1–3, δ, ρ1–2, with splice variants and homologues in nonmammalian species) (Barnard et al., 1992; Wisden and Seeburg, 1992; ffrench-Constant et al., 1993; Harvey et al., 1993; Macdonald and Olsen, 1994; Mohler et al., 1995), GABA$_A$ receptors are prime examples for receptor heterogeneity in the CNS. By investigating the physiological and pharmacological significance of GABA$_A$ receptor subtypes, in combination with their molecular architecture and cellular localization, it will be possible to identify the neuronal circuits regulating the above-mentioned CNS states and to provide

H. MOHLER, J. M. FRITSCHY, B. LÜSCHER, U. RUDOLPH, J. BENSON, and D. BENKE • Institute of Pharmacology, Federal Institute of Technology (ETH), and University of Zurich, CH-8057 Zurich, Switzerland.

Ion Channels, Volume 4, edited by Toshio Narahashi, Plenum Press, New York, 1996.

strategies for the development of subtype-specific drugs for selective therapies. The following is an overview of the structure, cellular localization, pharmacology, regulation and pathophysiology of the major $GABA_A$ receptors in the CNS.

2. SUBUNIT STRUCTURE AND STOICHIOMETRY

In line with the view that $GABA_A$ receptors are part of a structurally conserved superfamily that includes the nicotinic acetylcholine receptor and the glycine receptor (Betz, 1990; Barnard, 1992; Galzi and Changeux, 1994; Ortells and Lunt, 1995), a pentameric subunit structure had been predicted for $GABA_A$ receptors. This quaternary structure was verified in an electron microscopic analysis of $GABA_A$ receptors purified from porcine brain (Nayeem *et al.*, 1994). Most $GABA_A$ receptors are made up of three types of subunits: an α subunit, a β subunit and the $\gamma 2$ subunit. Concerning the stoi-chiometry of $GABA_A$ receptor subunits only indirect evidence is available. In an electrophysiological study on the recombinant receptor subtype $\alpha 3 \beta 2 \gamma 2$, the most likely subunit stoichiometry was found to be two α, one β, and two γ subunits (Backus *et al.*, 1993). This result is based on a correlation between changes in rectification and the mutation of charged amino acid residues near the channel. Biochemical studies support the view that the stoichiometry $2\alpha 1\beta 2\gamma$ may also occur in native receptors. Immuno-precipitation studies indicate that two α subunit variants can coexist within a single native receptor complex (Duggan *et al.*, 1991; Endo and Olsen, 1993; Mertens *et al.*, 1993; Pollard *et al.*, 1993). In addition, there is evidence for an association of two γ-subunits as shown for the $\gamma 2$ and $\gamma 3$ subunits in native receptors (Quirk *et al.*, 1994), and the two isoforms of the $\gamma 2$ subunit are frequently associated within a single receptor complex (Khan *et al.*, 1994). Finally, $GABA_A$ receptors appear to contain only a single β subunit variant (Benke *et al.*, 1994), which is consistent with the proposed stoichiometry. Thus, a combination of 2α, 1β, and 2γ subunits is likely to represent the subunit stoichiometry of at least some $GABA_A$ receptors. Since an interaction between an α and the $\gamma 2$ subunit is a prerequisite for the formation of benzodiazepine sites (see below), the α and γ subunits might be nearest neighbors. In addition, the β subunit might be flanked by the two α subunits based on their interaction in the formation of the GABA site (see below). Direct evidence, however, is still required to establish the definitive subunit stoichiometry.

3. FUNCTIONAL DOMAINS OF SUBUNITS

Molecular insight into the operation of the receptor as a signal transduction device has been obtained by mutational analysis of recombinant $GABA_A$ receptors. Structural domains that are critical for GABA-dependent gating and for drug-induced modulation have been identified (Fig. 1).

3.1. GABA Site

Two distinct homologous domains of the β subunit have been found to be essential for activation of the channel by GABA (Tyr[157], Thr[160] and Tyr[202], Thr[205] as determined

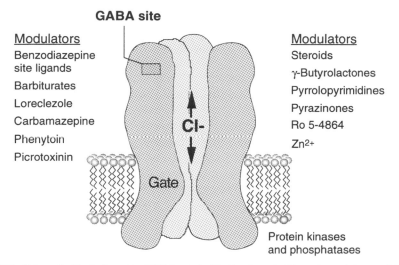

FIGURE 1. Schematic illustration of a GABA-gated chloride channel. Allosteric modulators of GABA_A receptor function are indicated.

for $\alpha1\beta2\gamma2$ receptors) (Amin *et al.*, 1994), consistent with previous photolabeling studies linking [^{3}H] muscimol binding to the β subunits (Casalotti *et al.*, 1986; Deng *et al.*, 1986). In addition, the GABA site receives a contribution from the α subunit, as shown in recombinant and native receptors (Sigel *et al.*, 1992; Smith and Olsen, 1994). In $\alpha1\beta2\gamma2$ and $\alpha5\beta2\gamma2$ recombinant receptors, mutation of Phe^{64} in the $\alpha1$ or $\alpha5$ subunit reduced the affinity of GABA and also reduced the cooperativity of GABA in gating the channel (Sigel *et al.*, 1992). The same residue, which is conserved among all GABA_A receptor α subunits, was targeted in the $\alpha1$ subunit of purified bovine receptors by [^{3}H] muscimol photolabeling (Smith and Olsen, 1994). Thus, the N-terminal extracellular domains of both α and β subunits appears to contribute to the agonist site. Neurotransmitter binding areas at subunit boundaries are a structural feature shared by other ligand-gated ion channels (Bertrand *et al.*, 1993; Unwin, 1993; Galzi and Changeux, 1994).

3.2. Benzodiazepine Site

The GABA response can be enhanced or reduced by agonists and inverse agonists, respectively, acting at the benzodiazepine (BZ) modulatory site. Photoaffinity labeling studies on native receptors originally located this site in the α subunits (for review, see Sieghart, 1985). Recombinant receptors, however, were responsive to classical BZ site ligands (agonists, antagonists, and inverse agonists) only when α and β subunits were coexpressed with the $\gamma2$ subunit, suggesting that the conformation of the α subunit is rendered BZ-sensitive by its interaction with the γ subunit. However, this holds true only when the α-subunit is $\alpha1$, $\alpha2$, $\alpha3$, or $\alpha5$ (Pritchett *et al.*, 1989; Pritchett and Seeburg, 1990), whereas receptors containing the $\alpha4$ or $\alpha6$ subunit remain insensitive to classical benzodiazepines (Luddens *et al.*, 1990; Wisden *et al.*, 1991; Knoflach *et al.*, 1996). This difference is due to a single amino acid residue, histidine in the BZ-sensitive and arginine in the BZ-insensitive α subunits (residue 100 in the α1 subunit) (Wieland *et al.*, 1992). Among the BZ-sensitive GABA$_A$ receptors, a change in a single amino acid residue in the α subunit results in the subtle differences between a type I pharmacological profile (receptors with the α1 subunit) and a type II profile (receptors with the subunits α2, α3), based on the substitution of glutamate in the α1 subunit by glycine in the α2 or α3 subunit (residue 225 in the α3 subunit) (Pritchett and Seeburg, 1991). In the γ subunit, the substitution of a single amino acid (Thr^{142}) profoundly alters the response of BZ site ligands, as shown for recombinant receptors containing the γ2 subunit in combination with α and β subunits (Mihic *et al.*, 1994). The γ3 subunit also imparts BZ sites to GABA$_A$ receptors when coexpressed with α and β subunits (α1βxγ3, α3β2γ3, α5βxγ2). While the bidirectional modulation of receptor function by agonists and inverse agonists is retained, the affinities of various BZ agonists are reduced compared to receptors containing the γ2 subunit (Knoflach *et al.*, 1991; Wilson-Shaw *et al.*, 1991; Herb *et al.*, 1992; Luddens *et al.*, 1994). In nicotinic acetylcholine or glycine receptors, a modulatory site homologous to the BZ site is so far unknown (Galzi and Changeux, 1994).

3.3.   Loreclezole Site

The GABA potentiating action of the anticonvulsant loreclezole depends on the type of β subunit in the receptor (Wafford *et al.*, 1994). Recombinant receptors containing the β2 or β3 subunit are responsive to loreclezole (e.g., α1β2, α1β3, α1β2γ2, α1β3γ2), while receptors containing the β1 subunit are insensitive. Loreclezole sensitivity critically depends on a defined amino acid residue (Asn^{289} in the β2 subunit; Asn^{290} in the β3 subunit) (Windgrove *et al.*, 1994).

3.4.   Barbiturate Site

The allosteric modulatory site by which barbiturates affect the function of GABA$_A$ receptors is less well defined. Barbiturate potentiation is seen on recombinant receptors even when the γ subunit, which is a prerequisite for the presence of the BZ site, is absent, and it is found even in homomeric receptors containing exclusively the α1 or β1 subunit (Pritchett *et al.*, 1988).

3.5.   Steroid Site

Various synthetic and natural steroids (e.g., alphaxolone, androsterone, pregnanolone) act as allosteric modulators of GABA$_A$ receptors (Harrison and Simmonds, 1984; Lambert *et al.*, 1990; Majewska, 1992; Paul and Purdy, 1992; Lambert *et al.*, 1995). GABA potentiation by steroids is found in both homomeric (β1) or heteromeric (α1β1, α1β1γ2) recombinant receptors (Puia *et al.*, 1990; Zaman *et al.*, 1992), where the type of α subunit influences the degree of potentiation (Lan *et al.*, 1991; Shingai *et al.*, 1991; Zaman *et al.*, 1992). The steroid site, which is likely to be different from the barbiturate site, may be of particular physiological significance, since the brain is capable of synthesizing steroids that affect GABA$_A$ receptor function (neurosteroids) (Beaulieu and Robel, 1990).

3.6.   Picrotoxinin Site

The convulsants picrotoxinin and TBPS (t-butylbicyclophosphorothionate), non-competitive GABA antagonists, interact with a site in the channel domain. Based on the resistance to cyclodiene pesticides found in the *Drosophila* mutant *Rdl*, a single amino acid in the putative channel lining domain of the *Drosophila* GABA$_A$ receptor subunit (Ala^{302} Ser) has been identified in homomeric recombinant receptors as the molecular determinant for the cyclodiene and picrotoxinin site (ffrench-Constant *et al.*, 1993).

3.7.   Novel Drug Binding Sites

Carbamazepine and phenytoin, in addition to their inhibitory action on voltage-gated sodium channels, were recently shown to enhance GABA$_A$-receptor function in neuronal culture and at a recombinant receptor subtype ($\alpha1\beta2\gamma2$, inactive on $\alpha3\beta2\gamma2$, $\alpha5\beta2\gamma2$, and $\alpha1\beta2$). This effect is not mediated via the benzodiazepine or barbiturate site (Granger *et al.*, 1995). Furthermore, pyrrolopyrimidines and pyrazinones enhance recombinant receptor responses at sites which are distinct from the loreclezole, the benzodiazepine site, and—at least for the pyrrolopyrimidines—from the barbiturate site (Im *et al.*, 1993, 1995).

3.8.   Phosphorylation Sites

GABA$_A$ receptor function can be modulated by phosphorylation in both native and recombinant receptors (for reviews, see Macdonald and Angelotti, 1993; Macdonald and Olsen, 1994). For instance, in mouse spinal cord neurons, acute intracellular application of protein kinase A (PKA) decreased the opening frequency of GABA$_A$ receptors (Porter *et al.*, 1990). Similarly, in recombinant receptors, the amplitude of the GABA response was reduced (Moss *et al.*, 1992b). Subunits amenable to phosphorylation by PKA include all β subunits and the $\alpha4$ and $\alpha6$ subunits. Besides PKA, activation of protein kinase C (PKC) decreased GABA-induced currents in oocytes injected with brain polyA mRNA (Sigel and Baur, 1988) and in other recombinant receptors (Sigel *et al.*, 1991; Leidenheimer *et al.*, 1992; Krishek *et al.*, 1994). Consensus sites for phosphorylation by PKC were found in the subunits $\alpha4$, $\alpha6$, $\beta1$–3, $\gamma2S$, $\gamma2L$, and $\rho2$ (Kellenberger *et al.*, 1992; Moss *et al.*, 1992a; Enz *et al.*, 1995). In the $\gamma2$ subunit, which occurs in two splice variants, the eight-amino acid insert of $\gamma2L$ contains a phosphorylation site for PKC (Whiting *et al.*, 1990), which has been suggested to be essential for the ethanol sensitivity of recombinant receptors (Wafford *et al.*, 1991), although this was disputed by Sigel *et al.* (1993).

In conclusion, the recognition of single amino acids as determinants for modulatory sites is of major relevance for drug development. Using gene targeting, it is now possible to generate animals that contain specifically altered GABA$_A$ receptors and are consequently unresponsive or responsive to a particular drug. Such animals will provide invaluable information for the assessment of the role of particular receptor subtypes in drug-induced behavior.

4. RECEPTOR DIVERSITY *IN VIVO* BY HETEROOLIGOMERIZATION

The structural, morphological, and pharmacological identity of $GABA_A$ receptor subtypes in the brain has been established mainly by immunobiological methods using antibodies selective for various subunits. Biochemically, the prevalence and main subunit repertoire of various receptor populations can be determined by subunit-selective immunoprecipitation (Fig. 2). Morphologically, the pattern of subunit expression allows receptor subtypes to be defined on the cellular level and allocated to particular neurons (Table I; Fig. 3). Pharmacologically, ligand affinities can be estimated from ligand binding to immunoprecipitated receptor populations (Table II). An assessment of ligand efficacy at different receptor subtypes *in vivo* requires refined electrophysiological analysis. The investigations mentioned, in combination with studies on recombinant receptors, will help to define the physiological and pharmacological relevance of $GABA_A$ receptor heterogeneity *in vivo*.

4.1. $GABA_A$ Receptors Containing the $\alpha 1\beta 2\gamma 2$ Subunits

The major $GABA_A$ receptor subtype in the brain is characterized by the subunit combination $\alpha 1\beta 2\gamma 2$, as shown by the prevalence of these subunits in immunoprecipitation (Fig. 1) (Benke *et al.*, 1991b) and their abundant colocalization in the same neurons demonstrated by triple immunohistochemical staining (Fritschy *et al.*, 1992; Fritschy and Mohler, 1995). The $\alpha 1$ subunit is the most abundant α subunit variant in the brain (Duggan and Stephenson, 1990; McKernan *et al.*, 1991; Endo and Olsen, 1993), in

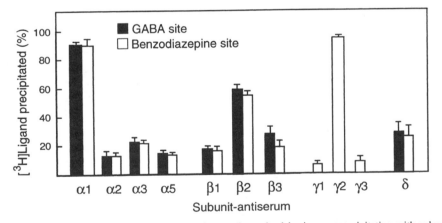

FIGURE 2. Prevalence of $GABA_A$ receptor subtypes determined by immunoprecipitation with subunit-specific antisera. Using purified receptor preparations (for α, β, and δ subunit antisera) or rat whole brain membrane extracts (γ subunit antisera), the percentage of immunoprecipitated receptors was measured by [^{3}H]muscimol binding (GABA site) or [^{3}H]flumazenil binding (BZ site). For the γ subunit antisera only the BZ sites were monitored by [^{3}H]flunitrazepam binding. The 100% values correspond to the sum of the sites present in the precipitate and the supernatant. Data are taken from Benke *et al.* (1991a,b,c; 1994; in preparation), Marksitzer *et al.* (1993), and Mertens *et al.* (1993).

TABLE I

GABA$_A$ Receptor Subtypes in Specified Neurons

Neurons	Subunits
Transmitter markers	
GABA (most brain areas)	α1β2γ2 and α1α3β2γ2
5-HT (raphe nuclei), ACh (basal forebrain)	α3β3γ2
NA (locus coeruleus)	α2α3γ2
DA (substantia nigra, pars compacta)	α2α3γ2
Olfactory bulb	
Mitral and tufted cells	α1α3β2γ2
Granule cells	α2α5β3γ2
Periglomerular cells	α2α5δ
Hippocampus	
Pyramidal cells	α2β3γ2 and α5β3γ2
Most interneurons	α1β2γ2
Thalamus, hypothalamus	
Relay neurons (ventrobasal complex)	α1β2γ2 δ
Reticular nucleus neurons	α3γ2
Supraoptic nucleus	α1α2β2,3γ2
Ventromedial and arcuate nuclei	α2β3γ2 and α5β3γ2
Brain stem	
Oculomotor, hypoglossal nucleus	α1α2γ2
Motor trigeminal nucelus	α1γ2
Reticular formation	α1β2γ2 and α1α3β2γ2
Cerebellum	
Purkinje cells, basket cells	α1β2,3γ2
Granule cells	α1α6β2,3γ2δ
Golgi type II cells	α2α3γ2

[a]The subunits analyzed are α1, α2, α3, α5, α6, β2,3, γ2, and δ. Colocalization of subunits was assessed in rat brain sections by double- and triple-immunofluorescence staining, using subunit-specific antisera and antibodies to neurotransmitter markers. For each neuron populuation, subunits that are not indicated were not detected. The presence of four or five subunits within a population of neurons was tested by multiple staining combinations (see Fritschy and Mohler, 1995, for details). 5-HT, serotonergic neurons; ACh, cholinergic neurons; NA, noradrenergic neurons; DA, dopaminergic neurons.

particular in the cortical mantle, thalamus, and brain stem (Benke *et al.*, 1991a). In subcortical regions, the α1 subunit displays conspicuous variations in distribution and/or staining intensity that closely follow distinct cytoarchitectural boundaries. Thus, the respective GABA$_A$ receptors may represent a useful tool for distinguishing nuclear boundaries, for example, in the thalamus, basal forebrain, and brain stem. Only a few regions are devoid of α1 subunit staining, notably the striatal complex, the granule cell layer of the olfactory bulb, and the reticular nucleus of the thalamus (Fritschy and Mohler, 1995). The β2 subunit is systematically associated with the α1 subunit within individual neurons, as shown by double immunofluorescence staining (Benke *et al.*, 1994). The γ2 subunit has the most widespread distribution of all subunits, being detected in early every brain region, albeit with variable intensity (Gutierrez *et al.*, 1994;

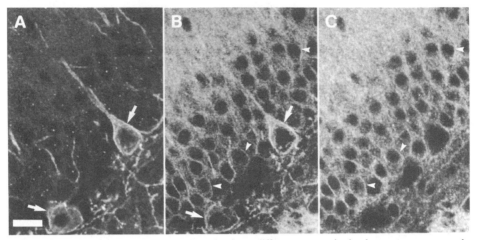

FIGURE 3. Allocation of GABA$_A$ receptor subunits to different neurons in the dentate gyrus, as seen by triple-immunofluorescence staining. Each panel depicts a different subunit immunoreactivity within the same section (A, α1; B, β2,3; C, α2), visualized by confocal laser microscopy (see Fritschy and Mohler, 1995, for details). The α1 subunit labels the soma and dendrites of presumptive interneurons while the α2 subunit is present in dentate gyrus granule cells. In contrast, the β2,3 subunits are colocalized with both the α1 subunit (arrows) and the α2 subunit (arrowheads), suggesting their contribution to two distinct receptor subtypes. Scale, 20 μm.

Fritschy and Mohler, 1995). The highest levels of γ2 subunit immunohistochemical staining are seen in the hippocampus and the lowest levels in the hypothalamus.

With regard to identified neurons, high levels of receptors containing the subunits α1β2γ2 are expressed in numerous populations of GABAergic neurons at all levels of the neuraxis (Table I). In particular, interneurons in cerebral cortex and hippocampus, cerebellar GABAergic neurons, and GABAergic neurons in the brain stem reticular formation, pallidum, substantia nigra, and basal forebrain are intensely immunoreactive for the α1 subunit (Fritschy *et al.*, 1992; Gao *et al.*, 1993, 1995; Gao and Fritschy, 1994). The α1β2γ2 subunit combination has also been allocated to identified non-GABAergic neurons, such as olfactory bulb mitral cells in association with the α3 subunit and relay neurons in the thalamus in combination with the δ subunit (Table I).

The ligand binding profile of the receptors containing α1β2γ2 subunits has been assessed by immunoprecipitation with either the α1 or β2 subunit antiserum (Mertens *et al.*, 1993; Benke *et al.*, 1994). In both cases, all classical ligands of the benzodiazepine (BZ) site tested, except CL 21887, displayed high nanomolar affinities (Table II). Thus, the neuronal circuits containing the GABA$_A$ receptor subtype α1β2γ2 contribute to the basic pharmacological and therapeutic spectrum of the classical BZ site ligands.

4.2. GABA$_A$ Receptors Containing α2β3γ2 Subunits or α3β3γ2 Subunits

Receptors containing the α2 subunit are most abundant in regions where the α1 subunit is absent or expressed at low levels (Marksitzer *et al.*, 1993; Benke *et al.*, 1994; Fritschy and Mohler, 1995). For instance, the α2 subunit is predominant in striatum, hippocampal

TABLE II
Drug Binding Profiles of Native GABA$_A$ Receptor Populations

Receptor population immuno-precipitated[a]	K$_i$[nM]; [^{3}H]flumazenil binding (except for α4 and γ1)						
	Flumazenil	Ro 15-4513	Fluni-trazepam	Diaze-pam	βCCM	Zolpidem	CL 218 872
α1	0.2 ± 0.07	0.9 ± 0.2	4.5 ± 2	18 ± 0.6	1.4 ± 0.8	8 ± 3	107 ± 27
α2	0.8 ± 0.2	—	4.9 ± 1.4	26 ± 3	5.7 ± 1	109 ± 43	1240 ± 280
α3	0.4 ± 0.1	2.9 ± 0.8	5.9 ± 2.3	21 ± 7	6.8 ± 2.9	75 ± 27	653 ± 67
α4[b]	130 ± 33	16 ± 3[d]	>10,000	>10,000	—	>10,000	>10,000
α5	0.6 ± 0.2	1.2 ± 0.4	3.1 ± 1.2	19 ± 7	2.7 ± 1.4	30 ± 6	276 ± 84
β1	1.3 ± 0.5[d]	—	7.7 ± 0.6	—	5.4 ± 0.8	133 ± 20	557 ± 58
β2	1.0 ± 0.2[d]	—	9.7 ± 0.6	—	3.1 ± 0.6	17 ± 4.0	283 ± 14
β3	1.0 ± 0.3[d]	—	9.3 ± 0.5	—	7.0 ± 1.6	90 ± 8.0	677 ± 117
γ1[c]	>10,000	>10,000	38 ± 5[d]	—	1550 ± 50	>10,000	400 ± 100
γ2	1.1 ± 0.1[d]	4.0 ± 1	12 ± 2	—	7.4 ± 0.5	32 ± 1	254 ± 47
γ3	1.1 ± 0.01[d]	6.9 ± 2	854 ± 552	—	11 ± 1	>10,000	227 ± 40
δ	0.5 ± 0.3	0.4 ± 0.13	2.1 ± 0.6	5 ± 3	0.2 ± 0.1	43 ± 23	396 ± 48

[a]Displacement potencies of various benzodiazepine receptor ligands were determined in radioligand binding to GABA$_A$ receptors immunoprecipitated with the respective subunit-specific antiserum from membrane extracts prepared from rat whole brain. Data taken from Marksitzer et al., 1993; Mertens et al., 1993; Benke et al., 1994; Benke et al., in preparation.
[b][^{3}H]-Ro 15-4513 binding.
[c][^{3}H]flunitrazepam binding.
[d]K$_D$ values derived from Scatchard analysis.

formation, and olfactory bulb. Like the α2 subunit, receptors containing the α3 subunit are frequently distributed in regions expressing only low or moderate levels of the α1 subunit, which include the lateral septum, reticular nucleus of the thalamus, and several brain stem nuclei (Gao et al., 1993; Fritschy and Mohler, 1995). On the cellular level, the α2 and α3 subunits are frequently coexpressed with the β3 and γ2 subunits (Gao et al., 1993, 1995; Benke et al., 1994). This is particularly evident in hippocampal pyramidal cells (α2β3γ2) and in cholinergic neurons of the basal forebrain (α3β3γ2) (Table I). Consistent with the presence of the γ2 subunit, the receptor populations immunoprecipitated by antisera specific for the α2, α3, or β3 subunit bind BZ site ligands (Table II). The binding profile of these receptors differs from that of the α1β2γ2 receptors in that βCCM displays a slightly lower displacing potency (four- to fivefold) and zolpidem and CL 218872 a considerably lower displacing potency (9- to 14-fold) (Table II).

During development, embryonic receptors containing the α2 subunit are increasingly replaced by receptors containing the α1 subunit, a phenomenon which is particularly prominent in the early postnatal period around the time of synapse formation (Fritschy et al., 1994). The functional relevance of this switch in subunit composition of GABA$_A$ receptors is as yet unknown.

4.3. GABA$_A$ Receptors Containing the α5 Subunit

Receptors containing the α5 subunit are of minor abundance in the brain (Fig. 2) and are concentrated mainly in the hippocampus, olfactory bulb, and trigeminal sensory nucleus (Fritschy and Mohler, 1995). They comprise various subtypes differentiated by the potency of zolpidem. For instance, in the hippocampus and spinal cord, the receptor population immunoprecipitated by the $\alpha 5$ antiserum displays micromolar affinity for zolpidem, while the receptor population in the cerebral cortex, striatum, and brain stem shows nanomolar affinity for zolpidem (Mertens *et al.*, 1993). Differential affinities for zolpidem were also observed in radioligand binding studies on cell membranes and tissue sections prepared from hippocampus and spinal cord (Ruano *et al.*, 1992; Benavides *et al.*, 1993). On recombinant receptors containing the $\alpha 5$ subunit, zolpidem is practically without effect ($K_i > 10$ μm; $\alpha 5\beta x\gamma x$) (Pritchett and Seeburg, 1990; Puia *et al.*, 1991; Hadingham *et al.*, 1993; Luddens *et al.*, 1994), suggesting that the receptors immunopurified from hippocampus and spinal cord by the $\alpha 5$ subunit antiserum might largely be represented by these zolpidem-insensitive subunit combinations. In contrast, in brain areas where native $\alpha 5$ receptors display nanomolar affinities for zolpidem, the $\alpha 5$ subunit might be associated with an additional α subunit, the $\alpha 1$ and $\alpha 3$ subunits being prime candidates (Mertens *et al.*, 1993).

4.4. GABA$_A$ Receptors Containing the $\alpha 4$ or $\alpha 6$ Subunits

The $\alpha 4$ and $\alpha 6$ subunits typify receptor populations of low abundance, with the $\alpha 4$ subunit being present mainly in thalamus, hippocampus, and olfactory bulb, as suggested by *in situ* hybridization histochemistry (Wisden *et al.*, 1991; Laurie *et al.*, 1992), while the $\alpha 6$ subunit is concentrated almost exclusively in cerebellar granule cells (Laurie *et al.*, 1992; Thompson *et al.*, 1992; Varecka *et al.*, 1994). Following their immunoprecipitation, both types of receptors lack high affinity for classical benzodiazepines (diazepam, flunitrazepam, clonazepam; Benke *et al.*, in preparation), which is in agreement with previous studies on recombinant receptors containing $\alpha 4$ or $\alpha 6$ subunits (Luddens *et al.*, 1990). However, the ligands bretazenil and Ro 15-4513, but not imidazenil, do potentiate the GABA response at these receptors ($\alpha 4\beta 2\gamma 2$; $\alpha 6\beta 2\gamma 2$) (Knoflach *et al.*, 1996). Likewise, *in vivo*, high affinity for Ro 15-4513 and its congeners has been demonstrated on diazepam-insensitive receptors in the cerebellum (Wong and Skolnik, 1992). Thus, while the $\alpha 4$ and $\alpha 6$ receptors do not contribute to the mode of action of classical BZ ligands, they can respond to certain nonclassical ligands (e.g., bretazenil, Ro 15-4513). Potentiation of $\alpha 6$ receptors may contribute to undesired drug side effects, since diazepam induces locomotor deficits in a rat mutant containing BZ-sensitive $\alpha 6$ receptors (Korpi *et al.*, 1993).

4.5. GABA$_A$ Receptors Containing the δ Subunit

The δ subunit characterizes a receptor population that is expressed mainly in the granule cells of the cerebellum and, to a minor extent, in other brain regions such as cerebral cortex, thalamus, olfactory bulb, and dentate gyrus, as shown by *in situ* hybridization histochemistry and by immunohistochemistry (Shivers *et al.*, 1989; Benke *et al.*, 1991c; Laurie *et al.*, 1992; Fritschy and Mohler, 1995). Receptors immunoprecipitated by δ subunit–specific antisera display high-affinity binding sites for [³H]flumazenil and [³H]muscimol (Benke *et al.*, 1991c), indicating that the δ subunit is a *bona fide* constituent of GABA$_A$ receptors and that at least some of these receptors also contain the $\gamma 2$ subunit. Biochemical evidence indicates that the δ subunit is potentially associated with the α1, α3, β2,3, and γ2 subunit (Mertens *et al.*, 1993). The pharmacological profile of the δ subunit–containing receptor population is unique in that the potency of βCCM and diazepam is higher than for all other receptors tested (Table II) (Mertens *et al.*, 1993). This has been substantiated by electrophysiological studies on recombinant receptors containing the α1β1γ2δ subunit combination. The GABA-induced currents were enhanced by diazepam more strongly than in the corresponding subunit combination lacking the δ subunit (Saxena and Macdonald, 1994). Furthermore, in comparison with the α1β1γ2 combination, receptors expressed from the subunits α1β1γ2δ displayed a larger main conductance level, a longer mean open time, and a slower rate of acute desensitization (Saxena and Macdonald, 1994), suggesting that the δ subunit influences not only the BZ site but also the kinetics of the GABA-gated channel to permit an enhanced GABA response.

4.6. GABA$_A$ Receptors with Unusual Subunit Combinations

In certain small populations of neurons, expression of unusual subunit combinations, lacking either the β2,3 or γ2 subunit, has been demonstrated immunohistochemically on the cellular level (Table I). For instance, in the reticular nucleus of the thalamus, immunoreactivity for the α3 and γ2 subunits is found, while antibodies recognizing the β2 and β3 subunits failed to be immunoreactive (Fritschy and Mohler, 1995). Thus the corresponding receptors contain the β1 subunit or lack β subunits completely. In addition, cerebellar granule cells are unusual in that at least five different types of subunit are expressed (α1, γ6, β2,3, γ2, δ) (Caruncho and Costa, 1994; Gao and Fritschy, 1995). This observation suggests that more than a single type of receptor is expressed in these cells. On the electron microscopic level, a distinction between synaptic and nonsynaptic receptors has been reported in cerebellar granule cells, based on the presence of either the α1 or the α6 subunit, respectively (Baude *et al.*, 1992). However, this finding is at variance with light microscopic evidence, which showed that the α1 and α6 subunits are generally colocalized on the subcellular level in granule cells (Gao and Fritschy, 1995). Nevertheless, a heterogeneity of receptors in certain types of neurons merits intense analysis.

The subunits γ1 and γ3 contribute to further GABA$_A$ receptor heterogeneity. Both subunits are of very low abundance in the adult rat brain (Fig. 2), and, so far, their distribution has been characterized only by *in situ* hybridization histochemistry (Araki *et al.*, 1992; Laurie *et al.*, 1992; Wisden *et al.*, 1992). There is biochemical evidence that while the γ1 and γ2 subunits do not coexist within single receptor complexes, the γ3 subunit may be associated with the γ2 subunit (Quirk *et al.*, 1994). The γ3 subunit appears to be assembled with most α subunit variants and with the β2,3 subunits to form multiple GABA$_A$ receptor subtypes (Togel *et al.*, 1994).

4.7. GABA Receptors Containing the ρ1 or ρ2 Subunits

The ρ1 or ρ2 subunits are expressed most abundantly in the retina (Cutting *et al.*, 1991, 1992), and have been shown by single cell *in situ* hybridization to be localized selectively to rod bipolar cells (Enz *et al.*, 1995). In recombinant expression systems, these subunits form homomeric, picrotoxinin-sensitive, GABA-gated chloride channels

that exhibit little desensitization and are almost insensitive to bicuculline, benzodiazepines, steroids, and barbiturates (Shimada *et al.*, 1992; Kusama *et al.*, 1993). GABA-
gated channels with similar properties have been identified electrophysiologically in
retina (Figenspan *et al.*, 1993; Qian and Dowling, 1993; Woodward *et al.*, 1993;
Feigenspan and Bormann, 1994). They also occur in higher visual centers and in the
cerebellum (Drew *et al.*, 1984; Drew and Johnston, 1992). Since these channels do not
respond to selective agonists and antagonists of $GABA_B$ receptors, they appear to
represent a pharmacologically distinct class of GABA receptors. They have been
tentatively named $GABA_C$ receptors (reviewed by Johnston, 1994), although based on
the fact that they represent GABA-gated chloride channels, they can nevertheless be
considered part of the $GABA_A$ receptor family. Clearly, receptors containing ρ subunits
display unusual properties.

5. DRUG RESPONSES OF $GABA_A$ RECEPTOR SUBTYPES

Medically, the most important drug modulatory site of $GABA_A$ receptors is the
benzodiazepine (BZ) site. The development of novel ligands aims to retain the therapeutic effectiveness of classical benzodiazepines while reducing unwanted side effects such
as tolerance, dependence liability, memory impairment, and ataxia (Costa *et al.*, 1995).
Two strategies are being followed to achieve this goal: 1) reduction of the efficacy of the
ligands at all or some receptors (partial agonists), and 2) targeting of the ligand to
particular receptor subtypes by selective affinity (selective agonists). Obviously a combination of both approaches is also possible. The following is a brief account of the
application of these principles in drug development.

Although there are individual variations (Ducic *et al.*, 1993), classical benzodiazepines such as diazepam interact with nearly all receptors with high affinity and high
efficacy, as demonstrated for both native and recombinant receptors. In contrast, partial
agonists, typified by bretazenil or imidazenil, display high affinity for most if not all
receptors but act with reduced efficacy compared to classical benzodiazepines (Haefely
et al., 1990; Puia *et al.*, 1992; Wafford *et al.*, 1993; Auta *et al.*, 1994). In addition, there are
BZ site ligands which display efficacies that vary depending on the $GABA_A$ receptor
subtype. For example, abecarnil acts as a partial agonist on recombinant receptors
containing $\alpha 2$ and $\alpha 5$ subunits but acts as a full agonist on receptors containing the $\alpha 1$
and $\alpha 3$ subunits (Knoflach *et al.*, 1993; Pribilla *et al.*, 1993). Abecarnil is currently
undergoing clinical trials as an anxiolytic. Finally, there are ligands which discriminate
among receptor subtypes exclusively by their affinity. The hypnotic zolpidem shows
high affinity for receptors containing the $\alpha 1$ subunit while displaying lower affinity for
receptors containing the $\alpha 2$ and $\alpha 3$ subunits and even lower affinity for most receptors
containing the $\alpha 5$ subunit (Pritchett and Seeburg, 1990; McKernan *et al.*, 1991; Mertens
et al., 1993). These examples indicate that favorable profiles of BZ site ligands can be
achieved either by reducing the efficacy of the ligand at all receptors (bretazenil,
imidazenil) or by avoiding a full activation of receptors containing $\alpha 2$ and $\alpha 5$ subunits
(abecarnil, zolpidem). The findings concerning abecarnil and zolpidem are of particular
interest with regard to receptor heterogeneity. Receptors containing $\alpha 2$ and $\alpha 5$ subunits
might be expected to be located in neuronal circuits which are involved in mediating

unwanted rather than desirable drug effects. Indeed, it is notable that receptors containing the α2 subunit are strongly expressed in brain areas mediating reward (e.g., nucleus accumbens), while receptors containing the α5 subunit are concentrated in certain areas linked to memory functions (e.g., hippocampus) (Fritschy and Mohler, 1995). In the future, it should be possible to determine the pharmacological relevance of certain GABA$_A$ receptor subtypes by using mutant mice lacking these receptors or containing pharmacologically inactive receptor subtypes. In animals with suitable receptor alterations, even a classical benzodiazepine such as diazepam would be expected to display fewer unwanted side effects than in normal animals.

6.   REGULATION OF RECEPTOR EXPRESSION

Adaptation to changing functional requirements is a key property of neuronal circuits. In the GABA system, a particular adaptational mechanism becomes apparent during ontogeny. It consist of a switch in subunit expression resulting in a change from one particular receptor subtype to another. This phenomenon has been studied using the α2 and α5 subunits as markers. Embryonic and perinatal GABA$_A$ receptors are characterized mainly by the presence of the α2 and α5 subunits, but not the α1 subunit. During postnatal development, these receptors are replaced by the adult pattern, in which the α1 subunit predominates and only minor receptor populations with α2 and α5 subunits remain (Paysan et al., 1993; Fritschy et al., 1994). It has been possible to show immunohistochemically that this switch does not involve receptor expression in different cells but rather occurs in the same neuron (Fritschy et al., 1994). This finding points to intricate developmental cues regulating receptor subtype expression.

To investigate which factors trigger the switch in gene expression, receptor maturation has been investigated in the cerebral cortex, where particular layers and areas display different developmental receptor patterns (Paysan et al., 1994a). The aim of these experiments was to determine whether receptor expression is governed by the afferent innervation. Thalamic fibers were prevented experimentally from innervating their cortical targets (mainly layers IV and VI) by unilateral lesioning of the lateral geniculate nucleus shortly after birth. In this case, the subunit expression pattern in the cerebral cortex was dramatically altered (Paysan et al., 1994b; Paysan et al., in preparation). On the lesioned side, the α1 subunit immunoreactivity in layers IV and VI of the visual cortex was strongly reduced in seven-day-old animals compared to the innervated control side of the cortex. In contrast, the expression of the α5 subunit, which is normally downregulated in these layers, was still prominent on the lesioned side. These results suggest that the expression of the α1 and α5 subunits in the neocortex can be regulated by factors derived from innervating neurons (epigenetic factors). The distribution of the α1 and α5 subunits was affected also in layers III and V in the deafferented visual cortex, resulting in a laminar distribution very similar to that seen in adjacent association areas. Thus, these experiments reveal the existence of an intrinsic developmental program in the neocortex, leading to a characteristic "default" laminar distribution of GABA$_A$ receptor subunits. In primary sensory areas, this default pattern is influenced early by the ingrowing thalamic afferents, leading to a different laminar distribution at P7, which can be prevented by a lesion of the thalamus. Since thalamic afferents are presumably glutamatergic, it appears that excitatory input to the neocortex has a determinative influence on its inhibitory counterpart during development.

Plasticity in $GABA_A$ receptor expression is retained in adulthood, as shown recently in the visual system (Hendry et al., 1994). Visual deprivation in monkeys induced by unilateral intraocular injection of tetrodotoxin (TTX) resulted in a downregulation of the $\alpha1$, $\beta2$, and $\gamma2$ subunits in layer IV of the primary visual cortex (area 17). The down-regulation was restricted to the ocular dominance columns corresponding to the deprived eye. This effect was evident at both the mRNA and the protein level (Huntsman et al., 1994), suggesting a transcriptional control of $GABA_A$ receptor expression. Neuronal activity is thus one of the determinants for the regulation of $GABA_A$ receptor gene expression.

7. ANALYSIS OF GENE PROMOTERS

Studies aimed at unraveling the mechanism of differential neuronal expression of $GABA_A$ receptor subunit genes have been initiated. The excellent correlation between immunohistochemical and in situ hybridization data on the levels of subunit expression suggests that differential expression of subunits is regulated largely at the level of tran-scription. The promoters of three subunit genes have been analyzed in some detail, re-vealing common and subunit gene–specific features. The $\beta3$ and δ subunit genes appear to be driven by similar promoters lacking a TATA box and very rich in GC content, and they both contain multiple clustered transcriptional initiation sites (Sommer et al., 1990; Kirkness and Fraser, 1993; Motejlek et al., 1994). A novel brain-specific DNA binding factor, BSF1, which recognizes a tandemly repeated purine-rich element in the 5′ flank-ing region of the δ subunit gene, has been described. The expression of BSF1 correlates well with the spatial δ subunit expression pattern in vivo and the temporal induction of this subunit in differentiating neurons in vitro, suggesting a role for BSF1 in cell–specific $GABA_A$ receptor subunit expression (Motejlek et al., 1994). Sequences with high homology to the BSF1 recognition site were also found in the $\beta3$ subunit gene promoter. In contrast to the δ and $\beta3$ subunits, the $\alpha1$ subunit appears to be driven by a more classical polymerase II promoter with a TATA-like sequence centered at position -25 (Bateson and Paetsch, 1993; Kang et al., 1994). A reporter gene construct driven by this promoter and transfected into primary neurons is strongly downregulated upon exposure to lorazepam, mimicking the downregulation of endogenous $\alpha1$ subunit mRNA observed in vivo in response to chronic exposure to this benzodiazepine (Kang and Miller, 1991). Thus, the drug-induced downregulation of $GABA_A$ receptors may be due at least in part to reduced transcription of subunit genes. Whether the changes in α subunit mRNAs induced by chronic ethanol treatment (Montpied et al., 1991; Mhatre and Ticku, 1992; Hirouchi et al., 1993) are similarly due to transcriptional control is so far unknown.

8. ANIMALS WITH MUTATIONS IN $GABA_A$ RECEPTOR SUBUNITS

A newly emerging method of assessing the physiological and pharmacological significance of receptor subtypes for brain function is the study of animals with muta-tions in specific subunit genes. Several deletions or point mutations in $GABA_A$ receptor subunits have been identified, in particular in the p-locus of the mouse chromosome 7, in the rat α6 subunit gene, and in a *Drosophila* GABA$_A$ receptor subunit gene.

Mice with the pink-eyed cleft palate (p^{cp}) mutation are characterized by hypopigmentation and an isolated cleft palate based on a deletion on chromosome 7. It extends from the second exon of the pink-eyed dilution (p) gene to the third exon of the GABA$_A$ receptor β3 subunit gene and includes both the γ3 and α5 subunit genes (Nakatsu *et al.*, 1993). Correspondingly, in prenatal hippocampus, which contains normally a high density of α5 and β3 subunits, the GABA$_A$ receptors were decreased by about 80%, as determined by [^{3}H]-Ro 15-1788 binding. The zolpidem-sensitive, low-affinity binding component, which is characteristic of α5 subunit–containing receptors, was absent. Homozygous p^{cp} mice usually die shortly after birth, presumably because of cleft palate–induced feeding problems (Nakatsu *et al.*, 1993). In addition, the cleft palate phenotype can be abolished by introducing a β3 subunit transgene (Culiat *et al.*, 1995).

By radiation-induced mutations at the p locus, Rinchik and coworkers derived a series of mice which retained the β3 subunit but lacked the α5 and γ3 subunits or only the γ3 subunit (Culiat *et al.*, 1993, 1994). These mice reached adulthood and did not exhibit cleft palate or any other overt phenotype, although drug-induced responses have not been tested. However, a complete concordance (for 20 deletions examined) between isolated cleft palate and the β3 subunit gene was found (Culiat *et al.*, 1993, 1994). Since treatment of developing fetal brains with diazepam or GABA may interfere with normal palate development (Wee and Zimmerman, 1983), β3 subunit–containing receptor complexes appear to play a role in palate development.

A point mutation in the α6 GABA$_A$ receptor subunit has been detected in the alcohol-nontolerant (ANT) rat line that displays impairment of postural reflexes following diazepam treatment. The molecular basis for this behavior is a naturally occurring arginine to glutamine substitution at position 100 of the α6 subunit (Korpi *et al.*, 1993), which renders the corresponding receptors BZ-sensitive. Since receptors containing the α6 subunit are present only in cerebellar granule cells, it is apparent that these neurons are involved in the regulation of postural reflexes.

In *Drosophila*, a point mutation in the channel-lining domain of the GABA$_A$ receptor (Ala 302 Ser; *Rdl* mutant) renders the channel insensitive to the convulsant picrotoxinin and cyclodiene pesticides, as shown for the respective homomeric recombinant GABA$_A$ receptor expressed from the *Drosophila Rdl* allele. This mutation may account for the widespread resistance of insects to cyclodiene pesticides (ffrench-Constant *et al.*, 1993).

9.   GABA$_A$ RECEPTORS IN DISEASE

Changes in GABA$_A$ receptor function may be of relevance for the pathophysiology of neurological diseases, in particular certain forms of epilepsy, Huntington's disease, and also Angelman's syndrome.

9.1.   Epilepsy

The "GABA hypothesis" of seizure disorders (Meldrum, 1979; Olsen *et al.*, 1986, 1992; Gale, 1992) suggests that a deficit in GABAergic inhibitory synaptic transmission may contribute to the synchronous hyperexcitable activity of epileptic brain. This hypothesis is supported by the effectiveness of anticonvulsant drugs which enhance GABAergic transmission (Gale, 1992). A primary role of the GABA system in the pathophysiology of epilepsy is, however, less well defined, which also concerns the role of $GABA_A$ receptors in epilepsy. Using positron emission tomography in patients with idiopathic partial epilepsy, it has been demonstrated that binding of the ^{11}C-labeled benzodiazepine receptor antagonist Ro 15-1788 was significantly lower in epileptic foci than in the contralateral homotopic reference region and in the remaining neocortex (Savic *et al.*, 1988). This reduction in $GABA_A$ receptor binding was detectable in the absence of overt cell loss. In neuropathological studies of surgical resections from patients with focal epilepsies, a loss of $GABA_A$ receptors, as well as a lack of receptor alterations, have been described (reviewed by Olsen *et al.*, 1990, 1992). At least in some cases $GABA_A$ receptor reduction was due to major cell loss (Olsen *et al.*, 1990; McDonald *et al.*, 1991; Wolf *et al.*, 1994). Thus, the pathophysiological role of $GABA_A$ receptors in seizure disorders is not clearly defined.

In animal models of epilepsy, a reduction in receptor binding has been found in seizure-susceptible gerbils (Olsen *et al.*, 1985). In kindling-induced seizures, increased $GABA_A$ receptor binding confined to the dentate gyrus of the hippocampus has been reported (Valdes *et al.*, 1982; Shin *et al.*, 1985; Nobrega *et al.*, 1990), and, in line with these findings, selective alterations of subunit mRNA levels have been observed exclusively in the dentate gyrus granule cells (Clark *et al.*, 1994; Kamphuis *et al.*, 1994; Kokaia *et al.*, 1994). In one of these studies (Kokaia *et al.*, 1994), the levels of the β3 and γ2 subunit mRNAs were initially reduced, 4 hours after the last electrical stimulation. During the next two days, the α1 and γ2 subunit mRNAs were increased, whereas the β3 subunit mRNA was not significantly altered. However, after five days the mRNA levels had returned to normal, indicating that the respective $GABA_A$ receptors may play a role in the initial events but do not appear to be involved in the maintenance of the kindled state.

9.2. Huntington's Disease

Huntington's disease is an inherited neurodegenerative disease characterized by progressive involuntary choreiform movements, psychopathological changes, and dementia (Hayden, 1981). Though a candidate gene for involvement in this disease has been identified (Huntington's Disease Collaborative Research Group, 1993; Strong *et al.*, 1993), the pathogenesis is still unknown. Neuropathologically, the hallmark of Huntington's disease is a profound loss of striatal projection neurons, leading to atrophy of the caudate nucleus and putamen. At early stages of the disease, a decrease in $GABA_A$ receptor binding has been observed in the caudate nucleus by positron emission tomography (Holthoff *et al.*, 1993). The loss of $GABA_A$ receptors has been confirmed immunohistochemically and autoradiographically (Faull *et al.*, 1993) and was shown to occur in the absence of overt neuropathological alterations. At late stages of Huntington's disease, there is a further decrease of $GABA_A$ receptors in the atrophied striatum, accompanied by a pronounced upregulation in the globus pallidus, pointing to hypo- and hypersensitivity of the GABA system in the respective regions (Walker *et al.*, 1984;

Whitehouse *et al.*, 1985; Faull *et al.*, 1993). These observations may be of pathophysiological significance and may explain why the beneficial effects of benzodiazepine therapy are not sustained at later stages of the disease.

9.3. Angelman's Syndrome

The human homologue of the region deleted in the mouse pink-eyed cleft palate mutation, which includes the α5, γ3, and β3 subunit genes, is associated with Angelman's syndrome (AS) (Saitoh *et al.*, 1992). This genetic disorder is characterized by mental retardation, seizures, ataxia, craniofacial abnormalities, and hypopigmentation. The smallest known deletion resulting in AS includes the β3 subunit gene (Knoll *et al.*, 1993; Sinnett *et al.*, 1993), although a single AS patient bearing a translocation possessed an apparently intact β3 subunit gene (Reis *et al.*, 1993). The GABA$_A$ receptor α5 subunit gene is not involved in AS. Although the mouse counterpart of the AS gene(s) is not imprinted, it is possible that the genes critical for AS may be within the chromosomal region corresponding to the *p^{cp}* deletion. It remains, however, to be determined what role, if any, the three GABA$_A$ receptor subunit genes play in the etiology of AS.

10. OUTLOOK

Since GABA$_A$ receptor subunits can be assembled in almost any desired combination in heterologous expression systems, potentially hundreds of different receptor subtypes might have been expected to exist in brain. However, *in vivo* the spectrum of GABA$_A$ receptor subtypes is limited. Three receptor subtypes, characterized by the subunit combinations α1β2γ2, α2β3γ2, and α3β3γ2, account for about 75% of all GABA$_A$ receptors, providing the basic repertoire for GABAergic signal transduction. Various additional minor populations of GABA$_A$ receptors have been identified, which appear to serve specialized functions in particular brain areas. These include receptors containing the α5 subunit (hippocampus), α6 subunit (cerebellar granule cells), or ρ subunit (retina), bringing the total number of GABA$_A$ receptors in the brain into the range of a few dozens of subtypes (see Table II).

Several receptor subtypes have been allocated to defined neuronal populations. This achievement opens the way for an analysis of the physiological significance of receptor subtypes in particular neuronal circuits. The necessary electrophysiological studies appear to be feasible since there is no indication of an extensive coexpression of various receptor subtypes in the same cell.

The role of receptor subtypes in the regulation of complex brain functions will be studied in animals that either lack certain GABA$_A$ receptors or contain specifically altered receptors. Such animal models, generated by subunit gene targeting, will help to identify the brain areas and neuronal circuits underlying GABA-mediated behavior and drug-induced responses.

Pharmacologically, receptor heterogeneity provides the opportunity to target drug actions to selected brain areas. Side effects which arise from the indiscriminate modulation of all GABA$_A$ receptors can thus be avoided. Both affinity and efficacy of ligands can be varied in a subtype-specific manner to advance the treatment of anxiety, insomnia, and epilepsy.

11. REFERENCES

Amin, J., Dickerson, I. M., and Weiss, D. S., 1994, The agonist binding site of the γ-aminobutyric acid type A channel is not formed by the extracellular cysteine loop, *Mol. Pharmacol.* **45**:317–323.

Araki, T., Kiyama, H., and Tohyama, M., 1992, The $GABA_A$ receptor γ1 subunit is expressed by distinct neuronal populations, *Mol. Brain Res.* **15**:121–132.

Auta, J., Giusti, P., Guidotti, A., and Costa, E., 1994, Imidazenil, a partial positive allosteric modulator of $GABA_A$ receptors, exhibits low tolerance and dependence liabilities in the rat. *J. Pharmacol. Exp. Ther.* **270**:1262–1269.

Backus, K. H., Arigoni, M., Drescher, U., Scheurer, L., Malherbe, P., Mohler, H., and Benson, J. A., 1993, Stoichiometry of a recombinant $GABA_A$-receptor deduced from mutation-induced rectification, *Neuroreport* **5**:285–288.

Barnard, E. A., 1992, Receptor classes and the transmitter-gated ion channels, *Trends Biochem. Sci.* **17**: 368–374.

Barnard, E. A., Bateson, A. N., Darlison, M. G., Glencorse, T. A., Harvey, R. J., Hicks, A. A., Lasham, A., Shingai, R., Usherwood, P. N. R., Verugdenhil, E., and Zaman, S. H., 1992, Genes for the $GABA_A$ receptor subunit types and their expression, in *GABAergic Synaptic Transmission* (G. Biggio, A. Concas, and E. Costa, eds.), Raven Press, New York, pp. 17–26.

Bateson, A. N., and Paetsch, P. R., 1993, Isolation and characterization of the rat $GABA_A$ receptor α1-subunit gene promoter, *Soc. Neurosci. Abstr.* **19**:476.

Baude, A., Sequier, J. M., McKernan, R. M., Oliver, K. R., and Somogyi, P., 1992, differential subcellular distribution of the α6 subunit versus the α1 and β2/3 subunits of the $GABA_A$/benzodiazepine receptor complex in granule cells of the cerebellar cortex, *Neuroscience* **51**:739–748.

Beaulieu, E. E., and Robel, P., 1990, Neurosteroids: A new brain function? *J. Steroid Biochem. Mol. Biol.* **37**:395–403.

Benavides, J., Peny, B., Ruano, D., Vitorica, J., and Scatton, B., 1993, Comparative autoradiographic distribution of central ω (benzodiazepine) modulatory site subtypes with high, intermediate and low affinity for zolpidem and alpidem, *Brain Res.* **604**:240–250.

Benke, D., Mertens, S., and Mohler, H., 1991a, Ubiquitous presence of $GABA_A$ receptors containing the α1-subunit in rat brain demonstrated by immunoprecipitation and immunohistochemistry, *Mol. Neuropharmacol.* **1**:103–110.

Benke, D., Mertens, S., Trzeciak, A., Gillessen, D., and Mohler, H., 1991b, $GABA_A$ receptors display association of γ2-subunit with α1- and β2/3 subunits, *J. Biol. Chem.* **266**:4478–4483.

Benke, D., Mertens, S., Trzeciak, A., Gillessen, D., and Mohler, H., 1991c, Identification and immuno-histochemical mapping of $GABA_A$ receptor subtypes containing the δ-subunit in rat brain, *FEBS Lett.* **283**:145–149.

Benke, D., Fritschy, J. M., Trzeciak, A., Bannwarth, W., and Mohler, H., 1994, Distribution, prevalence and drug-binding profile of $GABA_A$-receptors subtypes differing in β-subunit isoform, *J. Biol. Chem.* **269**:27100–27107.

Bertrand, D., Galzi, J. L., Devillers-Thiery, A., Bertrand, S., and Changeux, J. P., 1993, Stratification of the channel domain in neurotransmitter receptors, *Curr. Opin. Cell Biol.* **5**:688–693.

Betz, H., 1990, Ligand-gated ion channels in the brain: The amino acid receptor superfamily, *Neuron* **5**: 383–392.

Caruncho, H J., and Costa, E., 1994, Double-immunolabelling analysis of $GABA_A$ receptor subunits in label-fracture replicas of cultured rat cerebellar granule cells, *Recept. Channels* **2**:143–153.

Casalotti, S. O., Stephenson, F. A., and Barnard, E. A., 1986, Separate subunits for agonist and benzodiazepine binding in the γ-aminobutyric $acid_A$ receptor oligomer, *J. Biol. Chem.* **261**:15013–15016.

Clark, M., Massenburg, G. S., Weiss, S. R. B., and Post, R. M., 1994, Analysis of the hippocampal $GABA_A$

receptor system in kindled rats by autoradiographic and in situ hybridization techniques: Contingent tolerance to carbamazepine, *Mol. Brain Res.* **26:**309–319.

Costa, E., Auta, J., Caruncho, H., Guidotti, A., Impagnatiello, F., Pesold, C., and Thompson, D. M., 1995, A search for a new anticonvulsant and anxiolytic benzodiazepine devoid of side effects and tolerance liability, in: *GABA$_A$ Receptors and Anxiety: From Neurobiology to Treatment*, Advances in Biochemical Psychopharmacology, Vol. 48 (G. Biggio, E. Sanna, M. Serra, and E. Costa, eds.), Raven Press, New York, 75–92.

Culiat, C. T., Stubbs, L., Nicholls, R. D., Montgomery, C. S., Russell, L. B., Johnson, D. K., and Rinchik, E. M., 1993, Concordance between isolated cleft palate in mice and alterations within a region including the gene encoding the β3 subunit of the type A γ-aminobutyric acid receptor, *Proc. Natl. Acad. Sci. USA* **90:**5105–5109.

Culiat, C. T., Stubbs, L. J., Montgomery, C. S., Russel, L. B., and Rinchik, E. M., 1994, Phenotypic consequences of deletion of the γ3, α5, or β3 subunit of the type A γ-aminobutyric acid receptor in mice, *Proc. Natl. Acad. Sci. USA* **91:**2815–2818.

Culiat, C. T., Stubbs, L. J., Woychik, R. P., Russell, L. B., Johnson, D. K., and Rinchik, E. M., 1995, Deficiency of the β3 subunit of the type A γ-aminobutyric acid receptor causes cleft palate in mice, *Nature Genetics* **11:**344–346.

Cutting, G. R., Lu, L., O'Hara, B. F., Kasch, L. M., Montrose- Rafizadeh, C., Donovan, D. M., Shimada, S., Antonarakis, S. E., Guggino, W. B., Uhl, G. R., and Kazazian, H. H., 1991, Cloning of the γ-aminobutyric acid (GABA) ρl cDNA: A GABA receptor subunit highly expressed in the retina, *Proc. Natl. Acad. Sci. USA* **88:**2673–2677.

Cutting, G. R., Curristin, S., Zoghbi, H., O'Hara, B., Seldin, M. F., and Uhl, G. R., 1992, Identification of a putative γ-aminobutyric acid (GABA) receptor subunit ρ2 cDNA and colocalization of the genes encoding ρ2 (GABARR2) and ρl (GABARR1) to human chromosome 6q14-q21 and mouse chromosome 4, *Genomics* **12:**801–806.

Deng, L., Ransom, R. W., and Olsen, R. W., 1986, [^{3}H]muscimol photolabels the γ-aminobutyric acid receptor binding site on a peptide subunit distinct from that labeled with benzodiazepines, *Biochem. Biophys. Res. Commun.* **138:**1308–1314.

Drew, C. A., and Johnston, G., 1992, Bicuculline- and baclofen-insensitive γ-aminobutyric acid binding to rat cerebellar membranes, *J. Neurochem.* **58:**1087–1092.

Drew, C. A., Johnston, G., and Weatherby, R. P., 1984, Bicuculline-insensitive GABA receptors: Studies on the binding of (-)-baclofen to rat cerebellar membrane, *Neurosci. Lett.* **52:**317–321.

Ducic, I., Puia, G., Vicini, S., and Costa, E., 1993, Triazolam is more efficacious than diazepam in a broad spectrum of recombinant receptors, *Eur. J. Pharmacol.* **244:**29–35.

Duggan, M. J., and Stephenson, F. A., 1990, Biochemical evidence for the existence of γ-aminobutyrate$_A$ receptor iso-oligomers, *J. Biol. Chem.* **265:**3831–3835.

Duggan, M. J., Pollard, S., and Stephenson, F. A., 1991, Immunoaffinity purification of GABA$_A$ receptor α-subunit iso-oligomers. Demonstration of receptor populations containing α1α2, α1α3, and α2α3 subunit pairs, *J. Biol. Chem.* **266:**24778–24784.

Endo, S., and Olsen, R. W., 1993, Antibodies specific for α-subunit subtypes of GABA$_A$ receptors reveal brain regional heterogeneity, *J. Neurochem.* **60:**1388–1398.

Enz, R., Brandstatter, J. H., Hartveit, E., Wassle, H., and Bormann, J., 1995, Expression of GABA receptor ρl and ρ2 subunits in the retina and brain of the rat, *Eur. J. Neurosci.* **7:**1495–1501.

Faull, R. L. M., Waldvogel, H. J., Nicholson, L. F. B., and Synek, B. J. L., 1993, The distribution of GABA$_A$-benzodiazepine receptors in the basal ganglia in Huntington's disease and in the quinolinic acid-lesioned rat, *Prog. Brain Res.* **99:**105–123.

Feigenspan, A., and Bormann, J., 1994, Modulation of GABA$_C$ receptors in rat retinal bipolar cells by protein kinase C, *J. Physiol. (London)* **481:**325–330.

Feigenspan, A., Wassle, H., and Bormann, J., 1993, Pharmacology of GABA receptor Cl$^-$ channels in rat retinal bipolar cells, *Nature* **361:**159–162.

ffrench-Constant, R. H., Rocheleau, T. A., Steichen, J. C., and Chalmers, A. E., 1993, A point mutation in a Drosophila GABA receptor confers insecticide resistance, *Nature* **363:**449–451.

Fritschy, J. M., and Mohler, H., 1995, GABA$_A$-receptor heterogeneity in the adult rat brain: Differential regional and cellular distribution of seven major subunits, *J. Comp. Neurol.* **359:**154–194.

Fritschy, J. M., Benke, D., Mertens, S., Oertel, W. H., Bachi, T., and Mohler, H., 1992, Five subtypes of type A γ-aminobutyric acid receptors identified in neurons by double and triple immunofluorescence staining with subunit-specific antibodies, *Proc. Natl. Acad. Sci. USA* **89:**6726–6730.

Fritschy, J. M., Paysan, J., Enna, A., and Mohler, H., 1994, Switch in the expression of rat GABA$_A$-receptor subtypes during postnatal development: An immunohistochemical study, *J. Neurosci.* **14:**5302–5324.

Gale, K., 1992, GABA and epilepsy: Basic concepts from preclinical research, *Epilepsia* **33(5):**S3–S12.

Galzi, J. L., and Changeux, J. P., 1994, Neurotransmitter-gated ion channels as unconventional allosteric proteins, *Curr. Opin. Struct. Biol.* **4:**554–565.

Gao, B., and Fritschy, J. M., 1994, Selective allocation of GABA$_A$-receptors containing the α1-subunit to neurochemically distinct subpopulations of hippocampal interneurons, *Eur. J. Neurosci.* **6:**837–853.

Gao, B., and Fritschy, J. M., 1995, Cerebellar granule cells in vitro recapitulate the in vivo pattern of GABA$_A$-receptor subunit expression, *Dev. Brain Res.* **88:**1–16.

Gao, B., Fritschy, J. M., Benke, D., and Mohler, H., 1993, Neuron-specific expression of GABA$_A$-receptor subtypes: Differential associations of the α1- and α3-subunits with serotonergic and GABAergic neurons, *Neuroscience* **54:**881–892.

Gao, B., Hornung, J. P., and Fritschy, J. M., 1995, Identification of distinct GABA$_A$-receptor subtypes in cholinergic and parvalbumin-positive neurons of the rat and marmoset medial-septum-diagonal band complex, *Neuroscience* **65:**101–117.

Granger, P., Biton, X., Faure, C., Vige, X., Depportere, H., Graham, D., Langer, S. Z., Scatton, B., and Avenet, P., 1995, Modulation of the γ-aminobutyric acid type A receptor by the antiepileptic drugs carbamazepine and phenytoin, *Mol. Pharmacol.* **47:**1189–1196.

Gutierrez, A., Khan, Z. U., and De Blas, A. L., 1994, Immunocytochemical localization of γ2 short and γ2 long subunits of the GABA$_A$ receptor in the rat brain, *J. Neurosci.* **14:**7168–7179.

Hadingham, K. L., Wingrove, P. B., Wafford, K. A., Bain, C., Kemp, J. A., Palmer, K. J., Wilson, A. W., Wilcox, A. S., Sikela, J. M., and Whiting, P. J., 1993, Role of the β subunit in determining the pharmacology of human γ-aminobutyric acid type A receptors, *Mol. Pharmacol.* **44:**1211–1218.

Haefely, W., Martin, J. R., and Schoch, P., 1990, Novel anxiolytics that act as partial agonists at benzodiazepine receptors, *Trends Pharmacol. Sci.* **11:**452–456.

Harrison, N. L., and Simmonds, M. A., 1984, Modulation of the GABA receptor complex by a steroid anaesthetic, *Brain Res.* **323:**287–292.

Harvey, R. J., Kim, H. C., and Darlison, M. G., 1993, Molecular cloning reveals the existence of a fourth γ subunit of the vertebrate brain GABA$_A$ receptor, *FEBS Lett.* **331:**211–216.

Hayden, M. R., 1981, *Huntington's Chorea*, Springer Verlag, New York.

Hendry, S. H. C., Huntsman, M. M., Vinuela, A., Mohler, H., de Blas, A. L., and Jones, E. G., 1994, GABA$_A$ receptor subunit immunoreactivity in primate visual cortex: Distribution in macaques and humans and regulation by visual input in adulthood, *J. Neurosci.* **14:**2383–2401.

Herb, A., Wisden, W., Luddens, H., Puia, G., Vicini, S., and Seeburg, P. H., 1992, The third γ subunit of the γ-aminobutyric acid type A receptor family, *Proc. Natl. Acad. Sci. USA* **89:**1433–1437.

Hirouchi, M., Hashimoto, T., and Kuriyama, K., 1993, Alteration of GABA$_A$ receptor α1-subunit mRNA in mouse brain following continuous ethanol inhalation, *Eur. J. Pharmacol.* **247:**127–130.

Holthoff, V. A., Koeppe, R. A., Frey, K. A., Penney, J. B., Markel, D. S., Kuhl, D. E., and Young, A. B., 1993, Positron emission tomography measures of benzodiazepine receptors in Huntington's disease, *Ann. Neurol.* **34:**76–81.

Huntington's Disease Collaborative Research Group, 1993, A novel gene containing a trinucleotide repeat that is expanded and unstable on Huntington's disease chromosomes, *Cell* **72:**971–983.

Huntsman, M. M., Isackson, P. J., and Jones, E. G., 1994, Lamina-specific expression and activity-dependent regulation of seven GABA$_A$ receptor subunit mRNAs in monkey visual cortex, *J. Neurosci.* **14:**2236–2259.

Im, H. K., Im, W. B., Judge, T. M., Gamill, R. B., Hamilton, B. J., Carter, D. B., and Pregenzer, J. F., 1993, Substituted pyrazinones, a new class of allosteric modulators for γ-aminobutyric acid A receptors, *Mol. Pharmacol.* **44:**468–472.

Im, H. K., Im, W. B., Pregenzer, J. F., Carter, D. B., and Hamilton, B. J., 1995, U 89843 A represents a novel class of GABA$_A$-receptor agonists, *J. Pharmacol. Exp. Ther.* **275:**1390–1395.

Johnston, G. A. R., 1994, GABA$_C$ receptors, *Prog. Brain Res.* **100:**61–65.

Kamphuis, W., De Rijk, T. C., and Lopes da Silva, F. H., 1994, GABA$_A$ receptor β1-3 subunit gene expression in the hippocampus of kindled rats, *Neurosci. Lett.* **174:**5–8.

Kang, I., and Miller, L. G., 1991, Decreased GABA$_A$ receptor subunit mRNA concentrations following chronic lorazepam administration, *Br. J. Pharmacol.* **103:**1285–1287.

Kang, I., Lindquist, D. G., Kinane, T. B., Ercolani, L., Pritchard, G. A., and Miller, L. G., 1994, Isolation and characterization of the promoter of the human GABA$_A$ receptor α1 subunit gene, *J. Neurochem.* **62:**1643–1646.

Kellenberger, S., Malherbe, P., and Sigel, E., 1992, Function of the α1β2γ2S γ-aminobutyric acid type A receptor is modulated by protein kinase C via multiple phosphorylation sites, *J. Biol. Chem.* **267:**25660–25663.

Khan, Z. U., Gutierrez, A., and Deblas, A. L., 1994, Short and long form γ2 subunits of the GABA$_A$/benzodiazepine receptors, *J. Neurochem.* **63:**1466–1476.

Kirkness, E. F., and Fraser, C. M., 1993, A strong promoter element is located between alternative exons of a gene encoding the human γ-aminobutyric acid-type A receptor β3 subunit (GABRB3), *J. Biol. Chem.* **268:**4420–4428.

Knoflach, F., Rhyner, T., Villa, M., Kellenberger, S., Drescher, U., Malherbe, P., Sigel, E., and Mohler, H., 1991, The γ3- subunit of the GABA$_A$-receptor confers sensitivity to benzodiazepine receptor ligands, *FEBS Lett.* **293:**191–194.

Knoflach, F., Drescher, U., Scheurer, L., Malherbe, P., and Mohler, H., 1993, Full and partial agonism displayed by benzodiazepine receptor ligands at different recombinant GABA$_A$ receptor subtypes, *J. Pharmacol. Exp. Ther.* **266:**385–391.

Knoflach, F., Benke, D., Wang, Y., Scheurer, L., Hamilton, B. J., Carter, D. B., Luddens, H., Mohler, H., and Benson, J. A., 1996, Modulation of "diazepam-insensitive" α4β2γ2 and α6β2γ2 recombinant GABA$_A$ receptors by partial and inverse agonists.

Knoll, J H., Sinnett, D., Wagstaff, J., Glatt, K., Wilcox, A. S., Whiting, P. M., Wingrove, P., Sikela, J. M., and Lalande, M., 1993, FISH ordering of reference markers and of the gene for the α5 subunit of the γ-aminobutyric acid receptor (GABAR5) within the Angelman and Prader-Willi syndrome chromosomal regions, *Hum. Mol. Gen.* **2:**183–189.

Kokaia, M., Pratt, G. D., Elmer, E., Bengzon, J., Fritschy, J. M., Lindvall, O., and Mohler, H., 1994, Biphasic differential changes of GABA$_A$-receptor subunit mRNA levels in dentate gyrus granule cells following recurrent kindling induced seizures, *Mol. Brain Res.* **23:**323–332.

Korpi, E. R., Kleingoor, C., Kettenmann, H., and Seeburg, P. H., 1993, Benzodiazepine-induced motor impairment linked to point mutation in cerebellar GABA$_A$ receptor. *Nature* **361:**356–359.

Krishek, B. J., Xie, X., Blackstone, C., Huganir, R. L., Moss, S. J., and Smart, T. G., 1994, Regulation of GABA$_A$ receptor function by protein kinase C phosphorylation, *Neuron* **12:**1081–1095.

Kusama, T., Spivak, C. E., Whiting, P., Dawson, V. L., Schaffer, J. C., and Uhl, G. R., 1993, Pharmacology of GABA ρ1 and GABA α/β receptors expressed in Xenopus oocytes and COS cells, *Br. J. Pharmacol.* **109:**200–206.

Lambert, J. J., Peters J. A., Sturgess, N. C., and Hales, T. G., 1990, Steroid modulation of the GABA$_A$ receptor complex: Electrophysiological studies, in: *Steroids and Neuronal Activity*, Ciba Foundation Symposium 153, Wiley, Chichester, pp. 56–71.

Lambert, J. J., Belelli, D., Hill-Venning, C., and Peters, J. A., 1995, Neurosteroids and GABA$_A$ receptor function, *Trends Pharmacol. Sci.* **16:**295–303.

Lan, N. C., Gee, K. W., Bolger, M. B., and Chen, J. S., 1991, Differential responses of expressed recombinant human γ-aminobutyric acid$_A$ receptors to neurosteroids, *J. Neurochem.* **57:**1818–1821.

Laurie, D. J., Seeburg, P. H., and Wisden, W., 1992, The distribution of 13 GABA$_A$ receptor subunit mRNAs in the rat brain. II. Olfactory bulb and cerebellum, *J. Neurosci.* **12:**1063–1076.

Leidenheimer, N. J., McQuilkin, S. J., Hahner, L. D., Whiting, P., and Harris, R. A., 1992, Activation of protein kinase C selectively inhibits the γ-aminobutyric acid$_A$ receptor: Role of desensitization, *Mol. Pharmacol.* **41:**1116–1123.

Luddens, H., Pritchett, D. B., Kohler, M., Killisch, I., Keinanen, L., Monyer, H., Sprengel, R., and Seeburg, P. H., 1990, Cerebellar GABA$_A$-receptor selective for a behavioral alcohol antagonist, *Nature* **346:**648–651.

Luddens, H., Seeburg, P. H., and Korpi, E. R., 1994, Impact of β and γ variants on ligand-binding properties of γ-aminobutyric acid type A receptors, *Mol. Pharmacol.* **45:**810–814.

Macdonald, R. L., and Angelotti, T. P., 1993, Native and recombinant GABA$_A$ receptor channels, *Cell. Physiol. Biochem.* **3**:352–373.

Macdonald, R. L., and Olsen, R. W., 1994, GABA$_A$ receptor channels, *Annu. Rev. Neurosci.* **17**:569–602.

Majewska, M. D., 1992, Neurosteroids: Endogenous bimodal modulators of the GABA$_A$ receptor. Mechanism of action and physiological significance, *Prog. Neurobiol.* **38**:379–395.

Marksitzer, R., Benke, D., Fritschy, J. M., and Mohler, H., 1993, GABA$_A$-receptors: Drug binding profile and distribution of receptors containing the α2-subunit in situ, *J. Recept. Res.* **13**:467–477.

McDonald, J. W., Garofalo, E. A., Hood, T., Sackellares, J. C., Gilman, S., McKeever, P. E., Troncoso, J. C., and Johnston, M. V., 1991, Altered excitatory and inhibitory amino acid receptor binding in hippocampus of patients with temporal lobe epilepsy, *Ann. Neurol.* **29**:529–541.

McKernan, R. M., Quirk, K., Prince, R., Cox, P. A., Gillard, N. P., Ragan, C. I., and Whiting, P., 1991, GABA$_A$ receptor subtypes immunopurified from rat brain with α subunit-specific antibodies have unique pharmacological properties, *Neuron* **7**:667–676.

Meldrum, B., 1979, Convulsant drugs, anticonvulsants and GABA-mediated neuronal inhibition, in: *GABA-Neurotransmitters* (P. Krogsgard-Larsen, J. Scheell-Kruger, and H. Kofod, eds.), Munksgaard, Copenhagen, pp. 390–405.

Mertens, S., Benke, D., and Mohler, H., 1993, GABA$_A$ receptor populations with novel subunit combinations and drug binding profiles identified in brain by α5- and δ-subunit-specific immunopurification, *J. Biol. Chem.* **268**:5965–5973.

Mhatre, M. C., and Ticku, M. K., 1992, Chronic ethanol administration alters γ-aminobutyric acid$_A$ receptor gene expression, *Mol. Pharmacol.* **42**:415–422.

Mihic, S. J., McQuilkin, S. J., Eger, E. I., Ionescu, P., and Harris, R. A., 1994, Potentiation of γ-aminobutyric acid type A receptor-mediated chloride currents by novel halogenated compounds correlates with their abilities to induce general anesthesia, *Mol. Pharmacol.* **46**:851–857.

Mohler, H., Knoflach, F., Paysan, J., Motejlek, K., Benke, D., Luscher, B., and Fritschy, J. M., 1995, Heterogeneity of GABA$_A$-receptors: Cell-specific expression, pharmacology, and regulation, *Neurochem. Res.* **20**:559–564.

Montpied, P., Morrow, A. L., Karanian, J. W., Ginns, E. I., Martin, B. M., and Paul, S. M., 1991, Prolonged ethanol inhalation decreases γ-aminobutyric acid$_A$ receptor α subunit mRNAs in the rat cerebral cortex, *Mol. Pharmacol.* **39**:157–163.

Moss, S. J., Doherty, C. A., and Huganir, R. L., 1992a, Identification of the cAMP-dependent protein kinase and protein kinase C phosphorylation sites within the major intracellular domains of the β1, γ2S and γ2L subunits of the γ-aminobutyric acid type A receptor, *J. Biol. Chem.* **267**:14470–14476.

Moss, S. J., Smart, T. G., Blackstone, C. D., and Huganir, R. L., 1992b, Functional modulation of GABA$_A$ receptors by cAMP-dependent protein phosphorylation, *Science* **257**:661–665.

Motejlek, K., Hauselmann, R., Leitgeb, S., and Luscher, B., 1994, BSF1, a novel brain-specific DNA-binding protein recognizing a tandemly repeated purine DNA element in the GABA$_A$ receptor δ subunit gene, *J. Biol. Chem.* **269**:15265–15273.

Nakatsu, Y., Tyndale, R. F., DeLorey, T. M., Durham-Pierre, D., Gardner, J. M., McDanel, H. J., Nguyen, Q., Wagstaff, J., Lalande, M., Sikela, J. M., Olsen, R. W., Tobin, A. J., and Brilliant, M. H., 1993, A cluster of three GABA$_A$ receptor subunit genes is deleted in a neurological mutant of the mouse p locus, *Nature* **364**:448–450.

Nayeem, N., Green, T. P., Martin, I. L., and Barnard, E. A., 1994, Quaternary structure of the native GABA$_A$ receptor determined by electron microscopic image analysis, *J. Neurochem.* **62**:815–818.

Nobrega, J. N., Kish, S. J., and Burnham, W. M., 1990, Regional brain [^{3}H]muscimol binding in kindled rat brain: A quantitative autoradiographic examination, *Epilepsy Res.* **6**:102–109.

Olsen, R. W., Wamsley, J. K., McCabe, R. T., Lee, R. J., and Lomax, P., 1985, Benzodiazepine/γ-aminobutyric acid receptor deficit in the midbrain of the seizure-susceptible gerbils, *Proc. Natl. Acad. Sci. USA* **82**:6701–6705.

Olsen, R. W., Wamsley, J. K., Lee, R. J., and Lomax, P., 1986, Benzodiazepine/barbiturate/GABA receptor-chloride ionophore complex in a genetic model for generalized epilepsy, *Adv. Neurol.* **44**:365–378.

Olsen, R. W., Bureau, M., Houser, C. R., Delgado-Escueta, A. V., Richards, J. G., and Mohler, H., 1990, GABA/Benzodiazepine receptors in human focal epilepsy, in: *Neurotransmitters in Epilepsy, Advances in the Neurobiology of Epilepsy*, Vol. 1 (G. Avanzini *et al.*, eds.), Demos Publications, New York, pp. 515–527.

Olsen, R. W., Bureau, M., Houser, C. R., Delgado-Escueta, A. V., Richards, J. G., and Mohler, H., 1992, GABA/benzodiazepine receptors in human focal epilepsy, *Epilepsy Res. Suppl.* **8**:383–391.

Ortells, M. O., and Lunt, G. G., 1995, Evolutionary history of the ligand-gated ion-channel superfamily of receptors, *Trends Neurosci.* **18**:121–127.

Paul, S. M., and Purdy, R. H., 1992, Neuroactive steroids, *FASEB J.* **6**:2311–2322.

Paysan, J., Kossel, A., Thanos, R., Fritschy, J. M., Mohler, H., and Bolz, J., 1993, Area-specific regulation of the GABA_A- receptor α1-subunit in developing rat neocortex deprived of afferent innervation, *Soc. Neurosci. Abstr.* **19**:1107.

Paysan, J., Bolz, J., Mohler H., and Fritschy, J. M., 1994a, The GABA_A-receptor α1-subunit: An early marker of cortical parcellation, *J. Comp. Neurol.* **350**:133–149.

Paysan, J., Mohler, H., and Fritschy, J. M., 1994b, Switch in the expression of GABA_A-receptor subtypes during postnatal development, *Eur. J. Neurosci. Suppl.* **7**:81.

Pollard, S., Duggan, M. J., and Stephenson, F. A., 1993, Further evidence for the existence of α subunit heterogeneity within discrete γ-aminobutyric acid_A receptor subpopulations, *J. Biol. Chem.* **268**:3753–3757.

Porter, N. M., Twyman, R. E., Uhler, M. D., and Macdonald, R. L., 1990, Cyclic AMP-dependent protein kinase decreases GABA_A receptor current in mouse spinal neurons, *Neuron* **5**:789–796.

Pribilla, I., Neuhaus, R., Huba, R., Hillmann, M., Turner, J. D., Stephens, D. N., and Schneider, H. H., 1993, Abercanil is a full agonist at some, and a partial agonist at other recombinant GABA_A receptor subtypes, in: *Anxiolytic β-Carbolines* (D. N. Stephens, ed.), Springer Verlag, Berlin, pp. 50–61.

Pritchett, D. B., and Seeburg, P. H., 1990, γ-Aminobutyric acid_A receptor α5-subunit creates novel type II benzodiazepine receptor pharmacology, *J. Neurochem.* **54**:1802–1804.

Pritchett, D. B., and Seeburg, P. H., 1991, γ-Aminobutyric acid type A receptor point mutation increases the affinity of compounds for the benzodiazepine site, *Proc. Natl. Acad. Sci. USA* **88**:1421–1425.

Pritchett, D. B., Sontheimer, H., Gorman, C. M., Kettenmann, H., Seeburg, P. H., and Schofield, P. R., 1988, Transient expression shows ligand gating and allosteric potentiation of GABA_A receptor subunits, *Science* **242**:1306–1308.

Pritchett, D. B., Luddens, II., and Seeburg, P. H., 1989, Type I and type II GABA_A-benzodiazepine receptors produced in transfected cells, *Science* **245**:1389–1392.

Puia, G., Santi, M. R., Vicini, S., Pritchett, D. B., Purdy, R. H., Paul, S. M., Seeburg, P. H., and Costa, E., 1990, Neurosteroids act on recombinant human GABA_A receptors, *Neuron* **4**:759–765.

Puia, G., Vicini, S., Seeburg, P. H., and Costa, E., 1991, Influence of recombinant γ-aminobutyric acid_A receptor subunit composition on the action of allosteric modulators of γ-aminobutyric acid–gated Cl⁻ currents, *Mol. Pharmacol.* **39**:691–696.

Puia, G., Dudic, I., Vicini, S., and Costa, E., 1992, Molecular mechanisms of the partial allosteric modulatory effects of bretazenil at γ-aminobutyric acid type A receptor, *Proc. Natl. Acad. Sci. USA* **89**:3620–3624.

Qian, H., and Dowling, J. E., 1993, Novel GABA responses from rod-driven retinal horizontal cells, *Nature* **361**:162–164.

Quirk, K., Gillard, N. P., Ragan, C. I., Whiting, P. J., and McKernan, R. M., 1994, γ-aminobutyric acid type A receptors in the rat brain can contain both γ2 and γ3 subunits, but γ1 does not exist in combination with another γ subunit, *Mol. Pharmacol.* **45**:1061–1070.

Reis, A., Kunze, J., Ladanyi, L., Enders, H., Klein-Vogler, U., and Niemann, G., 1993, Exclusion of the GABA_A-receptor β3 subunit gene as the Angelman's syndrome gene, *Lancet* **341**:122–123.

Ruano, D., Vizuete, M., Cano, J., Machado, A., and Vitorica, J., 1992, Heterogeneity in the allosteric interaction between the γ-aminobutyric acid (GABA) binding site and three different benzodiazepine binding sites of the GABA_A/benzodiazepine receptor complex in the rat nervous system, *J. Neurochem.* **58**:485–493.

Saitoh, S., Kubota, T., Ohta, T., Jinno, Y., Niikawa, N., Sugimoto, T., Wagstaff, J., and Lalande, M., 1992, Familial Angelman syndrome cause by imprinted submicroscopic deletion encompassing GABA_A receptor β3-subunit gene, *Lancet* **339**:366–367.

Savic, I., Roland, P., Sedvall, G., Persson, A., Pauli, S., and Widen, L., 1988, In vivo demonstration of reduced benzodiazepine receptor binding in human epileptic foci, *Lancet* **2**:863–866.

Saxena, N. C., and Macdonald, R. L., 1994, Assembly of GABA_A receptor subunits: Role of the δ subunit, *J. Neurosci.* **14**:7077–7086.

Shimada, S., Cutting, G., and Uhl, G. R., 1992, γ-Aminobutyric acid A or C receptor? γ-aminobutyric acid ρl receptor RNA induces bicuculline-, barbiturate-, and benzodiazepine-insensitive γ-aminobutyric acid responses in Xenopus oocytes, *Mol. Pharmacol.* **41:**683–687.

Shin, C., Pedersen, H. B., and McNamara, J. O., 1985, γ-aminobutyric acid and benzodiazepine receptors in the kindling model of epilepsy: A quantitative radiohistochemical study, *J. Neurosci.* **5:**2696–2701.

Shingai, R., Sutherland, M. L., and Barnard, E. A., 1991, Effects of subunit types of the cloned GABA_A receptor on the response to a neurosteroid, *Eur. J. Pharmacol.* **206:**77–80.

Shivers, B. D., Killisch, I., Sprengel, R., Sontheimer, H., Kohler, M., Schofield, P. R., and Seeburg, P. H., 1989, Two novel GABA_A receptor subunits exist in distinct neuronal subpopulations, *Neuron* **3:**327–337.

Sieghart, W., 1985, Benzodiazepine receptors: Multiple receptors or multiple conformations, *J. Neural Transm.* **63:**191–208.

Sigel, E., and Baur, A, 1988, Activation of protein kinase C differentially modulates neuronal Na^+, Ca^{2+} and γ-aminobutyrate type A channels, *Proc. Natl. Acad. Sci. USA* **88:**6192–6196.

Sigel, E., Baur, A., and Malherbe, P., 1991, Activation of protein kinase C results in down-modulation of different recombinant GABA_A-channels, *FEBS Lett.* **291:**150–152.

Sigel, E., Baur, R., Kellenberger, S., and Malherbe, P., 1992, Point mutations affecting antagonist affinity and agonist dependent gating of GABA_A receptor channels, *EMBO J.* **11:**2017–2023.

Sigel, E., Baur, R., and Malherbe, P., 1993, Recombinant GABA_A receptor function and ethanol, *FEBS Lett.* **324:**140–142.

Sinnett, C., Wagstaff, J., Glatt, K., Woolf, E., Kirkness, E. J., and Lalande, M., 1993, High-resolution mapping of the γ-aminobutyric acid receptor subunit β3 and α5 gene cluster on chromosome 15qII-qI3, and localization of breakpoint in two Angelman syndrome patients, *Am. J. Hum. Genet.* **52:**1216–1229.

Smith, G. B., and Olsen, R. W., 1994, Identification of a [^{3}H]muscimol photoaffinity substrate in the bovine γ-aminobutyric acid_A receptor α subunit, *J. Biol. Chem.* **269:**20380–20387.

Sommer, B., Poustka, A., Spurr, N. K., and Seeburg, P. H., 1990, The murine GABA receptor δ subunit gene: Structure and assignment to human chromosome 1, *DNA Cell Biol.* **9:**561–568.

Strong, T. V., Tagle, D. A., Valdes, J. M., Elmer, L. W., Boehm, K., Swaroop, M., Kaatz, K. W., Collins, F. S., and Albin, R. L., 1993, Widespread expression of the human and rat Huntington's disease gene in brain and non-neural tissues, *Nature Genet.* **5:**259–265.

Thompson, C. L., Bodewitz, G., Stephenson, F. A., and Turner, J. D., 1992, Mapping of GABA_A receptor α5 and α6 subunit-like immunoreactivity in rat brain, *Neurosci. Lett.* **144:**53–56.

Togel, M., Mossier, B., Fuchs, K., and Sieghart, W., 1994, γ-aminobutyric acid_A receptors displaying association of γ3-subunits with β2/3 and different α-subunits exhibit unique pharmacological properties, *J. Biol. Chem.* **269:**12993–12998.

Unwin, N., 1993, Neurotransmitter action: Opening of a ligand-gated channel, *Neuron Suppl.* **10:**31–41.

Valdes, F., Daschiff, M. D., Birmingham, F., Crutcher, K. A., and McNamara, J. O., 1982, Benzodiazepine receptor increases after repeated seizures: Evidence for localization to dentate granule cells, *Proc. Natl. Acad. Sci. USA* **79:**193–197.

Varecka, L., Wu, C. H., Rotter, A., and Frostholm, A., 1994, GABA_A-benzodiazepine receptor α6 subunit mRNA in granule cells of the cerebellar cortex and cochlear nuclei: Expression in developing and mutant mice, *J. Comp. Neurol.* **339:**341–352.

Wafford, K. A., Burnett, D. M., Leidenheimer, N. J., Burt, D. R., Wand, J. B., Kofuji, P., Dunwiddie, T. V., Harris, R. A., and Sikela, J. M., 1991, Ethanol sensitivity of the GABA_A receptor expressed in Xenopus oocytes requires 8 amino acids contained in the γ2L subunit, *Neuron* **7:**27–33.

Wafford, K. A., Whiting, P. J., and Kemp, J. A., 1993, Differences in affinity and efficacy of benzodiazepine receptor ligands at recombinant γ-aminobutyric acid receptor subtypes, *Mol. Pharmacol.* **43:**240–244.

Wafford, K. A., Bain, C. J., Quirk, K., McKernan, R. M., Wingrove, P. B., Whiting, P. J., and Kemp, J. A., 1994, A novel allosteric modulatory site on the GABA_A receptor β subunit, *Neuron* **12:**775–782.

Walker, F. O., Young, A. B., Penney, J. B., Dovorini-Zis, K., and Shoulson, I., 1984, Benzodiazepine and GABA receptors in early Huntington's disease, *Neurology* **34:**1237–1240.

Wee, E. L., and Zimmerman, E. F., 1983, Involvement of GABA in palate morphogenesis and its relation to diazepam teratogenesis in two mouse strains, *Teratology* **28:**15–22.

Whitehouse, P. J., Trifeletti, R. R., Jones, B. E., Folslein, S., Price, D. L., Snyder, S. H., and Kuhar, M. J., 1985,

Neurotransmitter alterations in Huntington's disease: Autoradiographic and homogenate studies with special reference to benzodiazepine receptor complexes, *Ann. Neurol.* **18:**202–210.

Whiting, P., McKernan, R. M., and Iversen, L. L., 1990, Another mechanism for creating diversity in γ-aminobutyrate type A receptors: RNA splicing directs expression of two forms of γ2-subunit, one of which contains a protein kinase C phosphorylation site, *Proc. Natl. Acad. Sci. USA* **87:**9966–9970.

Wieland, H. A., Luddens, H., and Seeburg, P. H., 1992, A single histidine in GABA$_A$ receptors is essential for benzodiazepine agonist binding, *J. Biol. Chem.* **267:**1426–1429.

Wilson-Shaw, D., Robinson, M., Gambarana, C., Siegel, R. E., and Sikela, J. M., 1991, A novel γ subunit of the GABA$_A$ receptor identified using the polymerase chain reaction, *FEBS Lett.* **284:**211–215.

Windgrove, P. B., Wafford, K. A., Bain, C., and Whiting, P. J., 1994, The modulatory action of loreclezole at the γ-aminobutyric acid type A receptor is determined by a single amino acid in the β2 and β3 subunit, *Proc. Natl. Acad. Sci. USA* **91:**4569–4573.

Wisden, W., and Seeburg, P. H., 1992, GABA$_A$ receptor channels: From subunits to functional entities, *Curr. Opin. Neurobiol.* **2:**263–269.

Wisden, W., Herb, A., Wieland, H., Keinanen, K., Luddens, H., and Seeburg, P. H., 1991, Cloning, pharmacological characteristics and expression pattern of the rat GABA$_A$ receptor α4 subunit, *FEBS Lett.* **289:**227–230.

Wisden, W., Laurie, D. J., Monyer, H., and Seeburg, P. H., 1992, The distribution of 13 GABA$_A$ receptor subunit mRNAs in the rat brain. I. Telencephalon, diencephalon, mesencephalon, *J. Neurosci.* **12:**1040–1062.

Wolf, H. K., Spanle, M., Muller, M. B., Elger, C. E., Schramm, J., and Wiestler, O. D., 1994, Hippocampal loss of the GABA$_A$ receptor α1 subunit in patients with chronic pharmacoresistant epilepsies, *Acta Neuropathol.* **88:**313–319.

Wong, G., and Skolnik, P., 1992, Ro 15-4513 binding to GABA$_A$ receptors: Subunit composition determines ligand efficacy, *Pharmacol. Biochem. Behav.* **42:**107–110.

Woodward, R. M., Polenzani, L., and Miledi, R., 1993, Characterization of bicuculline/baclofen-insensitive (ρ-like) γ-aminobutyric acid receptors expressed in Xenopus oocytes. II. Pharmacology of γ-aminobutyric acid$_A$ and γ-aminobutyric acid$_B$ receptor agonists and antagonists, *Mol. Pharmacol.* **43:**609–625.

Zaman, S. H., Shingai, R., Harvey, R. J., Darlison, M. G., and Barnard, E. A., 1992, Effects of subunit types of the recombinant GABA$_A$ receptor on the response to a neurosteroid, *Eur. J. Pharmacol.* **225:**321–330.

CHAPTER 4

STRUCTURE AND REGULATION OF THE AMILORIDE-SENSITIVE EPITHELIAL SODIUM CHANNEL

PASCAL BARBRY and MICHEL LAZDUNSKI

1. INTRODUCTION

In high-resistance epithelia such as distal segments of the kidney tubule, distal colon, urinary bladder, skin, or airways, coupling of passive electrodiffusion of sodium through the apical membrane with active extrusion of intracellular sodium by $Na^+/K^+/ATPases$ present in the basolateral membrane generates active Na^+ reabsorption via a transcellular pathway (Palmer, 1992; Eaton and Hamilton, 1988; Garty and Benos, 1986). The rate-limiting factor for sodium reabsorption by the epithelial lining is formed by an apical Na^+ channel that can be blocked by the diuretic molecules triamterene and amiloride (Eigler and Crabbé, 1969). Different hormones control the activity of this channel. For instance, aldosterone regulates the sodium balance, blood volume, and blood pressure by stimulating channel activity in kidney and in the distal colon (Garty and Benos, 1986; Rossier and Palmer, 1992); similarly, glucocorticoids increase the reabsorption of sodium in lung, thereby regulating the amount of respiratory fluid (O'Brodovich, 1991; Champigny et al., 1994). In these different tissues, the epithelial Na^+ channel represents a major component of the total ionic permeability of the apical membrane.

Although the functional properties of the apical Na^+ channel have been extensively analyzed, its molecular structure has been elucidated only recently (Canessa et al., 1993, 1994b; Lingueglia et al., 1993b, 1994). Three homologous proteins 650–700 residues

PASCAL BARBRY and MICHEL LAZDUNSKI • Institute of Molecular and Cellular Pharmacology, CNRS, 06560 Valbonne, France.

Ion Channels, Volume 4, edited by Toshio Narahashi, Plenum Press, New York, 1996.

long are involved in the function of the channel. The absence of significant homology with other known ionic channel proteins, together with the existence of some conservations between its three subunits and proteins found in the nematode *Caenorhabditis elegans*, have suggested that they may constitute the first known members of a new gene superfamily. Recent identifications of other homologues demonstrate that this is indeed the case (Lingueglia *et al.*, 1995; Waldmann *et al.*, 1995b).

The electrophysiological description of a large number of distinct cationic channels which share certain common biophysical, pharmacological, or physiological properties with the amiloride-sensitive Na^+ channel raises the possibility of structural links among these different functional entities. Since several recent reviews (Palmer, 1992, Benos *et al.*, 1995) have discussed in detail the properties of these channels, the present paper will review the molecular properties of the highly Na^+-selective and highly amiloride-sensitive Na^+ channel in terms of its electrophysiological properties, in light of its recently elucidated structure.

Numerous studies have used radiolabeled amiloride derivatives to titrate the amiloride-sensitive Na^+ channel and to elucidate its structure (Benos *et al.*, 1986; Barbry *et al.*, 1987; Kleyman *et al.*, 1986). However, molecular cloning of several proteins has revealed the existence of high-affinity amiloride-binding proteins that are not necessarily involved in Na^+ transport. Their properties are described at the end of this paper.

2.   PHARMACOLOGY

Inhibition of Na^+ transport by the diuretic molecules triamterene and amiloride is the cornerstone of epithelial Na^+ channel studies. The diuretic properties of the two molecules were explained by the blockade of Na^+ permeability at the apical membrane of epithelial tissues (Bentley, 1968; Eigler and Crabbé, 1969). Triamterene, identified in 1954, was found to have a mild diuretic effect on the distal segments of rabbit nephron (Gross and Kokko, 1977). Amiloride was discovered in 1964 after screening for non-steroidal saliuretic agents with antikaliuretic properties in rat (Cragoe, 1979). Analysis of more than 20,000 compounds led to the synthesis of 3,5-diamino-N-(aminoimino-methyl)-6-chloropyrazinecarboxamide (for structure of the molecule, see Table I), which proved to be 5 to 20 times more potent that spironolactone (an antagonist of the mineralocorticoid receptor) and 20 times more potent than triamterene. The channel's sensitivity to these two diuretics, its insensitivity to the effectors of the voltage-dependent Na^+ channel found in excitable tissues, together with distinct biophysical properties demonstrated that apical Na^+ channels were not related to voltage-dependent Na^+ channels. The development of a large amiloride pharmacology and the high affinity of the channel for this molecule led to the extensive use of amiloride (reviewed by Kleyman and Cragoe, 1988). An acid-base equilibrium exists between the positively charged amine and the undissociated base of the guanidinium moiety ($pK_a = 8.7$). The positively charged form of the molecule rapidly and reversibly inhibits the Na^+ channel (Cuthbert, 1976), whereas the uncharged form is able to penetrate the biological membranes and enter the cytoplasm (Benos *et al.*, 1983), where its accumulation can lead to several nonspecific effects (see Section 8). Amiloride analogs inhibit several distinct transmembrane Na^+ transport systems: 1) the epithelial Na^+ channel, 2) the Na^+/H^+ exchange system, and 3) the Na^+/Ca^{2+} exchange system (reviewed by Frelin *et al.*, 1987). Structure-activity relationships using amiloride derivatives with selected modification of each of the functional groups of the molecule indicate that the three Na^+ transporting systems have distinct pharmacological profiles. Amiloride derivatives that are substituted on the guanidino moiety, such as phenamil or benzamil, are potent inhibitors of the epithelial Na^+ channel (with IC_{50}'s for inhibition of about 10 nM). Conversely, 5-N-disubstituted derivatives of amiloride, such as ethylisopropylamiloride, are the most potent inhibitors of the ubiquitous isoform of the Na^+/H^+ exchanger (Frelin *et al.*, 1987). The Na^+ channel is not sensitive to these derivatives. The Na^+/Ca^{2+} exchange system is poorly inhibited by amiloride, but some amiloride derivatives that are substituted on the guanidino moiety, such as dichlorobenzamil, inhibit it, although with a low affinity ($K_{0.5} \approx 10$ µM, Kaczarowski *et al.*, 1985). It is thus theoretically possible to selectively inhibit one or the other of the Na^+ transport systems by using specific amiloride derivatives. However, some recently characterized isoforms of the Na^+/H^+ antiporter are poorly inhibited by both amiloride and its 5-N-disubstituted derivatives (Tse *et al.*, 1993; Counillon *et al.*, 1993), suggesting that the inhibition (or absence of inhibition) of a Na^+ transport by a given amiloride derivative may not be sufficient to identify the correct Na^+ pathway.

A low concentration of amiloride induces a rapid and reversible inhibition of the Na^+ channel ($K_{0.5} \approx 100$ nM for the Na^+ channel expressed in *Xenopus* oocytes; see Table I), but has virtually no effect on the other Na^+ transport systems ($K_{0.5} \approx 10$ µM for the Na^+/H^+ antiporter; $K_{0.5} \geqslant 1$ mM for the Na^+/Ca^{2+} exchanger; see Table I). The mechanism of blockade of Na^+ permeation through the channel has been analyzed by Li *et al.* (1985, 1987) in the skin of the frog *Rana ridibunda*. These investigators proposed that the interaction between the positively charged side chain of the molecule and the channel, followed by its stabilization by an interaction between the chloride at the 6-position and another site, might lead to plugging of the channel. The existence of voltage dependence in amiloride blocking kinetics suggests that the amiloride molecule, which is positively charged, indeed penetrates the ion pore from the luminal side and binds at a site where it senses a part of the transmembrane electric field (Hamilton and Eaton, 1985a). Accordingly, Marunaka and Eaton (1988) observed that inhibition of the channel by an uncharged amiloride derivative was voltage-independent. This suggests that the molecule itself, rather than a conformational modification of the channel, senses the voltage. Amiloride treatment of an active channel induced more frequent closures and openings of the channel and a reduction in the mean open time (Hamilton and Eaton, 1985a; Palmer and Frindt, 1986).

Importantly, amiloride also interacts with many other proteins which are not involved in Na^+ transport (Kleyman and Cragoe, 1988). Although it is usually possible to discriminate Na^+ transport through the channel, through the Na^+/H^+ exchanger, or through the Na^+/Ca^{2+} exchanger by using a specific amiloride derivative, titration by so-called "specific" radiolabeled derivatives of amiloride would probably lead to the titration of one of these other amiloride binding proteins, usually much more abundant. The pharmacology of the epithelial Na^+ channel has therefore been established in several different tissues by measuring electrogenic Na^+ transport in the presence of

TABLE I

Binding Properties of Amiloride Derivatives on Na^+ Permeable Channels, Na^+/H^+ Exchangers, Na^+/Ca^{2+} Exchangers, L-type Ca^{2+} Channel, and Diamine Oxidase

	Phenamil	Benzamil	Dichlorobenzamil	I-NMBA
R_1	$-NH_2$	$-NH_2$	$-NH_2$	$-NH_2$
Na$^+$ channel				
In vivo	15nM	38nM	136nM	>10μM
Expression into oocytes of cloned subunits				
αRCNaCh	44nM			
αHLNaCh	50nM			
α + β + γRCNaCh		14nM		
Pig thyroid Na$^+$ channel	47nM			
Rat brain microvessel cationic channel		15nM		
Mouse cochlear transducer channel	12μM	5.5μM		1.8μM
Xenopus oocyte mechanosensitive channel		94μM		
Na$^+$/H$^+$ antiporter				
In vivo	≫100μM	>100μM	>100μM	
Expression into transfected cells of cloned subunits				
NHE1				
NHE2				
NHE3				
Na$^+$/Ca^{2+} exchanger	65μM	100μM	12μM	18μM
Ca^{2+} channel, L-type		25μM		4μM (NMBA)
Amiloride binding protein (diamine oxidase, E.C.1.4.3.6)				
[14C]putrescine oxidase activity	33nM	103nM		
[3H]phenamil binding				
Long form	10nM	43nM		
Short form	68nM	6.3μM		

TABLE I
(Continued)

$-NH_2$	$-N(CH_3)_2$	$-N\big\langle{}^{CH_3}_{(CH_2)_2CH_3}$	$-N\big\langle{}^{CH_2CH_3}_{CH(CH_3)_2}$	$-N\bigcirc$	
H−	H−	H−	H−	H−	
Amiloride	DMA	MPA	EIPA	HMA	Reference
340nM			>10μM		Kleyman and Cragoe, 1987
620nM			>300μM		Lingueglia et al., 1994
80nM			>10μM		Voilley et al., 1994
104nM					Canessa et al., 1994b
150nM			>10μM		Verier et al., 1989
	10μM				Vigne et al., 1989
53μM	41μM			4.3μM	Rüsch et al., 1994
500μM	357μM			34μM (Br-HMA)	Lane et al., 1991
3–84μM	0.04–7μM	20–240nM	7–400nM	670nM	Frelin et al., 1987; Kleyman and Cragoe, 1988 Counillon et al., 1993
3μM	100nM	80nM			
3μM	700nM	500nM			
100μM	11μM	10μM			
1100μM	550μM		129μM		Kaczorowski et al., 1995
100μM	33μM				Garcia et al., 1990
2μM	2μM	540nM	3.4μM		Novotny et al., 1994 Lingueglia et al., 1993b
1.4μM	1.6μM		1.6μM		
3.2μM					

amiloride derivatives, rather than by titration of the amiloride binding sites by radio-labeled amiloride derivatives. Inhibition of a short-circuit current in intact tissue (Eigler and Crabbé, 1969), of a current recorded by the patch-clamp technique (Vigne et al., 1989), or $^{22}Na^+$ uptake into intact cells (Verrier et al., 1989) or into membrane vesicles (Garty et al., 1983) have been used to determine the sensitivity of channels toward amiloride and its derivatives. In toad urinary bladder (Asher et al., 1987) or in rat distal colon (Bridges et al., 1989), measurements of short-circuit current and $^{22}Na^+$ uptake have provided similar results for a pathway with a high amiloride affinity, demonstrating that this is the route for active Na^+ reabsorption. A lower affinity pathway has also been identified in these two tissues (Asher et al., 1987; Bridges et al., 1989) and may correspond to other electrogenic Na^+ reabsorptive pathways, characterized by a distinct sensitivity to amiloride and its derivatives. The different pharmacological data obtained for Na^+-permeable channels, including the stretch-activated channel from Xenopus oocytes (Lane et al., 1991) and the transducer channel from mouse cochlea (Rüsch et al., 1994), or for other ion channels such as the L-type voltage-dependent Ca^{2+} channel (Garcia et al., 1990) are summarized in Table I.

In summary, the amiloride-sensitive Na^+ channel involved in active Na^+ transport in tight epithelia is characterized pharmacologically by its high sensitivity to amiloride, phenamil, and benzamil and by its insensitivity to 5-N-disubstituted amiloride derivatives such as ethylisopropylamiloride.

3. BIOPHYSICAL PROPERTIES

3.1. Macroscopic Measurements

Transepithelial Na^+ transport was first examined in frog skin and in toad urinary bladder using the short-circuit technique developed by Kœfœd-Johnsen and Ussing (1958). The amiloride-sensitive short-circuit current obtained when the epithelium is bathed on both sides in an identical medium and the transepithelial potential is clamped to 0 mV has permitted comparison of the net Na^+ influx in many different epithelia (Table II). Analysis of the transepithelial current noise induced by submaximal concentrations of amiloride has suggested that the apical Na^+ entry pathway is indeed an ionic channel characterized by a single-channel conductance value of 5.5 pS in frog skin (Lindemann and Van Driessche, 1977; Van Driessche and Lindemann, 1979) and 3.8 pS in rabbit colon (Zeiske et al., 1982). The selectivity sequence of the channel corresponds to Eisenmann's sequence XI (Eisenmann and Horn, 1983): $P_{Li^+} > P_{Na^+} \gg P_{K^+}$, with H^+ even more permeant than Li^+ (Palmer, 1990). Other experimental approaches have also clarified the role of amiloride-sensitive Na^+ channel in important physiological processes. For instance, by measuring the potential across the blastocyst trophectoderm, Powers et al. (1977) have shown that an amiloride-sensitive active Na^+ transport is responsible for generating the filling of the blastula cavity.

3.2. Patch-Clamp Studies

Hamilton and Eaton (1985a) were the first to successfully record single amiloride-sensitive Na^+ channels in the amphibian kidney cell line A6. Channels of 5 pS conduc-

TABLE II
Na+ Channel Densities in Different Epithelia[a]

Organism	Epithelium	Conditions	Na+ transport rate (μA/ cm^2)	Current	Reference
Human	Nasal epithelium	Normal	67	I_{Na^+}	Boucher et al., 1986
		Cystic fibrosis	150	I_{Na^+}	
	Trachea		28	I_{Na^+}	Yamaya et al., 1992
	Bronchi		49.6	I_{Na^+}	Knowles et al., 1984
	Sweat duct	Normal	990	I_{Eq}	Bijman and Frömter,
		Cystic fibrosis	769	I_{Eq}	1986
Rat	Distal colon	Control	0	I_{Na^+}	Lingueglia et al.,
		Low sodium diet	410	I_{Na^+}	1994
	Trachea		19	I_{Na^+}	Legris et al., 1982
	Cortical collect-	Control	6	Net flux	Schafer and Hawk,
	ing tubule	Vasopressin	58	Net flux	1992
		DOCA	80	Net flux	
		DOCA+vasopressin	410	Net flux	
Mouse	Nasal epithelium	Normal	11	I_{Na^+}	Grubb et al., 1994a
		CF mouse	46.2	I_{Na^+}	
	Trachea	Normal	8	I_{Na^+}	Grubb et al., 1994b
		CF mouse	11	I_{Na^+}	
	Inner medullary	Control	0.2	I_{Na^+}	Kizer et al., 1995
	collecting	Aldosterone	4.6	I_{Na^+}	
	duct (culture)				
Rabbit	Distal colon	Control	28	I_{Na^+}	Turnheim et al., 1986
		Low sodium diet	92	I_{Na^+}	
	Cortical collect-	Control	91	Net flux	Schwartz and Burg,
	ing tubule	DOCA	246	Net flux	1978
Chicken	Distal colon	Control	7	I_{Na^+}	Árnason and
		Low sodium diet	228	I_{Na^+}	Skadhauge, 1991
	Coprodeum	Control	0	I_{Na^+}	Árnason and
		Low sodium diet	413	I_{Na^+}	Skadhauge, 1991
Amphibian	Frog lung	Control	12	I_{Na^+}	Fischer and Clauss,
		Aldosterone	23	I_{Na^+}	1990
	A6 cells derived	Control	1.4	I_{SC}	Handler et al., 1981
	from Xenopus	Aldosterone	4	I_{SC}	
	kidney				
	Toad urinary	Control	6.5	I_{Na^+}	Palmer et al., 1982
	bladder	Aldosterone	15	I_{Na^+}	
	Toad skin	Control	16	I_{SC}	
		Aldosterone	43	I_{SC}	
Leech	Integument		33	I_{Na^+}	Weber et al., 1995
Crab	Gill		196	I_{Na^+}	Zeiske et al., 1992

[a]Na+ transport rate (expressed in μA/cm^2) has been evaluated as either amiloride-sensitive short-circuit current (I_{Na^+}); amiloride-sensitive equivalent current (I_{Eq}); current density calculated from net ^{22}Na+ flux per tubule length, assuming an inner diameter of 20 μm (net flux); or short-circuit current (I_{SC}). Vasopressin, tubule microperfused with vasopressin; DOCA, animal injected with desoxycorticosterone; CF mouse, murine model for cystic fibrosis, obtained after knock-out of the murine CFTR gene.

tance, a high Na^+ over K^+ selectivity, a high amiloride sensitivity ($K_{0.5} \sim 500$ nM), a mean open time of ~20 msec, and a mean closed time of ~ 280 msec were recorded when A6 cells were grown on a permeable support. The channel was not present when cells were grown on plastic, but Hamilton and Eaton (1985a,b) observed in such cases another cationic channel that was also sensitive to low concentrations of amiloride, but characterized by a lower Na^+ selectivity ($P_{Na^+}/P_{K^+} \sim 3$–4:1) and faster opening and closure kinetics ($\langle t_o \rangle \sim 50$ msec).

Sodium ion channels identified by Palmer and Frindt (1986) in the apical membrane of the rat cortical collecting tubule had a conductance of 5 pS with 140 mM NaCl in the pipette (cell-attached configuration). The conductance was a saturable function of external Na^+, with a maximal value of about 8 pS and a half saturation of about 75 mM Na^+. In excised inside-out patches, the selectivity of the channels for Na^+ over K^+ was estimated from reversal potentials to be at least 10:1. Open and closed states had mean lifetimes of 3–4 seconds. A direct interaction between the channel and H^+ ions was observed by Palmer and Frindt (1987), i.e., activity increased at alkaline pH whereas indirect regulation by internal Ca^{2+} was noted by these authors (see Section 7).

Electrophysiological measurements on fetal rat lung epithelial cells in primary cultures demonstrated the presence of an apparently identical Na^+ channel that is inhibited by amiloride ($K_{0.5} = 90$ nM) and some of its derivatives such as phenamil ($K_{0.5} = 19$ nM) and benzamil ($K_{0.5} = 14$ nM), but not by ethylisopropylamiloride (Voilley et al., 1994). An amiloride-sensitive Na^+ channel of 4 pS was recorded from outside-out patches excised from the apical membrane. This channel is highly selective for Na^+ ($P_{Na^+}/P_{K^+} \geqslant 10$).

Gögelein and Greger (1986) reported the presence of an amiloride-sensitive Na^+ channel ($P_{Na^+}/P_{K^+} > 19$) in the straight portion of the rabbit proximal tubule. The unitary conductance was 12 pS (at 37°C), with I-V curves being linear between -50 mV and $+50$ mV. Addition of 1 mM amiloride to the cytoplasmic side of the patch caused a complete block of the channel by reducing the mean open time of the channel. The high concentration of amiloride needed to block this channel is probably explained by the protocol of amiloride addition (i.e., to the cytosolic side of the patch). Since Frindt et al. (1993) have reported a doubling of the unitary conductance of the Na^+ channel from rat cortical collecting tubules between 25°C and 37°C, it might be that the channels described by Gögelein and Greger (1986) and by Palmer and Frindt (1986) correspond to the same molecular entity.

Frings et al. (1988) used the patch-clamp technique to investigate the activity of ion channels in the intact epithelium of the toad (Bufo marinus) urinary bladder, a classical tissue model of active Na^+ transport (Rossier, 1978; Palmer et al., 1982). Single-channel recordings revealed the activity of a highly Na^+-selective, amiloride-sensitive channel with a mean conductance of 4.8 pS. While this was the predominant channel, other channels with the same conductance but a different selectivity or amiloride sensitivity and channels with a different conductance were also recorded.

In confluent M-1 cells derived from a culture of murine cortical collecting duct cells, the apical amiloride-sensitive conductance was characterized by a high Na^+ to K^+ selectivity ratio (~15) (Korbmacher et al., 1993). Blockade of this conductance led to a

23-mV hyperpolarization of the cell membrane. Similar whole-cell properties have been described for a channel found in granular duct cells of mouse mandibular glands (Dinudom et al., 1993). In the inner ear, the endocochlear fluid is high in K^+ and low in Na^+. The marginal cells in the stria vascularis are considered to be responsible for producing the endocochlear potential and for secreting the endolymphatic fluid. An amiloride-sensitive Na^+-selective channel has been characterized at the apical membrane of these cells (Iwasa et al., 1994). This channel might be involved in the control of the Na^+ concentration in the endolymph (Ferrary et al., 1989).

The different channels described so far have in common their expression in epithelial tissues, their high selectivity for Na^+ over K^+, and their sensitivity to low concentrations of amiloride; unitary currents are characterized by a low conductance and by slow kinetics. Channels sharing most or some of these properties have been characterized in other tissue types.

In taste buds, a channel with a high selectivity for Na^+ and H^+ and a high amiloride sensitivity ($K_{0.5} = 200$ nM) has been identified (Avenet and Lindemann, 1991; Gilbertson et al., 1992). This channel, which is involved in salt and acid perception, shares similar amiloride analogue sensitivities (Schiffman et al., 1990) and ion selectivity (Heck et al., 1989) with the channel found in Na^+-reabsorbing epithelia, supporting the idea that it corresponds to the same entity.

Other amiloride-sensitive cationic channels have also been characterized. In human B lymphoid cells, whole-cell recordings have identified a highly Na^+-selective current which can be blocked by 2 μM amiloride and by 100 nM benzamil, i.e., pharmacological properties that are expected for epithelial Na^+ channels (Bubien and Warnock, 1993). The unitary properties of this channel were not reported, but the properties of the whole-cell currents suggest that the same channel would be expressed in epithelial tissues and in blood cells. Gaspar et al. (1992) have reported that 0.25–3 mM bretylium tosylate induced in human lymphocytes a sodium-dependent, amiloride-sensitive transient inward current reaching its maximum value approximately 20–30 sec after the administration of the drug and lasting 6–10 min. This current was activated by depolarization within 25 msec at around -42 mV, its inactivation took about 2 sec, and its reversal potential was 24 ± 5 mV.

Using the patch-clamp technique, Negulyaev and Vedernikova (1994) have described highly selective, non-voltage-dependent Na^+ channels in membranes of rat peritoneal macrophages. At 23°C, the channel had a unitary conductance of 10.2 pS in a 145 mM NaCl medium and 3.9 pS in a 145 mM LiCl medium. This channel was reversibly inhibited by external application of amiloride ($K_{0.5} = 0.87$ mM). The channel identified in the rat macrophage differs from the classical Na^+ channel in its strong insensitivity to amiloride and in its selectivity for alkali ($Na^+ > Li^+ \gg K^+$). The same channel was previously described by Van Renterghem and Lazdunski (1991) in primary cultures of rat vascular smooth muscle cells: 100 μM amiloride was unable to block the channel, whereas 50 μM phenamil prevented long openings of the channel. As for the Na^+ channel in rat cortical collecting tubules, large fluctuations were observed in the open probability (0.05–0.88). Slight modifications in the amiloride-sensitive Na^+ channel structure, resulting in transformation of the Eisenmann selectivity from sequence XI to sequence X (Eisenmann and Horn, 1983), together with an alteration of the amiloride binding site, may explain the properties of the channels described by Van Renterghem and Lazdunski (1993) and Negulyaev and Vedernikova (1994). Screening of the cells where this channel has been described with newly developed molecular probes raised against the apical Na^+ channel should make it possible to test for the presence of "epithelial-like," amiloride-insensitive channels in nonepithelial tissues.

In *Helix aspersa* neurons (Green *et al.*, 1994), as well as in *Aplysia californica* bursting (Ruben *et al.*, 1986) and motor (Belkin *et al.*, 1993) neurons, Phe-Met-Arg-Phe-NH_2 (FMRFamide) and structurally related peptides induce a fast excitatory depolarizing response due to direct activation of an amiloride-sensitive Na^+ channel (Green *et al.*, 1994). This current was carried through amiloride-sensitive, but tetrodotoxin- or lignocaine-insensitive, Na^+-selective channels. Gating of the channel was explained by direct binding of the peptide FMRFamide to the channel. The small unitary conductance (6.6 pS in 100 mM Na^+), together with the sensitivity of the channel to amiloride, and the high selectivity for Na^+ over K^+, suggest that these ionic channels may be structurally related to the amiloride-sensitive Na^+-selective channel from epithelial tissues (see Section 5.2).

Moderately or nonselective Na^+ channels are characterized by P_{Na^+}/P_{K^+} ratios below 6 and by larger unitary conductances (6.5–28 pS). They are sensitive to micromolar concentrations of amiloride (0.1 μM < $K_{0.5}$ < 10 μM). Hamilton and Eaton (1985b) reported the presence of a 7–10 pS amiloride-sensitive cationic channel characterized by a permeability ratio for Na^+ over K^+ of 3–5 in A6 cells. This channel coexisted with the highly Na^+-selective channel, but their respective amounts depended on culture conditions; when cells were grown on plastic, the moderately selective channel was predominant, whereas the reverse was observed when cells were grown on a permeable support. A similar 8–10 pS channel was reported by Cantiello *et al.* (1989) in A6 cells and by Ling *et al.* (1991) in primary cultures of rabbit cortical collecting tubule (P_{Na^+}/P_{K^+} = 5). A 15-pS channel was reported by Joris *et al.* (1989) in primary cultures of human sweat gland cells (P_{Na^+}/P_{K^+} = 3). In primary cultures of human nasal epithelium, Chinet *et al.* (1993) reported the presence of a cationic channel characterized by a higher unitary conductance (21.4 ± 1.5 pS in a 140 mM NaCl external solution). A permeability ratio P_{na^+}/P_{K^+} of 2–6 was evaluated by measurements of the inversion potential.

The last category of amiloride-sensitive cationic channels, described in primary cultures of blood-brain barrier cells (Vigne *et al.*, 1989), thyrocytes (Verrier *et al.*, 1989), type II pneumocytes (MacGregor *et al.*, 1994; Yue *et al.*, 1993; Orser *et al.*, 1991; Feng *et al.*, 1993), and inner medullary collecting tubule cells (Light *et al.*, 1988), is characterized by a low selectivity for Na^+ over K^+ (pNa^+/pK^+ ~ 1). Sensitivity to amiloride or its derivatives has been studied by electrophysiology in blood-brain barrier cells, where increasing concentrations of phenamil and amiloride decrease the open probability of the channel in a dose-dependent fashion (Vigne *et al.*, 1989). The concentrations required for half-maximal inhibition in 140 mM NaCl were 15 nM for phenamil and 10 μM for amiloride. The pharmacology of a Na^+ channel found in thyroid cells was established by $^{22}Na^+$ uptake experiments and appeared very similar to that reported in toad urinary bladder (Asher *et al.*, 1987), distal colon (Bridges *et al.*, 1989), or fetal rat lung epithelial cells (Voilley *et al.*, 1994). Surprisingly, the channel that was recorded during outside-out patch-clamp experiments was not selective for Na^+ over K^+ ($pNa^+/pK^+ = 1.2$) and had a very low conductance (2.6 pS in 140 mM NaCl at 25°C). Since whole-cell measurements have demonstrated the hyperpolarizing effects of micromolar concentrations of phenamil, i.e., that phenamil-sensitive conductance is selective for Na^+, two types of channels may have been present in the thyrocyte membrane: one that is prominent in $^{22}Na^+$ uptake experiments and in whole-cell recordings and corresponds to the classical Na^+ channel, and a second type which is observed during outside-out patch-clamp recording. A similar situation has been described by Matalon *et al.* (1993) in primary cultures of fetal lung epithelial cells, where they identified two different types of amiloride-sensitive channels. In rat inner medullary collecting tubules, a 28-pS cationic channel with short open and closed times (i.e., < 50 msec) was characterized by Light *et al.* (1988). The activity of this channel was inhibited by the atrial natriuretic factor (ANF) (Light *et al.*, 1989b), via cyclic GMP–dependent regulatory pathways (Light *et al.*, 1990), and was activated by the α_{i-3} subunit of the heterotrimeric G protein (Light *et al.*, 1989a).

In *Xenopus* oocytes, Lane *et al.* (1991) reported the presence of an amiloride-blockable mechanosensitive channel. The channel was inwardly rectified and weakly selective for Na^+, K^+, and Ca^{2+}. The amiloride derivative bromohexamethylene-amiloride, which has no effect on the epithelial Na^+ channel, was the most potent blocker of the channel found in the amiloride family. The channel, present in cochlear hair cells and involved in mechanoelectrical transduction, may be related to this last channel (Rüsch *et al.*, 1994).

In summary, although these different channels have usually been classified in the same family because of their common sensitivity to amiloride (with inhibition constants that differ by several orders of magnitude), there is no strong evidence that their structures are indeed related. Analysis of the different reports available suggests that active Na^+ transport occurs in kidney, lung, and colon via the Na^+ channel and is characterized by a high selectivity for Na^+ and K^+, a high sensitivity to amiloride, a low unitary conductance, and slow kinetics.

3.3.   Reconstitution into Artificial Membranes

Several groups have described the reconstitution of purified or semipurified Na^+ channels (see Benos *et al.*, 1995, for a more extensive description of these results). Sariban-Sohraby *et al.* (1992) used the patch-clamp technique to study the reincorporation of purified renal bovine Na^+ channels into a lipid bilayer formed on the tip of a glass pipette. A three-step purification technique consisting of affinity chromatography on a lectin column followed by size exclusion high performance liquid chromatography led to the purification of three polypeptides of 150, 40, and 31 kDa. In about 10% of the cases, the authors observed unitary conductance jumps ranging from 1.9 to 8.5 pS in 100 mM Na_2HPO_4. Transfer into a vial containing 0.1 μM amiloride induced rapid transitions between the open and blocked conductance states. Using different purification and reconstitution protocols, Oh and Benos (1993) reported the presence of two amiloride-sensitive conductance states in purified material: a state with a maximal conductance of 56 pS and Na^+ to K^+ permeability ratio of 6.7:1 and another state with a maximal conductance of 8.4 pS and Na^+ to K^+ permeability ratio of 7.8:1. Protein kinase A increased the open probability of the two channels, whereas protein kinase C decreased it (Oh *et al.*, 1993). The stimulatory effect of protein kinase A might be related to the well-known stimulation of amiloride-sensitive Na^+ channel activity by the antidiuretic hormone vasopressin (see section 7). The existence of several amiloride-sensitive conductances is intriguing, but has been reported by Frings *et al.* (1988) during patch-clamp experiments performed on intact toad bladder epithelial cells. This might be due to the existence of distinct channels, to different conductive substrates of the same channel, or to a modification of the channel properties after reconstitution in an artificial membrane or after formation of the seal. Partial degradation of the channel by proteases found at the apical membrane of the distal kidney tubule, such as kallikreins, is another possible explanation for the change observed in the conductive properties of the channel (Lewis and Alles, 1986). Although these different studies clearly demonstrate that amiloride-sensitive channel activity can be measured in artificial membranes, the high sensitivity of the reconstitution techniques may lead to detection of Na^+ channel activity borne by minor contaminants (Barbry *et al.*, 1990b). The identification of the minimal polypeptide complex forming the Na^+ channel requires therefore more extensive investigation, including molecular cloning of each putative subunit. A protocol such as that previously reported for the cAMP-activated chloride channel CFTR (Bear *et al.*, 1992), where homogeneity of the purification was checked by microsequencing of the purified protein, warrants investigation.

4.   EXPRESSION CLONING

Several groups (George *et al.*, 1989; Palmer *et al.*, 1990; Hinton and Eaton, 1989; Asher *et al.*, 1992a,b; Kroll *et al.*, 1989, 1991) have demonstrated that the epithelial Na^+ channel can be expressed in oocytes from *Xenopus laevis* after injection of mRNAs originating from A6 cells, toad urinary bladder, or bovine trachea. After injection of 1–50 ng messenger RNA into oocytes, Na^+ channel activity was assayed either as an amiloride-sensitive current under voltage-clamp conditions or as amiloride-sensitive $^{22}Na^+$ uptake, 1–3 days after injection. No activity was detected in noninjected or water-injected oocytes. The amiloride-sensitive pathway induced by the mRNAs had the same characteristics as the native channel in terms of ionic selectivity for Na^+ over K^+, sensitivity to amiloride and to its derivatives, and saturation by external Na^+. These studies demonstrated that the Na^+ channel could be expressed into oocytes and that the system was suitable for expression cloning. However, the level of expression usually remained rather low, since currents rarely exceeded 100 nA. Following treatment with steroids, distal segments of the mammalian colon (Edmonds, 1967; Frizzell *et al.*, 1976; Foster *et al.*, 1983; Will *et al.*, 1980) or bird colon and coprodeum (Clauss *et al.*, 1987) proved to be richer sources of functional Na^+ channels. This prompted our group to use dexamethasone-treated rat distal colon as starting material (Lingueglia *et al.*, 1993a) and independently led Canessa *et al.* (1993) to use a very similar model consisting of rat distal colon recovered from animals treated for at least 10 days with a low-sodium diet. Such a diet is known to largely increase the secretion of aldosterone by adrenal glands (Kanwar and Venkatachalam, 1992). Expression of mRNAs prepared from these tissues in *Xenopus* oocytes permitted the expression of amiloride-sensitive Na^+ currents of over 500 nA, i.e., ten times higher than with other tissues. Dexamethasone-treated rat distal colon mRNAs, previously verified as having high functional expression, were used to build a cDNA expression library. The first strand cDNA was reverse-transcribed after XhoI-$(dT)_{15}$ oligonucleotide priming in order to build a directional library. The second strand was synthesized, and the EcoRI adaptor was ligated, followed by a XhoI digestion. In order to protect internal XhoI sites from digestion 5-methyl-dCTP was used rather than dCTP during first strand cDNA synthesis. Since mRNA sizing experiments had shown that the rat mRNA(s) was larger than 2 kb, only cDNAs of more than 2 kb were ligated into the vector. The vector used for the library was a plasmid modified from pGEM-5Zf (-) (Promega). The main modifications were: 1) insertion of a T7 terminator to suppress the step of plasmid linearization prior to *in vitro* transcription, 2) addition of the flanking noncoding regions of *Xenopus laevis* globin 5' and 3' to the cDNAs in order to stabilize the cRNAs in oocyte cytoplasm (Cribbs *et al.*, 1990), and 3) presence of a poly$(A)_{27}$ tract followed by a poly$(C)_{13}$ immediately after the 3' noncoding sequence of *Xenopus* globin to permit efficient translation, even in mature oocytes (Krieg and Melton, 1984; Varnum and Wormington, 1990; Fox and Wickens, 1990).

Figure 1 describes the five steps that led us to identification of the first subunit of the rat colon amiloride-sensitive Na^+ channel. The functional properties obtained after injection of the cRNA into oocytes corresponded exactly to those expected from experiments performed on the native channel: the conductance was highly Na^+-selective and highly sensitive to amiloride and phenamil, but not to ethylisopropylamiloride. The level of the current, however, appeared excessively low. While typical currents recorded from oocytes injected with 50 ng of rat colon mRNAs were several hundreds of nanoamps, the current recorded for oocytes injected with 50 ng of the pure cRNA rarely exceeded 50 nA. Removal of the noncoding regions of the cDNA (unpublished data) and expression of the human homologue (Voilley *et al.*, 1994) failed to significantly increase the current. These results suggested that other factors translated from mRNA were required to maximize channel activity.

The clone that was characterized was 3081 nucleotides long and encoded a 699–amino acid protein. (The difference in the length between the clones reported by Lingueglia *et al.* and Canessa *et al.* is explained by the choice of a first or a second adjacent ATG as the first codon, resulting in proteins 699 or 698 residues long, respectively.) This protein, termed RCNaCh (Lingueglia *et al.*, 1993b) or αrENaC (Canessa *et al.*, 1993), contains two large hydrophobic domains potentially able to span the membrane. A human lung homologue of the α subunit was subsequently characterized (Voilley *et al.*, 1994; MacDonald *et al.*, 1994). The human gene for the α subunit of the human lung Na^+ channel was mapped on chromosome 12p13. The human protein is 669 residues long, is 83% identical to that of the rat, and has two large hydrophobic domains. When the human homologue was expressed into *Xenopus* oocytes, its functional properties were identical to those of the rat protein (Voilley *et al.*, 1994).

Molecular cloning of a first subunit of the rat colon amiloride-sensitive Na^+ channel

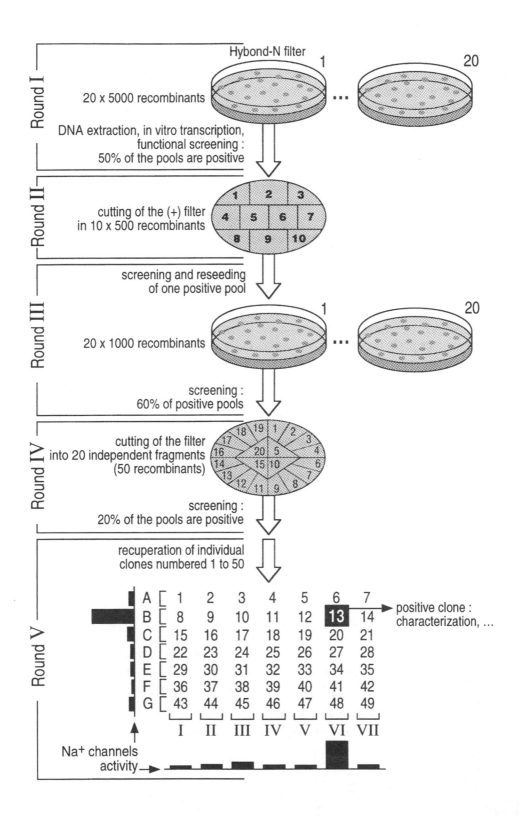

Round I

Hybond-N filter   1   20

20 x 5000 recombinants

DNA extraction, in vitro transcription,
functional screening :
50% of the pools are positive

Round II

cutting of the (+) filter
in 10 x 500 recombinants

1	2	3
4	5	6	7
8	9	10

screening and reseeding
of one positive pool

Round III

1   20

20 x 1000 recombinants

screening :
60% of positive pools

Round IV

cutting of the filter
into 20 independent fragments
(50 recombinants)

18 19 1 2 3
17 20 5 4
16 15 10 6
14 7
13 12 11 9 8

screening :
20% of the pools are positive

recuperation of individual
clones numbered 1 to 50

Round V

	1	2	3	4	5	6	7
A	1	2	3	4	5	6	7
B	8	9	10	11	12	**13**	14
C	15	16	17	18	19	20	21
D	22	23	24	25	26	27	28
E	29	30	31	32	33	34	35
F	36	37	38	39	40	41	42
G	43	44	45	46	47	48	49
	I	II	III	IV	V	VI	VII

positive clone :
characterization, ...

Na⁺ channels
activity

revealed the existence of a relationship between the channel and a set of proteins expressed in the mechanosensitive neurons of the nematode *Caenorhabditis elegans*. A significant sequence identity was found between the coned subunit of the Na$^+$ channel and three proteins, called degenerins, involved in mechanosensitivity in the nematode (Lingueglia *et al.*, 1993b; Canessa *et al.*, 1993; Chalfie *et al.*, 1993). Since two of these degenerins (i.e., mec-4 and mec-10) are expressed in the same cells (Mitani *et al.*, 1993; Huang and Chalfie, 1994), this suggested that homologue(s) of the first subunit might be necessary to complement Na$^+$ channel activity. Alignment of the sequences of the human and rat Na$^+$ channel proteins with the degenerin sequences revealed small stretches of identical residues (Fig. 2). Synthesis of degenerated oligonucleotides permitted amplification of a partial cDNA from rat distal colon mRNA by the polymerase chain reaction. This cDNA shared 54% similarity with the first subunit of the Na$^+$ channel. This fragment was used to screen a rat colon cDNA library, and a full length cDNA encoding a new homologous subunit was characterized (Lingueglia *et al.*, 1994). The deduced protein shared 34% identity with the α subunit and had no intrinsic activity when expressed alone in *Xenopus* oocytes, but its coexpression with the first subunit increased channel activity 18 ± 5-fold.

Canessa *et al.* (1994b) investigated the low level of expression induced by injection of the α subunit alone by a pure expression cloning approach. These investigators successfully complemented the Na$^+$ channel activity induced by expression of the α subunit by coexpressing the α subunit with different pools of a rat colon cDNA library.

←──

FIGURE 1. During the four first steps, the bacteria were grown on filters to avoid the selection of clones with a higher cell growth rate. Ø 150 mm Hybond-N filters were placed on LB-agar bacteriological plates and seeded with 5000 recombinants. Bacteria were grown overnight at 30° C. The next day, one replica was performed and grown for several hours until colonies were detected. The initial matrix was impregnated with 5% glycerol at 37° C, covered with a second filter, sealed with a plastic wrap, and stored at −70° C. The first replica was scraped with a rubber policeman and the corresponding bacteria were used for plasmid preparation. Plasmids were purified after extraction by alkaline lysis. cRNAs were transcribed *in vitro* using T7 RNA polymerase. A total of 100,000 clones, seeded on 20 Ø 150 mm filters, were analyzed during the first round. Filters were treated separately, but for convenience, the first functional screening was performed on ten pools (each pool representing two filters), designated I to X. Each pool of cRNAs was injected into *Xenopus* oocytes. Functional activity was measured as amiloride-sensitive ^{22}Na$^+$ uptake; positive pools were confirmed by measuring the sensitivity of the current measured at −70 mV to amiloride by the two-electrode voltage-clamp technique. At the end of the first round, five pools were positive for amiloride-sensitive ^{22}Na$^+$ uptake. Amiloride-sensitive channel activity was confirmed by electrophysiology for two of them. A second round of purification was performed for these two pools: their replicas were thawed, cultured at 37°C, cut in ten fragments (each containing 500 independent colonies). After this step, fraction 3a was seeded on 20 Ø 150 mm filters at a density of 1000 independent colonies per filter. Twelve out of 20 pools were positive for ^{22}Na$^+$ uptake. One of them was cut into 20 new fractions of 50 clones, numbered from 1 to 20. Pools 3, 4, 17, and 18 were positive for uptake and for electrophysiology. The 50 recombinants present on filter 3 were recovered independently. One clone was tested individually. The 49 other clones were arbitrarily put on a 7x7 matrix. This 7x7 matrix was analyzed by injecting the 14 different possible mixtures of seven clones. As expected, the clone of interest appeared in two pools that corresponded to the lane and the row of the 7x7 matrix where it was located. With this protocol, the positive clone was obtained by injecting only 15 different mixtures of cRNAs, rather than 50 different pure cRNAs. The activity of the clone was confirmed by a new injection, and the amiloride-sensitive Na$^+$ current subsequently characterized.

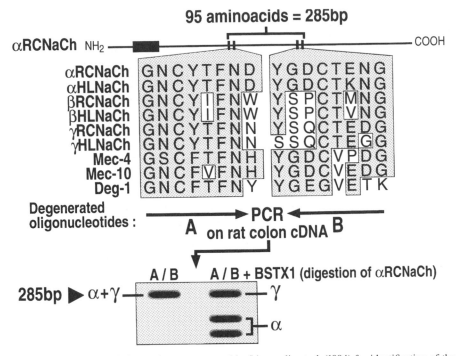

FIGURE 2. Polymerase chain reaction strategy used by Lingueglia *et al.* (1994) for identification of the γ subunit (initially called RCNaCh2), with an alignment of the human and rat Na⁺ channel subunits and of the degenerins around the segments that have been used to synthesize the degenerate oligonucleotides.

This functional complementation strategy led to identification of two homologous cDNAs, called β and γ (the latter corresponding to the second subunit cloned by Lingueglia *et al.*, termed RCNaCh2), sharing 35% and 34% identities with the α subunit, respectively. The degenerate oligonucleotides used by Lingueglia *et al.* (1994) did not match properly with the β cDNA, explaining *a posteriori* why only fragments of the α and γ subunits were amplified (Fig. 2). The ion-selective permeability, the gating properties, and the pharmacological profile of the channel formed by coexpressing the three subunits in oocytes are similar to those of the native channel (Fig. 3). The high channel activity measured by the two-electrode, voltage-clamp technique permitted Canessa *et al.* (1994b) to study the unitary properties of the channel expressed in oocytes by the patch-clamp technique. They demonstrated the presence of a voltage-independent channel selective for Na⁺ (characterized by a unitary conductance of 4.5 pS in 140 mM NaCl) and for Li⁺ (characterized by a unitary conductance of 6.5 pS in 140 mM LiCl), but not for K⁺ (Fig. 3). This channel was identical to the channels previously described in intact tissues (Hamilton and Eaton, 1985b; Palmer and Frindt, 1986; Voilley *et al.*, 1994). Nearly identical properties have been reported for the homologues cloned from other

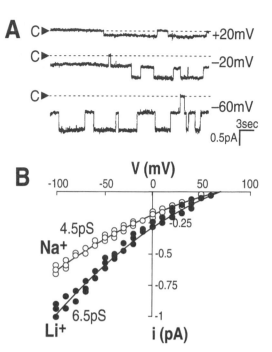

FIGURE 3. Functional expression of the amiloride-sensitive Na^+ channel in *Xenopus* oocytes after expression of the three subunits of the rat colon Na^+ channel (courtesy of Dr. Guy Champigny). (A) Single-channel properties of the amiloride-sensitive Na^+ channel. (B) Current-voltage relationships.

species. Expression in oocytes of the three Na^+ channel subunits cloned from a human lung cDNA library (Voilley *et al.*, 1994, 1995) generates a voltage-independent channel selective for Na^+ (characterized by a unitary conductance of 5.6 pS in 140 mM NaCl) and for Li^+ (characterized by a unitary conductance of 7.8 pS in 140 mM LiCl), but not for K^+ (Waldmann *et al.*, 1995b). The rat and human subunits can be readily exchanged without affecting the functional properties of the channels expressed into oocytes (Macdonald *et al.*, 1995). Molecular cloning of the amphibian epithelial Na^+ channel led Puoti *et al.* (1995) to determine the slope conductance of the channel expressed in oocytes, which was equal to 5.8 pS for Na^+ and 9.74 pS for Li^+. The highly positive reversal potential was consistent with a high selectivity for Na^+ over K^+. These different molecular clonings also revealed the existence of distinct forms of the subunits. A splice variant of the rat α chain encodes a nonfunctional truncated protein that retains some capacity to bind amiloride (Li *et al.*, 1995). In *Xenopus* kidney, several sizes of β mRNA have been detected (Puoti *et al.*, 1995). Screening of a human lung cDNA library has also revealed two forms for the β subunit, which differ by the presence or absence of a 464 bp fragment in the 3' region (Voilley *et al.*, 1995). A frameshift in the short form modifies the COOH terminal sequence of the corresponding protein. The existence of this COOH-truncated β chain might be of physiological importance, since several similar frameshifts mutations have recently been reported in patients affected by a rare form of hypertension (Shimkets *et al.*, 1994).

5. BIOCHEMICAL PROPERTIES OF THE CLONED SUBUNITS

5.1. Transmembrane Topology of the α Na$^+$ Channel Subunit

The transmembrane topology of the protein was first established by Renard *et al.* (1994) using a biochemical approach: rabbit reticulocyte lysate was used to translate cRNAs into proteins in the presence of canine pancreatic microsomal membranes. The proteins translated in this system usually adopt a correct secondary structure across the phospholipid bilayer and can even undergo full maturation permitting functional expression (Rosenberg *et al.*, 1992; Awayda *et al.*, 1995). This has been demonstrated for the *Shaker* K$^+$ channel (Rosenberg and East, 1992): after cell-free protein translation, microsomal membrane processing of nascent channel proteins, and reconstitution of newly synthesized *Shaker* K$^+$ channels into planar lipid bilayers, Rosenberg and East (1992) recorded *in vitro* the activity of functional voltage-dependent potassium channels.

The protein inserted into canine pancreatic microsomes is protected from complete degradation by proteases that do not have access to the protein fragment that is located toward the lumen of the microsomes. The principle of the experiment performed by Renard *et al.* (1994) is illustrated in Fig. 4. After treatment of the microsomes by proteinase K, these investigators observed a protected 50-kDa fragment which they identified as the large loop located between the two hydrophobic domains. Four distinct polyclonal antibodies raised against four distinct parts of the channel were used to immunoprecipitate the full-length protein expressed by *in vitro* translation. Only one of these sera, directed against an immunogenic peptide located between the two hydrophobic domains, reacted with the protease-protected 50-kDa fragment. This result,

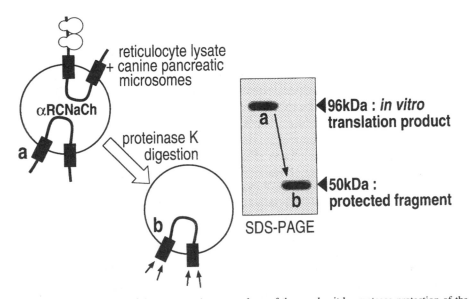

FIGURE 4. Determination of the transmembrane topology of the α subunit by protease protection of the protein after *in vitro* translation and insertion into canine pancreatic microsomes (Renard *et al.*, 1994).

together with several other findings obtained with the entire or truncated forms of the α subunit, including sensitivity to proteolytic enzymes cleaving the C-terminal part of the proteins and the size of the protease-protected fragments obtained with the different truncated forms, led to a model where the protein contains a large extracellular loop located between two transmembrane α-helices. The NH_2- and COOH-terminal domains are cytoplasmic. A cysteine-rich domain is found in the external segment, but its functional role remains unclear. A last point of interest concerns the existence of slight differences between the theoretical and experimental molecular weights, which suggest the presence of a segment in the vicinity of one of the two transmembrane α-helices that may play a role similar to the H5 segment forming the ionic pore of several voltage-dependent ionic channels (see also Chapter 1 of this volume).

Using an approach similar to the one initially developed by Rosenberg et al. (1992) (who recorded the in vitro activity of functional Shaker K^+ channels after cell-free protein translation, microsomal membrane processing of nascent channel proteins, and reconstitution into planar lipid bilayers), Awayda et al. (1995) have reconstituted into liposomes the bovine renal α subunit after an in vitro translation. Awayda et al. (1995) observed incorporation into planar lipid bilayers of voltage-independent Na^+ channel activity with a single-channel conductance of 40 pS, a high sensitivity to amiloride, and a cation selectivity ($Li^+:Na^+:K^+$ perm selectivity of 2:1:0.14). Although the unitary conductance of this channel differs from the one recorded in Xenopus oocytes after expression of the three cloned subunits, this work documents the expression of a functional ionic channel using an in vitro translation system and validates the use of this expression system for analyzing the structural properties of the Na^+ channel subunits.

Snyder et al. (1994) subsequently used the in vitro translation approach to analyze the properties of the α subunit: six of the eight putative N-glycosylation sites were effectively found to be occupied, corresponding to Asn^{190}, Asn^{259}, Asn^{320}, Asn^{339}, Asn^{424}, and Asn^{538}, i.e., the asparagine residues located in the 50-kDa extracellular fragment first identified by Renard et al. (1994). Canessa et al. (1994a) confirmed these results using chimeras between the α subunit of the Na^+ channel and the β subunit of the $Na^+/K^+/ATPase$. Mutagenesis of the six asparagine residues into glutamine had no effect on functional expression of the channel measured in oocytes.

5.2. An SH3 Binding Region Is Present in the α Na^+ Channel Subunit

A unique proline-rich sequence identified in the COOH-terminal region of the α subunit is found in the α, β, and γ subunits of the human, rat, and Xenopus Na^+ channel. Coimmunoprecipitation experiments demonstrated that this segment is responsible for binding of the α subunit to α-spectrin (Rotin et al., 1994). When microinjected into the cytoplasm of polarized rat alveolar epithelial cells, a recombinant fusion protein containing the proline-rich region of the α subunit localized exclusively to the apical area of the plasma membrane. This localization paralleled that of α-spectrin. Rotin et al. (1994) proposed that cytoplasmic interactions via SH3 domains mediate the apical localization of the Na^+ channel and would provide a novel mechanism for retaining proteins in specific membranes of polarized epithelial cells. Experiments must now be designed to determine whether insertion into the apical membrane is explained only by the interaction with cytoskeletal via SH3 binding domains or whether other factors also play a role. The functional importance of the COOH-terminal segment (including the SH3 binding domain) of the Na^+ channel β and γ subunits was recently highlighted by the demonstration that mutations in the human β and γ subunit gene are responsible for Liddle's syndrome, a rare form of hypertension (Shimkets *et al.*, 1994): mutations with insertion of a stop codon or shift of the open reading frame modified the COOH-terminal sequence of the human β and γ subunits and removed the putative SH3 binding domain. These truncated subunits increases Na^+ channel activity expressed in *Xenopus* oocytes, mainly through increase of the number of active channels (Schild *et al.*, 1995), but it is not yet demonstrated whether the SH3 binding domains play a role in this regulation.

5.3.   Cellular Specificity of Expression of the α, β, and γ Na^+ Channel Subunits

The biophysical properties of the channels recorded in rat cortical collecting ducts (Palmer and Frindt, 1986) and fetal lung epithelial cells (Voilley *et al.*, 1994) appear quite similar, although the physiological roles of the channel appear somewhat different in kidney, where it controls the amount of Na^+ to be reabsorbed by the distal nephron, and in lung, where it regulates hydration of the mucus. Functional evidence indicates that the same structure is present in these different tissues. These functional observations have been confirmed by direct detection of the three subunits with different molecular probes raised against them.

Transcripts of rat α, β, and γ subunit mRNAs have been detected by Northern blot analysis as unique bands of 3.7, 2.2, and 3.2 kb, respectively. The rat transcripts were specifically detected at a high level in epithelial tissues such as the renal cortex and medulla, distal colon, urinary bladder, lung, placenta, and salivary glands (Canessa *et al.*, 1994b; Voilley *et al.*, 1994, 1996). In the tongue, Northern blots and *in situ* hybridization analyses revealed the presence of the α subunit in the adjacent epithelial layers rather than in taste buds (Li *et al.*, 1994). Low levels of transcripts were also identified in proximal colon, uterus, thyroid, and intestine. No signal was detected by Northern blot analysis of total RNA from liver, stomach, duodenum, muscle (smooth and striated), heart, brain, or blood-brain barrier microvessels. In humans, Voilley *et al.* (1994) reported the presence of the α subunit in lung, colon, kidney, thyroid, liver, placenta, and pancreas. Some of the differences observed between species might be accounted for by a different hormonal status, since hormones such as aldosterone control the transcription of Na^+ channel subunits (see Section 6).

Li *et al.* (1994) identified expression of the α subunit in epithelial cells of colon and lung by *in situ* hybridization. The same pattern of expression was observed for α, β, and γ RNAs (Duc *et al.*, 1994). The three subunits were immunodetected by Duc *et al.* (1994) and Renard *et al.* (1995) who used specific antipeptide antibodies raised against each of the three subunits. The three subunits of the Na^+ channel were colocalized at the apical membrane of the surface epithelium from distal colon, the secretory ducts of the salivary and sweat glands, and airway epithelium. The three proteins were found in distal convoluted tubule, connecting tubule, and cortical and outer medullary collecting tubules. Duc *et al.* (1994) and Renard *et al.* (1995) provided the first evidence that all three subunits are found at the same site as the functional Na^+ channel. It is therefore likely that in these different tissues, their participation in a heteromultimeric structure fully accounts for the formation of a channel identical to that expressed in *Xenopus* oocytes. In lung, a high level of expression of the α, β, and γ subunits was observed in the airways (Renard *et al.*, 1995). No labeling was detected in lung alveoli, suggesting that the major site of expression of the Na^+ channel in lung corresponds to the airways. This conclusion would be in agreement with the very low amiloride-sensitive Na^+ absorption across the alveolar epithelium reported by Basset *et al.* (1987) and with the distinct pharmacological properties of the Na^+ conductive pathways in alveolar type II pneumocytes (Matalon *et al.*, 1991). The low level of expression of the Na^+ channel subunits in the distal epithelium would fit with a model where fluid comes from distal secretion near or at the alveoli and then moves toward airways, where active Na^+ transport prevents flooding and obstruction (Kilburn, 1968). Since Goodman and Crandall (1982) and Mason *et al.* (1982) have suggested that alveolar epithelium, at least in primary cultures, absorbs water and salt, it remains possible that an active Na^+ transport also exists in distal lung.

6. THE Na^+ CHANNEL GENE SUPERFAMILY

6.1. The Degenerins of the Nematode *Caenorhabditis elegans*

A unique feature of the nematode nervous system is the conservation of the number of cells and the interactions between a given set of neurons. For instance, the neural pathways for touch-induced movement in *C. elegans* contain six touch receptors, five pairs of interneurons, and 69 motor neurons (Chalfie *et al.*, 1985). The synaptic relationships among these cells have been deduced from reconstructions of serial section electron micrographs; the roles of the cells were assessed by examining the behavior of animals after selective killing of cell precursors by laser microsurgery. The pathways for touch-mediated movement have been investigated by genetic screening and this has led to identification of the genes required for function of the six touch receptor neurons. Eighteen different genes were identified (Chalfie and Au, 1989); they potentially encode components of the mechanosensory apparatus. Two of them, called mec-4 (Driscoll and Chalfie, 1991) and mec-10 (Huang and Chalfie, 1994), encode homologous proteins and are directly involved in the function of touch cells. The mec-4 and mec-10 genes are expressed in the same cells, although each has a unique function. The deg-1 gene, expressed in distinct cell types, is homologous to mec-4 and mec-10 (Chalfie and Wolinski, 1990). Two other *C. elegans* genes, namely mec-6, expressed specifically in the six touch cells, and unc-105, a muscle-affecting gene in which rare dominant alleles cause a hypercontracted phenotype, are also suspected as belonging to the degenerin family (Jentsch, 1994). The other mec genes may be involved in the generation, specification, or maintenance of the different cells that mediate mechanosensation (Chalfie and Au, 1989).

Degenerins are structurally related to the three subunits of the Na^+ channel (Lingueglia *et al.*, 1993b; Canessa *et al.*, 1993; Chalfie *et al.*, 1993). Degenerins contain two hydrophobic domains, two cysteine-rich regions in the extracellular portion located in positions similar to the Na^+ channel subunits. The ~ 12% identity between the three subunits of the Na^+ channel and the three degenerins mec-4, mec-10, and deg-1 is consistent with a model where degenerins are subunits of an ionic channel.

The genetic analysis of the different mutated degenerins has provided important insights about the function of these proteins. Two different kinds of mutations occur in degenerin genes, inducing gain-of-function or loss-of-function phenotypes. When mec-4, deg-1, and mec-10 genes are modified by gain-of-function mutations, the corresponding toxic gene product(s) can cause neurodegeneration that results in specific swelling of the cells in which the mutated degenerin is expressed, leading to their vacuolization and subsequent lysis (a degeneration mechanism that differs from necrosis and apoptosis; Driscoll and Chalfie, 1991). The various mutations that lead to degeneration of the small number of mechanosensory neurons have been extensively analyzed. Three dominant mutations of mec-4 that cause degeneration of the touch receptor neurons affect the alanine in position 442, which is located immediately before the second hydrophobic domain of mec-4 (Driscoll and Chalfie, 1991). *In vitro* mutagenesis analysis has shown that alanine modification by residues with a large lateral chain, such as valine, threonine, phenylalanine, aspartic acid, arginine, leucine, or proline, induces degeneration, whereas modification of the alanine by residues with a small lateral chain, such as glycine, serine, or cysteine, has no functional effect. Mutation of the equivalent alanine residue into valine in deg-1 induces degeneration of the posterior ventral cord interneurons; a similar mutation in mec-10 induces degeneration of the touch cells (Huang and Chalfie, 1994).

After introduction of the A673V mutation into mec-10, Huang and Chalfie (1994) analyzed the corresponding degenerations for different genetic backgrounds. Several loss-of-function missense mutations have been characterized in mec-10. One mec-10 missense mutation, G676R, found before the second hydrophobic domain, suppresses the mec-10–induced degeneration, as do other mutations affecting mec-4 and mec-6 genes. This suggests that mec-10, mec-4, and mec-6 belong to the same heteromultimeric structure involved in mechanosensation. Moreover, several copies of mec-4 and mec-10 may participate in formation of the active complex (Huang and Chalfie, 1994; Hong and Driscoll, 1994). A mutation (S105F) that affects a residue located before the first hydrophobic domain of mec-10 and three other mutations (L679R, G680E, and G684R) that are found near or in the second hydrophobic domain enhance mec-10–induced degeneration caused by mutations of Ala^{673}. This surprising observation may indicate that these four mutant mec-10 proteins are unable to associate, leading to the formation of a complex with only gain-of-function mec-10 mutants. Quantification of the expressed mutant proteins will be necessary to address this question directly.

Hong and Driscoll (1994) also showed that several loss-of-function mec-4 mutations modified three residues located within the second hydrophobic domain (corresponding to the missense mutations S455F, T458I, and E461K). These positions, every three residues, would be consistent with a model in which Ser^{455}, Thr^{458}, and Glu^{461} are on the same side of a transmembrane α-helix and delineate a portion of the ionic pore. These authors proposed that modification of these residues into Phe, Ileu, or Lys, respectively, induces loss of function of the system. The existence of a significant sequence conservation between the second transmembrane segment of mec-4, deg-1, and the α subunit of the Na^+ channel was used by Hong and Driscoll (1994) to study the role of the conserved amino acids in the second membrane-spanning domain in mec-4 activity. Specific substitutions in the second membrane-spanning domain, whether encoded *cis* or *trans* to the A442V mutation, block or delay the onset of degeneration. Remarkably, chimeric proteins obtained by domain swapping of the second membrane-spanning domain from mec-4 with the corresponding sequences from deg-1 or the α subunit of the Na^+ channel still participate in an intact mechanosensory function when expressed in the nematode.

Importantly, up until today, the functional properties of the degenerins have been determined solely by genetic studies, since no electrophysiological studies have yet been conducted on nematode touch cells. Direct comparison of the touch cell channel and the amiloride-sensitive Na^+ channel has thus been impossible. However, several important structural properties are shared by all the members of this new gene superfamily. They include: 1) the residues located before the second transmembrane α-helical segment (Driscoll and Chalfie, 1991), and 2) a nontypical transmembrane segment in the vicinity of the α-helical transmembrane segment of the α subunit of the Na^+ channel (Renard *et al.*, 1994). Moreover, exchange of the second hydrophobic domain of the Na^+ channel α subunit by that of mec-4 results in a functional ion channel with changed pharmacology for amiloride and benzamil and changed selectivity, conductance, gating, and voltage dependence. Similar results were obtained when Ser^{589} and Ser^{593} of the second hydrophobic domain were mutated, suggesting that these two residues are essential for the correct ion permeation through the channel (Waldmann *et al.*, 1995).

The genetic approach in *C. elegans* has already provided important insight into the properties of the proteins belonging to the same gene superfamily. Characterization of other mec genes may reveal new proteins involved in correct function of the Na^+ channel. Stoichiometry of the different degenerins in a functional complex may provide information about the quaternary structure of the epithelial Na^+ channel. It is essential however to avoid complete analogy between the two systems, which are carrying out clearly distinct physiological roles. This is revealed by the existence of segments specific to the degenerins or to the Na^+ channel proteins. This is the case for the SH3 binding domain (Rotin *et al.*, 1994), found in all Na^+ channel subunits, but not present in degenerins. Conversely, a missense mutation within an extracellular 22–amino acid region found in all the *C. elegans* degenerins, but not in the mammalian proteins, causes cell death similar to that caused by the dominant mutations affecting the predicted pore lining (Garcia-Anoveros *et al.*, 1995). This clearly shows that if the two systems obviously have a common backbone, the existence of distinct regulatory regions leads to proteins of very different activities.

C. elegans genome sequencing project has permitted the identification of new worm proteins, such as C41C4.5 or T28D9.7, which display clear homology with the other members of the family, especially within the second hydrophobic region and the cysteine-rich region. Their respective physiological roles are unknown.

6.2. Toward Other Homologues

Divergence between chordates (including man, rat, frog and chick) and nematodes occurred at least 650 millions years ago, suggesting that proteins related to this super-family may occur in numerous other species. The amiloride-sensitive Na^+ channel has indeed been studied in the leech integument (Weber *et al.*, 1995), and amiloride-sensitive Na$^+$ channels have been characterized in the nervous system of the snail *Helix aspersa* (Green *et al.*, 1994). A cDNA has now been isolated from *Helix* nervous tissue (Lingueglia *et al.*, 1995). It encodes an FMRFamide-activated Na$^+$ channel (FaNaCh) that can be blocked by amiloride. Although the corresponding protein shares a low sequence identity with the previously cloned proteins of the family, it displays the same overall structural organization as the epithelial Na$^+$ channel subunits and *C. elegans* degenerins. This suggests that peptides may directly control the cellular excitability by direct gating of channels from that family.

Comparison of the α, β, and γ subunits with the database of expressed sequence tags has revealed the existence of a mammalian homologue (Genebank Accession Number T19320). A full length cDNA that encodes a protein of 638 amino acids was cloned (Waldmann *et al.*, 1995a). The identity with α, β, and γ is 27–37%, and the mRNA was mainly detected in brain, pancreas, testis, and ovary. When expressed alone in *Xenopus* oocytes, this new protein generates a small amiloride-sensitive Na$^+$ current, as does the α subunit alone. When expressed with the β and γ subunits, it generates a large amiloride-sensitive Na$^+$ current, with biophysical and pharmacological properties distinct from those observed after expression of the α, β, and γ subunits (higher unitary conductance, lower sensitivity to amiloride derivatives, different selectivity sequence).

Adenosine 5′-triphosphate (ATP) is the principal transmitter from sympathetic nerves to certain smooth muscles (arterioles, bladder, vas deferens), where it also activates a ligand-gated cation channel (P2X receptor). Two cDNAs encoding proteins of ~400 residues have been isolated by expression cloning from vas deferens and from rat pheochromocytoma PC12 cells. They encode an ATP-gated channel when expressed in oocytes or in HEK 293 cells (Brake *et al.*, 1994; Valera *et al.*, 1994). A partial cDNA clone obtained by subtraction hybridization in thymocytes is selectively expressed in cells induced to die. This clone corresponds to an incompletely processed P2X receptor gene that differs from P2X receptor only in its first 37 nucleotides. The P2X receptor and related molecules may have functional roles in synaptic transmission and programmed cell death. The overall topology in the membrane is similar to that of the amiloride-sensitive Na$^+$ channel, with two hydrophobic regions, a pore-forming motif which resembles that of potassium channels, and a large extracellular domain containing numerous cysteins, but no significant homology was found between the P2X receptor and the Na$^+$ channel subunits, or any other member of the family. It is therefore unlikely that they belong in a strict sense to the same family. The existence of the same membrane topology in the different proteins, i.e., a large extracytoplasmic region containing a cysteine-rich domain, however, suggests some kind of structural relationships. Identifications of new homologues will surely clarify the exact relationships between these distinct proteins and the exact role of their common domains.

7. REGULATION OF THE Na$^+$ CHANNEL BY STEROIDS

A fundamental aspect of Na$^+$ channel function concerns its regulatory properties involved in the control of extracellular osmolarity and volume. Numerous external and internal factors participate in the adaptation of the Na^+ channel activity. Aldosterone is the major sodium-retaining hormone in kidney and colon (Rossier and Palmer, 1992). Its effects are pleiotropic along with increasing apical Na^+ permeability: it inhibits electro-neutral sodium transport in rat colon (Bastl and Hayslett, 1992) and in distal segments of the nephron (Tomita et al., 1985), reduces tight junction conductance (Reif et al., 1986), modifies the morphology of the renal (Stanton et al., 1985) and intestinal (Elbrond et al., 1993) cells, and induces H^+ secretion in rabbit cortical and medullary collecting tubules (Koeppen and Helman, 1982; Stone et al., 1983) and K^+ secretion in cortical collecting tubules and colon (Field and Giebisch, 1985; Rechkemmer and Halm, 1989). In amphibians as well as terrestrial vertebrates, aldosterone increases the amiloride-sensitive apical Na^+ permeability (Crabbé, 1963). However, large differences in experimental results have been observed among heterogeneous experimental preparations. Some effects that have been documented are mediated by transcriptional events, whereas others seem to be mediated by nontranscriptional events.

7.1. Transcription-Independent Activation by Corticosteroid Hormones

The increase in apical Na^+ permeability by aldosterone may involve the activation of silent channels which are already present in the cells. The activation of silent channels might be the result of direct effects of the steroids on the plasma membrane (reviewed by Wehling et al., 1993). The acute increase in Na^+ permeability that occurs before any neosynthesis suggests the existence of such silent channels (Palmer et al., 1982). This process could be produced by changes in internal pH (Harvey et al., 1988), by methylation reactions (Sariban-Sohraby et al., 1984), and/or by changes in internal calcium (Petzel et al., 1992). A fast response independent of neosynthesis is also suggested by the impaired aldosterone response obtained when the apical membrane of toad urinary bladder cells is pretreated with proteolytic enzymes or carboxyl-reactive reagents (Garty and Edelman, 1983; Kipnowski et al., 1983) and by the observation that aldosterone primarily increases the open probability of the Na^+ channels expressed in frog kidney cell line A6, with a minor effect on the number of channels per patch (Kemendy et al., 1992).

7.2. Mechanism of Transcriptional Activation by Corticosteroid Hormones

Transcriptional activation by aldosterone or any other steroid hormone involves free diffusion of the hormone across the membrane, binding to a cytosolic receptor (Arriza et al., 1987), nuclear translocation of the hormone-receptor complex, and binding to steroid-responsive elements located in the regulatory region of target gene promoters (Wahli and Martinez, 1991). There are two types of high-affinity aldosterone receptors. The type I or mineralocorticoid receptor has a high affinity for aldosterone while the type II or glucocorticoid receptor has a lower affinity for the hormone but better affinities for synthetic glucocorticoids such as dexamethasone (Rossier and Palmer, 1992). Both receptors belong to the same family as thyroid hormone, vitamin D, retinoic acid receptor, etc. (Wahli and Martinez, 1991): they are divided into several distinct cassettes which have distinct functional roles such as ligand binding, nuclear localization, DNA binding, and transactivation. The nucleotide sequence 5'-AGGACA(N)$_3$TGTCCT-3' has been identified as the hormonal responsive element for several distinct steroid hormones, including glucocorticoid hormone, progesterone, androgen, and aldosterone. The estrogen responsive element differs by a 2-bp modification of the 15-bp palindromic DNA sequence (5'-AGGTCA(N)TGACCT-3' in place of 5'-AGGACA(N)$_3$TGTCCT-3'), but existing experimental methods have so far not permitted discrimination among glucocorticoid, progesterone, androgen, and aldosterone responsive elements. The effects of the hormones are inhibited by blockers of transcription or translation and involve *de novo* RNA and protein syntheses.

Regulation of the amiloride-sensitive Na$^+$ channel by corticosteroids has been investigated in detail in toad urinary bladder, mammalian and chicken intestine, kidney, and lung, where different mechanisms have been identified.

7.3.   Toad Bladder

In toad bladder, Palmer *et al.* (1982) showed that the aldosterone-mediated increase in Na$^+$ channel activity was due to an increase in the number of active channels. Asher *et al.* (1988, 1992a) analyzed the aldosterone-induced augmentation of Na$^+$ transport by comparing the hormone's effect on the transepithelial short-circuit current, on the amiloride-sensitive ^{22}Na$^+$ uptake in isolated membrane vesicles, and more recently on the amiloride-blockable Na$^+$ channel expressed in *Xenopus* oocytes after injection of urinary bladder RNA. Short-term aldosterone treatment caused an increase in the amiloride-sensitive short-circuit current, but had no effect on amiloride-sensitive Na$^+$ transport in apical vesicles derived from treated tissue or on the ability of isolated RNA to express functional channels in oocytes. Long-term aldosterone treatment produced an additional augmentation of the short-circuit current, together with an increase in the channel activity in isolated membranes and in the channel activity expressed in oocytes. Asher *et al.* (1988, 1992a) suggested that aldosterone increases the apical Na$^+$ permeability in the toad bladder by two different mechanisms: a relatively fast effect (less than or equal to 3 hr), which is not sustained by the isolated membrane, and a later response (greater than 3 hr), mediated by the increased transcription of specific genes which is preserved by the isolated membrane and after injection of the RNAs into *Xenopus* oocytes. Their data also suggest that the short-term effect in toad urinary bladder is mediated by mineralocorticoid receptors, whereas the long-term effect is of a glucocorticoid type.

7.4.   Colon

The observation that colonic Na$^+$ channels are strongly regulated at the RNA level by dietary salt intake in rat and chicken has been one of the keys to successful expression cloning of the Na$^+$ channel (Canessa *et al.*, 1993, 1994b; Lingueglia *et al.*, 1993b, 1994). A low sodium diet stimulates the renin-angiotensin system and leads to large increases in aldosterone secretion and Na$^+$ channel activity (Pácha *et al.*, 1993). Oocytes injected with mRNA from chicken lower intestinal epithelium (colon + coprodeum) showed no detectable sodium currents when they were prepared from control animals, whereas oocytes injected with mRNA prepared from animals fed a low-sodium diet had high amiloride-blockable sodium currents (Weber *et al.*, 1992). Similar results were obtained with RNAs prepared from distal colon of rats infused with dexamethasone or fed a low-sodium diet (unpublished data).

In rat distal colon, aldosterone both induces amiloride-sensitive electrogenic Na^+ transport and inhibits electroneutral Na^+ absorption (Sandle and Binder, 1987). The Na^+ transport observed in nonstimulated control rats involves coupled transport of NaCl via a Na^+/H^+ antiporter and a Cl^-/HCO_3^- antiporter (Charney and Feldman, 1984; Bastl and Hayslett, 1992). The same mechanism is responsible for the Na^+ reabsorption observed in distal and proximal segments of the colon from control animals. In chicken colon, besides stimulation of electrogenic Na^+ transport, a low-sodium diet also inhibits Na^+-driven cotransporters of amino acids and sugars (Årnason and Skadhauge, 1991). Steroid stimulation of electrogenic amiloride-sensitive Na^+ transport in distal colon is mediated by type I mineralocorticoid receptors, as shown by the inhibitory effect of spironolactone, a specific type I mineralocorticoid receptor antagonist (Bastl, 1988), and by the control of electrogenic Na^+ absorption by nanomolar concentrations of aldosterone (Fromm *et al.*, 1993), occurring via *de novo* protein synthesis. Synthetic steroids with a high specificity for glucocorticoid receptors stimulate the electroneutral transport of NaCl but have no effect on the electrogenic Na^+ transport (Bastl, 1987; Turnamian and Binder, 1989).

Linguelia *et al.* (1994) and Renard *et al.* (1995) have analyzed the effects of a low-sodium diet and dexamethasone treatment on expression of the Na^+ channel subunits at the mRNA and protein levels in rat distal colon. They showed that steroids control Na^+ channel activity via stimulation of the transcription of the β and γ subunits; the level of expression of the α subunit mRNA is not significantly affected. Using specific antipeptide antibodies raised against each subunit of the Na^+ channel, Renard *et al.* (1995) failed to detect any α, β, or γ proteins in the apical membrane of colonocytes from control animals, whereas all three proteins were detected in colon from animals kept on a low-sodium diet or treated with dexamethasone. The strong elevation in transcription of the β and γ subunits by aldosterone (or dexamethasone) caused an increase in the level of β and γ subunit proteins detected by immunocytochemistry. The level of expression of the α subunit, which was detected only in colon from steroid-treated animals, appears to be indirectly controlled by the steroid levels. Renard *et al.* (1995) proposed that this regulation occurs at a posttranslational level as a result of multimerization of the three subunits after steroid treatment: formation of a stabilized complex between newly synthesized β and γ subunits and the α subunit would permit maturation of the complex toward the apical membrane. A similar property has been observed for the nicotinic receptor in transfected cells expressing its α, β, γ, and δ forming subunits (Green and Claudio, 1994). The transcriptional (for β and γ subunits) and posttranslational (for the α subunit) regulations proposed by Renard *et al.* (1995) would lead to very strong control and to high stimulation factors of the Na^+ channel activity by the circulating steroids.

7.5. Kidney

Frindt *et al.* (1990) showed that a low-sodium diet causes a ~100-fold increase in the amiloride-sensitive whole-cell current measured in rat cortical collecting tubule cells. The effects of a low-sodium diet on the activity of the Na^+ channel from rat cortical collecting tubules have been demonstrated by cell-attached patch-clamp recordings. Pácha *et al.* (1993) reported an increase in the number of active Na^+ channels but did not detect any modification in the open probability of the channels, suggesting an all-or-nothing expression of the channel in rat collecting tubules. Reif *et al.* (1986) noted the synergistic effects of antidiuretic hormone and desoxycorticosterone on Na^+ transport in perfused rat cortical collecting tubules. Amiloride failed to inhibit the lumen-to-bath flux of $^{22}Na^+$ under basal conditions, in agreement with Pácha *et al.* (1993). Addition of antidiuretic hormone to the bath produced a stable, amiloride-sensitive lumen-to-bath $^{22}Na^+$ flux that was potentiated by pretreatment of the animals by desoxycorticosterone. The rapid activation of the Na^+ channels by vasopressin, observed within minutes (Reif *et al.*, 1986), suggests that a pool of silent channels can be activated, even though Pácha *et al.*, 1993) failed to detect such activation in their cell-attached patch-clamp recordings performed in the absence of cAMP-elevating agents. These different studies clearly indicate that steroid treatment considerably increases amiloride-sensitive Na^+ channel activity in kidney, but the mechanism of action seems distinct from that observed in distal colon: 1) steroid treatments do not alter significantly the levels of α, β, and γ subunits RNA in kidney, whereas a dramatic increase in the levels of β and γ subunits RNA is observed in distal colon (Renard *et al.*, 1995); 2) a pool of silent channels is present in membranes of nontreated rat kidney cells (possibly in an internal pool), as revealed after vasopressin treatment (Reif *et al.*, 1986). In distal colon, cyclic AMP and vasopressin do not affect the amiloride-sensitive Na^+ transport (Bridges *et al.*, 1984). Differences have also been noted between species. For instance, a high basal activity of the Na^+ channel is observed in cortical collecting tubules and distal colon from control rabbits (Chen *et al.*, 1990; Turnheim *et al.*, 1986).

In the amphibian kidney cell line A6 grown on a permeable support, Kemendy *et al.* (1992) observed that aldosterone alters the open probability of amiloride-blockable Na^+ channels, mainly via modulation of the open time of the channel. The number of Na^+ channels was not altered in a statistically significant manner. Thus, Na^+ transport into mRNA-injected oocytes did not increase whether the A6 cells used to isolate mRNA had been treated with aldosterone or not, suggesting that the number of messenger RNAs for Na^+ channels was not modified (George *et al.*, 1989; Palmer *et al.*, 1990; also see Hinton and Eaton, 1989).

A nonzero expression of the renal Na^+ channel in control rat has been suggested after transcription and immunohistochemical analyses by Renard *et al.* (1995). Northern blot analysis failed to reveal any significant modification in the level of the mRNAs for the three subunits after steroid treatment (i.e., a low-sodium diet and dexamethasone infusion). Since the α, β, and γ subunits are detected in several distinct segments of the kidney tubule besides the cortical collecting tubule, the effects of steroid treatments may indeed depend on the site of expression. More sensitive analysis, such as RNAse mapping or quantitative PCR in each segment of the nephron, are necessary to address this question more accurately. Immunolabeling experiments are, however, consistent with a constitutive expression of the Na^+ channel subunits in distal segment of the kidney tubule: no effect of steroid treatments was observed for the expression of the α subunit in medullary rays or of the γ subunit in the distal convoluted tubules. Due to cryosection preparation of the tissues, it remains unclear whether labeling was limited to the apical membrane or whether it also concerned internal vesicles. More accurate experiments using paraffin sections or electron microscopy will make it possible to evaluate whether steroid treatments alter the targeting of the subunits up to the apical membrane. The existence of two distinct regulation mechanisms involving glucocorticoid and/or mineralocorticoid receptors in different segments of the renal tubule may also explain some of the effects observed in kidney. Thus, regulatory mechanisms specific to the kidney seem to play an important role and permit rapid adaptation of Na^+ excretion to changes in environmental conditions.

7.6. Regulation of the Lung Na^+ Channel by Glucocorticoids

Amiloride-sensitive Na^+ channels control the quantity and composition of the respiratory tract fluid and play a key role in the transition from a fluid-filled to an air-filled lung at the time of birth (Strang, 1991; O'Brodovich, 1991). The mRNAs for the α, β, and γ subunits have been identified in the lung (Renard et al., 1995), and the biophysical properties reported by Voilley et al. (1994) appear identical to those of the renal Na^+ channel, suggesting that the structure present in lung is the same as in colon and kidney. Around birth, an increase in Na^+ channel transcription and expression results in a switch of the ionic transport in lung from active Cl^- secretion to active Na^+ reabsorption (Voilley et al., 1994; O'Brodovich et al., 1993; MacDonald et al., 1994; Voilley et al., 1996). This results in clearance of the pulmonary fluid as the lung switches to an air-conducting system. Combined administration of glucocorticoid and thyroid hormones has previously been demonstrated to induce a Na^+-absorptive capacity in the immature fetal lung (Barker et al., 1991), suggesting that this modification may be due to stimulation by glucocorticoids. Champigny et al. (1994) analyzed the mechanisms of this stimulation in primary cultures of fetal rat lung epithelial cells and showed that the activity of the amiloride-sensitive Na^+ channel is controlled by corticosteroids. Dexamethasone (0.1 μM) or aldosterone (1 μM) increased amiloride-induced hyperpolarization and the amiloride-sensitive current. An increased transcription of the three Na^+ channel subunits was observed (Champigny et al., 1994; Voilley et al., 1996). A parallel increase in the amount of α subunit mRNA was explained by a stimulation of the transcription. Using synthetic specific agonists and antagonists for mineralocorticoid and glucocorticoid receptors, Champigny et al. (1994) showed that the steroid action on lung Na^+ channel expression is mediated via glucocorticoid receptors. In lung epithelial cells, triiodothyronine, which is known to modulate steroid action in several epithelial tissues (Barlet-Bas et al., 1988, but see also Geering et al., 1982), had no effect on the amiloride-sensitive Na^+ current or the level of the mRNA for the Na^+ channel protein, but potentiated the stimulatory effect of dexamethasone. The increase in Na^+ channel activity observed in the lung around birth can thus be explained by a direct increase in transcription of the Na^+ channel genes (similar results have subsequently been obtained for the β and γ subunits, unpublished experiments), in good agreement with the role played by corticosteroids in the maturation of fetal lung before birth (Strang, 1991; O'Brodovich, 1991). As glucocorticoids and triiodothyronine levels are known to increase in vivo around birth (Barker et al., 1991), their inducing effect on the Na^+ absorptive capacity in immature fetal lung can be explained by their action on Na^+

channel transcription. Besides their well-known effects on the secretion of the surfactant by type II pneumocytes, it seems therefore that glucocorticoids also alter *in vivo* the expression of other important pulmonary systems such as the epithelial Na^+ channel. Although a rise in plasma epinephrine concentration also takes place during labor and delivery and plays an important role in the reabsorption of fetal lung liquid (Strang, 1991), no effect of cAMP-raising agents were observed in fetal rat lung epithelial cells in primary cultures either at the level of the RNA or at the level of the amiloride-sensitive Na^+ current (Voilley and Champigny, unpublished data).

Renard *et al.* (1995) detected the three subunits at the apical membrane of airway epithelial cells using an immunohistochemical technique. Neither a low-sodium diet nor dexamethasone treatment altered mRNA (as revealed by Northern blot analysis) or protein expression (as revealed by immunolabeling) for any of the three subunits, whereas adrenalectomy drastically reduced expression of the three subunits, Distinct regulation of the Na^+ channel by steroids has thus been observed in lung and in colon. While β and γ mRNAs are not expressed in the colon from control animals, their level is high in lung tissue from the same animals. The α subunit, which is constitutively transcribed in control colon, is regulated by steroids in fetal lung epithelial cells. Steroid treatment of fetal lung epithelial cells in primary culture alters the level of transcription of the α subunit (Champigny *et al.*, 1994), as well as the β and γ subunits (Voilley *et al.*, 1996). These effects, together with the decreased transcription of the three subunits after adrenalectomy, suggest that glucocorticoids are the major regulators of the expression of the α, β, and γ subunits in lung, and that, under normal conditions, their level is usually high enough to ensure maximal activation of α, β, and γ subunit transcription in this tissue. The capacity of distinct corticosteroid receptors to recognize identical hormone responsive elements and to modulate the expression of the corresponding genes may allow a regulation of the Na^+ channel expression adapted to each tissue.

8.   REGULATION OF THE Na^+ CHANNEL BY INTRACELLULAR SIGNALING PATHWAYS

8.1.   Vasopressin and Cyclic AMP

Vasopressin (the antidiuretic hormone) controls body water balance by regulating renal water excretion and acts on several distinct transport systems located along the excretory tract. In the collecting ducts, transepithelial-osmotic water permeability (Verkman, 1989), Na^+ reabsorption (Reif *et al.*, 1986), and K^+ secretion (Schafer *et al.*, 1990) are increased. The effects of the hormone on water and Na^+ transport are usually rapid, i.e., within minutes, and appear to be mediated by an increase in cyclic AMP (Schafer and Troutman, 1990). Vasopressin treatments increase the expression of a 29-kDa protein called AQP-CD (for aquaporin-collecting duct) at the apical membrane and in the subapical vesicles of collecting duct principal cells and inner medullary collecting duct cells (Fushimi *et al.*, 1993; Nielsen *et al.*, 1993). Functionally, the AQP-CD protein behaves like a "water-selective" channel. According to the "shuttle" hypothesis, subapical vesicles contain a reservoir of water channels. Vasopressin increases the apical membrane water permeability by triggering exocytosis of these vesicles to the apical membrane (Chevalier *et al.*, 1974). Long-term water deprivation, which stimulates vasopressin secretion, also increases water permeability as well as the amount of AQP-CD detected by immunolabeling (Nielsen *et al.*, 1993). An increase in water channel expression may therefore be additive to the acute effect of antidiuretic hormone on water permeability.

The vasopressin-induced increase in Na^+ permeability may be due to the same type of mechanism, although effects of the hormone on Na^+ and water permeability are clearly different: 1) the increase in water permeability caused by antidiuretic hormone is sensitive to pretreatment of the tissues by colchicine, cytochalasin B, trifluoperazine, and methohexital sodium, whereas the increase in Na^+ permeability is not (reviewed by Garty and Benos, 1986); 2) lower concentrations of the hormone suffice to activate Na^+ permeability. Stimulation of amiloride-sensitive Na^+ channels by vasopressin in vasopressin-responsive tissues, or by its second messenger cyclic AMP in cells devoid of vasopressin receptors, is not constant. In rat, for instance, cyclic AMP stimulates the Na^+ channel in cortical collecting tubules (Schafer and Hawk, 1992) but not in colon (Bridges *et al.*, 1984), even though the three subunits that make up the channel are present in these two tissue types. Interspecies differences have also been observed. For instance, in rat cortical collecting tubules and in frog skin, antidiuretic hormone causes stable stimulation of amiloride-sensitive Na^+ transport, whereas in toad urinary bladder and in rabbit cortical collecting tubules, the stimulatory effect is transient, and the tissues become desensitized to a second application of the hormone (Chen *et al.*, 1990; Garty and Benos, 1986). Since vasopressin has been shown to increase both cytoplasmic cyclic AMP and intracellular calcium in rat and rabbit cortical collecting ducts (Breyer and Ando, 1994), complex responses to the hormone, such as those observed in rabbit, may explain these differences.

After *in vitro* treatment of toad urinary bladder membrane vesicles with cyclic AMP–dependent protein kinase, no stimulation of the amiloride-sensitive $^{22}Na^+$ uptake was noticed with respect to controls, whereas a two- to fourfold increase was observed when vesicles were prepared from vasopressin-treated urinary bladder (Lester *et al.*, 1988). This suggests that direct posttranslational modification of the channel is probably not the cause of its activation. Frings *et al.* (1988), in excised patch-clamp studies performed on toad urinary bladder, were able to stimulate a Na^+ channel in three out of nine patches by adding protein kinase A to the cytoplasmic surface of the membrane. The fact that these investigators succeeded in activating Na^+ channels in some excised patches suggests that, in toad urinary bladder, the Na^+ channel is already present in the apical membrane prior to activation by protein kinase A.

An increase in the number of conductive Na^+ channels with little or no change in the open probability of individual Na^+ channels has been observed in A6 cells pretreated with vasopressin (Marunaka and Eaton, 1991). The effect was mimicked by treatment with a permeant analog of cyclic AMP and with cholera toxin. The large increase in the number of conductive Na^+ channels per patch (i.e., from a mean of 2 in controls to 9 after treatment by vasopressin) suggests that these three agents may act by activating clusters of previously silent Na^+ channels. A similar increase in the number of active Na^+ channels without alteration of the open probability has been observed after treatment with antidiuretic hormone (ADH) in frog skin (Helman *et al.*, 1983), frog colon (Krattenmacher *et al.*, 1988), and toad urinary bladder (Li *et al.*, 1982). Stimulation of the Na^+ transport by ADH appears to be a complex mechanism requiring an intact system for membrane biogenesis, as recently suggested by studies in toad bladder by Weng and Wade (1994) using brefeldin A, an inhibitor of the vesicular transport from the endoplasmic reticulum. In the presence of brefeldin A, these authors have shown that the stimulation of Na^+ transport by ADH becomes increasingly insensitive to the hormone during successive challenges. Surprisingly, stimulations by forskolin and vasopressin might not be equally sensitive to brefeldin A (Kleyman *et al.*, 1994; Coupaye-Gérard *et al.*, 1994).

Additional types of positive regulation by cyclic AMP–elevating agents have been reported in other tissues. In hamster fungiform taste cells, amiloride-sensitive Na^+ and H^+ currents can be increased by treatment with arginine-8-vasopressin or 8-Br-cAMP (Gilbertson *et al.*, 1993). These Na^+ and H^+ currents are related to amiloride's ability to inhibit responses to a variety of taste stimuli, including salty, sweet, and sour (acid). In lung, beta-adrenergic agonists increase lung fluid clearance in anesthetized ventilated adult sheep; the effect is inhibited by propranolol a beta blocker, and by amiloride (Berthiaume *et al.*, 1987; Strang, 1991).

Molecular cloning of the α, β, and γ subunits of the epithelial Na^+ channel provides interesting data about channel regulation by cyclic AMP. No conserved consensus sites for phosphorylation by protein kinase A are found in the cytoplasmic domains of human and rat α, β, or γ subunits (Renard *et al.*, 1994). Cyclic AMP has no fast effect in oocytes injected with the three subunits. Thus the vasopressin-induced activation of the Na^+ channel activity is probably not mediated by a phosphorylation of one of these three pore-forming subunits. This is in agreement with detection of the same three subunits in kidney, i.e., a vasopressin-responsive tissue, and in colon, i.e., a tissue where Na^+ channel activity is not regulated by cyclic AMP (Bridges *et al.*, 1984). Clearly vasopressin regulation of the Na^+ channel requires other components that remain to be identified. Stutts *et al.* (1955) have reported that CFTR (cystic fibrosis transmembrane conductance regulator) can act as a cAMP-dependent regulator of Na^+ channel: the amiloride-sensitive short-circuit current generated by the heterologous expression of the three Na^+ channel subunits in Madin Darby canine kidney cells (MDCK) was stimulated by forskolin in the absence of CFTR, but heterologous expression of CFTR reverted the forskolin-effect, which became inhibitory on the amiloride-sensitive short-circuit current. This observation might be consistent with the well known increase of the Na^+ channel activity observed during cystic fibrosis, i.e., when the CFTR activity is severely reduced (Boucher *et al.*, 1988).

8.2. Regulation by Protein Kinase C

Activation of protein kinase C reduces the activity of the Na^+ channel in A6 cells grown on a permeable support (Yanase and Handler, 1986), in rabbit cortical collecting tubules (Hays *et al.*, 1987), and in mammalian airway epithelia (Graham *et al.*, 1992). Protein kinase C activators (PMA and OAG) reverse the increases of the open probability and of the number of active channels observed in a cell-attached patch when the luminal concentration of sodium (i.e., the concentration of Na^+ outside the pipette) is reduced from 129 mM to 3 mM (Ling and Eaton, 1989). The effects of decreasing the luminal Na^+ concentration were mimicked by a protein kinase C inhibitor (D-sphingosine). In rat collecting tubules, the activity of the Na^+ channel is decreased by an increase in $[Ca^{2+}]_i$ (Palmer and Frindt, 1987). This effect is observed in cell-attached patches but not in excised patches, suggesting an indirect rather than a direct effect of Ca^{2+} ions on the channel. In rat and rabbit cortical collecting tubule, an increase in the cytoplasmic concentration of Na^+ produces an increase in $[Ca^{2+}]_i$, presumably *via* activation of a basolateral Na^+/Ca^{2+} exchange (Silver *et al.*, 1993; Frindt and Windhager, 1990). The so-called feedback inhibition of the Na^+ channel prevents the cell swelling expected from the presumed increase in cytoplasmic Na^+ activity (MacRobbie and Ussing, 1961). A decrease in the luminal entry of Na^+ induced by the addition of amiloride reduces $[Ca^{2+}]_i$ and increases the number of active channels and their open probability. These results may be explained by modifications of the apical membrane voltage, since amiloride treatment increases the electrical driving force for Na^+ entry, *via* the hyper-polarization of the membrane. An increase in $[Na^+]_i$ (for instance, after blockade of $Na^+/K^+/ATPase$) increases $[Ca^{2+}]_i$ and decreases the number of active channels and their open probability (Frindt *et al.*, 1993; Silver *et al.*, 1993). Besides effects on the apical membrane potential, it seems, however, that several distinct mechanisms modulate the activity of the Na^+ channel during feedback inhibition (Frindt *et al.*, 1993; Silver *et al.*, 1993). In A6 cells and in distal segments of the nephron, this could involve an activation of a protein kinase C (Ling and Eaton, 1989; Frindt *et al.*, 1993). A different situation may exist in frog skin, where a Ca^{2+}-independent isoform of protein kinase C activates the Na^+ channel (Civan *et al.*, 1991). In this last tissue, the increase in $[Na^+]_i$ acidifies the cytoplasm due to activation of a basolateral Na^+/H^+ antiporter (Harvey *et al.*, 1988). This results in a decrease in apical Na^+ permeability, since its activity is reduced at acidic pH (Harvey *et al.*, 1988; Palmer and Frindt, 1987). In rat cortical collecting tubules, no alteration of the internal pH was noticed under the same experimental conditions (Silver *et al.*, 1993). Other mechanisms, such as the reorganization of the cytoskeleton may also be involved in feedback inhibition (frog skin; Els and Chou, 1993). Ling and Eaton (1989) have proposed that the activation of protein kinase C may also influence Na^+ "self-inhibition," corresponding to the reduction in apical Na^+ permeability observed when the luminal Na^+ concentration is increased (Fuchs *et al.*, 1977; VanDriessche and Lindemann, 1979). The Na^+ self-inhibition reported by Fuchs *et al.* (1977) occurs rapidly in frog skin (i.e., within seconds), whereas the time course of inhibition by protein kinase C lasts for minutes (Ling and Eaton, 1989). In frog skin, direct interaction of the Na^+ ion with an external site of the channel might thus be responsible for the inhibition observed.

Several factors are known to affect antidiuretic action of vasopressin (reviewed by Morel and Doucet, 1986). Prostaglandin E2 (Holt and Lechene, 1981; Stokes and Kokko, 1977), bradykinin (Tomita *et al.*, 1985), and acetylcholine (Wiesmann *et al.*, 1978) inhibit Na^+ transport in the distal nephron and in the toad urinary bladder. Although occupancy of the corresponding receptors by these effectors activates several distinct intracellular signaling pathways, it seems that Na^+ channel inhibition correlates with transient increases in intracellular inositol-1,3,5-triphosphate and to an activation of a Ca^{2+}-dependent protein kinase C, at least for prostaglandin E2 (Kokko *et al.*, 1994). Whereas acute treatment of A6 cells with prostaglandin E2 inhibited amiloride-sensitive Na^+ transport *via* a pertussis toxin–insensitive pathway, a long-term treatment by prostaglandin E2 resulted in stimulation of Na^+ transport, but this late effect of prostaglandin E2 was explained by a slow rise in cyclic AMP induced by the agent (Kokko *et al.*, 1994).

It is currently unknown whether regulation by protein kinase C is due to direct phosphorylation of one channel subunit. Reconstitution experiments in planar bilayers have revealed an inhibitory effect of protein kinase C on an amiloride-sensitive channel (Oh *et al.*, 1993b) which clearly differed from the 4-pS, highly selective Na^+ channel (unitary conductance: 45 pS; pNa^+/pK^+:7). Alignment of rat and human α, β, and γ subunits reveals that five putative phosphorylation sites for protein kinase C are conserved between human and rat ([Ser or Thr]-X-[Arg or Lys], where X is any amino acid): one site is found in the α chain, two sites are found in the β chain, and two sites are found in the γ chain. It will be interesting to evaluate the role of the corresponding residues in channel function. Interestingly, mutations that modify these sites in the human β and γ chains have been observed in Liddle's syndrome, a rare form of hypertension characterized by hyperactivity of the amiloride-sensitive Na^+ channel; suppression of one of these sites may explain the increased channel activity which is the hallmark of the disease (Shimkets *et al.*, 1994; Hansson *et al.*, 1995).

8.3. Regulation by Tyrosine Kinases

Several hormones that interact with tyrosine kinases are active on Na^+ transport, although *a priori* by distinct mechanisms. Epidermal growth factor (EGF), which is detected at high concentrations in the distal nephron, inhibits the amiloride-sensitive Na^+ channel (Muto *et al.*, 1991; Warden and Stokes, 1993). It is currently unclear whether the effects of EGF on Na^+ transport can be mediated by activation of protein kinase C, as is the case for EGF-induced inhibition of water transport (Breyer *et al.*, 1988), or whether a distinct regulatory mechanism is involved.

Functional measurements in toad urinary bladder (Herrera, 1965), A6 cells (Fidelmann *et al.*, 1982; Walker *et al.*, 1984), frog skin (Civan *et al.*, 1988), and mammalian kidney (DeFronzo *et al.*, 1976; Stenvinkel *et al.*, 1992) have demonstrated that insulin has a minor stimulatory effect on the amiloride-sensitive Na^+ channel and enhances the effect of aldosterone in a synergistic manner. Patch-clamp studies have shown that after a 5–10 min lag time insulin treatment of A6 cells results in an increase of the open probability of an amiloride-sensitive Na^+ channel, without alteration of the single-channel conductance or the current/voltage relationship (Marunaka *et al.*, 1992). Rodriguez-Commes *et al.* (1994) have shown that the insulin-mediated increase of the short-circuit current is inhibited by the calcium chelator BAPTA and is also sensitive to dihydroxychlorpromazine, trifluoperazine, and genistein, inhibitors of protein kinase C, Ca^{2+}-dependent/calmodulin-dependent protein kinase, and tyrosine kinase, respectively. Insulin-mediated activation of the Na^+ channel thus seems to be a complex process involving several distinct regulatory pathways.

No consensus site for phosphorylation by tyrosine kinases is found in the cytoplasmic segments of the α, β, and γ subunits of the Na^+ channel.

8.4. G Proteins and Atrial Natriuretic Factor

A role for G proteins in the regulation of epithelial Na^+ channels was first suggested by Mohrmann et al. (1987). However, the effects were described for a distinct ionic channel characterized by a unitary conductance of 9 pS and a pNa^+/pK^+ selectivity ratio of 5. Other effects, such as activation of the Na^+ channel in toad bladder vesicles after treatment with GTPγS (Garty et al., 1989) or activation of the Na^+ channel by GDPβS and pertussis toxin in A6 cells (Ohara et al., 1993), involve the highly Na^+-selective channel. Elucidation of the physiological role of this regulatory pathway deserves further investigation.

The atrial natriuretic factor (ANF) increases urinary Na^+ excretion, but no effect has been observed on the highly selective Na^+ channel (Maack et al., 1985). ANF and its cellular effector, cyclic GMP, inhibit the activity of an amiloride-sensitive, 28-pS nonselective cation channel found in inner medullary collecting tubules (Light et al., 1989a,b).

9. BIOCHEMISTRY OF THE AMILORIDE BINDING PROTEINS

Functional studies have shown that amiloride binds to the Na^+ channel in a 1:1 stoichiometry. Cuthbert and Edwardson (1981) were the first to titrate amiloride binding sites with [^{3}H]benzamil in kidney and colon, although not in muscle, brain, or heart. These binding sites were characterized by a higher sensitivity to amiloride derivatives substituted at their carboxy guanidinium by lipophilic molecules such as phenamil or benzamil than to amiloride itself. Similar results were subsequently reported by Kleyman et al. (1986) in bovine kidney with [^{3}H]benzamil and by Barbry et al. (1986, 1990b) in pig kidney with [^{3}H]phenamil and [^{3}H]bromobenzamil. Since similar inhibition constants were obtained during binding experiments with radiolabeled amiloride derivatives and during functional measurements of the Na^+ channel activity (Kleyman et al., 1986; Barbry et al., 1986), it was proposed that the Na^+ channel itself was actually being titrated, i.e., that the amiloride binding site was associated with the Na^+ channel. The [^{3}H]benzamil binding site was estimated to have a size of 600 kDa by the target-size inactivation technique (Edwardson et al., 1981), but Barbry et al. (1987) subsequently showed that the [^{3}H]phenamil binding site actually had a size of 90 kDa under experimental conditions where the number, but not the inhibition constant, of the binding sites was altered by irradiation. Barbry et al. (1990b) later demonstrated the specific covalent labeling of [^{3}H]bromobenzamil in a protein of ~105 kDa associated into a homodimer of 185 kDa by sulfhydryl groups. This was the only labeled protein in pig kidney, a result substantially different from the report by Kleyman et al. (1986), where polypeptides of 176 kDa, 77 kDa, and 47 kDa from bovine kidney membranes were labeled by [^{3}H]bromobenzamil. In both cases, however, the number of amiloride binding sites ($\geq$ 10 pmol/ mg protein) exceeded the predicted number of Na^+ channels by several orders of magnitude, suggesting that other proteins were also titrated by the radiolabeled compounds. Kleyman et al. (1989) have used 2'-methoxy-5'-nitrobenzamil (NMBA), a photoreactive amiloride analog, and anti-amiloride antibodies to identify photolabeled polypeptides. NMBA specifically labels a 130-kDa polypeptide in bovine kidney microsomes and in the A6 cell line. NMBA photolabeling and [^{3}H]benzamil binding were used to examine the cellular pool of binding sites following aldosterone treatment of A6 cells, during which a 3.8-fold stimulation was observed in transepithelial Na$^+$ transport. No noticeable alteration was caused by the mineralocorticoid agonist aldosterone or the antagonist spironolactone. A chronic low-sodium diet, known to stimulate Na$^+$ channel activity, did not modify the number of [^{3}H]phenamil binding sites in chicken colon (Goldstein et al., 1993) or pig kidney (P. Barbry and O. Chassande, unpublished data).

[^{3}H]methylbromoamiloride was synthesized by Lazorick et al. (1985) in order to titrate the Na$^+$ channel. The affinity of this compound for the Na$^+$ channel was around 100 nM (Sariban-Sohraby and Benos, 1986). The thermodynamic properties of the interaction between the different radiolabeled amiloride derivatives and the membranes appeared very similar: the time course of association varied between 5 and 10 min for 140 nM [^{3}H]methylbromoamiloride and for 0.05–90 nM [^{3}H]phenamil at 0°C, whereas dissociation was effective after 20 sec for [^{3}H]methylbromoamiloride and after 10 min for [^{3}H]phenamil. In comparison with the functional inhibition of the Na$^+$ channel by amiloride and its derivatives, which occurs within seconds, these time courses appear very slow (Goldstein et al., 1993). Careful analysis of [^{3}H]phenamil and [^{3}H]benzamil binding and Na$^+$ transport inhibition has revealed substantial differences between the high-affinity [^{3}H]phenamil binding site and the site whose occupancy by phenamil blocks Na$^+$ transport: 1) ethylisopropylamiloride was found to displace bound [^{3}H]phenamil at concentrations that are at least tenfold lower than those needed to block the channel; 2) the rates at which [^{3}H]phenamil associates and dissociates from this site are lower than the rates at which Na$^+$ channels are inhibited and reactivated under similar conditions (Goldstein et al., 1993).

The 105-kDa protein shown by photoaffinity labeling to be the main amiloride binding site in pig kidney was purified and cloned (Barbry et al., 1987, 1990a,b). Stable transfection of human and rat cDNA in 293 cells (Barbry et al., 1990a; Lingueglia et al., 1993a) led to the expression of a high-affinity binding site for amiloride and phenamil. The ~750-residue-long protein appears very hydrophilic, possesses a signal peptide, and is found in the cell culture supernatant. The protein is detected in epithelial tissues such as kidney, intestine, placenta, and skin, as well as in hematopoietic tissues such as thymus. This tissue specificity of expression differs from that expected for a subunit of the Na$^+$ channel: for instance, in rat, the protein is more abundant in proximal colon than in distal colon (Lingueglia et al., 1993a; Verity and Fuller, 1994). Its expression is not regulated by corticosteroids (Verity and Fuller, 1994), but in seminal vesicles, androgens exert strong control (Izawa, 1991). Novotny et al. (1994) have demonstrated that the protein corresponds to an amiloride-blockable diamine oxidase. This enzyme (also called histaminase, E.C.1.4.3.6) oxidatively deaminates putrescine and histamine (Novotny et al., 1994; see also Mu et al., 1994). While the protein's exact physiological role remains unclear, it might be important in some aspect of cell growth. Diamine oxidase is anchored to the basolateral membrane of epithelial cells via an interaction with glycosaminoglycans (Haddock et al., 1987) and can be released into the serum by heparin infusion (D'Agostino et al., 1989). This explains a posteriori why amiloride binding sites were titrated in plasma membranes. Circulating levels of diamine oxidase are raised during pregnancy (Kobayashi, 1967). In rat, two distinct forms have been cloned from colon and lung, corresponding to transcripts of 2.7 kb and 1.2 kb. The short form corresponds to the 3' terminus of the longer one and encodes a nonfunctional enzyme (Lingueglia *et al.*, 1993a), i.e., without the region containing topaquinone, the cofactor of the enzyme (Janes *et al.*, 1992). It might be formed by alternative transcription under the control of an internal promoter (Lingueglia *et al.*, 1993a; Chassande *et al.*, 1994). Both the long and short forms of the protein bind amiloride and some of its derivatives, but they have distinct pharmacologies (Lingueglia *et al.*, 1993a).

The exact relationships between the identification of diamine oxidase as ABP and other purifications of the amiloride-sensitive Na^+ channel remain unclear. Using [^{3}H]methylbromoamiloride as a tracer, Benos and coworkers (1986, 1987) used a two-step technique to purify a complex revealed by nonreducing sodium dodecyl sulfate gel electrophoresis analysis as a band of approximately 730 kDa (Benos *et al.*, 1986, 1987). Upon reduction, this band was resolved into five major polypeptide bands with apparent average M_r values of 300, 150, 95, 70, 55, and 40 kDa. Assuming a molecular mass of 703 kDa, the methylbromoamiloride receptor was expected to be near homogeneity. [^{3}H]methylbromoamiloride photoincorporated into a 150-kDa polypeptide and this incorporation was blocked by addition of excess amiloride (Benos *et al.*, 1987). The 300-kDa subunit is a substrate for phosphorylation by cyclic AMP–dependent protein kinase (Sariban-Sohraby *et al.*, 1988); GTP-dependent incorporation of methyl groups is detected in the 95-kDa subunit after treatment of A6 cells by aldosterone (Sariban-Sohraby *et al.*, 1993); the 40-kDa subunit, which can be ADP-ribosylated by the pertussis toxin, corresponds to the α_{i-3} subunit of the heterotrimeric G proteins (Ausiello *et al.*, 1992). Recently, purifications by the same authors led to isolation of a complex formed only by the 150-kDa and 40-kDa subunits (Sariban-Sohraby *et al.*, 1992; Oh *et al.*, 1993a). The exact stoichiometry and interaction of the different proteins of this purified fraction are not known and may very well result from copurification of independent proteins. Information about the primary structures of these different proteins will help to explain their exact relationships with the other characterized amiloride binding proteins. Due to the very similar biochemical approaches developed by Benos and coworkers (1987) and by Barbry and coworkers (1987), it might also be particularly important to test for the presence of diamine oxidase activity in the different purified fractions. For instance, the truncated form of diamine oxidase (Lingueglia *et al.*, 1993a) and an amiloride binding protein which is immunologically related to the complex purified by Benos and coworkers (Oh *et al.*, 1992) are found in rat lung. Since these two proteins share very similar pharmacological properties, both may be related to diamine oxidase.

Lin *et al.* (1994) developed an antiidiotypic approach to generate antibodies directed against the amiloride binding domain of the channel. First, monoclonal antibodies were raised against amiloride. One of them, mAb BA7.1, has been used as a model system to analyze the three-dimensional conformation of an amiloride binding site. The nucleotide sequences of the variable regions of the heavy and light chains of mAb BA7.1 were determined and amino acid sequences deduced to analyze the structure of the amiloride binding site (Lin *et al.*, 1994). Antiidiotypic monoclonal antibody RA6.3 mimicked the effect of amiloride by inhibiting Na^+ transport across A6 cell monolayers when applied to the apical cell surface (Kleyman *et al.*, 1991). Inhibition of transport required pretreatment of the apical cell surface with trypsin in the presence of amiloride in order to enhance accessibility of the antibody to the amiloride binding site. This antibody specifically immunoprecipitated a large 750–700-kDa protein from [^{35}S]methionine-labeled A6 cell cultures, which was resolved further under reducing conditions as a set of polypeptides with apparent molecular masses of 260–230, 180, 140–110, and 70 kDa. The antibody recognized the 140-kDa subunit on immunoblots of purified A6 cell Na$^+$ channel. Immunocytochemical localization in A6 cultures revealed apical membrane as well as cytosolic immunoreactive sites, but also sites at or near the basolateral plasma membrane (Kleyman *et al.*, 1991). The interaction of amiloride and the so-called "specific blockers of the Na$^+$ channel" with unrelated proteins, which are usually much more abundant, is the major limitation of any approach using titration of an amiloride binding site. In the absence of a specific blocker of the Na$^+$ channel that does not interact with the other amiloride binding proteins, titration techniques, whether direct by a radiolabeled amiloride derivative or indirect by interaction with an antiidiotypic antibody, lead to the obligatory identification of the most abundant forms, i.e., non-Na$^+$ channel proteins. Although functional reconstitution of amiloride-sensitive cationic conductance in a planar lipid bilayer was achieved with the material purified by Oh *et al.* (1993a,b), and by Sariban-Sohraby *et al.* (1992), immunodetection with the antibodies raised by Brown *et al.* (1989) and Kleyman *et al.* (1989, 1991) of antigens in the renal inner medulla, rather than in the cortical and outer medullary collecting ducts (Renard *et al.*, 1995; Duc *et al.*, 1994) strongly suggests that the antigens revealed by these antibodies differ from the Na$^+$ channel subunits. This clearly implies that these purified materials are not directly related to the cloned subunits of the Na$^+$ channel.

In summary, none of the different amiloride binding proteins identified so far have been proven unambiguously to be related to one of the cloned subunits of the amiloride-sensitive Na$^+$ channel. The existence of enzymes, such as diamine oxidase, that share some common pharmacological properties with the Na$^+$ channel has been a major problem that has impeded the successful biochemical characterization of this channel.

10. PATHOLOGIES ASSOCIATED WITH Na$^+$ CHANNEL DYSFUNCTION

Cystic fibrosis (CF) is the major autosomal recessive disease in the Caucasian population. Disruption of the exocrine function of the pancreas, biliary cirrhosis, chronic bronchopulmonary infection with emphysema, high sweat electrolyte, and abnormalities in the transport ducts of the male genital system are the main manifestations of the disease (Quinton, 1990). Mutations in the cystic fibrosis transmembrane conductance regulator (CFTR) gene are responsible for the defective activity of a Cl$^-$ channel activated by the cyclic AMP–dependent protein kinase (Tsui, 1992). The CFTR protein, which corresponds to this Cl$^-$ channel, is a member of the ATP binding cassette superfamily (Riordan *et al.*, 1989). Pathological mutations in CFTR lead to the loss or reduction of cyclic AMP–dependent chloride secretion into the affected tissues (Welsh and Smith, 1993). In airway epithelium, the reduction in Cl$^-$ secretion is accompanied by an increase in active Na^+ absorption (an approximately twofold increase in Na^+ permeability, documented by measurements on intact tissues and on primary cultures of respiratory epithelial cells; see Table II), which is mediated at the apical membrane by the amiloride-sensitive Na^+ channel. The increase in Na^+ reabsorption, together with the decrease in Cl^- secretion, result in altered fluid transport characterized by excessive fluid absorption, leading *in vivo* to dehydration of the mucus (Jiang *et al.*, 1993). This has suggested that amiloride aerosol therapy might be beneficial for CF patients. Indeed, amiloride inhalation has been shown to increase mucus clearance and to retard the decline in lung function (Knowles *et al.*, 1990). These modifications might be related to changes in ion content, hydration, and rheology of sputum (Tomkiekwicz *et al.*, 1993). Surprisingly, hyper-Na^+ reabsorption has only been observed in the airways and not in other CF-affected tissues such as the ductal cells of sweat glands (Quinton, 1983; Bijman and Frömter, 1986). Transgenic mice, obtained by knock-out of the CFTR murine gene, have reduced cyclic AMP–activated Cl^- secretion, and some strains also develop some hyper-Na^+ reabsorption, among other symptoms of the disease (Grubb *et al.*, 1994a).

Few reports describing the unitary properties of the Na^+ permeable channel in the human airway epithelium have been published. Chinet *et al.* (1993) have characterized amiloride-blockable cationic channels in the apical membrane of these cells. Comparing the properties of this channel in normal and CF-affected cells, these authors reported an increase in the open probability (Chinet *et al.*, 1994). Molecular cloning of the α, β, and γ subunits of the highly selective and highly amiloride-sensitive Na^+ channel from a human lung cDNA library (Voilley *et al.*, 1994, 1995) suggests however that active Na^+ reabsorption in airways is mediated by this latter channel rather than by the channel characterized by Chinet *et al.* (1993). Accordingly, α, β, and γ subunit transcripts have been detected in human nasal, tracheal, and bronchial epithelia (unpublished data).

The mechanism leading from a mutation within the gene of a Cl^- channel to the increased Na^+ reabsorption is still unknown but could theoretically be explained by an increase of either the number of active channels or their open probability. Modifications in membrane recycling have been described during cystic fibrosis (Bradbury *et al.*, 1992). This may lead to increased expression of the Na^+ channel at the apical membrane of airway epithelial cells or altered targeting of an important regulatory protein at this same apical membrane.

Another human disease is also associated with enhanced activity of the Na^+ channel. Liddle *et al.* (1963) reported a case of hypertension associated with hypo-kaliemic alkalosis not due to hyperaldosteronism. The index case was characterized by hypoaldosteronism, hypokalemia, and decreased renin and angiotensin. A renal trans-plant following development of renal failure resulted in normalization of the aldosterone and renin responses to salt restriction (Botero-Velez *et al.*, 1994). Rodriguez *et al.* (1981), Wang *et al.* (1981), and Nakada *et al.* (1987) later demonstrated that amiloride and triamterene were effective treatments so long as dietary Na^+ intake was restricted. Very recently, Shimkets *et al.* (1994) have demonstrated a link between the disease locus and the β subunit of the Na^+ channel, both located in the region p13.11 of human chromosome 16. A C-to-T transition at the first nucleotide of codon Arg^{564} introduces a stop codon in the gene. This mutation truncates the cytoplasmic COOH-terminus in affected subjects from Liddle's original kindred. Analysis of subjects with the disorder from four additional kindreds (Gardner *et al.*, 1971) demonstrated either premature termination or frameshift mutations in the same region. More recently, Hansson *et al.* (1995) have identified a W574X mutation that affects the γ chain gene and removes most of the cytoplasmic COOH terminal segment of the protein. *A priori*, one would expect that these mutations would remove two possibly important sites in these two subunits. The first possible one is an SH3 binding domain, whose sequence is found not only in the α subunit (Rotin *et al.*, 1994) but also in β and γ subunits. However, since this motif was proposed to target the proteins at the apical membrane, its absence would be expected to reduce the activity of the Na^+ channel (Rotin *et al.*, 1994). A second possibility would be that the domain removed by the mutations is involved in an interaction with a negative regulator of the channel. A third explanation would involve the putative site for phosphorylation by protein kinase C found in the vicinity of this SH3 binding region. In this case, the absence of phosphorylation in the amputated β subunit could indeed lead to hyperactivity of the channel. After having introduced such mutations in the rat β and γ chains, Schild *et al.* (1995) have observed an increased activity of the channel expressed in oocytes, mainly due to an increased number of active channels.

Other human pathologies may also involve the Na^+ channel directly. For instance, loss of electrogenic Na^+ absorption in the distal colon is an important factor in the pathogenesis of diarrhea in ulcerative colitis (Sandle *et al.*, 1990).

11.   CONCLUSIONS AND PERSPECTIVES

The identification of the α, β, and γ subunits of the epithelial Na^+ channel has allowed the elucidation of some of its properties. The interaction of each pore-forming subunit with the membrane or with the adjacent cytoskeleton is beginning to be understood. Transcriptional regulations by corticosteroid hormones have already been studied, and most of the different tools needed for a systematic study are now available. The direct or indirect involvement of these proteins in human pathologies (primary hypertension, cystic fibrosis) is now firmly established. This scientific progress has not only solved some old questions, but has also revealed new unknowns: 1) the quaternary structure of the Na^+ channel complex is still not known, since expression into *Xenopus* oocytes has not yet determined the stoichiometry. Conjunction of biochemical, genetic, and electrophysiological approaches will be necessary to determine it. Important domains involved for instance in the amiloride binding, in the selectivity filter, or in the interaction between subunits have not yet been completely localized; 2) the regulation of the Na^+ channel expression by corticosteroid hormones, especially in kidney, seems to be more complex than initially thought and cannot be explained simply by a transcriptional regulation of the Na^+ channel; 3) control of the Na^+ channel activity by other hormones, such as vasopressin, and more generally by intracellular cascades, is till poorly understood. Progress in this area is particularly important for establishing the molecular mechanisms of Na^+ channel dysfunction in human pathologies; 4) identification of new homologues, especially in excitable tissues, will allow us to assess the physiological importance of the gene superfamily.

Electrophysiological characterization of the channels encoded by *C. elegans* degen erins may provide important information. Identification of the functional, structural, and pharmacological relationships between channels previously characterized in epithelial and nonepithelial tissues (smooth muscle cells, blood cells) will probably become easier with the cloning of new members of the family.

ACKNOWLEDGMENTS. We are grateful to our many colleagues who have contributed to our research work and to Guy Champigny and Eric Lingueglia for a careful reading of the manuscript. This work has been supported by grants from the Centre National de la Recherche Scientifique (CNRS), Rhône-Poulenc-Rorer, and the Association Française de Lutte contre la Mucoviscidose (AFLM). The expert technical assistance of Catherine Roulinat and Franck Aguila is greatly acknowledged.

12. REFERENCES

Alliegro, M. C., Alliegro, M. A., Cragoe, E. J., and Glaser, B. M., 1993, Amiloride inhibition of angiogenesis *in vitro*, *J. Exp. Zool.* **267**:245–252.

Árnason, S. S., and Skadhauge, E., 1991, Steady-state sodium absorption and chloride secretion of colon and coprodeum, and plasma levels of osmoregulatory hormones in hens in relation to sodium intake, *J. Comp. Physiol.* **B161**:1–14.

Arriza, J. L., Weinberger, C., Cerelli, G., Glaser, T. M., Handelin, B. L., Housman, D. E., and Evans, R. M., 1987, Cloning of human mineralocorticoid receptor complementary DNA: Structural and functional kinship with the glucocorticoid receptor, *Science* **237**:268–275.

Asher, C., and Garty, H., 1988, Aldosterone increases the apical Na^+ permeability of toad bladder by two different mechanisms, *Proc. Natl. Acad. Sci. USA* **85**:7413–7417.

Asher, C., Cragoe, Jr., E. J., and Garty, H., 1987, Effects of amiloride analogues on Na^+ transport in toad bladder membrane vesicles. Evidence for two electrogenic transporters with different affinities towards pyrazine carboxamides, *J. Biol. Chem.* **262**:8566–8573.

Asher, C., Eren, R., Kahn, L., Yeger, O., and Garty, H., 1992a, Expression of the amiloride-blockable Na^+ channel by RNA from control versus aldosterone-stimulated tissue, *J. Biol. Chem.* **267**:16061–16065.

Asher, C., Singer, D., Eren, R., Yeger, O., Dascal, N., and Garty, H., 1992b, NaCl-dependent expression of an amiloride-blockable Na^+ channel in *Xenopus* oocytes, *Am. J. Physiol.* **262**:G244–G248.

Ausiello, D. A., Stow, J. L., Cantiello, H. F., de, A. J., and Benos, D. J., 1992, Purified epithelial Na^+ channel complex contains the pertussis toxin–sensitive G alpha i-3 protein, *J. Biol. Chem.* **267**:4759–4765.

Avenet, P., and Lindemann, B., 1991, Noninvasive recording of receptor cell action potentials and sustained currents from single taste buds maintained in the tongue: The response to mucosal NaCl and amiloride, *J. Membr. Biol.* **124**:33–41.

Awayda, M. S., Ismailov, I. I., Berdiev, B. K., and Benos, D. J., 1995, A cloned renal epithelial Na^+ channel protein displays stretch activation in planar lipid bilayers, *Am. J. Physiol.* **268**:C1450–C1459.

Barbry, P., Frelin, C., Vigne, P., Cragoe Jr., E. J., and Lazdunski, M., 1986, [^{3}H]Phenamil, a radiolabelled diuretic for the analysis of the amiloride-sensitive Na^+ channels in kidney membranes, *Biochem. Biophys. Res. Commun.* **135**:25–32.

Barbry, P., Chassande, O., Vigne, P., Frelin, C., Ellory, C., Cragoe Jr., E. J., and Lazdunski, M., 1987, Purification and subunit structure of the [^{3}H]phenamil receptor associated with the renal Na^+ channel, *Proc. Natl. Acad. Sci. USA* **84**:4836–4840.

Barbry, P., Champe, M., Chassande, O., Munemitsu, S., Champigny, G., Lingueglia, E., Maes, P., Frelin, C., Tartar, A., Ullrich, A., and Lazdunski, M., 1990a, Human kidney amiloride-binding protein: cDNA structure and functional expression, *Proc. Natl. Acad. Sci. USA* **87**:7347–7351.

Barbry, P., Chassande, O., Marsault, R., Lazdunski, M., and Frelin, C., 1990b, [^{3}H]Phenamil binding protein of the renal epithelium Na^+ channel. Purification, affinity labeling, and functional reconstitution, *Biochemistry* **29**:1039–1045.

Barker, P. M., Walters, D. V., Markiewicz, M., and Strang, L. B., 1991, Development of the lung liquid reabsorptive mechanism in fetal sheep: Synergism of triiodothyronine and hydrocortisone, *J. Physiol. (London)* **433:**435–449.

Barlet-Bas, C., Khadouri, C., Marsy, S., and Doucet, A., 1988, Sodium-independent *in vitro* induction of Na⁺,K⁺-ATPase by aldosterone in renal target cells: Permissive effect of triiodothyronine, *Proc. Natl. Acad. Sci. USA* **85:**1707–1711.

Basset, G., Crone, C., and Saumon, G., 1987, Fluid absorption by rat lung in situ: Pathways for sodium entry in the luminal membrane of alveolar epithelium, *J. Physiol. (London)* **384:**325–345.

Bastl, C. P., 1987, Regulation of cation transport by low doses of glucocorticoids in *in vivo* adrenalectomized rat colon, *J. Clin. Invest.* **80:**348–356.

Bastl, C. P., 1988, The effect of spironolactone on glucocorticoid-induced colonic cation transport, *Am. J. Physiol.* **255:**F1235–F1242.

Bastl, C. P., and Hayslett, J. P., 1992, The cellular action of aldosterone in target epithelia, *Kidney Int.* **42:** 250–264.

Bear, C. E., Li, C., Kartner, N., Bridges, R. J., Jensen, T. J., Ramjeesingh, M., and Riordan, J. R., 1992, Purification and functional reconstitution of the cystic fibrosis transmembrane conductance regulator (CFTR), *Cell* **68:**809–818.

Belkin, K. J., and Abrams, T. W., 1993, FMRFamide produces biphasic modulation of the LFS motor neurons in the neural circuit of the siphon withdrawal reflex of *Aplysia* by activating Na⁺ and K⁺ currents, *J. Neurosci.* **13:**5139–5152.

Benos, D. J., Reyes, J., and Shoemaker, D. G., 1983, Amiloride fluxes across erythrocyte membranes, *Biochim. Biophys. Acta* **734:**99–104.

Benos, D. J., Saccomani, G., Brenner, B. M., and Sariban-Sohraby, S., 1986, Purification and characterization of the amiloride-sensitive sodium channel from A6 cultured cells and bovine renal papilla, *Proc. Natl. Acad. Sci. USA* **83:**8525–8529.

Benos, D. J., Saccomani, G., and Sariban-Sohraby, S., 1987, The epithelial sodium channel. Subunit number and location of the amiloride binding site, *J. Biol. Chem.* **262:**10613–10618.

Benos, D. J., Awayda, S., Ismailov, I. I., and Johnson, J. P., 1995, Structure and function of amiloride-sensitive Na⁺ channels, *J. Membr. Biol.* **143:**1–18.

Bentley, P. J., 1968, Amiloride: A potent inhibitor of sodium transport across the toad bladder, *J. Physiol.* **195:**317–330.

Berthiaume, Y., Staub, N. C., and Matthay, M. A., 1987, β adrenergic agonists increase lung liquid clearance in anesthetized sheep, *J. Clin. Invest.* **79:**335–343.

Bijman, J., and Frömter, E., 1986, Direct demonstration of high transepithelial chloride-conductance in normal human sweat duct which is absent in cystic fibrosis, *Pflügers. Arch.* **407:**S123–S127.

Botero-Velez, M., Curtis, J. J., and Warnock, D. G., 1994, Liddle's syndrome revisited: A disorder of sodium reabsorption in the distal tubule, *N. Engl. J. Med.* **330:**178–181.

Boucher, R. C., Stutts, M. J., Knowles, M. R., Cantley, L., and Gatzy, J. T., 1986, Na⁺ transport in cystic fibrosis respiratory epithelia. Abnormal basal rate and response to adenylate cyclase activation, *J. Clin. Invest.* **78:**1245–1252.

Bradbury, N. A., Jilling, T., Berta, G., Sorscher, E. J., Bridges, R. J., and Kirk, K. L., 1992, Regulation of plasma membrane recycling by CFTR, *Science* **256:**530–532.

Brake, A. J., Wagenbach, M. J., and Julius, D., 1994, New structural motif for ligand-gated ion channels defined by an ionotropic ATP receptor, *Nature* **371:**519–523.

Breyer, M. D., and Ando, Y., 1994, Hormonal signaling and regulation of salt and water transport in the collecting duct, *Annu. Rev. Physiol.* **56:**711–739.

Breyer, M. D., Jacobson, H. R., and Breyer, J., 1988, Epidermal growth factor inhibits the hydroosmotic effect of vasopressin in the isolated perfused rabbit cortical collecting tubule, *J. Clin. Invest.* **82:**1313–1320.

Bridges, R. J., Rummel, W., and Wollenberg, P., 1984, Effects of vasopressin on electrolyte transport across isolated colon from normal and dexamethasone-treated rats, *J. Physiol. (London)* **355:**11–23.

Bridges, R. J., Cragoe, E. J. J., Frizzell, R. A., and Benos, D. J., 1989, Inhibition of colonic Na⁺ transport by amiloride analogues, *Am. J. Physiol.* **256:**C67–C74.

Brown, D., Sorscher, E. J., Ausiello, D. A., and Benos, D. J., 1989, Immunocytochemical localization of Na⁺ channels in rat kidney medulla, *Am. J. Physiol.* **256:**F366–F369.

Bubien, J. K., and Warnock, D. G., 1993, Amiloride-sensitive sodium conductance in human B lymphoid cells, *Am. J. Physiol.* **265:**C1175–C1183.

Canessa, C. M., Horisberger, J. D., and Rossier, B. C., 1993, Epithelial sodium channel related to proteins involved in neurodegeneration, *Nature* **61:**467–470.

Canessa, C. M., Merillat, A. M., and Rossier, B. C., 1994a, Membrane topology of the epithelial sodium channel in intact cells, *Am. J. Physiol.* **267:**C1682–C1690.

Canessa, C. M., Schild, L., Buell, G., Thorens, B., Gautschi, I., Horisberger, J. D., and Rossier, B., 1994b, Amiloride-sensitive epithelial Na$^+$ channel is made of three homologous subunits, *Nature* **367:**463–467.

Cantiello, H. F., Patenaude, C. R., and Ausiello, D. A., 1989, G-protein subunit, αi-3, activates a pertussis toxin–sensitive Na$^+$ channel from the epithelial cell line, A6, *J. Biol. Chem.* **264:**20867–20870.

Chalfie, M., and Au, M., 1989, Genetic control of differentiation of the *C. elegans* touch receptor neurons, *Science* **243:**1027–1033.

Chalfie, M., and Wolinski, E., 1990, The identification and suppression of inherited neurodegeneration in *Caenorhabditis elegans*, *Nature* **345:**410–416.

Chalfie, M., Sulston, J. E., White, J. G., Southgate, E., Thomson, J. N., and Brenner, S., 1985, The neural circuit for touch sensitivity in *Caenorhabditis elegans*, *J. Neurosci.* **5:**956–964.

Chalfie, M., Driscoll, M., Huang, M., 1993, Degenerin similarities, *Nature* **361:**504.

Champigny, G., Voilley, N., Lingueglia, E., Friend, V., Barbry, P., and Lazdunski, M., 1994, Regulation of expression of the lung amiloride-sensitive Na$^+$ channel by steroid hormones, *EMBO J.* **13:**2177–2181.

Charney, A., and Feldman, G., 1984, Systemic acid-base disorders and intestinal electrolyte transport, *Am. J. Physiol.* **247:**G1–G12.

Chassande, O., Renard, S., Barbry, P., and Lazdunski, M., 1994, The human gene for diamine oxidase, an amiloride binding protein: Molecular cloning, sequencing and characterization of the promoter, *J. Biol. Chem.* **269:**14484–14486.

Chen, L., Wiliams, S. K., and Schafer, J. A., 1990, Differences in synergistic actions of vasopressin and deoxycorticosterone in rat and rabbit CCD, *Am. J. Physiol.* **259:**F147–F156.

Chevalier, J., Bourguet, J., and Hugon, J. S., 1974, Membrane associated particles: Distribution in frog urinary bladder epithelium at rest and after oxytocin treatment, *Cell Tissue Res.* **152:**129–140.

Chinet, T. C., Fullton, J. M., Yankaskas, J. R., Boucher, R. C., and Stutts, M. J., 1993, Sodium-permeable channels in the apical membrane of human nasal epithelial cells, *Am. J. Physiol.* **265:**C1050–C1060.

Chinet, T. C., Fullton, J. M., Yankaskas, J. R., Boucher, R. C., and Stutts, M. J., 1994, Mechanism of sodium hyperabsorption in cultured cystic fibrosis nasal epithelium: A patch clamp study, *Am. J. Physiol.* **265:**C1061–C1068.

Civan, M. M., Peterson-Yantorno, K., and O'Brien, T. G., 1988, Insulin and phorbol ester stimulate conductive Na$^+$ transport through a common pathway, *Proc. Natl. Acad. Sci. USA* **85:**963–967.

Civan, M. M., Oler, A., Peterson-Yantorno, K., Georges, K., and O'Brien, T. G., 1991, Ca^{2+}-independent form of protein kinase C may regulate Na$^+$ transport across frog skin, *J. Membr. Biol.* **121:**37–50.

Clauss, W., Dürr, J. E., Guth, D., and Skadhauge, E., 1987, Effects on adrenal steroids on Na$^+$ transport in the lower intestine (coprodeum) of the hen, *J. Membr. Biol.* **96:**141–152.

Counillon, L., Scholz, W., Lang, H. J., and Pouysségur, J., 1993, Pharmacological characterization of stably transfected Na$^+$/H$^+$ antiporter isoforms using amiloride analogs and a new inhibitor exhibiting antiischemic properties, *Mol. Pharmacol.* **44:**1041–1045.

Coupaye-Gérard, B., Kim, H. J., Singh, A., and Blazer-Yost, B. L., 1994, Differential effects of brefeldin A on hormonally regulated Na$^+$ transport in a model renal epithelial cell line, *Biochim. Biophys. Acta* **1190:**449–456.

Crabbé, J., 1963, Site of action of aldosterone on the bladder of the toad, *Nature* **200:**787–788.

Cragoe Jr., E. J., 1979, Structure activity relationships in the amiloride series, in: *Amiloride and Epithelial Sodium Transport*, (A. W. Cuthbert, G. M. Fanelli, Jr., and A. Scrabini, eds.), Urban and Schwarzenberg, Baltimore, pp. 1–20.

Cribbs, L. L., Satin, J., Fozzard, H. A., and Rogard, R. B., 1990, Functional expression of the rat heart I Na$^+$ channel isoform. Demonstration of properties characteristic of native cardiac Na$^+$ channels, *FEBS Lett.* **275:**195–200.

Cuthbert, A. W., 1976, Importance of guanidinium groups for blocking sodium channels in epithelia, *Mol. Pharmacol.* **12:**945–957.

Cuthbert, A. W., and Edwardson, J. M., 1981, Benzamil binding to kidney cell membranes, *Biochem. Pharmacol.* **30:**1175–1183.

D'Agostino, L., Pignata, S., Danièle, B., Ventriglia, R., Ferrari, G., Ferraro, C., Spagnuolo, S., Luchelli, P. E., and Mazzacca, G., 1989, Release of diamine oxidase into plasma by glycosaminoglycans in rats, *Biochim. Biophys. Acta* **993:**228–232.

DeFronzo, R. A., Goldberg, M., and Agus, Z. S., 1976, The effects of glucose and insulin on renal electrolyte transport, *J. Clin. Invest.* **58:**83–90.

Dinudom, A., Young, J. A., and Cook, D. I., 1993, Amiloride-sensitive Na$^+$ current in the granular duct cells of mouse mandibular glands, *Pflügers Arch.* **423:**164–166.

Driscoll, M., and Chalfie, M., 1991, The mec-4 gene is a member of a family of *Caenorhabditis elegans* genes that can mutate to induce neuronal degeneration, *Nature* **349:**588–593.

Duc, C., Farman, N., Canessa, C. M., Bonvalet, J. P., and Rossier, B. C., 1994, Cell-specific expression of epithelial sodium channel α, β, and γ subunits in aldosterone-responsive epithelia from the rat: Localization by in situ hybridization and immunocytochemistry, *J. Cell Biol.* **127:**1907–1921.

Eaton, D. C., and Hamilton, K. L., 1988, The amiloride-blockable sodium channel of epithelial tissue, in: *Ion Channels*, Vol. 1 (T. Narahashi, ed.), Plenum, New York, pp. 151–182.

Edmonds, C., 1967, Transport of sodium and secretion of potassium and bicarbonate by the colon of normal and sodium depleted rats, *J. Physiol. (London)* **193:**589–602.

Edwardson, J. M., Fanestil, D. D., Ellory, J. C., and Cuthbert, A. W., 1981, Extraction of a [^{3}H]benzamil binding component from kidney cell membranes, *Biochem. Pharmacol.* **30:**1185–1189.

Eigler, J., and Crabbé, J., 1969, Effect of diuretics on active Na-transport in amphibian membranes, in: *Renal Transport and Diuretics* (K. Thurau and H. Jahrmärker, eds.), Springer-Verlag KG, Berlin, pp. 195–208.

Eisenmann, G., and Horn, R., 1983, Ion selectivity revisited: The role of kinetic and equilibrium processes in ion permeation through channels, *J. Membr. Biol.* **76:**197–225.

Elbrond, V. S., Dantzer, V., Mayhew, T. M., and Skadhauge, E., 1993, Dietary and aldosterone effects on the morphology and electrophysiology of the chicken coprodeum, in: *Avian Endocrinology* (P. J. Sharp, ed.), Journal of Endocrinology Ltd., Bristol, England, pp. 217–226.

Els, W. J., and Chou, K.-Y., 1993, Sodium-dependent regulation of epithelial sodium channel densities in frog skin; a role for the cytoskeleton, *J. Physiol.* **462:**447–464.

Feng, Z. P., Clark, R. B., and Berthiaume, Y., 1993, Identification of nonselective cation channels in cultured adult rat alveolar type II cells, *Am. J. Respir. Cell. Mol. Biol.* **9:**248–254.

Ferrary, E., Bernard, C., Oudar, O., Sterkers, O., and Amiel, C., 1989, Sodium transfer from endolymph through a luminal amiloride-sensitive channel, *Am. J. Physiol.* **257:**F182–F189.

Fidelmann, M. L., May, J. M., Biber, T. U. L., and Watlington, C. O., 1982, Insulin stimulation of Na$^+$ transport and glucose metabolism in cultured renal cells, *Am. J. Physiol.* **242:**C121–C123.

Field, M. J., and Giebisch, G. J., 1985, Hormonal control of renal potassium excretion, *Kidney Int.* **27:**379–387.

Fischer, H., and Clauss, W., 1990, Regulation of Na$^+$ channels in frog lung epithelium: A target tissue for aldosterone action, *Pflügers Arch.* **416:**62–67.

Foster, E. S., Zimmerman, T. W., Hayslett, J. P., and Binder, H. J., 1983, Corticosteroid alteration of active electrolyte transport in rat distal colon, *Am. J. Physiol.* **245:**G668–G675.

Fox, C. A., and Wickens, M., 1990, Poly(A) removal during oocytes maturation: A default reaction selectively prevented by specific sequences in the 3'UTR of certain maternal mRNAs, *Genes Dev.* **4:**2287–2298.

Frelin, C., Vigne, P., Barbry, P., and Lazdunski, M., 1987, Molecular properties of amiloride and of its Na$^+$ transporting targets, *Kidney Int.* **32:**785–792.

Frindt, G., and Windhager, E. E., 1990, Ca^{2+}-dependent inhibition of sodium transport in rabbit cortical collecting tubules, *Am. J. Physiol.* **258:**F568–F582.

Frindt, G., Sackin, H., and Palmer, L. G., 1990, Whole-cell currents in rat cortical collecting tubule: Low Na$^+$ diet increases amiloride-sensitive conductance, *Am. J. Physiol.* **258:**F562–F567.

Frindt, G., Silver, R. B., Windhager, E. E., and Palmer, L. G., 1993, Feedback regulation of Na channels in rat CCT. II. Effects of inhibition of Na entry, *Am. J. Physiol.* **264:**F565–F574.

Frings, S., Purves, R. D., and Macknight, A. D. C., 1988, Single channel recordings from the apical membrane of the toad urinary bladder epithelial cell, *J. Membr. Biol.* **106:**157–172.

Frizzell, R. A., Koch, M. J., and Schultz, S. G., 1976, Ion transport by rabbit colon. I. Active and passive components, *J. Membr. Biol.* **27:**297–316.

Fromm, M., Schulzke, J. D., and Hegel, U., 1993, Control of electrogenic Na$^+$ absorption in rat late distal colon by nanomolar aldosterone added *in vitro*, *Am. J. Physiol.* **264**:E68–E73.

Fuchs, W., Hviid Larsen, E., and Lindemann, B., 1977, Current-voltage curve of sodium channels and concentration dependence of sodium permeability in frog skin, *J. Physiol. (London)* **267**:137–166.

Fushimi, K., Uchida, S., Hara, Y., Hirata, Y., Marumo, F., and Sasaki, S., 1993, Cloning and expression of apical membrane water channel of rat kidney collecting tubule, *Nature* **361**:549–552.

Garcia, M. L., King, V. F., Shevell, J. L., Slaughter, R. S., Suarez, K. G., Winquist, R. J., and Kaczorowski, G. J., 1990, Amiloride analogs inhibit L-type calcium channels and display calcium entry blocker activity, *J. Biol. Chem.* **265**:3763–3771.

Garcia-Anoveros, J., Ma, C., and Chalfie, M., 1995, Regulation of *Caenorhabditis elegans* degenerin proteins by a putative extracellular domain, *Curr. Biol.* **5**:441–448

Gardner, J. D., Lapey, A., Simopoulos, A. P., and Bravo, E. L., 1971, Abnormal membrane sodium transport in Liddle's syndrome, *J. Clin. Invest.* **50**:2253–2258.

Garty, H., and Benos, D. J., 1986, Characteristics and regulatory mechanisms of the amiloride-blockable Na$^+$ channel, *Phys. Rev.* **68**:309–373.

Garty, H., and Edelman, I. S., 1983, Amiloride-sensitive trypsinization of apical sodium channels. Analysis of hormonal regulation of sodium transport in toad bladder, *J. Gen. Physiol.* **81**:785–803.

Garty, H., Rudy, B., and Karlish, S. J. D., 1983, A simple and sensitive procedure for measuring isotope fluxes through ion-specific channels in heterogeneous populations of membrane vesicles, *J. Biol. Chem.* **258**:13094–13099.

Garty, H., Yeger, O., Yanovsky, A., and Asher, C., 1989, Guanosine nucleotide-dependent activation of the amiloride-blockable Na$^+$ channel, *Am. J. Physiol.* **256**:F965–F969.

Gaspar, R. J., Krasznai, Z., Marian, T., Tron, L., Recchioni, R., Falasca, M., Moroni, F., Pieri, C., and Damjanovich, S., 1992, Bretylium-induced voltage-gated sodium current in human lymphocytes, *Biochim. Biophys. Acta* **1137**:143–147.

Geering, K., Girardet, M., Bron, C., Kraehenbuhl, J. P., and Rossier, B. C., 1982, Hormonal regulation of (Na$^+$,K$^+$)-ATPase biosynthesis in the toad bladder. Effects of aldosterone and 3,5,3-triiodo-t-thyronine, *J. Biol. Chem.* **257**:10338–10343.

George Jr., A. L., Staub, O., Geering, K., Rossier, B., Kleyman, T. R., and Kraehenbuhl, J. P., 1989, Functional expression of the amiloride-sensitive sodium channel in *Xenopus* oocytes, *Proc. Natl. Acad. Sci. USA* **86**:7295–7298.

Gilbertson, T. A., Avenet, P., Kinnamon, S. C., and Roper, S. D., 1992, Proton currents through amiloride-sensitive Na channels in hamster taste cells: Role in acid transduction, *J. Gen. Physiol.* **100**:803–824.

Gilbertson, T. A., Roper, S. D., and Kinnamon, S. C., 1993, Proton currents through amiloride-sensitive Na$^+$ channels in isolated hamster taste cells: Enhancement by vasopressin and cAMP, *Neuron* **10**:931–942.

Gögelein, H., and Greger, R., 1986, Na$^+$ selective channels in the apical membrane of rabbit late proximal tubules (pars recta), *Pflügers Arch.* **406**:198–203.

Goldstein, O., Asher, C., Barbry, P., Cragoe Jr., E. J., Clauss, W., and Garty, H., 1993, An epithelial high-affinity amiloride-binding site, different from the Na$^+$ channel, *J. Biol. Chem.* **268**:7856–7862.

Goodman, B. E., and Crandall, E. D., 1982, Dome formation in primary cultured monolayers of alveolar epithelial cells, *Am. J. Physiol.* **243**:C96–C100.

Graham, A., Steel, D. M., Alton, E. W., and Geddes, D. M., 1992, Second-messenger regulation of sodium transport in mammalian airway epithelia, *J. Physiol.* **453**:475–491.

Green, K. A., Falconer, S. W. P., and Cottrell, G. A., 1994, The neuropeptide Phe-Met-Arg-Phe-NH$_2$ (FMRFamide) directly gates two ion channels in an identified helix neurone, *Pflügers Arch.* **428**:232–240.

Green, W. N., and Claudio, T., 1994, Acetylcholine receptor assembly: Subunit folding and oligomerization occur sequentially, *Cell* **74**:54–69.

Gross, J. B., and Kokko, J. P., 1977, Effects of aldosterone and potassium-sparing diuretics on electrical potential differences across the distal nephron, *J. Clin. Invest.* **59**:82–89.

Grubb, B. R., Paradiso, A. M., and Boucher, R. C., 1994a, Anomalies in ion transport in CF mouse tracheal epithelium, *Am. J. Physiol.* **267**:C293–C300.

Grubb, B. R., Vick, R. N., and Boucher, R. C., 1994b, Hyperabsorption of Na$^+$ and raised Ca^{2+}-mediated Cl$^-$ secretion in nasal epithelia of CF mice, *Am. J. Physiol.* **266**:C1478–1483.

Haddock, R. C., Mack, P., Fogerty, F. J., and Baenziger, N. L., 1987, Role of receptors in metabolic interaction of histamine with human vascular endothelial cells and skin fibroblasts. An ordered sequence of enzyme action, *J. Biol. Chem.* **262:**10220–10228.

Hamilton, K. L., and Eaton, D. C., 1985a, Single-channel recordings from two types of amiloride-sensitive epithelial Na^+ channels, *Membr. Biochem.* **6:**149–171.

Hamilton, K. L., and Eaton, D. C., 1985b, Single-channel recordings from amiloride-sensitive epithelial sodium channel, *Am. J. Physiol.* **249:**C200–C207.

Handler, J. S., Preston, A. S., Perkins, F. M., Matsumura, M., Johnson, J. P., and Watlington, C. O., 1981, The effect of adrenal steroid hormones on epithelia formed in culture by A6 cells, *Ann. N.Y. Acad. Sci.* **372:**442–454.

Hansson, J. H., Nelson-Williams, C., Suzuki, H., Schild, L., Shimkets, R. A., Lu, Y., Canessa, C. M., Iwasaki, T., Rossier, B. C., and Lifton, R. P., 1995, Hypertension caused by a truncated epithelial sodium channel γ subunit: genetic heterogeneity of Liddle syndrome, *Nature Genet.* **11:**76–82.

Harvey, B. J., Thomas, S. R., and Ehrenfeld, J., 1988, Intracellular pH controls cell membrane Na and K conductances and transport in frog skin epithelium, *J. Gen. Physiol.* **92:**767–791.

Hays, S. R., Baum, M., and Kokko, J. P., 1987, Effects of protein kinase C activation on sodium, potassium, chloride, and total Co_2 transport in the rabbit cortical collecting tubule, *J. Clin. Invest.* **80:**561–570.

Heck, G. L., Persan, K. C., and DeSimone, J. A., 1989, Direct measurement of translingual epithelial NaCl and KCl currents during the chorda tympani taste response, *Biophys. J.* **55:**843–857.

Helman, S. I., Cox, T. C., and Van Driessche, W., 1993, Hormonal control of apical membrane Na transport in epithelia. Studies with fluctuation analysis, *J. Gen. Physiol.* **82:**201–220.

Herrera, F. C., 1965, Effect of insulin on short-circuit and sodium transport across toad urinary bladder, *Am. J. Physiol.* **209:**819–824.

Hinton, C. F., and Eaton, D. C., 1989, Expression of amiloride-blockable sodium channels in *Xenopus* oocytes, *Am. J. Physiol.* **257:**C825–C829.

Holt, W. F., and Lechene, C., 1981, ADH-PGE2 interactions in cortical collecting tubule. I. Depression of sodium transport, *Am. J. Physiol.* **241:**F452–F460.

Hong, K., and Driscoll, M., 1994, A transmembrane domain of the putative channel subunit mec-4 influences mechanotransduction and neurodegeneration in *C. elegans*, *Nature* **367:**470–473.

Huang, M., and Chalfie, M., 1994, Gene interactions affecting mechanosensory transduction in *Caenorhabditis elegans*, *Nature* **367:**467–470.

Iwasa, K. H., Mizuta, K., Lim, D. J., Benos, D. J., and Tachibana, M., 1994, Amiloride-sensitive channels in marginal channels in the stria vascularis of the guinea pig cochlea, *Neurosci. Lett.* **172:**163–166.

Izawa, M., 1991, Nucleotide sequences of cDNA clones, pSv-1 and pSv-2, hybridizing to androgen-stimulated mRNAs in rat seminal vesicles, *Endocrinol. Jpn.* **38:**577–581.

Janes, S. M., Palcic, M. M., Scaman, C. H., Smith, A. J., Brown, D. E., Dooley, D. M., Mure, M., and Klinman, J. P., 1992, Identification of topaquinone and its consensus sequence in copper amine oxidases, *Biochemistry* **31:**12147–12154.

Jentsch, T. J., 1994, Trinity of cation channels, *Nature* **367:**412–413.

Jiang, C., Finkbeiner, W. E., Widdicombe, J. H., McCray Jr., P. B., and Miller, S. S., 1993, Altered fluid transport across airway epithelium in cystic fibrosis, *Science* **262:**424–427.

Joris, L., Krouse, E., Hagiwara, G., Bell, C. L., and Wine, J. J., 1989, Patch-clamp study of cultured human sweat duct cells: Amiloride-blockable Na^+ channel, *Pflügers Arch.* **414:**369–372.

Kaczorowski, G. J., Barros, F., Dethmers, J. K., Trumble, M. J., and Cragoe Jr., E. J., 1985, Inhibition of Na^+/Ca^{2+} exchange in pituitary plasma membrane vesicles by analogues of amiloride, *Biochemistry* **24:**1394–1403.

Kanwar, Y. S., and Venkatachalam, M. A., 1992, Ultrastructure of glomerulus and juxtaglomerular apparatus, in *Renal Physiology*, Vol. I (E. E. Windhager, ed.), Oxford University Press, New York, pp. 3–40.

Kemendy, A. E., Kleyman, T. R., and Eaton, D. C., 1992, Aldosterone alters the open probability of amiloride-blockable sodium channels in A6 epithelia, *Am. J. Physiol.* **263:**C825–C837.

Kilburn, K. H., 1968, A hypothesis for pulmonary clearance and its implications, *Am. Rev. Respir. Dis.* **98:**449–463.

Kipnowski, J., Park, C. S., and Fanestil, D. D., 1983, Modification of carboxyl of Na^+ channel inhibits aldosterone action on Na^+ transport, *Am. J. Physiol.* **245:**F726–F734.

Kizer, N. L., Lewis, B., and Stanton, B. A., 1995, Electrogenic sodium absorption and chloride secretion by an inner medullary collecting duct cell line (mIMCD-K2), *Am. J. Physiol.* **268**:F347–F355.

Kleyman, T. R., and Cragoe, E. J., 1988, Amiloride and its analogs as tools in the study of ion transport, *J. Membr. Biol.* **105**:1–21.

Kleyman, T. R., Yulo, T., Ashbaugh, C., Landry, D., Cragoe Jr., E. J., and Al-Awqati, Q., 1986, Photoaffinity labeling of the epithelial sodium channel, *J. Biol. Chem.* **261**:2839–2843.

Kleyman, T. R., Cragoe Jr., E. J., and Kraehenbuhl, J. P., 1989, The cellular pool of Na^+ channels in the amphibian cell line analysis using a new photoactive amiloride analog in combination with anti amiloride antibodies, *J. Biol. Chem.* **264**:11995–12000.

Kleyman, T. R., Kraehenbuhl, J. P., and Ernst, S. A., 1991, Characterization and cellular localization of the epithelial Na^+ channel. Studies using an anti-Na^+ channel antibody raised by an antiidiotypic route, *J. Biol. Chem.* **266**:3907–3915.

Kleyman, T. R., Ernst, S. A., and Coupaye-Gérard, B., 1994, Arginine vasopressin and forskolin regulate apical cell surface expression of epithelial Na^+ channels in A6 cells, *Am. J. Physiol.* **266**:F506–F511.

Knowles, M., Murray, G., Shallal, J., Askin, F., Ranga, V., Gatzy, J., and Boucher, R., 1984, Bioelectric properties and ion flow across excised human bronchi, *J. Appl. Physiol.* **56**:868–877.

Knowles, M. R., Church, N. L., Waltner, W. E., Yankaskas, J. R., Gilligan, P., King, M., Edwards, L. J., Helms, R. W., and Boucher, R. C., 1990, A pilot study of aerolized amiloride for the treatment of lung disease in cystic fibrosis, *N. Engl. J. Med.* **322**:1189–1194.

Kobayashi, Y., 1967, Plasma diamine oxidase titres of normal and pregnant rats, *Nature* **203**:146.

Kœfœd-Johnsen, V., and Ussing, H. H., 1958, The nature of the frog skin potential, *Acta Physiol. Scand.* **42**:298–308.

Koeppen, B. M., and Helman, S. I., 1982, Acidification of luminal fluid by the rabbit cortical collecting tubule perfused *in vitro*, *Am. J. Physiol.* **242**:F521–531.

Kokko, K. E., Matsumoto, P. S., Ling, B. N., and Eaton, D. C., 1994, Effects of prostaglandin E2 on amiloride-blockable Na^+ channels in a distal nephron cell line (A6), *Am. J. Physiol.* **267**:C1414–C1425.

Korbmacher, C., Segal, A. S., Fejes-Tóth, G., Giebisch, G., and Boulpaep, E. L., 1993, Whole-cell currents in single and confluent M-1 mouse cortical collecting duct cells, *J. Gen. Physiol.* **102**:761–793.

Krattenmacher, R., Fischer, H., Van Driessche, W., and Clauss, W., 1988, Noise analysis of cAMP-stimulated Na^+ current in frog colon, *Pflügers Arch.* **412**:568–573.

Krieg, P. A., and Melton, D. A., 1984, Functional messenger RNAs produced by SP6 *in vitro* transcription of cloned cDNAs, *Nucleic Acids Res.* **12**:7057–7070.

Kroll, B., Bautsch, W., Bremer, S., Wilke, M., Tümmler, B., and Frömter, E., 1989, Expression of Na^+ channel from mRNA respiratory epithelium in *Xenopus* oocytes, *Am. J. Physiol.* **257**:L284–L288.

Kroll, B., Bremer, S., Tümmler, B., Kottra, G., and Frömter, E., 1991, Sodium dependence of the epithelial sodium conductance expressed in *Xenopus laevis* oocytes, *Pflügers Arch.* **419**:101–107.

Lane, J. W., MacBride, D. W., and Hamill, O. P., 1991, Amiloride block of the mechanosensitive cation channel in *Xenopus* oocytes, *J. Physiol. (London)* **441**:347–366.

Lazorick, K., Miller, C., Sariban-Sohraby, S., and Benos, D. J., 1985, Synthesis and characterization of methyl-bromoamiloride, a potential biochemical probe of epithelial Na^+ channels, *J. Membr. Biol.* **86**:69–77.

Legris, G. J., Will, P. C., and Hopfer, U., 1982, Inhibition of amiloride-sensitive sodium conductance by indoleamines, *Proc. Natl. Acad. Sci. USA* **79**:2046–2050.

Lester, D. S., Asher, C., and Garty, G., 1988, Characterization of cAMP-induced activation of epithelial sodium channels, *Am. J. Physiol.* **254**:C802–C808.

Lewis, S. A., and Alles, W. P., 1986, Urinary kallikrein: A physiological regulator of epithelial Na^+ absorption, *Proc. Natl. Acad. Sci. USA* **83**:5345–5348.

Li, J. H.-Y., Palmer, L. G., Edelman, I. S., and Lindemann, B., 1982, The role of sodium channel density in the natriferic response of the toad urinary bladder to an antidiuretic hormone, *J. Membr. Biol.* **64**:77–84.

Li, J. H.-Y., Cragoe Jr., E. J., and Lindemann, B., 1985, Structure-activity relationship of amiloride analogs as blockers of epithelial NA channels. 1. Pyrazine-ring modifications, *J. Membr. Biol.* **83**:45–56.

Li, J. H.-Y., Cragoe Jr., E. J., and Lindemann, B., 1987, Structure-activity relationship of amiloride analogs as blockers of epithelial Na^+ channels. 2. Side-chain modifications, *J. Membr. Biol.* **95**:171–185.

Li, X.-J., Blackshaw, S., and Snyder, S. H., 1994, Expression and localization of amiloride-sensitive sodium channel indicate a role for non taste cells in taste perception, *Proc. Natl. Acad. Sci. USA* **91**:1814–1818.

Li, X.-J., Xu, R. H., Guggino, W. B., and Snyder, S. H., 1995, Alternatively spliced forms of the α subunit of the epithelial sodium channel: distinct sites for amiloride binding and channel pore, *Mol. Pharmacol.* **47:**1133–1140.

Liddle, G. W., Bledsoe, T., and Coppage Jr., W. S., 1963, A familial renal disorder simulating primary aldosteronism but with negligible aldosterone secretion, *Trans. Assoc. Am. Physicians* **76:**199–213.

Light, D. B., McCann, F. V., Keller, T. M., and Stanton, B. A., 1988, Amiloride-sensitive cation channel in apical membrane of inner medullary collecting duct, *Am. J. Physiol.* **255:**F278–F286.

Light, D. B., Ausiello, D. A., and Stanton, B. A., 1989a, Guanine nucleotide-binding protein, alpha i-3, directly activates a cation channel in rat renal inner medullary collecting duct cells, *J. Clin. Invest.* **84:**352–356.

Light, D. B., Schwiebert, E. M., Karlson, K. H., and Stanton, B. A., 1989b, Atrial natriuretic peptide inhibits cation channel in renal inner medullary collecting duct cells, *Science* **243:**383–385.

Light, D. B., Corbin, J. D., and Stanton, B. A., 1990, Dual ion-channel regulation by cyclic GMP and cyclic GMP-dependent protein kinase, *Nature* **344:**336–339.

Lin, C., Kieber, E. T., Villalobos, A. P., Foster, M. H., Wahlgren, C., and Kleyman, T. R., 1994, Topology of an amiloride-binding protein, *J. Biol. Chem.* **269:**2805–2813.

Lindemann, B., and Van Driessche, W., 1977, Sodium-specific membrane channels of frog skin are pores: Current fluctuations reveal high turnover, *Science* **195:**292–294.

Ling, B. N., and Eaton, D. C., 1989, Effects of luminal Na$^+$ on single Na$^+$ channels in A6 cells, a regulatory role of protein kinase C, *Am. J. Physiol.* **256:**F1094–F1103.

Ling, B. N., Hinton, C. F., and Eaton, D. C., 1991, Amiloride-sensitive sodium channels in rabbit cortical collecting tubule primary cultures, *Am. J. Physiol.* **261:**F933–F944.

Lingueglia, E., Renard, S., Voilley, N., Waldmann, R., Chassande, O., Lazdunski, M., and Barbry, P., 1993a, Molecular cloning and functional expression of different molecular forms of rat amiloride binding proteins, *Eur. J. Biochem.* **216:**679–687.

Lingueglia, E., Voilley, N., Waldmann, R., Lazdunski, M., and Barbry, P., 1993b, Expression cloning of an epithelial amiloride-sensitive Na$^+$ channel. A new channel type with homologies to *C. elegans* degenerins, *FEBS Lett.* **318:**95–99.

Lingueglia, E., Renard, S., Waldmann, R., Voilley, N., Champigny, G., Plass, H., Lazdunski, M., and Barbry, P., 1994, Different homologous subunits of the amiloride-sensitive Na$^+$ channel are differently regulated by aldosterone, *J. Biol. Chem.* **269:**13736–13739.

Lingueglia, E., Champigny, G., Lazdunski, M., and Barbry, P., 1995, Cloning of a neuronal FMRFamide receptor, a peptide-gated sodium channel, *Nature*, in press.

Maack, T., Camargo, M. J. F., Kleinert, H. D., Laragh, J. H., and Atlas, S. A., 1985, Atrial natriuretic factor: Structure and functional properties, *Kidney Int.* **27:**607–615.

MacDonald, F. J., Snyder, P. M., McCray, P. J., and Welsh, M. J., 1994, Cloning, expression, and tissue distribution of a human amiloride-sensitive Na$^+$ channel, *Am. J. Physiol.* **266:**L728–L734.

MacDonald, F. J., Price, M. P., Snyder, P. M., and Welsh, M. J., 1995, Cloning and expression of the α- and γ-subunits of the human epithelial Na$^+$ channel, *Am. J. Physiol.* **268:**C1157–C1163.

MacGregor, G. G., Olver, R. E., and Kemp, P. J., 1994, Amiloride-sensitive Na$^+$ channels in fetal type II pneumocytes are regulated by G proteins, *Am. J. Physiol.* **267:**L1–L8.

MacRobbie, E. A. C., and Ussing, H. H., 1961, Osmotic behavior of the epithelial cells of frog skin, *Acta Physiol. Scand.* **53:**348–365.

Marunaka, Y., and Eaton, D. C., 1988, Effects of CDPC on single channel currents of the amiloride-sensitive NA channel from cultured renal cells, *Biophys. J.* **53:**522a.

Marunaka, Y., and Eaton, D. C., 1991, Effects of vasopressin, adenosine 3′-5′-cyclic monophosphate and cholera toxin on single amiloride-blockable Na channels in renal cells, *Am. J. Physiol.* **260:**C1071–C1084.

Marunaka, Y., Hagiwara, N., and Tohda, H., 1992, Insulin activates single amiloride-blockable Na channels in a distal nephron cell line (A6), *Am. J. Physiol.* **263:**F392–F400.

Mason, R. J., Williams, M. C., Widdicombe, J. H., Sanders, M. J., Misfeldt, D. S., and Berry, L. C., 1982, Transepithelial transport by pulmonary alveolar type II cells in primary culture, *Proc. Natl. Acad. Sci. USA* **79:**6033–6037.

Matalon, S., Bridges, R. J., and Benos, D. J., 1991, Amiloride-inhibitable Na$^+$ conductive pathways in alveolar type II pneumocytes, *Am. J. Physiol.* **260:**L90–L96.

Matalon, S., Bauer, M. L., Benos, D. J., Kleyman, T. R., Lin, C., Cragoe Jr., E. J. and O'Brodovich, H., 1993, Fetal lung epithelial cells contain two populations of amiloride sensitive Na^+ channels, *Am. J. Physiol.* **264**:L357–L364.

Mitani, S., Du, H., Hall, D. H., Driscoll, M., and Chalfie, M., 1993, Combinatorial control of touch receptor neuron expression in *Caenorhabditis elegans*, *Development* **119**:773–783.

Mohrmann, M., Cantiello, H. F., and Ausiello, D. A., 1987, Inhibition of epithelial Na^+ transport by atriopeptin, protein kinase C, and pertussis toxin, *Am. J. Physiol.* **253**:F372–F376.

Morel, F., and Doucet, A., 1986, Hormonal control of kidney functions at the cell level, *Phys. Rev.* **66**: 377–468.

Mu, D., Medzihradszky, K. F., Adams, G. W., Mayer, P., Hines, W. M., Burlingame, A. L., Smith, A. J., Cai, D., and Klinman, J. P., 1994, Primary structures for a mammalian cellular and serum copper amine oxidase, *J. Biol. Chem.* **269**:9926–9932.

Muto, S., Furaya, H., Tabei, K., and Asano, Y., 1991, Site and mechanism of action of epidermal growth factor in rabbit cortical collecting duct, *Am. J. Physiol.* **260**:F163–F169.

Nagel, W., and Crabbé, J., 1980, Mechanism of action of aldosterone on active sodium transport across toad skin, *Pflügers Arch.* **385**:181–187.

Nakada, T., Koike, H., Akiya, T., Katayama, T., Kawamata, S., Takaya, K., and Shigematsu, H., 1987, Liddle's syndrome, an uncommon form of hyporeninemic hypoaldosteronism: Functional and histopathological studies, *J. Urol.* **137**:636–640.

Negulyaev, Y. A., and Vedernikova, E. A., 1994, Sodium-selective channels in membranes of rat macrophages, *J. Membr. Biol.* **138**:37–45.

Nielsen, S., DiGiovanni, S. R., Christensen, E. I., Knepper, M. A., and Harris, H. W., 1993, Cellular and subcellular immunolocalization of vasopressin-regulated water channel in rat kidney, *Proc. Natl. Acad. Sci. USA* **90**:11663–11667.

Novotny, W. F., Chassande, O., Baker, M., Lazdunski, M., and Barbry, P., 1994, Diamine oxidase is the amiloride binding protein and is inhibited by amiloride analogues, *J. Biol. Chem.* **269**:9921–9925.

O'Brodovich, H., 1991, epithelial ion transport in the fetal and perinatal lung, *Am. J. Physiol.* **261**:C555–C564.

O'Brodovich, H., Canessa, C., Ueda, J., Rafii, B., Rossier, B. C., and Edelson, J., 1993, Expression of the epithelial Na^+ channel in the developing rat lung, *Am. J. Physiol.* **265**:C491–C496.

Oh, Y., and Benos, D. J., 1993, Single channel characteristics of a purified bovine renal amiloride-sensitive Na^+ channel in planar lipid bilayers, *Am. J. Physiol.* **264**:C1489–C1499.

Oh, Y., Matalon, S., Kleyman, T. R., and Benos, D. J., 1992, Biochemical evidence for the presence of an amiloride binding protein in adult alveolar type II pneumocytes, *J. Biol. Chem.* **267**:18498–18504.

Oh, Y. S., Smith, P. R., Bradford, A. L., Keeton, D., and Benos, D. J., 1993, Regulation by phosphorylation of purified epithelial Na^+ channels in planar lipid bilayers, *Am. J. Physiol.* **265**:C85–C91.

Ohara, A., Matsunaga, H., and Eaton, D. C., 1993, G protein activation inhibits amiloride-blockable highly selective sodium channels in A6 cells, *Am. J. Physiol.* **264**:C352–C360.

Orser, B. A., Bertlik, M., Fedorko, L., and O'Brodovich, H., 1991, Cation selective channel in fetal alveolar type II epithelium, *Biochim. Biophys. Acta* **1094**:19–26.

Pácha, J., Frindt, G., Antonian, L., Silver, R. B., and Palmer, L. G., 1993, Regulation of Na channels of the rat cortical collecting tubule by aldosterone, *J. Gen. Physiol.* **102**:25–42.

Palmer, L. G., 1990, Epithelial Na^+ channels: The nature of the conducting pore, *Renal Physiol. Biochem.* **13**:51–58.

Palmer, L. G., 1992, Epithelial Na channels: Function and diversity, *Annu. Rev. Physiol.* **54**:51–66.

Palmer, L. G., and Frindt, G., 1986, Amiloride-sensitive Na channels from the apical membrane of the rat cortical collecting tubule, *Proc. Natl. Acad. Sci. USA* **83**:2767–2770.

Palmer, L. G., and Frindt, G., 1987, Effects of cell Ca^{2+} and pH on Na^+ channels from rat cortical collecting tubules, *Am. J. Physiol.* **253**:F333–F339.

Palmer, L. G., Li, J. H.-Y., Lindemann, B., and Edelman, I. S., 1982, Aldosterone control of the density of sodium channels in the toad urinary bladder, *J. Membr. Biol.* **57**:59–71.

Palmer, L. G., Corrthesy-Theulaz, M., Gaeggeler, H. P., Kraehenbuhl, J. P., and Rossier, B., 1990, Expression of epithelial Na channels in *Xenopus* oocytes, *J. Gen. Physiol.* **96**:23–46.

Petzel, D., Ganz, M. B., Nestler, E. J., Lewis, J. J., Goldenring, J., Akcicek, F., and Hayslett, J. P., 1992,

Correlates of aldosterone-induced increases in Ca_{i2+} and Isc suggest that Ca_{i2+} is the second messenger for stimulation of apical membrane conductance, *J. Clin. Invest.* **89:**150–156.

Powers, R. D., Borland, R. M., and Biggers, J. D., 1977, Amiloride-sensitive rheogenic Na^+ transport in rabbit blastocyst, *Nature* **270:**603–604.

Puoti, A., May, A., Canessa, C. M., Horisberger, J. D., Schild, L., and Rossier, B. C., 1995, The highly selective low-conductance epithelial Na channel of *Xenopus laevis* A6 kidney cells, *Am. J. Physiol.* **269:**C188–C197.

Quinton, P. M., 1983, Chloride impermeability in cystic fibrosis, *Nature* **301:**421–422.

Quinton, P. M., 1990, Cystic fibrosis: A disease in electrolyte transport, *FASEB J.* **4:**2709–2717.

Rechkemmer, G., and Halm, D. R., 1989, Aldosterone stimulates K secretion across mammalian colon independent of Na absorption, *Proc. Natl. Acad. Sci. USA* **86:**397–401.

Reif, M. C., Troutman, S. L., and Schafer, J. A., 1986, Sodium transport by rat cortical collecting tubule. Effects of vasopressin and desoxycorticosterone, *J. Clin. Invest.* **77:**1291–1298.

Renard, S., Lingueglia, E., Voilley, N., Lazdunski, M., and Barbry, P., 1994, Biochemical analysis of the membrane topology of the amiloride-sensitive Na^+ channel, *J. Biol. Chem.* **269:**12981–12986.

Renard, S., Voilley, N., Bassilana, F., Lazdunski, M., and Barbry, P., 1995, Localization and regulation by steroids of the α, β, and γ subunits of the amiloride-sensitive Na^+ channel in colon, lung and kidney, *Pflügers Arch., Eur. J. Physiol.* **430:**299–307.

Riordan, J. R., Rommens, J. M., Kerem, B.-S., Alon, N., Rohzmahel, R., Grzelczak, Z., Zielenski, J., Lok, S., Plavsic, N., Chou, J.-L., Drumm, M. L., Iannuzzi, M. C., Collins, F. S., and Tsui, L.-C., 1989, Identification of the cystic fibrosis gene: Cloning and characterization of complementary DNA, *Science* **245:**1066–1073.

Rodriguez, J. A., Biglieri, E. G., and Schambelan, M., 1981, Pseudohyperaldosteronism with renal tubular resistance to mineralocorticoid hormones, *Trans. Assoc. Am. Physicians* **94:**172–182.

Rodriguez-Commes, J., Isales, C., Kalghati, L., Gasalla-Herraiz, J., and Hayslett, J. P., 1994, Mechanism of insulin-stimulated electrogenic sodium transport, *Kidney Int.* **46:**666–674.

Rosenberg, R. L., and East, J. E., 1992, Cell-free expression of functional *Shaker* potassium channels, *Nature* **360:**166–169.

Rossier, B. C., 1978, Role of RNA in the action of aldosterone on Na^+ transport, *J. Membr. Biol.* **40:**187–197.

Rossier, B. C., and Palmer, L. G., 1992, Mechanism of aldosterone action on sodium and potassium transport, in: *The Kidney: Physiology and Pathophysiology*, 2nd Ed. (D. W. Seldin and G. Giebsich, eds.), Raven Press, New York, pp. 1373–1409.

Rotin, D., Bar-Sagi, D., O'Brodovich, H., Merilainen, J., Lehto, V. P., Canessa, C. M., Rossier, B. C., and Downey, G. P., 1994, An SH3 binding region in the epithelial Na^+ channel (α rENaC) mediates its localization at the apical membrane, *EMBO J.* **13:**4440–4450.

Ruben, P., Johnson, J. W., and Thompson, S., 1986, Analysis of FMRF-amide effects on *Aplysia* bursting neurons, *J. Neurosci.* **6:**252–259.

Rüsch, A., Kros, C. J., and Richardson, G. P., 1994, Block by amiloride and its derivatives of mechano-electrical transduction in outer hair cells of mouse cochlear cultures, *J. Physiol. (London)* **474:**75–86.

Sandle, G. I., and Binder, H. J., 1987, Corticosteroids and intestinal ion transport, *Gastroenterology* **93:** 188–196.

Sandle, G. I., Higgs, N., Crowe, P., Marsh, M. N., Venkatesan, S., and Peters, T. J., 1990, Cellular basis for defective electrolyte transport in inflamed human colon, *Gastroenterology* **99:**97–105.

Sariban-Sohraby, S., and Benos, D. J., 1986, Detergent solubilization, functional reconstitution, and partial purification of epithelial amiloride-binding protein, *Biochemistry* **25:**4639–4646.

Sariban-Sohraby, S., Burg, M., Wiesmann, W. P., Chiang, P. K., and Johnson, J. P., 1984, Methylation increases sodium transport into A6 apical membrane vesicles: Possible mode of aldosterone action, *Science* **225:**745–746.

Sariban-Sohraby, S., Sorscher, E. J., Brenner, B. M., and Benos, D. J., 1988, Phosphorylation of a single subunit of the epithelial Na^+ channel protein following vasopressin treatment of A6 cells, *J. Biol. Chem.* **263:**13875–13879.

Sariban-Sohraby, S., Abramow, M., and Fisher, R. S., 1992, Single-channel behavior of a purified epithelial Na^+ channel subunit that binds amiloride, *Am. J. Physiol.* **263:**C1111–C1117.

Sariban-Sohraby, S., Fisher, R. S., and Abramow, M., 1993, Aldosterone-induced and GTP-stimulated methylation of a 90-kDa polypeptide in the apical membrane of A6 epithelia, *J. Biol. Chem.* **268:**26613–26617.

Schafer, J. A., and Hawk, C. T., 1992, Regulation of Na^+ channels in the cortical collecting duct by AVP and mineralocorticoids, *Kidney Int.* **41:**255–268.

Schafer, J. A., and Troutman, S. L., 1990, cAMP mediates the increase in apical membrane Na^+ conductance produced in the rat CCD by vasopressin, *Am. J. Physiol.* **259:**F823–F831.

Schafer, J. A., Troutman, S. L., and Schlatter, E., 1990, Vasopressin and mineralocorticoid increase apical membrane driving force for K^+ secretion in rat CCD, *Am. J. Physiol.* **258:**F199–F210.

Schiffman, S. S., Frey, A. E., Suggs, M. S., Cragoc Jr., E. J., and Erickson, R. P., 1990, The effect of amiloride analogs on taste responses in gerbil, *Physiol. Behav.* **47:**435–441.

Schild, L., Canessa, C. M., Shimkets, R. A., Gautschi, I., Lifton, R. P., and Rossier, B. C., 1995, A mutation in the epithelial sodium channel causing Liddle disease increases channel activity in the *Xenopus laevis* oocyte expression system, *Proc. Natl. Acad. Sci. USA* **92:**5699–5703.

Schwartz, G. J., and Burg, M. B., 1978, Mineralocorticoid effects on cation transport by cortical collecting tubules *in vitro*, *Am. J. Physiol.* **235:**F576–F585.

Shimkets, R. A., Warnock, D. G., Bositis, C. M., Nelson-Williams, C., Hansson, J. H., Schambelan, M., Gill Jr., J. R., Ulick, S., Milora, R. V., Findling, J. W., Canessa, C. M., Rossier, B. C., and Lifton, R. P., 1994, Liddle's syndrome: Heritable human hypertension caused by mutations in the β subunit of the epithelial sodium channel, *Cell* **79:**407–414.

Silver, R. B., Frindt, G., Windhager, E. E., and Palmer, L. G., 1993, Feedback regulation of Na channels in rat CCT. I. Effects of inhibition of Na pump, *Am. J. Physiol.* **264:**F557–F564.

Snyder, P. M., McDonald, F. J., Stokes, J. B., and Welsh, M. J.,, 1994, Membrane topology of the amiloride-sensitive epithelial sodium channel, *J. Biol. Chem.* **269:**24379–24383.

Stanton, B., Janzen, A., Wade, J., DeFronzo, R., and Giebisch, G., 1985, Ultrastructure of rat initial collecting tubule: Effect of adrenal corticosteroid treatment, *J. Clin. Invest.* **75:**1317–1326.

Stenvinkel, P., Bolinder, J., and Alvestrand, A., 1992, Effects of insulin on renal hemodynamics and the proximal and distal tubular sodium handling in healthy subjects, *Diabetologia* **35:**1042–1048.

Stokes, J. B., and Kokko, J. P., 1977, Inhibition of sodium transport by prostaglandin E2 across the isolated, perfused rabbit collecting tubule, *J. Clin. Invest.* **59:**1099–1104.

Stone, D. K., Seldin, D. W., Kokko, J. P., and Jacobson, N. R., 1983, Mineralocorticoid modulation of rabbit medullary collecting duct acidification, *J. Clin. Invest.* **72:**77–83.

Strang, L. B., 1991, Fetal lung liquid: Secretion and reabsorption, *Physiol. Rev.* **71:**991–1133.

Stutts, M. J., Canessa, C. M., Olsen, J. C., Hamrick, M., Cohn, J. A., Rossier, B. C., and Boucher, R. C., 1995, CFTR as a cAMP-dependent regulator of sodium channels, *Science* **269:**847–850.

Tomita, K., Pisano, J. J., and Knepper, M. A., 1985, Control of sodium and potassium transport in the collecting duct of the rat. Effects of bradykinin, vasopressin, and desoxycorticosterone, *J. Clin. Invest.* **76:**132–136.

Tomkiewicz, R. P., App, E. M., Zayas, J. G., Ramirez, O., Church, N., Boucher, R. C., Knowles, M. R., and King, M., 1993, Amiloride inhalation therapy in cystic fibrosis: Influence on ion content, hydration and rheology of sputum, *Am. Rev. Respir. Dis.* **148:**1002–1007.

Tse, C. M., Levine, S. A., Yun, C. H., Brant, S. R., Pouysségur, J., Montrose, M. H., and Donowitz, M., 1993, Functional characteristics of a cloned epithelial Na^+/H^+ exchanger (NHE3): Resistance to amiloride and inhibition by protein kinase C, *Proc. Natl. Acad. Sci. USA* **90:**9110–9114.

Tsui, L.-C., 1992, Mutations and sequence variations detected in the cystic fibrosis transmembrane conductance regulator (CFTR) gene: A report from the Cystic Fibrosis Genetic Analysis Consortium, *Hum, Mutat.* **1:**197–203.

Turnamian, S. G., and Binder, H. J., 1989, Regulation of active sodium and potassium transport in the distal colon of the rat, *J. Clin. Invest.* **84:**1924–1929.

Turnheim, K., Plass, H., Grasl, M., Krinavek, P., and Wiener, H., 1986, Sodium absorption and potassium secretion in rabbit colon during sodium deficiency, *Am. J. Physiol.* **250:**F235–F245.

Valera, S., Hussy, N., Evans, R. J., Adami, N., North, R. A., Surprenant, A., and Buell, G., 1994, A new class of ligand-gated ion channel defined by P2x receptor for extracellular ATP, *Nature* **371:**516–519.

Van Driessche, W., and Lindemann, B., 1979, Concentration dependence of currents through single sodium-selective pores in frog skin, *Nature* **282:**519–520.

Van Renterghem, C., and Lazdunski, M., 1991, A new non-voltage-dependent, epithelial-like Na$^+$ channel in vascular smooth muscle cells, *Eur. J. Physiol.* **419:**401–408.

Varnum, S. M., and Wormington, W. M., 1990, Deadenylation of maternal mRNAs during *Xenopus* oocyte maturation does not require specific cis-sequences: A default mechanism for translational control, *Genes Dev.* **4:**2278–2286.

Verity, K., and Fuller, P. J., 1994, Isolation of a rat amiloride-binding protein cDNA clone: Tissue distribution and regulation of expression, *Am. J. Physiol.* **266:**C1505–C1512.

Verkman, A. S., 1989, Mechanisms and regulation of water permeability in renal epithelial, *Am. J. Physiol.* **257:**C837–C850.

Verrier, B., Champigny, G., Barbry, P., Gérard, C., Mauchamp, J., and Lazdunski, M., 1989, Identification and properties of a novel type of Na$^+$-permeable amiloride-sensitive channel in thyroid cells, *Eur. J. Biochem.* **183:**499–505

Vigne, P., Champigny, G., Marsault, R., Barbry, P., Frelin, C., and Lazdunski, M., 1989, A new type of amiloride-sensitive cationic channel in endothelial cells of brain microvessels, *J. Biol. Chem.* **264:**7663–7668.

Voilley, N., Lingueglia, E., Champigny, G., Mattéi, M.-G., Waldmann, R., Lazdunski, M., and Barbry, P., 1994, The amiloride-sensitive Na$^+$ channel in lung epithelial cells: Biophysical properties, pharmacology, ontogenesis, molecular cloning, expression and chromosomic localization of the human channel, *Proc. Natl. Acad. Sci. USA* **91:**247–251.

Voilley, N., Bassilana, F., Mignon, C., Merscher, S., Mattéi, M.-G., Carle, G. F., Lazdunski, M., and Barbry, P., 1995, Cloning, chromosomal localization and physical linkage of the β and γ subunits of the human epithelial amiloride-sensitive sodium channel, *Genomics* **28:**560–565.

Voilley, N., Galibert, A., Bassilana, F., Renard, S., Lingueglia, E., Le Néchet, S., Champigny, G., Hoffman, P., Lazdunski, M., and Barbry, P., 1996, From primary structure to function of the amiloride-sensitive Na$^+$ channel, *Comp. Biochem. Physiol*, in press.

Wahli, W., and Martinez, E., 1991, Superfamily of steroid nuclear receptors: Positive and negative regulators of gene expression, *FASEB J.* **5:**2243–2249.

Waldmann, R., Champigny, G., and Lazdunski, M., 1995, Functional degenerin-containing chimeras identify residues essential for amiloride-sensitive Na$^+$ channel function, *J. Biol. Chem.* **270:**11735–11737.

Waldmann, R., Champigny, G., Bassilana, F., Voilley, N., Lazdunski, M., 1995b, Molecular cloning and functional expression of a novel amiloride-sensitive Na$^+$ channel, *J. Biol. Chem.*, in press.

Walker, T. C., Fidelman, M. L., Watlington, C. O., and Biber, T. U. L., 1984, Insulin decreases apical cell membrane resistance in cultured kidney cells (A6), *Biochem. Biophys. Res. Commun.* **124:**614–618.

Wang, C., Chan, T. K., Yeung, R. T. T., Coghlan, J. P., Scoggins, B. A., and Stockigt, J. R., 1981, The effect of triamterene and sodium intake on renin, aldosterone, and erythrocyte sodium transport in Liddle's syndrome, *J. Clin. Endocr. Metab.* **52:**1027–1032.

Warden, D. H., and Stokes, J. B., 1993, EGF and PGE2 inhibit rabbit CCD Na$^+$ transport by different mechanisms: PGE2 inhibits Na$^+$-K$^+$ pump, *Am. J. Physiol.* **264:**F670–F677.

Weber, W. M., Asher, C., Garty, H., and Clauss, W., 1992, Expression of amiloride-sensitive Na$^+$ channels of hen lower intestine in *Xenopus* oocytes: Electrophysiological studies on the dependence of varying NaCl intake, *Biochim. Biophys. Acta* **1111:**159–164.

Weber, W. M., Blank, U., and Clauss, W., 1995, Regulation of electrogenic Na$^+$ transport across leech skin, *Am. J. Physiol.* **268:**R605–R613.

Wehling, M., Christ, M., and Gerzer, R., 1993, Aldosterone-specific membrane receptors and related rapid, non-genomic effects, *Trends Pharmacol. Sci.* **14:**1–4.

Welsh, M. J., and Smith, A. E., 1993, Molecular mechanisms of CFTR chloride channel dysfunction in cystic fibrosis, *Cell* **73:**1251–1254.

Weng, K., and Wade, J. B., 1994, Effect of brefeldin A on ADH-induced transport of toad bladder, *Am. J. Physiol.* **266:**C1069–C1076.

Wiesmann, W., Sinha, S., Yates, J., and Klahr, S., 1978, Cholinergic agents inhibit sodium transport across the isolated toad bladder, *Am. J. Physiol.* **235:**F564–F569.

Will, P. C., Lebowitz, J. L., and Hopfer, U., 1980, Induction of amiloride-sensitive sodium transport in the rat colon by mineralocorticoids, *Am. J. Physiol.* **238:**F261–F268.

Yamaya, M., Finkbeiner, W. E., Chun, S. Y., and Widdicombe, J. H., 1992, Differentiated structure and function of cultures from human tracheal epithelium, *Am. J. Physiol.* **262:**L713–L724.

Yanase, M., and Handler, J. S., 1986, Activators of protein kinase C inhibit sodium transport in A6 epithelia, *Am. J. Physiol.* **250:**C517–C523.

Yue, G., Hu, P., Oh, Y., Jilling, T., Shoemaker, R. L., Benos, D. J., Cragoe Jr., E. J., and Matalon, S., 1993, Culture-induced alterations in alveolar type II cell Na^+ conductance, *Am. J. Physiol.* **265:**C630–C640.

Zeiske, W., Wills, N. K., and Van Driessche, W., 1982, Sodium channels and amiloride-induced noise in the mammalian colon epithelium, *Biochim. Biophys. Acta* **688:**201–210.

Zeiske, W., Onken, H., Schwarz, H. J., and Graszynski, K., 1992, Invertebrate epithelial Na^+ channels: Amiloride-induced current-noise in crab gill, *Biochim. Biophys. Acta* **1105:**245–252.

CHAPTER 5

VDAC, A CHANNEL IN THE OUTER MITOCHONDRIAL MEMBRANE

MARCO COLOMBINI, ELIZABETH BLACHLY-DYSON, and MICHAEL FORTE

1. BACKGROUND

Proteins that form aqueous channels in membranes generate conduction pathways with a variety of shapes and sizes. Perhaps the largest channel-forming protein is the 2-MDa ryanodine receptor while the smallest may be gramicidin. However, the size of the conducting pathway is not correlated with the amount of protein mass needed to make up the structure, as demonstrated by the fact that some of the narrowest conducting pathways are produced by very large amounts of protein (e.g., 0.3 MDa for the $Na^+/K^+/Ca^{2+}$ channel family). In contrast, the focus of this review, the voltage-dependent anion channel (VDAC) of the mitochondrial outer membrane, produces one of the largest aqueous pathways from a single 30-kDa protein. VDAC also demonstrates that functional complexity does not seem to correlate well with the amount of protein used to form a channel. VDAC has a small amount of protein mass but displays complex behavior. It has two voltage-gating processes, can be controlled by metabolites and regulatory proteins, is able to form complexes with other proteins and enzymes, and responds to the protein concentration of the cytoplasm (Colombini, 1994). Thus, many functions are packed into a single, relatively small VDAC protein.

The molecular structure of channels is also quite variable. The only structures known with atomic resolution are gramicidin and some members of the bacterial porin family. The former forms a variety of structures but the membrane channel is a β-helix

MARCO COLOMBINI • Department of Zoology, University of Maryland, College Park, Maryland 20742. ELIZABETH BLACHLY-DYSON and MICHAEL FORTE • Vollum Institute, Oregon Health Sciences University, Portland, Oregon 97201.

Ion Channels, Volume 4, edited by Toshio Narahashi, Plenum Press, New York, 1996.

(for discussion, see Durkin *et al.*, 1990); the latter forms β-barrels (Weiss *et al.*, 1990). Other channel-forming proteins seem to form the transmembrane aqueous pathway in different ways, using α-helices (the cystic fibrosis chloride channel; Akabas *et al.*, 1994b), a combination of α-helices and extended regions (the nicotinic acetylcholine receptor; Akabas *et al.*, 1994a), extended regions only (K^+ channels; Yellen *et al.*, 1991), and β-barrel/α-helix combinations (VDAC). This variety of structures is likely to increase as more detailed structural information is obtained about additional channels.

One goal of studying membrane channels is to interrelate electrical properties of the channel, the molecular rearrangements that underlie these properties, and the *in vivo* physiological function of individual channel types. In the case of VDAC, as with other channels, more is known about the properties of the channel than about its physiological role. Most investigators agree however that VDAC is the main pathway by which metabolites cross the outer mitochondrial membrane. These metabolites can reach a molecular weight of almost 1000, and therefore the permeability pathway must be large. If VDAC functions to control the outer membrane's permeability for these molecules and ions, it may work quite differently than channels whose function is to control the permeability of small alkali-metal ions. The goal of this review is to outline our current understanding of VDAC: the properties of the channel as assayed following reconstitution into planar phospholipid membranes, the molecules that generate these channels, the nature of the conformational transitions underlying basic channel properties, and the regulatory interactions that influence VDAC's function and that may represent important functional interactions *in vivo*.

2. FUNDAMENTAL CHANNEL PROPERTIES

All VDACs described to date have a remarkably conserved set of biophysical properties. Whether isolated from yeast or humans, VDAC proteins form channels with a roughly similar single-channel conductance (4 nS in 1 M KCl), open channel selectivity (2:1 preference for the conduction of Cl^- over K^+ for a gradient of 1 M versus 0.1 M KCl), and voltage-dependent conductance (Colombini, 1989). In addition, each channel closes to lower-conducting or "closed" states when either positive or negative potentials are applied. Thus, although the details listed below are generated primarily from study of VDACs isolated from fungi like yeast or *Neurospora crassa*, the same general properties are observed for VDAC isolated from essentially all eukaryotes.

2.1. Two Gating Processes

When reconstituted into planar phospholipid membranes, VDACs are open most of the time at low voltages (~ 10 mV), although they undergo rare transitions to closed states. Thus, when channels insert into the membrane from the aqueous phase at low potentials, one typically observes current increments that form a staircase effect (Fig. 1). There is open-channel noise which manifests itself more and more as the number of reconstituted channels increases. Typically, channels insert into such membranes as single, discrete conducting units defined as single channels. It is unlikely that the single

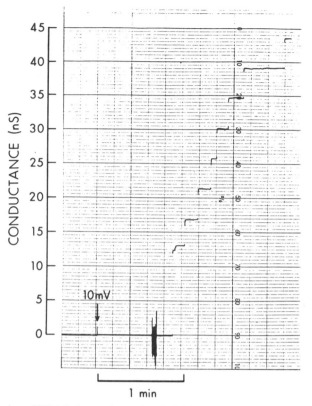

FIGURE 1. Insertion of VDACs into a planar phospholipid membrane. The membrane was made by the monolayer method of Montal and Mueller as per Schein *et al.* (1976) using soybean phospholipids. The aqueous solution was 1.0 M KCl, 5 mM $CaCl_2$. The transmembrane voltage was clamped at 10 mV beginning where indicated. Five microliters of Triton X100 solubilized *N. crassa* VDAC were added at the stirring artifact. The first conductance increment is the insertion of a triplet of VDACs, a common occurrence for *N. crassa* VDAC. Reproduced with permission from Colombini (1980a).

channel is actually the insertion of two channels side by side because when a single channel closes, the drop in conductance is usually more than half the original conductance. Smaller insertion events are interpreted as channels inserting in one of the closed states. Channels inserting with low conductance have been observed to open and then behave like normal, fully open channels. When channels do not behave like the majority, they are assumed to be damaged, modified, or improperly folded.

When the membrane potential is increased, typically above 30 mV, VDACs undergo transitions to closed states (Schein *et al.*, 1976; Colombini, 1989). Although both the open and closed states are permeable to simple salts (the former more permeable than the latter), their permeability to mitochondrial metabolites (mostly organic anions) is dramatically different (see Table I). Thus, from a physiological perspective, the terms open and closed are quite appropriate. Closed channels will reopen, but in VDAC from most species they tend to remain closed until the voltage is reduced. In addition, although

TABLE I
Ion Selectivity of VDAC[a]

Salt	P_{anion}/P_{cation} [b]	
	Open state	Closed state
K^+ Cl^-	5.1	0.15
Na^+ $H_2PO_4^-$	1.9	0.093
Na^+ HPO_4^{2-}	0.53	0.035
Na^+ Succinate^{2-}	0.61	0.021
Na^+ Citrate^{3-}	0.54	0.044

[a]From Hodge and Colombini, in preparation.
[b]The $H_2PO_4^-$ and HPO_4^{2-} experiments were run at pH 5.5 and 8, respectively. The rest were run at pH 7.

there is a single open state, there does not appear to be a single closed state but rather a variety of possible closed states, some more stable than others. For example, immediately after closing, channels reopen more readily than after they have been closed for a long time, suggesting that structural rearrangements take place in order to achieve more stable closed conformations. These rearrangements are sometimes visible as transitions between states shortly after the membrane potential is elevated. Indeed, while the voltage is kept constant, the channel often returns to the highest conducting state (the open state) before closing again to a closed state with a different conductance (Fig. 2). In addition, there are electrically silent changes that can be detected only by the fact that the rate of reopening of the channels is reduced with increasing time in the closed state. In contrast to other channel types, VDAC closure is observed at both positive and negative potentials (Schein et al., 1976). This is true not only in multichannel membranes but also in single-channel membranes (Fig. 3), demonstrating that a single channel forms one conductance pathway that responds to both positive and negative potentials.

2.2. Asymmetric Structure Yields Symmetrical Behavior

The ability of individual channels to behave in a symmetrical manner seems best explained by a symmetrical structure. Initial indications supported the conclusion that VDAC was a homodimer of 30-kDa subunits. The detergent-solubilized VDAC from rat liver seemed to be a dimer (Linden and Gellerfors, 1983), and there is a linear dependence of the number of channels reconstituted into a planar membrane on the amount of detergent-solubilized protein added to the aqueous phase (Roos et al., 1982). However, more precise experiments indicate that a single VDAC is formed by a single VDAC protein.

The ability of VDAC from N. crassa to make two-dimensional crystals was exploited to estimate the mass of one channel. Figure 4 shows the surface topography of such a crystal (upper panel) and a view of its internal structure after freezing in vitreous ice (lower panel). Mannella estimated that there was not enough mass per pore in these crystals for each channel to be composed of a dimer of 30-kDa proteins (Mannella, 1986, 1987). Consistent with this conclusion, Thomas and coworkers (1991) measured the mass

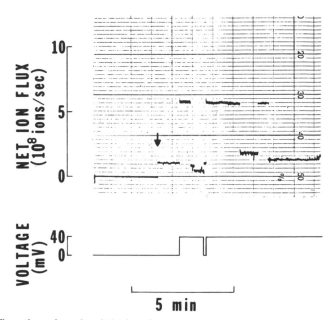

FIGURE 2. The voltage-dependent behavior of a single VDAC in the presence of a salt gradient. The conditions were as in Fig. 1 except that one side of the membrane contained ten times less KCl. The indicated voltage refers to the value on the high-salt side. The recorded current was converted to a net ion flux. The arrow indicates the point in time at which a single channel inserted. When the voltage was raised to 40 mV, channel closure caused the current to decline. The different current levels indicate different conducting states of this one channel. Four different low-conductance (closed) states are visible. These states may vary in either conductance and/or selectivity. Reproduced with permission from Colombini (1980b).

of an area of the crystalline array (using tobacco mosaic virus as a standard) by scanning transmission electron microscopy. Since the number of channels per unit area was known, it was possible to estimate that each channel in an array had an effective mass of 44 kDa. Since one VDAC polypeptide is 30 kDa, one channel can only contain one polypeptide. The extra mass is likely due to the association of phospholipids and/or sterols with the VDAC protein. Consistent with this idea, there is evidence that sterols are part of the basic structure of VDAC (Pfaller et al., 1985).

Taking a different approach, Peng and coworkers (1992b) tried to produce VDACs composed of two different VDAC polypeptides by expressing wild-type genes and genes containing site-directed mutations that result in channels with altered selectivity together in yeast. If VDAC channels are dimers, one would expect that three kinds of channels would be present in these cells: 1) wild-type dimers, 2) mutant dimers, and 3) mixed dimers with an intermediate selectivity. When individual channels from such cells were examined, channels with wild-type selectivity and channels with mutant selectivity were observed with almost equal frequency. No channels were found with intermediate selectivity. If dimers had formed either as two adjacent channels, two channels in tandem, or two semi-cylindrical halves forming one large pore, channels with intermediate selectivity would have been detected. It is possible that the failure to observe

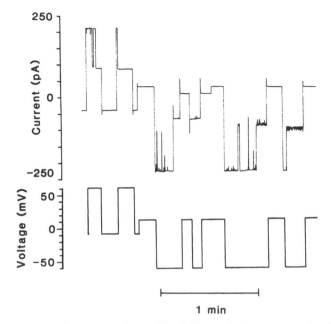

FIGURE 3. Two voltage-gating processes in one VDAC. The conditions were as in Fig. 1 except that the VDAC was isolated from the yeast *S. cerevisiae*. The aqueous phase was 1.0 M KCl, 1 mM $CaCl_2$, 5 mM MES at pH 5.8. Channel closure occurs at both positive and negative applied voltages. The channels occupy the high-conducting, open state at low potentials.

"hybrid" channels reflects some unknown process related to the translation of VDAC transcripts and the assembly and targeting of newly formed VDAC proteins to mito-chondria which prevents "mixing" of different VDAC proteins, although this is un-likely. Statistically, enough channels were examined to reduce the probability of having missed hybrid channels by random chance to 1 in 10^7.

Thus, these findings agree with the electron microscopy experiments in the conclu-sion that one channel is composed of only one 30-kDa polypeptide chain. Since the primary structure of VDAC proteins does not contain palindromic sequences or any significant structural repeats, it is very likely that the molecule is fundamentally asym-metric. In fact, the inherent structural asymmetry of the channel can be readily observed as functional asymmetry when channels are reconstituted in unusual lipids such as diphytanoylphosphatidylcholine (Liu, unpublished observations) or when point muta-tions are engineered at particular sites (Zizi *et al.*, 1995). Thus, while the presence of two symmetric gating processes in planar phospholipid membranes has been conserved throughout evolution, it is possible to establish experimental conditions in which gating becomes more asymmetric.

2.3. Selectivity of Open and Closed States

The open channels differ from closed channels not only in conductance but in selectivity. The open state prefers anions, e.g., Cl^- over K^+ by a factor of 5 (Table I) at 0.1

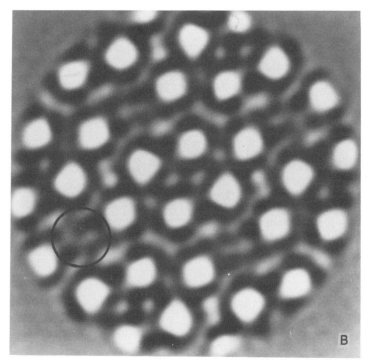

FIGURE 4. Electron microscopy of two-dimensional crystals of VDACs induced in the outer membranes of *N. crassa* mitochondria by slow phospholipase treatment. Both figures have been computer filtered and averaged with respect to the six-channel repeating unit. The upper panel shows a surface view obtained by freeze-drying and shadowing these crystals (courtesy of Lorie Thomas). The openings of the channels are seen as dark depressions. The lower figure shows the crystals after freezing in a thin layer of vitreous ice (courtesy of C. Mannella). The white areas are the channel lumens and the dark areas are the protein structures. The circle indicates the center of a location surrounded by six channels. This location corresponds to the elevated (white) region in the upper panel.

M salt concentrations (measurements made at higher salt concentrations and steeper salt gradients yield lower values; cf. Colombini, 1989). The closed states generally favor cations (Colombini, 1980b; Benz *et al.*, 1990), although the selectivities vary depending on the closed state that is achieved (Zhang and Colombini, 1990). When organic anions are examined, channel closure results in large drops in the selectivity for these ions (Table I). The selectivities for succinate and citrate drop dramatically, and thus VDAC becomes a poor conduit for these anions. Higher voltages or the addition of agents that induce VDAC closure (see Section 6) result in channels entering states of much lower conductance. These lower-conducting states have not been well studied but there is evidence that the selectivity for cations is even greater in these states (Peng and Colombini, unpublished). This large change in selectivity for small ions and organic compounds is consistent with the idea that the transition from open to closed channel is associated with a large change in the overall structure of the protein (see Section 4).

2.4. Biophysical Basis for Selectivity

Traditionally, the study of channel selectivity has focused on the ability of these molecules to distinguish among similar small ions such as the alkali metals, alkaline earths, and halides. The work of Eisenmann and Krasne (1975), however, has demonstrated the importance of ion dehydration on selectivity; the energy needed to dehydrate the ion is just as important as the binding energy to specific sites on the protein. More recently the discussion has focused on how one can get specific binding and, at the same time, fast throughput.

In the case of proteins that form large aqueous pores, selectivity among ions that are much smaller than the pore must rest largely on more long-range forces than direct atomic interactions (electrostatics, dipole, or van der Waals). In the case of VDAC, the large pore size (2.5–3 nm in diameter) means that ions located in the center of the pore may feel a rather different electrostatic environment compared to those moving closer to the walls of the channel. In fact, the distribution of electrostatic charge on the walls of the pore may be uneven, resulting in preferred conduits. The extent of the inhomogeneity depends on the ionic strength, and the effect depends on the size of the ion. Thus, ion flow through large channels is likely to represent a composite of a number of distinctly different processes when compared to ion flow through narrow channels.

The selectivity of a channel for different ions is often estimated by determining the zero-current potential (reversal potential) in the presence of an ion gradient. By modifying a theory developed by Teorell (1953) for ion flow through ion-exchange membranes, a theoretical description of ion flow through VDAC was achieved (Zambrowicz and Colombini, 1993). This large channel theory provides values for the reversal potential under various conditions of ion activities and activity gradients. The theory accounts at least qualitatively and often quantitatively for the reversal potentials observed under different conditions. Hence, the model is likely to reflect to some extent the way the ions actually flow through VDAC.

Large channel theory divides the cylindrical plug of solution in the channel into two compartments: an outer shell of solution close to the wall of the channel that contains immobile charge and a central cylindrical compartment that is totally devoid of the effects of fixed charge. These two pathways are in parallel, and current may even flow in opposite directions in the two compartments. The physical size of each compartment depends on the ionic strength. The thickness of the cylindrical shell is defined by the cylindrical equivalent of the debye length, i.e., the distance at which the surface potential decays to 1/e of the value at the surface.

This rather crude framework accounts for some unusual observations. The reversal potential does not increase monotonically with increase in activity gradient of the salt but goes through a maximum (Fig. 5). The reversal potential varies with ionic strength even if the activity ratio does not. It goes from one value at low ionic strength to another at high ionic strength. At low ionic strength it is dominated by the properties of the cylindrical shell and thus reflects mainly the charge on the walls of the channel. At high ionic strength it reflects the properties of the central cylinder and thus the properties of the ions making up the salt (e.g., the difference in ion mobility).

Thus, ion flow through VDAC's large aqueous pore is probably more complex than one might have imagined at first. The selectivity change accompanying channel closure is therefore a result of not only a change in the net charge on the wall of the channel, but also a change in the diameter of the pore, causing the effect of the cylindrical shell to become more dominant. This model is consistent with many of the results obtained by analysis of VDACs containing site-directed mutations as described below.

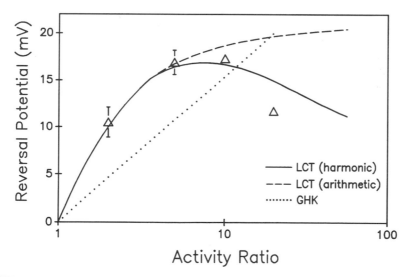

FIGURE 5. The dependence of the channel selectivity on the salt gradient across the membrane. Ion selectivity was quantified as the zero-current potential (reversal potential) in the presence of a KCl salt gradient. The low-salt-side was always at a KCl activity of 0.06 while the high-salt side was varied as indicated. The lines represent the different theoretical predictions of different theories: GHK is Goldman/Hodgkin/Katz; LCT (arithmetic) and LCT (harmonic) are two versions of the large channel theory. Reproduced with permission from Zambrowicz and Colombini (1993).

3. STRUCTURE OF VDAC

3.1. Genes Encoding VDAC

VDAC genes have now been isolated from mammals and plants, as well as from fungi. While originally only single sequences were reported from the fungi *Neurospora crassa* and *Saccharomyces cerevisiae*, it is now clear that many species have multiple VDAC genes. The spectrum of genes encoding VDAC isoforms has been most completely characterized in humans. Human cDNAs representing two different VDAC genes (HVDAC1 and HVDAC2; Blachly-Dyson *et al.*, 1993) have been identified. HVDAC1 encodes the protein purified from human B-lymphocytes by Kayser *et al.* (1989). The proteins encoded by HVDAC1 and HVDAC2 are only 75% identical at the amino acid level, and HVDAC2 contains a 11-residue amino-terminal extension relative to HVDAC1. Both human VDACs have less than 30% sequence identity with yeast VDAC. In addition, a cDNA that differs from HVDAC2 at the 5' end (Ha *et al.*, 1993) has been described. This transcript may represent a splicing variant of HVDAC2, or perhaps an incompletely spliced transcript, since it contains the 5' end sequences of HVDAC2 within its 5' untranslated region. In addition, it contains a single nucleotide deletion relative to HVDAC2 near the 3' end of the coding region, resulting in a frameshift which alters the translation of the C-terminal end of the protein. Two additional human sequences highly homologous to HVDAC1 have been identified by polymerase chain reaction (Blachly-Dyson *et al.*, 1994), but it is not known whether these represent expressed genes. A protein highly homologous to HVDAC1 and a cDNA highly homologous to HVDAC2 have been isolated from rat brain (Bureau *et al.*, 1992). In addition, an HVDAC1-like cDNA has been isolated from bovine brain (Dermietzel *et al.*, 1994). Three mouse VDAC genes, with 65–75% sequence identity to each other have also been cloned (Craigen *et al.*, 1994).

VDAC genes have also been isolated from several plants. VDAC genes from pea and maize were found to have 58% sequence homology with each other, while each of them was only about 25% identical to fungal or human VDAC (Fischer *et al.*, 1994). cDNAs representing two different potato VDAC genes have been characterized (Heins *et al.*, 1994). These two isoforms have about 75% identity to each other, 25% with human VDAC, and less than 25% with fungal VDACs. A wheat VDAC sequence is also available in the database. In addition to these genes, a VDAC gene has been cloned from *Dictyostelium discoideum* (Troll *et al.*, 1992), and expressed sequence-tagged sites from *C. elegans* appear to encode a VDAC gene from this species.

3.2. Conservation of Primary and Secondary Structure

Although, as indicated above, the VDAC molecules of different species have very little sequence conservation, the channel gating and selectivity properties are highly conserved. Both HVDAC1 and HVDAC2 (expressed in yeast cell lacking the endogenous VDAC gene) have single-channel conductances and ion selectivities (K^+ versus Cl^-) that are virtually indistinguishable from those of yeast VDAC (Blachly-Dyson *et al.*, 1993). Likewise, the plant VDACs (Blumenthal *et al.*, 1993) and *Dictyostelium*

VDAC (Troll *et al.*, 1992) form channels with very similar properties, although the sequences are highly divergent.

VDAC sequences contain no long stretches of hydrophobic amino acid residues that could form transmembrane α-helices. However, when these sequences are examined for secondary structure motifs, a common pattern emerges. In all the VDAC sequences analyzed, the N-terminal end contains a sequence that can form an amphiphilic α-helix. Downstream sequences all have 12 or more peaks of strong alternating hydrophobic/hydrophilic segments, which could contribute to the formation of a "sided" β-barrel structure with a hydrophilic inner surface facing the pore and a hydrophobic outer surface buried in the membrane. This can be easily quantitated and visualized in a plot called a β pattern (Fig. 6). These homologous sequence patterns have led to a general model for the structure of the VDAC, in which the pore of the channel is formed by the hydrophilic side of a cylindrically curved β-sheet that has a hydrophobic side facing the membrane interior (Blachly-Dyson *et al.*, 1989; Song and Colombini, 1995).

3.3. Site-Directed Mutagenesis of Charged Residues in Yeast VDAC

To test this model, extensive site-directed mutagenesis was performed, changing the charge of amino acid residues throughout the sequence of the molecule. Charge changes at positions lining the pore would be expected to alter the selectivity of the open channel; charge increases should increase anion selectivity, while charge decreases (or increases in negative charge) should decrease or reverse the anion selectivity. While a number of the mutations had no effect on selectivity, residues were found throughout the molecule which had the expected effect on selectivity when the charge was changed (Fig. 7). These residues were found within the putative amino-terminal α-helix and 12 of the proposed transmembrane β-strands (Blachly-Dyson *et al.*, 1990). This led to a refinement of the general model outlined above, in which the VDAC is formed by the amino-terminal α-helix and a β-barrel consisting of the 12 β-strands whose position is now constrained by these functional studies of mutant channels (Fig. 8). (Note: While residues whose charge affects selectivity *must* line the pore and thus must reside in transmembrane strands, transmembrane strands that do not line the pore may not affect selectivity). Thus, many charged residues distributed throughout the length of the protein sequence contribute to the overall charge of the cylindrical shell in a roughly additive manner to determine the selectivity of the open channel.

4. MOLECULAR ANALYSIS OF THE GATING PROCESS

By definition, voltage-gated channels must undergo a voltage-dependent conformational change. This can be achieved by coupling the gating process to the motion of charges ("gating charges") relative to the electric field or the alignment of dipoles with the field. Two approaches have been used to identify the portion of the VDAC molecule that moves in response to an applied field. First, the site-directed mutations were used to determine whether the same residues that contributed to the selectivity of the open channel also contributed to the selectivity of the closed state of the channel. Residues that

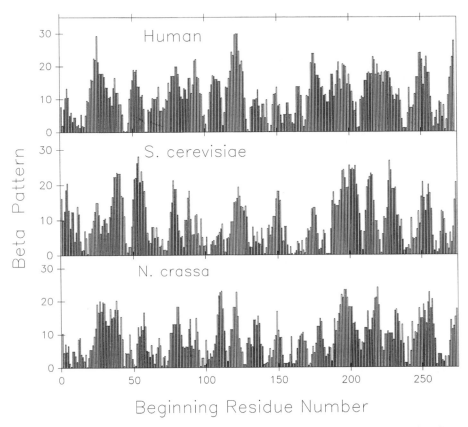

FIGURE 6. An evaluation of the potential of stretches of amino acids in the VDAC sequences from human (HVDAC1), *S. cerevisiae* (YVDAC1), and *N. crassa* to form β-strands lining the walls of a water-filled pore. The hydropathy values (Kyte-Doolittle) of each group of ten amino acids was summed in such a way that every other value was multiplied by −1. The absolute value of these sums is plotted versus the number from the N-terminus of the first amino acid in the group. Large values (peaks) indicate a good alternating polar/nonpolar pattern expected for a transmembrane β-strand separating a polar environment from a nonpolar one.

contributed to the selectivity of one state, but not the other, were candidates for being part of the "gate" that moves when the channel opens. Second, these mutations were examined for their effect on the gating of the channels. Changing a charged residue that contributes to the gating charge (i.e., a charge that "senses" the applied field during gating) should affect channel gating parameters in predictable ways. Both of these approaches indicated that a large portion of the protein moves during channel gating.

4.1. Magnitude of the Structural Change Associated with Voltage Gating

VDAC is a large pore that undergoes large changes in both size and selectivity during channel closure (see above). Thus, it should not be surprising that large conforma-

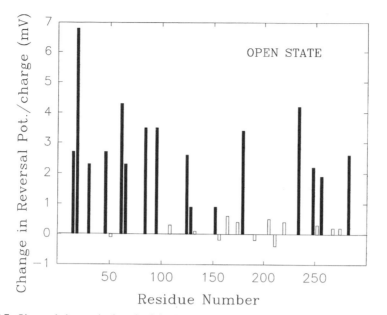

FIGURE 7. Observed changes in the selectivity (reversal potential) in yeast VDACs following single site-directed mutations. The reversal potentials were divided by the magnitude and sign of the charge change at the site produced by the amino acid substitution. They are plotted against the position of the residue from the N-terminus. The reversal potentials are compilations of previously published data (Blachly-Dyson *et al.*, 1990; Peng *et al.*, 1992a). They were collected on single channels after reconstitution into a planar membrane (as in Fig. 1) in the presence of a tenfold KCl gradient. Positive values indicate that the selectivity change was in the direction expected from the charge change. Solid bars represent significant changes in selectivity.

tional changes may be involved in channel gating. While the large conductance change associated with VDAC closure suggests a large structural change, there is often a poor correlation between conductance and pore size (Finkelstein, 1985), so it is important to determine the size of the open and closed state pores by other methods. A good assessment of open pore size is obtained by using nonelectrolytes to probe the steric barrier. The ability of inulin, polyethylene glycol (PEG 3400), and dextran (Zalman *et al.*, 1980) to pass through the open state of VDAC (Colombini, 1980b) indicate a large open pathway for the narrowest portion of the channel. The flexibility of PEG and dextran may tend to overestimate the pore size of the open state, but inulin is not as flexible. Electron microscopy of negatively stained images of VDACs in two-dimensional crystalline arrays indicates pore diameters of 2.4 to 3 nm. Estimates based on access resistance considerations (Vodyanoy *et al.*, 1992) yield a pore diameter of 2.4 nm after correction for the size of the permeating ion. Hence, a reasonable estimate of the pore diameter is 2.5 to 3 nm.

Closer agreement exists for the pore diameter of the low-conducting closed state.

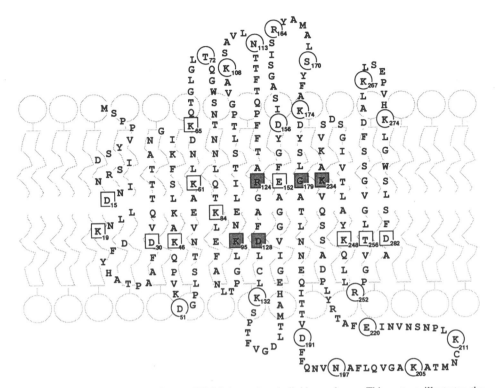

FIGURE 8. Folding pattern of yeast VDAC in a phospholipid membrane. This pattern illustrates the transmembrane strands proposed to form the walls of the pore of the VDAC. The extended β-strands would be hydrogen-bonded together and to the α-helix, forming a cylindrical structure. The boxed residues are sites that affect the ion selectivity of the channel. Circled residues are sites that had no effect on ion selectivity.

Functionally, this state is just barely permeable to gamma cyclodextrin (Colombini *et al.*, 1987). This rigid spheroid is 1.9 nm in diameter. This is almost the same estimate obtained by electron microscopy of negatively stained arrays (Mannella and Guo, 1990). These estimates must be viewed with caution however since the closed state for both estimates was induced by the addition of König's polyanion, a synthetic polymer that induces channel closure at very low concentrations. While all indications favor the conclusion that the states induced by this polymer are the same as those induced by the electric field, this may not be totally correct.

To estimate the change in pore volume during closure, Zimmerberg and Parsegian (1986) used macromolecules that could not penetrate the pore of VDAC to induce a tension within the channel that favored the closed state. From the energy change induced in the molecule, these investigators calculated a change in the volume of the channel upon closure of 20–40 nm^3. This is consistent with a large, global change in the structure of the pore as opposed to a local, shutter-like, constriction at one point in the channel.

Together, these studies lead to the conclusion that closure of the channel results in a rather large change in effecting pore size.

4.2. Location of the Mobile Domain

Functional experiments on channels reconstituted into planar membranes indicated that regions of the channel move across the membrane during the gating process. Addition of the reagents aluminum hydroxide (Zhang and Colombini, 1990) or succinic anhydride (Doring and Colombini, 1985) to one side or the other of the membrane caused asymmetric changes in the behavior of the channels in the membrane. The changes observed depended on the state of the channel (open or closed) when the reagent was added. These results were best explained by the translocation of some groups all the way through the membrane.

In general, the motion of a protein domain can be detected if, in its new environment, its effect on a property of the protein is quite different. Since the selectivity of the channel changes rather dramatically upon closure, the electrostatic nature of the ion-conducting pathway must change radically. This could result from a change in the residues forming the protein wall of the pore; new residues could be introduced or existing residues could be removed. Since channel closure also results in a reduction in the effective diameter of the pore, a removal of residues forming the wall of the open channel seems more likely.

Having identified regions that form the wall of the pore in the open state, Peng and coworkers sought to determine if some of these no longer affected the channel in the closed state. Such regions would be part of the mobile domain. Using the battery of mutant proteins with the single amino acid substitutions at locations either within the pore or outside the pore, Peng et al. (1992a) identified residues that influenced the selectivity of the channel in the open state and not in the closed state (Fig. 9). In addition, there were residues that still affected the selectivity of the closed state but to a lesser extent than on the selectivity of the open state. Both of these classes of mutations were considered to identify mobile regions in the channel. Most of these mutations fell in the N-terminal end of the channel, specifically the α-helix and nearby three β-strands in the model in Fig. 8. One mutation was at position 282 at the C-terminus. Since in a barrel-like structure this transmembrane segment is expected to be adjacent to the α-helix, this mutation could still be part of a contiguous mobile domain. However, a mutation at position 152, in the middle of a nonmobile region, also affected the selectivity of the channels in the open state but not the closed state. Subsequent studies have indicated that this region of the protein contributes in unique ways to the gating process and, in more recent refinements of the general structural model, has been repositioned to reside in mobile regions (see Fig. 10).

Thus, these results are consistent with the movement of regions that form part of the wall of the channel in the open state out of the channel proper or at least away from the ion stream. If the region that moves has a net positive charge, this change would result in a reduction in the net positive charge on the walls of the pore and a reduction in the effective pore radius. It would also result in a reduction in the volume of water within the pore. All these predictions are consistent with experimental observations (see previous sections).

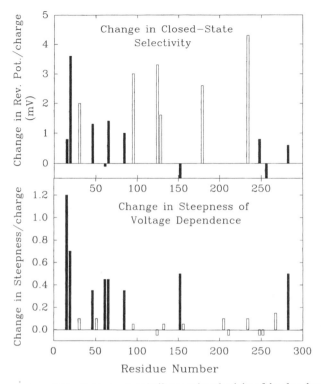

FIGURE 9. A comparison of sites that had reduced effects on the selectivity of the closed state with those that affected the steepness of the voltage dependence. The upper panel shows the changes in the closed-state reversal potential (normalized for the sign and magnitude of the engineered charge change) of the sites that had significant effect on the open-state selectivity (Fig. 7). The open bars are sites that had an effect on the selectivity of the closed state that was similar in magnitude to that of the open state. The solid bars are sites that had either no significant effect on the selectivity of the closed state or an effect that was smaller than expected. The lower panel shows changes in the steepness of the voltage dependence (normalized as well) with significant changes in solid bars and nonsignificant changes in open bars. Positive values indicate changes that were in the direction expected from the sign of the engineered charge change.

4.3. Identification and Localization of the Voltage Sensor

Based on these results, we hypothesized that the domains that affect selectivity in the open but not the closed VDAC channel correspond to or overlap the voltage sensor and that this voltage sensor is a moiety with a net positive charge that moves perpendicular to the membrane out of the channel wall during channel closure. The hypothesis that this mobile domain contains the voltage sensor makes very clear predictions: 1) If the charge on any part of this domain is altered it must affect the *steepness* of the voltage dependence of the channel. Steepness is reflected in the parameter n in a two-state model defined by Eq. (1).

$$\ln[(G_{max} - G)/(G - G_{min})] = (nFV - nFV_0)/RT \tag{1}$$

where V_0 represents the voltage at which half the channels are closed; G, G_{max}, and G_{min} are the conductance at any voltage V, the maximum conductance, and the minimum conductance, respectively; and F, R, and T are the Faraday constant, the gas constant, and the absolute temperature, respectively; 2) The parameter n should be increased if the charge in sensing domains is made more positive and decreased if it is made more negative; 3) The change in n induced by such a change in charge must be proportional to the magnitude of the charge change and the fraction of the electric field through which the charge moves.

These predictions were tested by examining the effects on the parameter n of mutations that changed the charge of specific residues. Amino acid substitutions that changed the charge at eight positions in the N-terminal α-helix, three nearby β-strands, or the C-terminal β-strand increased or decreased n if the site was made more positive or more negative, respectively (Thomas et al., 1993). These sites matched (dark bars in Fig. 9) the sites that seem to move, based on selectivity changes (Section 4.2 above). This correspondence by two different approaches provides strong evidence that these regions are moving through the field during the gating process. The lack of correspondence at position 248, a site close to the mouth of the channel, could be explained by the residue moving out of the ion stream without moving through a significant portion of the electric field.

There are, however, difficulties in understanding the motion of the sensor. The substitution of lysine for aspartate at position 30 (D30K) had no significant effect on n. This substitution also affected the selectivity of VDAC in both the open and closed states, indicating that this residue is not moving out of the channel upon channel closure. Thus, although these results are consistent with each other, it is hard to see how the nearby regions of the protein can move out of the channel without this strand also moving. D51K also did not generate the expected changes in n. Located in the loop region between the second and third β-strands, this residue is positioned outside the pore in current models since D51K does not affect open-state selectivity and would not have to traverse the electric field if the mobile domain moved toward it. Changes in a nearby residue, K46E, decrease the voltage dependence of gating processes at both positive and negative potentials, indicating that the domain containing K46 and D51 may move in both directions. D51K would be expected to increase the voltage dependence of at least one of the processes. The D51K mutation had no effect on n, however. D51K thus would be expected to increase the voltage dependence of at least one of the processes. These findings point out the fact that although current models are valuable tools for understanding and summarizing results obtained in functional studies, the models are almost certainly simplified representations of structures and changes that are undoubtedly more complex.

Amino acid substitutions that changed the charge at other locations generally had no effect on the steepness of the voltage dependence. A notable exception was glutamate 152 on β-strand 7 (Fig. 8). This region is in the middle of the molecule between strands that are proposed to remain fixed during the gating process. E152K increased the steepness of the voltage dependence but did so in an asymmetric manner, i.e., it increased the voltage dependence of only one of the two gating processes. The asymmetry can be explained if E152 is located near one end of a transmembrane strand. More difficult to

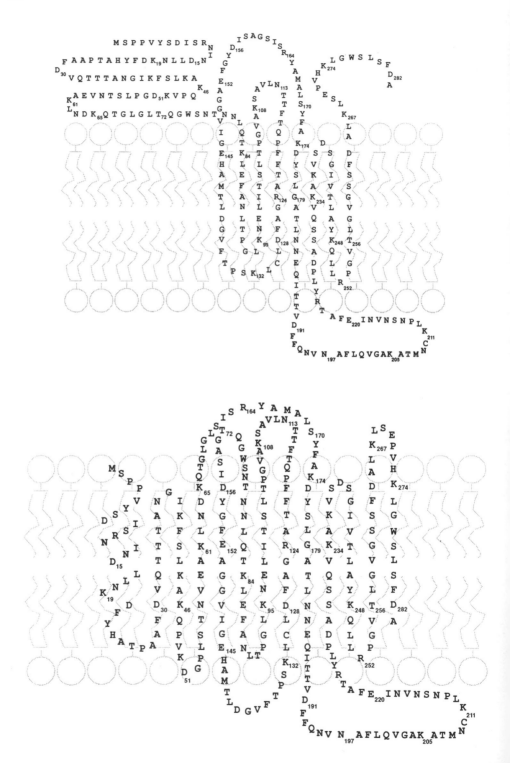

explain is the fact that this strand has a net negative charge and reducing this negative charge by the glutamate to lysine substitution increases the voltage dependence. This means that normally this negatively charged strand moves toward the negative side of the membrane. This seems possible only if it is coupled to the motion of positively charged regions moving in the same direction. However, the adjacent β-strands in Fig. 8 do not move based on selectivity and voltage-dependence measurements on charge substitutions in these regions. It was therefore proposed (Zizi et al., 1995) that this strand might be located elsewhere, perhaps within the mobile region (as in Fig. 10, upper panel). This relocation is possible because of the long loop regions connecting this strand to the nearby transmembrane segments. More importantly, the relocation of the strand containing residue 152 into the mobile domain provides constraints on the movement of this domain. Because the strand is linked to nonmobile regions, it may act as a tether that is stressed by channel closure and this stress would be released by the reopening process (Fig. 10, lower panel). Such a proposal would account for the rapid, submillisecond rates of channel opening (Schein et al., 1976; Colombini, 1979).

If the strand containing residue 152 does in fact move normal to the membrane upon channel closure, the nearby glutamate 145 is likely to move as well. Changing this residue to a positive charge should increase the voltage dependence of closure. When glutamate 145 was mutated to lysine (E145K), an asymmetrical increase in voltage dependence was observed, similar to the effect of the E152K mutation (Fig. 11). Furthermore, if residues 145 and 152 are located at opposite ends of a transmembrane strand, E145K and E152K should increase the voltage dependence of the two different gating processes, respectively. This was tested (Zizi et al., 1995) by generating the double mutant. E145K/E152K showed elevated steepness of voltage dependence in response to both positive and negative potentials (Fig. 11), consistent with this strand moving in one direction at positive potentials and the other at negative potentials.

Thus, while the details of the gating process need to be clarified, the voltage sensor is clearly a region of the protein-forming part of the wall of the pore. This region moves out of the channel upon channel closure and in the process traverses part of the electric field, resulting in a structural change that accounts for the voltage dependence of the conductance.

4.4. Activation Energy and the Gating Mechanism

The large conformation change proposed for the gating process raises the question of the size of the activation energy. While VDAC closure is a slow process, the opening is fast (submillisecond). Can such a large conformational change occur in such a short time?

The movement of the domain illustrated in Fig. 12 requires the cleavage of the hydrogen bonds connecting transmembrane β-strands. On closure, polar side chains

FIGURE 10. A new folding pattern for yeast VDAC in a membrane and the proposed motion of regions of the protein upon channel closure. In the upper panel β-strand 7 from the N-terminus (Fig. 8) was moved so as to lie between β-strands 3 and 4. In the lower panel, regions proposed to move out of the channel in one of the two gating processes (positive potential below the membrane).

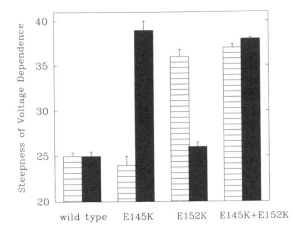

FIGURE 11. Effect of glutamate (E) to lysine (K) substitutions at positions 145 (E145K) and 152 (E152K) on the voltage dependence of yeast VDAC. The steepness of the voltage dependence was measured in the presence of 5.75×10^{-2} mg/ml dextran sulfate (500 kDa), a value that magnifies the steepness of the voltage dependence by a factor of 10 (see also Fig. 15). The two bars indicate the voltage dependence of each of the two gating processes. This is a replotting of the data of Zizi *et al.*, 1995.

facing the aqueous phase within the pore would face a similar environment on the membrane surface. Apolar groups facing the hydrocarbon chains of the phospholipids need not be removed from this environment but can slide along to a location near the membrane surface. Thus it is not unreasonable to postulate that breaking the hydrogen bonds between the mobile domain and the rest of the structure may be the major contribution to the energy barrier.

There is no general agreement as to the stabilizing effect of hydrogen bonds on protein folding. Although estimates are very crude, it is generally agreed that hydrogen bonds need to be formed within the protein structure, but the strength of these intramolecular hydrogen bonds as compared to that of hydrogen bonds with water is not very different, perhaps 4 kJ/mol more stable. Since there are two edges where hydrogen bonds between β-strands must be broken in order to move these strands relative to each other during the closing process, the proposed VDAC gating mechanism would require approximately 20 hydrogen bonds to be broken for channel closure. Only half as many need to be broken for channels to open from the closed state since only one set of hydrogen bonds need to be cleaved to reinsert the portion that moved out into the wall of

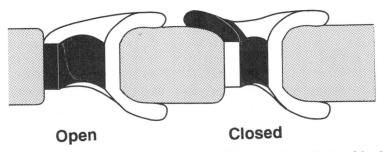

Open   **Closed**

FIGURE 12. Model of one of the VDAC gating processes. The open and closed states of the channel are illustrated as if bisected. The domain in black is the mobile domain, the voltage sensor. It forms part of the wall of the channel in the open state and is moved out of the channel upon closure.

the channel. Consistent with this mechanism, estimates of the enthalpy change for the transition from the open to the closed states are 35 to 40 kJ/mol (Pavlin and Colombini, in preparation), or roughly ten hydrogen bonds. Thus, this difference in activation energy can account for the 1000-fold difference in the rate of channel closure compared to channel opening (Colombini, 1979) (taking reasonable values for the energy to break a hydrogen bond).

4.5. Ion Flow and the Gating Process

Another unusual observation that can be explained by the proposed gating mechanism is the observed shift along the voltage axis of the switching region of the voltage-gating processes in the presence of a salt gradient. When the salt concentration on both sides of the membrane is the same, the two gating processes generally occur at the same magnitude of the membrane potential (VDAC reconstituted into soybean phospholipid membranes). However, in the presence of a salt gradient (1 M versus 0.1 M) there is a pronounced shift in that higher negative potentials and lower positive potentials on the high-salt side are required to close the channels. It appears that moving the sensor against the salt gradient requires more energy than moving it down the gradient. These observations lead to the hypothesis that some of the kinetic energy from the flow of salt through the channel was imparted to the mobile domain favoring motion down the gradient. This was tested and confirmed by using salts of different mass and size (Zizi, Byrd, and Colombini, in preparation). Thus, as illustrated in Fig. 13, the bias that ion flow imparts to the gating process can be explained by the transfer of kinetic energy to the mobile

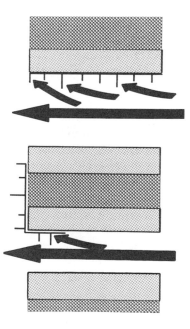

FIGURE 13. An illustration of how the flow of salt through the channel could impart kinetic energy to the wall of the channel and favor the movement of the sensor out of the channel. The sensor is depicted as a strand with bars representing the amino acid side chains extending into the lumen of the channel. The arrows indicate the direction of flow of the salt and its collision with the extended amino acid side chains. The upper figure shows the open state and the lower figure the closed state.

domain. This notion is consistent with the proposed gating mechanism since residues forming the ion conduction pathway also form part of the voltage sensor or gate.

5. COMPARISON OF PROPOSED GATING MECHANISMS

Section 4 focused on one proposed mechanism for voltage gating in VDAC. Very different mechanisms have been proposed by Mannella and others (Mannella, 1990; Mannella *et al.*, 1992; Adams and McCabe, 1994). In addition, the actual mechanism may differ from any yet proposed. Here, we consider how the current experimental evidence may exclude other possible mechanisms (see Fig. 14).

5.1. Corking the Bottle

In analogy to the mechanism that seems to underlie slow inactivation of Na^+ and K^+ channels, the idea of plugging the channel with a protein domain from the surface seems simple and straightforward. To be consistent with the conductivity of closed VDACs, this would need to be a porous cork. However, even with such a modification, this mechanism has difficulty accounting for a number of experimental observations:

1. the large volume change associated with channel closure
2. the fact that certain residues that influence selectivity in the open state no longer do so in the closed state
3. the ability of residues that influence channel selectivity in the open state to influence the steepness of the voltage dependence
4. the voltage dependence of the gating process.

A modified proposal by Mannella (1990) tries to circumvent some of these problems. He proposes that α-helix enters the channel, resulting in a reduced pore volume, thereby overlying regions of the wall of the pore and masking the effect of charges at these sites on the selectivity of the resulting closed channel. There are difficulties with this proposal:

1. The α-helix is proposed to lie on the surface of the membrane in the open channel, away from the ion stream. Yet charge substitutions at two sites in the helix (D15K and K19E) affect the selectivity of the open state of the channel, indicating that this region is in intimate contact with the ion stream. In the closed state, these sites on the α-helix have less, not more, effect on selectivity, in contrast to the expected results if the helix enters the channel during closure.
2. While the α-helix of VDAC from some species has a net charge, in some cases there is no net charge (e.g., in yeast VDAC). Thus there is no apparent way for enough charge to move through the field to account for the voltage dependence of the gating process.
3. Since one surface of the α-helix is hydrophobic, the movement of this strand from the membrane surface to the inside of a polar channel would require a high activation energy.
4. It is unclear how putting an α-helix in the channel would mask the effect of

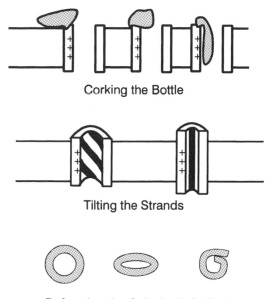

Corking the Bottle

Tilting the Strands

Deforming the Cylinder/Jelly Roll

FIGURE 14. Various alternative (incorrect) schemes by which the VDAC might close. In the upper figure, a domain extends into the channel opening or into the channel proper. In the middle scheme, the β-strands are tilted in the open state and the tilt is reduced in the closed state, producing a longer channel with narrower diameter. In the bottom figure, the channel is shown in cross section. Closure here results from either the partial collapse of the cylinder or the overlap of the free ends of the cylinder.

charged residues on ion flow. The formation of ion pairs would require residues of opposite charge to interact with the existing residues on the walls of the pore. In order for the α-helix to dramatically change the selectivity of the channel (favoring cations), it would have to be a negatively charged structure. In fact, the α-helices of VDACs from a wide variety of species are either neutral or positively charged (Song and Colombini, 1995). Thus, this mechanism could not account for the change in selectivity associated with the gating process.

5. This mechanism cannot account for two distinct gating processes in each VDAC.

5.2. Forming a Jelly Roll

The possibility that the β-barrel could be reduced in size by overlapping the ends of the wall has been proposed (Mannella, 1990). Since the N- and C-terminal ends seem to be involved in the structural change, this has some appeal. However, it is unclear how such a structural change could result in charge motion through the electric field and thus be voltage-dependent. It is also difficult to see how the apolar groups facing the hydrocarbon portion of the membrane could be induced to now face the polar protein surface that faced the aqueous pore.

5.3. Tilting the Strands

Following the mechanism proposed for gating of gap junction channels (Unwin and Zampighi, 1980), one mechanism of closure might involve the straightening of transmembrane β-strands. In the open channel, transmembrane β-segments must be severely tilted (roughly 60° in the current model) given the dimensions of the channel formed by a single VDAC protein. In addition to not accounting for voltage dependence and the selectivity change associated with gating in VDAC, such a conformational change would require that every hydrogen bond between each of the staves of the barrel be broken. This would be an unlikely event and should have a high enthalpic activation energy.

5.4.   Deforming the Cylinder

One might imagine that the cylindrical channel might be deformed in such a way that the cross section would look like an ellipse. However this proposal suffers from some of the same shortcomings encountered by the tilting of strands. In addition, the high curvatures produced by such a change would result either in strained hydrogen bonds between some strands or actual broken bonds. If broken, it is hard to see how these could be satisfied by the apolar membrane interior.

While the mechanism illustrated in Fig. 12 is consistent with the vast majority of the experimental data and explains many of the special properties of VDACs, there are a few apparent inconsistencies. In addition to those already mentioned, Mannella and Guo (1990) have shown that closure results in only small changes in the packing of the channels into two-dimensional arrays. One might expect that the reduction in the overall diameter of the pore upon closure would result in a tighter packing. However, it is known that a considerable amount of phospholipid is present in the ordered arrays, and because lipids move laterally at a very rapid rate, it is not unreasonable to expect that phospholipids have filled in any space generated by the closing process. The spacing of the channels themselves may be determined more by the nature of the surface regions of the channel than simply the packing of cylindrical proteins. Indeed, the packing of VDACs is not a simple, hexagonal, close packing, indicating specific interactions between individual channels.

6.   REGULATION

In addition to two voltage-gating processes, there are a number of other agents and factors that affect the probability of finding VDACs in either an open or closed state. Whether cells actually use these to regulate VDAC and thus the permeability of the mitochondrial outer membrane has yet to be demonstrated directly. However, a number of these regulatory interactions have been found to be highly conserved in VDACs from widely different sources, implying that these mechanisms are maintained by selective pressure. Thus, it is likely that these observations define important regulatory interactions *in vivo*.

6.1.   Amplification of the Voltage Dependence

The parameter n, reflecting the steepness of the voltage dependence of VDAC reconstituted into planar membranes, varies from 2 to 5 depending on the source of

VDAC and the experimental conditions. This is comparable to the voltage dependence of voltage-gated channels responsible for action potentials. The n value of VDAC can be dramatically augmented by treatment with small amounts of polyanions. Thus, n is increased tenfold by the presence of 100 nM dextran sulfate (500 kDa) and it can be increased by another factor of 2 at higher concentrations (Mangan and Colombini, 1987) (Fig. 15). A variety of polyanions, from polyaspartic acid to RNA, have similar effects (Colombini *et al.*, 1989). Stronger effects are observed with simple polymers, and the potency of the effect increases with increasing charge density on individual polymers, indicating that the structure of specific polyanions may determine the extent of interaction with VDAC.

The polyanions appear to act by favoring the closed state of the channel rather than by a voltage-dependent blockage. The drop in conductance upon channel closure is virtually the same whether voltage alone is used to close the channel or voltage-dependent closure is augmented by the addition of dextran sulfate. There is no sign of a decrease in single-channel conductance or of channel flickering in the presence of polyanions.

The ultra-sheep voltage dependence induced by the polyanions is consistent with the voltage-dependent increase in the partitioning of the polyanion into the access-resistance region at the mouth of the channel (Mangan and Colombini, 1987). There, the polyanion presumably interacts with the positively charged voltage sensor to increase the likelihood of these domains translocating out of the membrane and closing the channel. The fact that the polyanion acts preferentially from the negative side of the membrane is consistent with this proposal.

Polyanions with hydrophobic regions show much more complex behavior. The best studied is König's polyanion, a copolymer of methacrylate, maleate, and styrene in a 1:2:3 ratio with average molecular weight of 10,000. In addition to increasing the voltage dependence as dextran sulfate does, it binds and induces closure even in the absence of a membrane potential (Colombini *et al.*, 1987). The effect seems to be progressive in that, with time, the channels tend to enter states of lower and lower conductance. When König's polyanion was used to increase the probability of closure, VDAC-containing vesicles showed a reduction in permeability to nonelectrolytes, consistent with a reduction in channel diameter, but the channel was still permeable to cyclodextrin (1.9 nm in diameter). This is consistent not with voltage-dependent block by a leaky plug but simply an increase in the probability of the channel entering a closed state.

To date, there are no examples of other channels responding to polyanions in this way. Thus the structure of VDAC and its gating process may be specifically inclined to respond to these polyanions. The possibility that these polyanions mimic the action of a natural substance led to the discovery of a protein (the VDAC modulator) that acts somewhat like the polyanions.

6.2. Action of the VDAC Modulator

The VDAC modulator is a soluble protein that induces VDAC closure in planar phospholipid membranes when added at very low concentrations (Holden and Colombini, 1988). It is very sensitive to proteases, requires the presence of dithiothreitol (DTT)

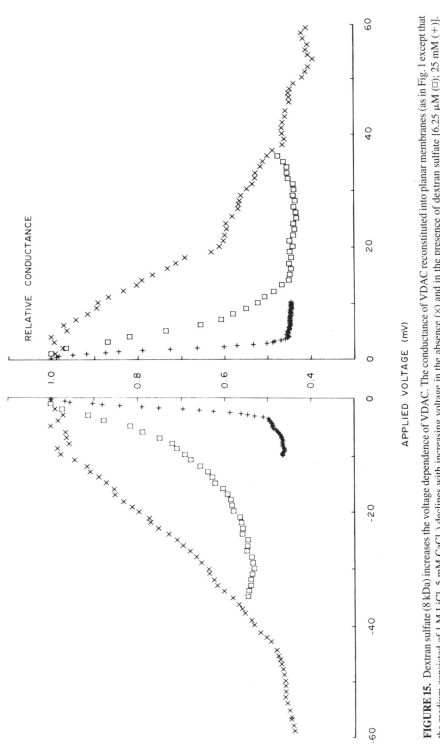

FIGURE 15. Dextran sulfate (8 kDa) increases the voltage dependence of VDAC. The conductance of VDAC reconstituted into planar membranes (as in Fig. 1 except that the medium consisted of 1 M LiCl, 5 mM CaCl₂) declines with increasing voltage in the absence (×) and in the presence of dextran sulfate [6.25 μM (□); 25 mM (+)]. Reproduced with permission from Mangan and Colombini (1987).

for stability (Liu *et al.*, 1993), and migrates with an apparent molecular weight of 100,000 by gel filtration chromatography. It has been identified in mitochondria from mammals, fungi, and higher plants (Liu and Colombini, 1991) and localized to the intermembrane space of *N. crassa* mitochondria (Holden and Colombini, 1993). There also appears to be more than one VDAC modulator activity in calf liver, one more potent than the other (Liu *et al.*, 1993). When added to VDAC reconstituted into planar membranes, the modulator binds very tightly since it is not washed out easily by perfusing the aqueous compartment. Functionally, the modulator not only increases the voltage dependence of the channels (Liu and Colombini, 1992b) but also tends to keep the channels in closed states, inducing channels to enter states of very low conductance which are usually achieved only at high voltages. The modulator acts from both sides of VDAC incorporated into planar membranes. As was the case with dextran sulfate, the modulator acts when the modulator-containing side is made negative. When added to mitochondria with intact outer membranes, the VDAC modulator reduces the permeability of the outer membrane and inhibits mitochondrial metabolic activities that require the flux of metabolites through the outer membrane (Liu and Colombini, 1992a). When the outer membrane is damaged, the inhibition is eliminated.

If the modulator is located only in the intermembrane space, its ability to bind from outside the mitochondria seems puzzling. It is possible that the modulator or related molecules may also be present in the cytoplasm, although this has yet to be demonstrated. However, since VDAC can gate in response to either positive or negative potentials, the modulator might be able to bind to it on either side of the channel if the same domain is exposed during each gating process.

6.3. Action of Nucleotides

When nucleotides (ATP, ADP, cAMP, cGMP, NADH, NADPH, and NAD^+) were added to VDACs reconstituted into phospholipid membranes, only NADH and NADPH had significant effects on the function of the channel, NADH being an order of magnitude more potent (Xu, unpublished observations). NADH doubles the voltage dependence of the channel with a K_D estimated to be in the low micromolar range (Zizi *et al.*, 1994). ATP has been reported to bind to human VDAC (Florke *et al.*, 1994) and even to a peptide consisting of only the 35 N-terminal amino acids. However, it has no effect on the function of VDAC reconstituted into phospholipid membranes (Zizi *et al.*, 1994). The addition of NADH and NADPH to mitochondria with intact outer membranes results in a sixfold reduction in the permeability of the outer membrane to ADP (Lee *et al.*, 1994). The effect of this on mitochondrial respiration is most pronounced at physiologically relevant ADP concentrations. These results are consistent with NADH inducing VDAC closure.

6.4. Colloidal Osmotic Pressure

The presence in the medium of macromolecules that cannot enter VDAC's aqueous pore results in an imbalance in the activity of water that can only be corrected by decreasing the hydrostatic pressure in the pore. This reduced pressure favors channel closure if the volume of the pore in the closed state is reduced. This was the line of reasoning used to estimate the volume change upon channel closure (Zimmerberg and Parsegian, 1986). Thus, VDAC is sensitive to the concentration of macromolecules and will be influenced by the concentration of these molecules in the cell. The influence of the colloidal osmotic pressure on isolated mitochondria is consistent with the closure of VDACs by increasing osmotic pressure. Mitochondrial functions are greatly inhibited by the addition of 10% dextran (Gellerich *et al.*, 1993).

6.5. Inhibition by Metal Hydroxides

Neutral metal trihydroxides strongly inhibit the ability of reconstituted VDACs to close in response to a membrane potential (Zhang and Colombini, 1989). The best studied is aluminum trihydroxide (Dill *et al.*, 1987), formed by the addition of aluminum chloride to a buffered solution, and is by far the major species at physiological pH. It is, however, metastable and should eventually precipitate. Other metals in the same group (Ga, In) have a similar effect, as do transition metals that form a neutral trihydroxide. The potency of the different metal trihydroxides depends on the relative amount of trihydroxide formed under given conditions (e.g., a particular pH) versus other possible hydrated species.

The action of aluminum hydroxide resembles what one would expect when the voltage sensor is neutralized. However, aluminum trihydroxide does not affect the selectivity of the channel in the open conformation, indicating that the net charge on the wall of the pore (including that in the mobile domain) is unchanged (Dill *et al.*, 1987). In addition, experiments indicate that the aluminum binding site moves through the membrane in the opposite direction to that of the voltage sensor (Zhang and Colombini, 1990). Thus, the aluminum binding may result in the effective movement of positive charge in the opposite direction to the sensor, thus reducing the net charge transferred.

7. AUTO-DIRECTED INSERTION

Experimental observations demonstrating that VDACs insert into phospholipid membranes in an oriented manner has led to the proposal that these channels have the property of "auto-directed insertion" (Zizi *et al.*, 1995). While a preferential insertion direction is not surprising, the evidence indicates that for VDAC the direction of insertion of the first channel determines the direction of insertion of virtually all other channels. If so, an inserted VDAC must: 1) interact with the other channels as they insert, 2) provide directional information that constrains the direction of insertion, and 3) accelerate the rate at which channels insert.

Experimental evidence in support of auto-directed insertion was obtained by examining the voltage-gating properties of two yeast VDAC mutants, each possessing a single amino acid substitution that rendered the channel asymmetric in its behavior. The asymmetric behavior allowed one to determine the direction of insertion. The degree of asymmetry was the same whether one or 100 channels were present in the membrane, indicating that, in the multichannel membrane, virtually all the channels were inserted in the same direction. More important, the direction of the insertion varied from membrane to membrane in an apparently random fashion. This led to the hypothesis that the first insertion is random and subsequent insertions are determined by the direction of the first insertion. Various possible causes for the apparent randomness of the membrane-to-membrane variation in the direction of insertion were investigated. Neither the sign of the membrane potential during the insertion process nor the side of the membrane to which the VDAC-containing sample was added had any significant influence on the direction of insertion. Thus, the most likely cause of this variation is the random nature of the first insertion.

The proposal that the pre-inserted channels interact with channels attempting to insert into the membrane requires that insertions at the location of the pre-inserted channel be highly favored over insertions at other locations. Since the entire membrane had a surface area 10^9 times greater than that of a single VDAC, in order for channels to preferentially insert next to the pre-inserted channel over anywhere else on the membrane, the preference must be greater than 10^9. To explain the observation that the insertion direction in a single experiment is almost perfect, the preferential insertion should be at least an order of magnitude greater. Thus, the most straightforward interpretation of the experiments is that VDACs both catalyze and determine the direction of insertion of fellow channels. This property of auto-directed insertion has important implications in protein targeting and the maintenance of oriented VDACs in the outer membrane of mitochondria.

8. ROLE OF VDAC IN CELL FUNCTION

Several enzymes (hexokinase, glucokinase, and glycerol kinase) that use ATP produced by mitochondria have been found to associate with the outer surface of the mitochondrial outer membrane by binding to VDAC protein. This localization presumably gives the enzyme preferential access to mitochondrially generated ATP. In addition, creatine kinase of the mitochondrial intermembrane space has been found to associate with VDAC on the inner surface of the outer membrane.

Hexokinase associates with mitochondria to varying extents, depending on tissue type, metabolic state, and developmental stage, with especially high levels of mitochondrially bound hexokinase seen in highly glycolytic tumor cells (reviewed in Adams et al., 1991). The outer membrane binding site for hexokinase has been identified as the VDAC protein, and hexokinase binds to purified VDAC reconstituted into liposomes (Linden et al., 1982; Fiek et al., 1982). This binding is inhibited by glucose 6-phosphate, the product of hexokinase activity (Fiek et al., 1982), although the fraction of mitochondrially bound hexokinase that can be released with glucose-6-phosphate treatment varies between species (Kabir and Wilson, 1994). The binding of hexokinase to outer membranes is thought to increase the accessibility of mitochondrially generated ATP to the enzyme, and hexokinase bound to mitochondria preferentially uses mitochondrial, as opposed to exogenous, ATP (Rasschaert and Malaisse, 1990). Binding of hexokinase and glucokinase to the mitochondria in pancreatic β cells and adipocytes may be involved in regulation of blood sugar. Hexokinase in adipocytes and glucokinase in an insulinoma cell line (RINm5F) coimmunoprecipitate with a VDAC isoform found in these cells, and purified VDAC from adipocytes and pancreatic β cells mediates the binding of liver hexokinase to liposomes in a glucose-6-phosphate–dependent manner. Treatment of RINm5F cells or adipocytes with the drug glimepiride, which lowers blood sugar *in vivo*, reduces the fraction of glucokinase or hexokinase that can be coimmunoprecipitated with VDAC from the respective cell types and also reduces the ability of VDAC purified from the treated cells to bind liver hexokinase (Müller *et al.*, 1994).

The N-terminal 15 amino acid residues of hexokinase are required for binding to mitochondria (reviewed in Adams *et al.*, 1991). The region of VDAC required for binding hexokinase probably includes glutamate 72. Treatment of mitochondria with N,N′-dicyclohexylcarbodiimide (DCCD), which covalently modifies VDAC and no other outer membrane proteins, inhibits hexokinase binding (Nakashima *et al.*, 1986). Recently, glutamate 72 has been identified as the site of DCCD modification (dePinto *et al.*, 1993). The fact that the lipid-soluble reagent DCCD reacted with E72 but water-soluble reagents did not (Nakashima *et al.*, 1986) indicates that E72 is located in a hydrophobic environment. The amino-terminal domain of hexokinase is hydrophobic and is accessible to hydrophobic reagents only when hexokinase is bound to mitochondria (Xie and Wilson, 1990). Thus it is likely that both this N-terminal domain and VDAC's E72 are either buried in the membrane or in a hydrophobic pocket of the protein.

A portion of cellular glycerol kinase has also been shown to associate with mitochondria (Yilmaz *et al.*, 1987). Glycerol kinase from rat liver cytosol binds to purified VDAC in liposomes, but only when hexokinase binding has been inhibited by addition of glucose-6-phosphate (Fiek *et al.*, 1982; Müller *et al.*, 1994). This suggests that the two enzymes bind to the same site or to overlapping sites on VDAC.

The mitochondrial form of creatine kinase, which phosphorylates creatine using mitochondrial ATP, is located in the intermembrane space and is concentrated in contact sites between the inner and outer membranes. Mitochondrial creatine kinase exists as dimers which can further associate into cube-like homooctamers of 45-kDa subunits (Schlegel *et al.*, 1988). The octameric form can mediate binding between inner mitochondrial membrane lipids and outer membrane lipids (Rojo *et al.*, 1991). Mitochondrial creatine kinase also binds VDAC, and binding to VDAC favors formation of octamers from dimers (Brdiczka *et al.*, 1994). Binding to VDAC may increase the access of creatine kinase to cytoplasmic creatine, which presumably enters the intermembrane space through the VDAC. In addition, binding to creatine kinase alters the properties of VDACs; when VDAC-octamer complexes insert into lipid bilayers, the single-channel conductance was reduced to about half the level of uncomplexed VDAC (Brdiczka *et al.*, 1994). It is unclear whether these are closed channels or whether association with creatine kinase partially blocks the pore.

9. CONCLUSIONS

Study of the VDAC has shown that this small molecule can perform a wide variety of functions. It forms large anion-selective pores that can close to smaller cation-selection pores in two distinct voltage-dependent gating processes. Thus, this molecule can serve as a model of how the relevant charged groups forming a voltage "sensor" couple to other regions of a protein to produce the atomic rearrangements that generate channel gating in response to voltage changes. One remarkable feature of all VDACs is the high conservation of basic channel properties. VDAC proteins from yeast, plants, or humans are not highly conserved at the primary sequence level yet form channels with essentially identical single-channel conductance, ion selectivity, and voltage dependence in phospholipid membranes. Using yeast VDAC, the functional consequences of a large number of charge changes introduced by site-directed mutagenesis have been investigated. These functional studies have allowed the development of a model of the transmembrane topology of the yeast VDAC protein in the open state, defined regions of the protein that are removed from the pore during channel closure, and identified specific residues forming the voltage sensor. Additional studies have demonstrated that gating can be modulated by a variety of factors, including NADH and a protein found in the mitochondrial intermembrane space. The VDAC protein also specifically binds to cellular kinase enzymes, presumably giving them preferential access to metabolites transported through the VDAC. Thus, it is likely that VDAC forms the coordination point for a complex consisting of a number of molecules, and regulation of VDAC's gating properties is likely to depend not only on external factors, such as NADH, but on the dynamic association with a variety of other proteins that constitute a significant metabolic regulatory interaction. Clearly, understanding the functional implications of these associations is critical to understanding VDAC's function in the cell. These issues, as well as further refinements of existing models of the transmembrane topology of the VDAC and the conformational transitions associated with voltage gating, are likely to form the focus of future analysis of this unique protein.

ACKNOWLEDGMENTS. This work was supported by grants from the Office of Naval Research (N00014-90-J-1024) and the National Institutes of Health (GM 35759).

10. REFERENCES

Adams, V., and McCabe, E. R. B., 1994, Role of porin-kinase interactions in disease, in: *Molecular Biology of Mitochondrial Transport Systems, NATO ASI Series*, Vol. 83 (M. Forte and M. Colombini, eds.), Springer-Verlag, Berlin, pp. 357–377.

Adams, V., Griffin, L., Towbin, J., Gelb, B., Worley, K., and McCabe, E. R. B., 1991, Porin interaction with hexokinase and glycerol kinase: Metabolic microcompartmentation at the outer mitochondrial membrane, *Biochem. Med. Metabol. Biol.* **45:**271–291.

Akabas, M. H., Kaufmann, C., Archdeacon, P., and Karlin, A., 1994a, Identification of acetylcholine receptor channel-lining residues in the entire M2 segment of the alpha subunit, *Neuron* **13:**919–927.

Akabas, M. H., Kaufmann, C., Cook, T. A., and Archdeacon, P., 1994b, Amino acid residues lining the chloride channel of the cystic fibrosis transmembrane conductance regulator, *J. Biol. Chem.* **269:**14865–14868.

Benz, R., Kottke, M., and Brdiczka, D., 1990, The cationically selective state of the mitochondrial outer membrane pore: A study with intact mitochondria and reconstituted mitochondrial porin, *Biochim. Biophys. Acta* **1022:**311–318.

Blachly-Dyson, E., Peng, S., Colombini, M., and Forte, M., 1989, Probing the structure of the mitochondrial channel, VDAC, by site-directed mutagenesis: A progress report, *J. Bioenerg. Biomembr.* **21:**471–483.

Blachly-Dyson, E., Peng, S., Colombini, M., and Forte, M., 1990, Selectivity changes in site-directed mutants of the VDAC ion channel: Structural implications, *Science* **247:**1233–1236.

Blachly-Dyson, E., Zambrowicz, E. B., Yu, W.-H., Adams, V., McCabe, E. R. B., Adelman, J., Colombini, M., and Forte, M., 1993, Cloning and functional expression in yeast of two human isoforms of the outer mitochondrial membrane channel, the voltage-dependent anion channel, *J. Biol. Chem.* **268:**1835–1841.

Blachly-Dyson, E., Baldini, A., Litt, M., McCabe, E. R. B., and Forte, M., 1994, Human genes encoding the voltage-dependent anion channel (VDAC) of the outer mitochondrial membrane: Mapping and identification of two new isoforms, *Genomics* **20:**62–67.

Blumenthal, A., Kahn, K., Beja, O., Galun, E., Colombini, M., and Breiman, A., 1993, Purification and characterization of the voltage-dependent anion-selective channel protein from wheat mitochondrial membranes, *Plant Physiol.* **101:**579–587.

Brdiczka, D., Kaldis, P., and Wallimann, T., 1994, *In vitro* complex formation between the octamer of mitochondrial creatine kinase and porin, *J. Biol. Chem.* **269:**27640–47644.

Bureau, M. H., Khrestchatisky, M., Heeren, M. A., Zambrowicz, E. B., Kim, H., Grisar, T. M., Colombini, M., Tobin, A. J., and Olsen, T. W., 1992, Isolation and cloning of a voltage-dependent anion channel-like M_r36,000 polypeptide from mammalian brain, *J. Biol. Chem.* **267:**8679–8684.

Colombini, M., 1979, A candidate for the permeability pathway of the outer mitochondrial membrane, *Nature* **279:**643–645.

Colombini, M., 1980a, Pore size and properties of channels from mitochondria isolated from *Neurospora crassa*, *J. Membr. Biol.* **53:**79–84.

Colombini, M., 1980b, Structure and mode of action of a voltage-dependent anion-selective channel (VDAC) located in the outer mitochondrial membrane, *Ann. N.Y. Acad. Sci.* **341:**552–563.

Colombini, M., 1989, Voltage gating in the mitochondrial channel, *J. Membr. Biol.* **111:**103–111.

Colombini, M., 1994, Anion channels in the mitochondrial outer membrane, in: *Current Topics in Membranes*, Vol. 42 (W. Guggino, ed.), Academic Press, San Diego, pp. 73–101.

Colombini, M., Yeung, C. L., Tung, J., and König, T., 1987, The mitochondrial outer membrane channel, VDAC, is regulated by a synthetic polyanion, *Biochim. Biophys. Acta* **905:**279–286.

Colombini, M., Holden, M. J., and Mangan, P. S., 1989, Modulation of the mitochondrial channel VDAC by a variety of agents, in: *Anion Carriers of Mitochondrial Membranes* (A. Azzi *et al.*, eds.), Springer-Verlag, Berlin, pp. 215–224.

Craigen, W. J., Lovell, R. S., and Sampson, M. J., 1994, Structure and expression of mouse mitochondrial voltage dependent anion channel genes, *Am. J. Hum. Genet.* **55:**(Abstr. 3) supplement.

dePinto, V., Jamal, J. A., and Palmieri, F., 1993, Location of the dicyclohexylcarbodiimide-reactive glutamate residue in the bovine heart mitochondrial porin, *J. Biol. Chem.* **268:**12977–12982.

Dermietzel, R., Hwang, T.-K., Buettner, R., Hofer, A., Dotzler, E., Kremer, M., Deutzmann, R., Thinnes, F. P., Fishman, G. I., Spray, D. C., and Siemen, D., 1994, Cloning and in situ localization of a brain-derived porin that constitutes a large-conductance anion channel in astrocytic plasma membranes, *Proc. Natl. Acad. Sci. USA* **91:**499–503.

Dill, E. T., Holden, M. J., and Colombini, M., 1987, Voltage gating in VDAC is markedly inhibited by micromolar quantities of aluminum, *J. Membr. Biol.* **99:**187–196.

Doring, C., and Colombini, M., 1985, The mitochondrial voltage-dependent channel, VDAC, is modified asymmetrically by succinic anhydride, *J. Membr. Biol.* **83:**87–94.

Durkin, J. T., Koeppe II, R. E., and Anderson, O. S., 1990, Energetics of gramicidin hybrid channel formation: A test for structural equivalence. Side-chain substitutions in the native sequence, *J. Mol. Biol.* **211:**221–234.

Eisenman, G., and Krasne, S., 1975, The ion selectivity of carrier molecules, membranes and enzymes, in: *MTP International Review of Science, Biochemistry Series*, Vol. 2 (C. F. Fox, ed.), Butterworths, London, pp. 27–59.

Fiek, C., Benz, R., Roos, N., and Brdiczka, D., 1982, Evidence for identity between the hexokinase-binding protein and the mitochondrial porin in the outer membrane of rat liver mitochondria, *Biochim. Biophys. Acta* **688:**429–440.

Finkelstein, A., 1985, The ubiquitous presence of channels with wide lumen and their gating by voltage, *Ann. N.Y. Acad. Sci.* **456:**26–32.

Fischer, K., Weber, A., Brink, S., Arbinger, B., Schünemann, D., Borchert, S., Heldt, H. W., Popp, B., Benz, R., Link, T.-A., Eckerskorn, C., and Flügge, U., 1994, Porins from plants: Molecular cloning and functional characterization of two new members of the porin family, *J. Biol. Chem.* **269:**25754–25760.

Florke, H., Thinnes, F. P., Winkelbach, H., Stadtmuller, U., Paetzold, G., Morys-Wortmann, C., Hesse, D., Sternbach, H., Zimmermann, B., and Kaufmann-Kolle, P., 1994, Channel active mammalian porin, purified from crude membrane fractions of B lymphocytes and bovine skeletal muscle, reversibly binds adenosine triphosphate (ATP), *Biol. Chem. Hoppe-Seyler* **375**:513–520.

Gellerich, F. N., Wagner, M., Kapischke, M., Wicker, U., and Brdiczka, D., 1993, Effect of macromolecules on the regulation of the mitochondrial outer membrane pore and the activity of adenylate kinase in the inter-membrane space, *Biochim. Biophys. Acta* **1142**:217–227.

Ha, H., Hajek, P., Bedwell, D. M., and Burrows, P. D., 1993, A mitochondrial porin cDNA predicts the existence of multiple human porins, *J. Biol. Chem.* **268**:12143–12149.

Heins, L., Mentzel, H., Schmid, A., Benz, R., and Schmitz, U. K., 1994, Biochemical, molecular and func-tional characterization of two different porins from potato mitochondria, *J. Biol. Chem.* **264**:26402–26410.

Holden, M. J., and Colombini, M., 1988, The mitochondrial outer membrane channel, VDAC, is modulated by a soluble protein, *FEBS Lett.* **241**:105–109.

Holden, M. J., and Colombini, M., 1993, The outer mitochondrial membrane channel, VDAC, is modulated by a protein localized in the intermembrane space, *Biochim. Biophys. Acta* **1144**:396–402.

Kabir, F., and Wilson, J. E., 1994, Mitochondrial hexokinase in brain: Coexistence of forms differing in sensitivity to solubilization by glucose-6-phosphate on the same mitochondria, *Arch. Biochem. Biophys.* **310**:410–416.

Kayser, H., Kratzin, H. D., Thinnes, F. P., Götz, H., Schmidt, W. E., Eckart, K., and Hilschmann, N., 1989, Charakterisierung und primärstruktur eines 31-kDa-porins aus menschlichen B-Lumphozyten (porin 31HL), *Biol. Chem. Hoppe-Seyler* **370**:1265–1278.

Lee, A.-C., Zizi, M., and Colombini, M., 1994, β-NADH decreases the permeability of the mitochondrial outer membrane to ADP by a factor of 6, *J. Biol. Chem.* **269**:30974–30980.

Linden, M., and Gellerfors, P., 1983, Hydrodynamic properties of porin isolated from outer membrane of rat liver mitochondria, *Biochim. Biophys. Acta* **736**:125–129.

Linden, M., Gellerfors, P., and Nelson, B. D., 1982, Pore protein and the hexokinase-binding protein from the outer membrane of rat liver mitochondria are identical, *FEBS Lett.* **141**:189–192.

Liu, M. Y., and Colombini, M., 1991, Voltage gating of the mitochondrial outer membrane channel VDAC is regulated by a very conserved protein, *Am. J. Physiol.* **260**:C371–C374.

Liu, M. Y., and Colombini, M., 1992a, Regulation of mitochondrial respiration by controlling the permeability of the outer membrane through the mitochondrial channel, VDAC, *Biochim. Biophys. Acta* **1098**:255–260.

Liu, M. Y., and Colombini, M., 1992b, A soluble protein increases the voltage dependence of the mitochondrial channel, VDAC, *J. Bioenerg. Biomembr.* **24**:41–46.

Liu, M. Y., Torgrimson, A., and Colombini, M., 1993, Characterization and partial purification of the VDAC-channel-modulating protein from calf liver mitochondria, *Biochim. Biophys. Acta* **1185**:203–212.

Mangan, P., and Colombini, M., 1987, Ultrasteep voltage dependence in a membrane channel, *Proc. Natl. Acad. Sci. USA* **84**:4896–4900.

Mannella, C. A., 1986, Mitochondrial outer membrane channel (VDAC, porin): Two-dimensional crystals from *Neurospora*, *Methods Enzymol.* **125**:595–610.

Mannella, C. A., 1987, Electron microscopy and image analysis of the mitochondrial outer membrane channel, VDAC, *J. Bioenerg. Biomembr.* **19**:329–340.

Mannella, C. A., 1990, Structural analysis of mitochondrial pores, *Experientia* **46**:137–145.

Mannella, C. A., and Guo, X. W., 1990, Interaction between the VDAC channel and a polyanionic effector, *Biophys. J.* **57**:23–31.

Mannella, C. A., Forte, M., and Colombini, M., 1992, Toward the molecular structure of the mitochondrial channel, VDAC, *J. Bioenerg. Biomembr.* **24**:7–19.

Müller, G., Korndörfer, A., Kornak, U., and Malaisse, W. J., 1994, Porin proteins in mitochondria from rat pancreatic islet cells and white adipocytes: Identification and regulation of hexokinase binding by the sulfonylurea glimepiride, *Arch. Biochem. Biophys.* **308**:8–23 (erratum: *Ibid.* **313**:382).

Nakashima, R. A., Mangan, P. S., Colombini, M., and Pedersen, P. L., 1986, Hexokinase receptor complex in hepatoma mitochondria: Evidence from N, N′-dicyclohexylcarbodiimide-labeling studies for the in-volvement of the pore-forming protein VDAC, *Biochemistry* **25**:1015–1021.

Peng, S., Blachly-Dyson, E., Forte, M., and Colombini, M., 1992a, Large scale rearrangement of protein domains is associated with voltage gating of the VDAC channel, *Biophys. J.* **62**:123–135.

Peng, S., Blachly-Dyson, E., Forte, M., and Colombini, M., 1992b, Determination of the number of polypeptide subunits in a functional VDAC channel from *Saccharomyces cerevisiae*, *J. Bioenerg. Biomembr.* **24**:27–31.

Pfaller, R., Freitag, H., Harmey, M. A., Benz, R., and Neupert, W., 1985, A water-soluble form of porin from the mitochondrial outer membrane of *Neurospora crassa*, *J. Biol. Chem.* **260:**8188–8193.

Rasschaert, J., and Malaisse, W. J., 1990, Hexose metabolism in pancreatic islets: Preferential utilization of mitochondrial ATP for glucose phosphorylation, *Biochim. Biophys. Acta* **1015:**353–360.

Rojo, M., Hovius, R., Demel, R. A., Nicolay, K., and Willimann, T., 1991, Mitochondrial creatine kinase mediates contact formation between mitochondrial membranes, *J. Biol. Chem.* **266:**20290–20295.

Roos, N., Benz, R., and Brdiczka, D., 1982, Identification and characterization of the pore-forming protein in the outer membrane of rat liver mitochondria, *Biochim. Biophys. Acta* **686:**204–214.

Schein, S. J., Colombini, M., and Finkelstein, A., 1976, Reconstitution in planar lipid bilayers of a voltage-dependent anion-selective channel obtained from Paramecium mitochondria, *J. Membr. Biol.* **30:**99–120.

Schlegel, J., Wyss, M., Schürch, U., Schnyder, T., Quest, A., Wegmann, G., Eppenberger, H. M., and Wallimann, T., 1988, Mitochondrial creatine kinase from cardiac muscle and brain are two distinct iso-enzymes but both form octameric molecules, *J. Biol. Chem.* **263:**16963–16969.

Song, J. M., and Colombini, M., 1996, Indications of a common folding pattern for VDAC channels from all sources, *J. Bioenerg. Biomembr.*, in press.

Teorell, T., 1953, Transport processes and electrical phenomena in ionic membranes, *Prog. Biophys. Biophys. Chem.* **3:**305–369.

Thomas, L., Kocsis, E., Colombini, M., Erbe, E., Trus, B. L., and Steven, A. C., 1991, Surface topography and molecular stoichiometry of the mitochondrial channel, VDAC, in crystalline arrays, *J. Struct. Biol.* **106:**161–171.

Thomas, L., Blachly-Dyson, E., Colombini, M., and Forte, M., 1993, Mapping of residues forming the voltage sensor of the VDAC channel, *Proc. Natl. Acad. Sci. USA* **90:**5446–5449.

Troll, H., Malchow, D., Müller-Taubenberger, A., Humbel, B., Lottspeich, F., Ecke, M., Gerisch, G., Schmid, A., and Benz, R., 1992, Purification, functional characterization, and cDNA sequencing of mitochondrial porin from dictyostelium discoideum, *J. Biol. Chem.* **267:**21072–21079.

Unwin, P. N. T., and Zampighi, G., 1980, Structure of the junction between communicating cells, *Nature* **283:**545–549.

Vodyanoy, I., Bezrukov, S. M., and Colombini, M., 1992, Measurement of ion channel access resistance, *Biophys. J.* **61:**A114.

Weiss, M. S., Wacker, T., Weckesser, J., Welte, W., and Schulz, G. E., 1990, The three-dimensional structure of porin from *Rhodobacter capsulatus* at 3 Å resolution, *FEBS Lett.* **267:**268–272.

Xie, G., and Wilson, J. E., 1990, Rat brain hexokinase: The hydrophobic N-terminus of the mitochondrially bound enzyme is inserted in the lipid bilayer, *Arch. Biochem. Biophys.* **276:**285–293.

Yellen, G. M., Jurman, M., Abramson, T., and MacKinnon, R., 1991, Mutations affecting internal TEA blockade identify the probable pore-forming region of a K^+ channel, *Science* **251:**939–942.

Yilmaz, M. T., Sener, A., and Malaisse, W. J., 1987, Glycerol phosphorylation and oxidation in pancreatic islets, *Mol. Cell. Endocrinol.* **52:**251–256.

Zalman, L. S., Nikaido, H., and Kagawa, Y., 1980, Mitochondrial outer membrane contains a protein producing nonspecific diffusion channels, *J. Biol. Chem.* **255:**1771–1774.

Zambrowicz, E. B., and Colombini, M., 1993, Zero-current potentials in a large membrane channel: A simple theory accounts for complex behavior, *Biophys. J.* **65:**1093–1100.

Zhang, D. W., and Colombini, M., 1989, Inhibition by aluminum hydroxide of the voltage-dependent closure of the mitochondrial channel, VDAC, *Biochim. Biophys. Acta* **991:**68–78.

Zhang, D. W., and Colombini, M., 1990, Group IIIA-metal hydroxides indirectly neutralize the voltage sensor of the voltage-dependent mitochondrial channel, VDAC, by interacting with a dynamic binding site, *Biochim. Biophys. Acta* **1025:**127–134.

Zimmerberg, J., and Parsegian, V. A., 1986, Polymer inaccessible volume changes during opening and closing of a voltage-dependent ionic channel, *Nature* **323:**36–39.

Zizi, M., Forte, M., Blachly-Dyson, E., and Colombini, M., 1994, NADH regulates the gating of VDAC, the mitochondrial outer membrane channel, *J. Biol. Chem.* **269:**1614–1616.

Zizi, M., Thomas, L., Blachly-Dyson, E., Forte, M., and Colombini, M., 1995, Oriented channel insertion reveals the motion of a transmembrane beta strand during voltage gating of VDAC, *J. Membr. Biol.* **144:**121–129.

CHAPTER 6

ION CHANNELS AND MEMBRANE RECEPTORS IN FOLLICLE-ENCLOSED *XENOPUS* OOCYTES

ROGELIO O. ARELLANO, RICHARD M. WOODWARD,
and RICARDO MILEDI

1. INTRODUCTION

The follicle-enclosed *Xenopus* oocyte (follicle) is a well-studied system that has been utilized in investigations of cell division, ion channels, membrane receptors, second messenger systems, protein synthesis, cell-to-cell interactions and development, cellular polarity, fertilization, *etc.* (Masui and Markert, 1971; Gurdon *et al.*, 1971; Kusano *et al.*, 1977, 1982; Robinson, 1979; Kado *et al.*, 1981; Miledi, 1982; Miledi and Parker, 1984; Parker and Miledi, 1986; Browne and Werner, 1984; Webb and Nuccitelli, 1985). Much of this knowledge helped to develop the oocyte as a system for expressing exogenous mRNA or cRNA and studying the transcribed proteins (Gurdon *et al.*, 1971; Barnard *et al.*, 1982; Miledi *et al.*, 1982; Gundersen *et al.*, 1983, 1984; Colman, 1984a,b). Because of all the background knowledge accumulated, and because *Xenopus* oocytes offer many advantages, they are now commonly used in laboratories interested in diverse areas of the biomedical sciences.

Researchers have been studying the electrophysiological characteristics of the *Xenopus* follicle for over 25 years. While several reviews have appeared on the physiology of the oocyte and its use in expression studies (Dascal, 1987; Miledi *et al.*, 1989a;

ROGELIO O. ARELLANO, RICHARD M. WOODWARD, and RICARDO MILEDI • Laboratory of Cellular and Molecular Neurobiology, Department of Psychobiology, University of California, Irvine, Irvine, California 92717. *Present address of R.M.W.*: Acea Pharmaceuticals, Irvine, California 97218.

Ion Channels, Volume 4, edited by Toshio Narahashi, Plenum Press, New York, 1996.

Goldin, 1991), there has been no specific compendium on the electrical properties of the native follicle, that is, an oocyte still surrounded by its enveloping ovarian tissue. The present is an attempt to correct, at least partially, this deficiency. Recent studies show that some types of follicular electrical responses to neurotransmitters and hormones, monitored by electrodes inserted in the oocyte, require maintenance of the electrical communication between the oocyte and its surrounding follicular cells. This in turn suggests that the responses originate in the membrane of the follicular cells and that these cells may contain all the elements for response generation, i.e., membrane receptors, several types of ion channels, and the intracellular machinery involved in channel gating and modulation. Studies of these elements are very important for understanding follicular physiology, its cellular compartmentation, and interrelations and raise issues that should be considered when oocytes are used as a heterologous protein expression system or as a model for studying other cellular processes.

2. THE FOLLICLE-ENCLOSED *XENOPUS* OOCYTE

2.1. Morphology

Detailed morphological characterization of *Xenopus* follicle-enclosed oocytes was prompted when it was realized that they could be very useful to study oocyte and early embryonic development. The description of oogenesis most commonly used is that of Dumont (1972), who defined six developmental stages, from the previtellogenic period (stage I), in which the follicle-enclosed oocyte is still a small transparent cell (50–300 μm in diameter), to the fully developed state (stave VI) following the vitellogenic process, where the oocyte reaches its maximum size (1200–1400 μm in diameter) and acquires its characteristic opaque and polarized appearance. The enclosed oocyte is then ready to continue its maturation, which has been halted in the first prophase of the meiotic division (Smith, 1989). Another important feature is that oogenesis in *Xenopus* is continuous and asynchronous so follicles in every developmental stage can be obtained from a single ovary. Furthermore, the frogs can be stimulated by hormones (i.e., gonadotropins) to produce a plentiful supply of fully grown oocytes.

At any stage of development the follicles are a complex multicellular system consisting of the oocyte itself surrounded by several cellular and noncellular envelopes (Fig. 1). The layer immediately adjacent to the oocyte membrane is the vitelline envelope, an acellular fibrous sheath that begins to form during stage II and completely surrounds the oocyte by stage III (Dumont, 1972). Adjacent to the outer surface of the vitelline envelope is a discontinuous monolayer of follicular cells (also termed "follicle cells"). In stage II the follicular cells begin to develop thin cellular processes (macrovilli) which extend, through pores in the vitelline envelope, toward the underlying oocyte membrane. At the same stage, the oocytes develop their own membrane processes (microvilli). Follicular cell macrovilli and oocyte microvilli make numerous contacts, forming a complex network that grows and is maintained throughout follicular stages III to VI (1000–1400 μm). Lying directly above the follicular cells is a basement membrane, formed mainly of collagen, which separates the follicular cell layer from overlying

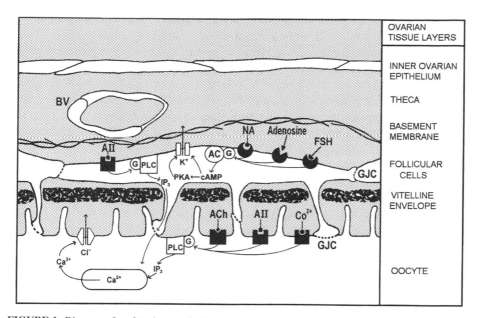

FIGURE 1. Diagram of surface layers of a follicle-enclosed *Xenopus* oocyte. The follicle consists of the oocyte surrounded by five distinct ovarian layers. Electrical and metabolic coupling is maintained between follicular cells and oocyte via gap junctions. Also illustrated are examples of ion channels, membrane receptors, and receptor-channel coupling mechanisms present in the follicular complex. Shading indicates extracellular spaces. BV, blood vessel; GJC, gap junction channels; G, G protein; AC, adenylate cyclase; PKA, protein kinase A; PLC, phospholipase C; FSH, follicle stimulating hormone; NA, noradrenaline; AII, angiotensin II; ACh, acetylcholine.

tissues (see plate 1, Miledi and Woodward, 1989a). In the original description (Dumont, 1972), the basement membrane of the follicular cells is grouped together with the thecal layer which also contains fibroblasts and blood vessels. The final external layer of the follicle is the inner ovarian epithelium, also referred to by Dumont (1972) as "surface epithelium." Perhaps the most intriguing among the follicle's cellular interactions is that which occurs between the oocyte and its surrounding layer of follicular cells. Their close relationship, maintained through oogenesis, is a prerequisite for the proper development of the gamete (Eppig, 1979; Buccione *et al.*, 1990a). Follicular cells also help to maintain an adequate microenvironment for the oocyte. In particular, follicular cells are responsible for the production and secretion of steroids and other factors which promote the growth and maturation of the oocyte (for reviews, Hsueh *et al.*, 1983; Dorrington *et al.*, 1983; Shultz, 1985). Communication between oocyte and follicular cells is not only paracrine, through secreted substances (Buccione *et al.*, 1990a,b), but also direct, via gap junctions (Anderson and Albertini, 1976; Dumont and Brummett, 1978; Browne *et al.*, 1979; Browne and Werner, 1984; van den Hoef *et al.*, 1984). These membrane structures are electrical and metabolic channels of communication that establish direct continuity between the cytoplasm of the coupled cells. From studies in other systems it is known that gap junctions are permeable to molecules with molecular weights up to *ca.* 1000 Da

(Lowenstein, 1981), although there are differences between junctional systems (Robinson *et al.*, 1993; Brissette *et al.*, 1994). A detailed electrophysiological characterization of gap junction channels in *Xenopus* follicles has not been reported, but they seem to be permeable to molecules such as carboxyfluorescein (Browne *et al.*, 1979) and various nucleotides (Heller and Shultz, 1980). The gap junctions also allow passage of several important second messengers, such as cAMP, cGMP, and inositol-1,4,5-triphosphate (IP$_3$) (Miledi and Woodward, 1989a; Sandberg *et al.*, 1992), and other substances such as tetraethylammonium, sucrose, and polyethylene glycols (up to ~600 Da) (Arellano and Miledi, 1995).

In a variety of cellular systems, coupling via gap junctions is under short-term regulation by neurotransmitters and hormones (Piccolino *et al.*, 1984; Lasater and Dowling, 1985; Neyton and Trautmann, 1986). In *Xenopus* follicles this aspect of gap junction modulation has yet to be investigated properly, though there is evidence suggesting long-term regulation by hormones during oogenesis (Browne *et al.*, 1979). At present, it appears that gap junction channels in *Xenopus* follicles at stages V and VI maintain a constant and high level of physiological coupling between follicular cells and oocyte, without need of a specific chemical or electrical stimuli. This can be assessed from the rapid rates at which follicular electrical responses develop following application of agonists (e.g., fast inward currents, or F_{in} currents) or injection of substances into the oocytes (e.g., cAMP, cGMP, sucrose; see below), and by the fast diffusion of carboxyfluorescein from the oocyte into the follicular cell compartments (R. O. Arellano, unpublished results), and the abundance of follicular gap junctions (van den Hoef *et al.*, 1984) also support this conclusion.

With respect to follicular electrophysiology, the electrical coupling between follicular cells and the oocyte and between the follicular cells themselves means that some membrane current responses monitored by microelectrodes inserted into the oocyte may be arising in the membrane of the follicular cells and not in the oocyte membrane itself. Furthermore, when characterizing follicular membrane currents one needs to remember that a change in these currents could be a consequence of indirect effects at the level of oocyte–follicular cell coupling (i.e., modulation of gap junction channels).

2.2. Electrophysiological Methods

Techniques for electrophysiological studies in *Xenopus* follicles were essentially the same as those developed for oocytes. These have been described in detail (Miledi, 1982; Miledi *et al.*, 1989a; Sumikawa *et al.*, 1989; Goldin, 1991), and many procedures useful for studies on oocytes have been fully described (Colman, 1984; Kay and Peng, 1991; Rudy and Iverson, 1992). For studies on follicles it is important that the constituent cells retain their native characteristics *in vitro*. We consistently find that follicles progressively lose their endogenous electrical responses (e.g., K$^+$-cAMP currents, Section 4), although in some cases there may be an initial enhancement (e.g., F_{in} currents, Section 5). For this reason, we usually work with fresh follicles (1–3 days after dissection) maintained at 16–18°C in unsupplemented Barth's medium (88 mM NaCl, 1 mM KCl, 2.4 mM NaHCO$_3$, 0.33 mM Ca (NO$_3$)$_2$, 0.41 mM CaCl$_2$, 0.82 mM MgSO$_4$, 5 mM N-(2-hydroxyethyl)piperazine-N'-(2-ethanesulfonic acid) (HEPES) adjusted to pH

7.4, with gentamicin 70 μg/ml) (Barth and Barth, 1959). When we wanted to maintain follicles for a longer time *in vitro*, we used a sterile modified Barth's solution (88 mM NaCl, 0.2 mM KCl, 2.4 mM NaHCO$_3$, 0.33 mM Ca(NO$_3$)$_2$, 0.41 mM CaCl$_2$, 0.82 mM MgSO$_4$, 0.88 mM KH$_2$PO$_4$, 2.7 mM Na$_2$HPO$_4$, 5 mM glucose, pH 7.4, with gentamicin 70 μg/ml) supplemented with 0.1% fetal bovine serum. Under the latter conditions the follicles retained their original electrical responses for about ten days.

Essentially, two different preparations of follicles were routinely used for electro-physiological recordings. In one, the native follicle was simply "plucked" from the ovary, thus retaining all the enveloping layers. In the other, the inner epithelia of the follicle, together with the thecal blood vessels, were dissected with sharp forceps. In this procedure, the basement membrane was not removed to provide some protection for the underlying follicular cells. In the present paper we will refer to this preparation as "epithelium-removed" follicles. This procedure simplified the interpretation of results by eliminating the possibility of thecal tissue involvement in follicular responses. In addition, removal of the inner ovarian epithelium and external thecal tissue facilitated the insertion of microelectrodes. Preparatory studies indicated that the follicular basement membrane was important to preserve the coupling and/or the electrical properties of the follicular cells. In addition to the follicle preparations we also used oocytes deprived of all surrounding follicular tissues except the vitelline layer. These are here termed "defolliculated" oocytes (see below), or simply oocytes.

Electrical responses in follicles were usually monitored with a two-electrode volt-age clamp (Miledi, 1982). The follicles were continuously superfused with either normal frog Ringer's (NR) solution containing 115 mM NaCl, 2 mM KCl, 1.8 mM CaCl$_2$, and 5 mM HEPES at pH 7.0; or a hypoosmotic Ringer solution (HR) containing 88 mM NaCl, 2 mM KCl, 1.8 mM CaCl$_2$, and 5 mM HEPES at pH 7.0. Unless otherwise stated, follicles and oocytes were voltage-clamped at −60 mV so as to be away from the equilibrium potentials for K$^+$, Na$^+$, Ca^{2+}, and Cl$^-$ (Kusano *et al.*, 1982), and the connotation of inward and outward ionic currents is given with respect to this potential. Drugs were applied by superfusion. For studies on follicular inward currents (e.g., Sections 5 and 6) we routinely blocked K$^+$ currents by adding 1–2 mM BaCl$_2$ to the superfusate, or tetraethylammonium (TEA$^+$), either substituting NaCl in the bath solu-tion (20–28 mM) or injected (0.2–2 nmol) into the oocyte (Arellano and Miledi, 1995).

Voltage-clamped follicles were injected intracellularly by pressure pulses applied to a third micropipette inserted into the oocyte (Miledi and Parker, 1984). Solutions for injection usually contained 5 mM HEPES and 50–200 μM EGTA [ethylene glycol-bis(β-amino-ethylether)N,N,N',N'-tetraacetic acid] and were adjusted to pH 7.0 with KOH. Substances injected included divalent cations, EGTA, BAPTA [1,2-bis(2-aminophenoxy)ethane-N,N,N',N'-tetraacetic acid], cAMP, cGMP, IP$_3$, TEA$^+$, protein kinase inhibitors, sucrose, and PEGs (polyethylene glycols) with different average molecular weights, e.g., PEG 200, PEG 300. These solutions were prepared at concentra-tions which were dependent on their activity in the cytosol, taking into consideration a dilution factor in the whole follicle and assuming homogeneous distribution. The final concentration in the follicle was estimated from the diameter of ejected droplets mea-sured in air before and after the intra-oocyte injection, and assuming an oocyte volume of 1 μl.

We have also used patch-clamp recording techniques (Hamill *et al.*, 1981) to study follicular cell channels directly. In these experiments, epithelium-removed follicles were further dissected a few minutes before recording. In order to gain access to the surface of the follicular cell membrane, a small "window" was opened through the basement membrane and the recording patch-clamp pipette was positioned above this site. Note that during the cell-attached recordings follicular cells remained *in situ* and were presumably still coupled to the oocyte.

2.3. Defolliculation Methods

Follicular electrical responses can be simply divided into currents that require the presence of follicular cells and currents that are independent of them, i.e., currents that originate wholly in the oocyte itself. For many years the most common method of defolliculation has been treatment with collagenase, and this continues to be the preferred method for preparing oocytes for expression studies. For collagenase treatment, isolated follicles were placed in vials containing 0.5–2 mg/ml collagenase (Sigma type I) in NR solution for 1–2 hr. Vials were then briskly shaken by hand, which usually loosens and removes inner ovarian epithelial as well as most of the follicular cells. Treated follicles were then repeatedly washed and returned to sterile Barth's medium. Before placing in an incubator, the collagenase-treated follicles can be checked microscopically to remove residues of external layers which often remain stuck to the vitelline envelope.

Unfortunately, there remains some uncertainty as to whether collagenase treatment consistently removes all the follicular cells (Miledi and Woodward, 1989a). Moreover, there are always concerns that, in addition to removing the external layers, the enzymatic digestion may also disrupt membrane proteins needed for generating responses (e.g., Dascal, 1987; Lupu-Meiri *et al.*, 1990). This possibility is enhanced by the fact that the most commonly used collagenase preparations, which are also the most effective for defolliculation, contain proteases that may exert nonspecific effects on membrane proteins. Considerations such as these prompted us to develop mechanical methods of defolliculation which not only ensure complete removal of follicular cells, but also do not involve enzymatic or chemical treatments. To estimate the efficacy of the procedures we used electrophysiological assays on the same oocytes before and after defolliculation and also quantified directly the extent of defolliculation using scanning electron microscopy (Miledi and Woodward, 1989a). For manual defolliculation oocytes were dissected from the ovary by peeling away the inner ovarian epithelium with sharp forceps. In contrast to the technique used for preparing epithelium-removed follicles, which retains the basement membrane of the follicular cells, the epithelial covering was removed together with the associated blood vessels and the basement membrane. Complete removal of the basement membrane, thecal layer, and epithelium only occurs when dissection produces a characteristic "unzipping" effect, where the oocyte extrudes itself through a rip in the enclosing layers. After this dissection the oocytes were still surrounded by most of their follicular layer, as shown directly by scanning electron microscopy studies and indirectly by the maintenance of several follicular electrical responses (Miledi and Woodward, 1989a; see Fig. 6B). Elimination of the follicular cells in these "unzipped follicles" was achieved by rolling the oocytes over glass slides coated with poly-L-lysine, either in Barth's or Ca^{2+}-free Ringer's solution. The follicular cells became stuck to the slide and were stripped from the oocyte, either singly or as ragged sheets. The "rolled oocytes" were then washed and incubated for 1–2 hr in Barth's solution before recording. Another approach to assure defolliculation without enzymatic treatment was to simply store the "unzipped follicles" in sterile normal Barth's solution for three or more days. Under these conditions the follicles either spontaneously lose the follicular monolayer or it can be removed using forceps.

3. FOLLICLE ION CURRENTS: AN OVERVIEW

3.1. Oocyte-Based Ca^{2+}-Dependent Cl^- Currents

Follicles respond to numerous neurotransmitters, peptides, and hormones that generate a variety of membrane current responses. The first responses described were those elicited by cholinergic and catecholaminergic agonists (Kusano *et al.*, 1977, 1982). These studies showed that activation of muscarinic acetylcholine (ACh) receptors generates multiphasic membrane currents in which the amplitudes of the different components vary greatly between follicles from different frogs. One component of these ACh responses is a characteristic oscillatory current that is preserved after defolliculation (Figs. 2A,B; 3B). It is now clear that these oscillatory membrane currents are mediated by ACh receptors and chloride-selective channels, both of which are located in the membrane of the oocyte itself (cf. Fig. 1). The oscillatory Cl^- response and an associated smooth current (see below) have a mechanism of activation that involves an increase in intracellular Ca^{2+} via receptor-stimulated activation of phospholipase C (PLC), IP_3 production, mobilization of intracellular Ca^{2+}, and opening of Ca^{2+}-dependent Cl^- channels (denoted here as $I_{Cl(Ca)}$) (Miledi, 1982; Miledi and Parker, 1984; Oron *et al.*, 1985; Parker *et al.*, 1985a,b; Parker and Miledi, 1986, 1987; Takahashi *et al.*, 1987). In oocytes from some frogs, the ACh-elicited oscillatory current includes an initial spike-like component (Fig. 2A) that may be over 20 μA in amplitude. The other muscarinic component of $I_{Cl(Ca)}$ is a smooth, slow Cl^- current, which develops and deactivates over several minutes and is due to an IP_3-dependent calcium influx from the external medium (Parker *et al.*, 1985a; Parker and Miledi, 1987). During the smooth component, hyperpolarization steps produce a further increase in the Ca^{2+}-dependent Cl^- current (denoted *transient* inward or T_{in} current; Parker and Miledi, 1987; Yao and Parker, 1993) (not shown). Calcium influx and opening of $I_{Cl(Ca)}$ can also be achieved by membrane depolarization, which opens voltage-dependent Ca^{2+} channels (denoted *transient* outward or T_{out} current; Miledi, 1982; Barish, 1983) (not shown).

Coexistence of all the ACh-elicited components of $I_{Cl(Ca)}$ in native oocytes is rare, but it is common in oocytes expressing exogenous receptors (Parker *et al.*, 1987). These multiphasic responses can also be elicited by activation of other native receptors located in the oocyte membrane, for example, receptors to divalent cations, a serum factor, and angiotensin II (AII) (Miledi *et al.*, 1989b; Tigyi *et al.*, 1990; Woodward and Miledi, 1991). The characteristics of these responses and the complex mechanisms involved in their activation have been discussed previously and will not be considered here (for reviews,

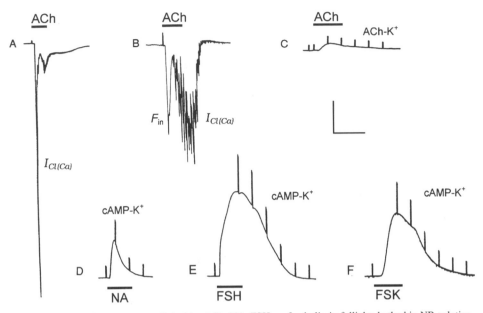

FIGURE 2. Membrane currents elicited by ACh, NA, FSH, or forskolin in follicles bathed in NR solution. (A–C) Various responses to ACh (100 μM) in follicles from different frogs, showing the variability due to the different proportions of current components; (A) a follicle which displays mainly the oocyte-based oscillatory $I_{Cl(Ca)}$; (B) a mixture of the fast and transient inward current (F_{in}), which originates in the follicular cells, along with the $I_{Cl(Ca)}$; (C) a follicular response consisting mainly of ACh-K$^+$ current. (D–F) cAMP-K$^+$ current responses recorded in a single follicle in response to NA (100 μM), FSH (0.5 μM), or forskolin (FSK, 0.5 μM). In this and following figures, drugs were applied by bath superfusion for the time indicated by the bars, and except where indicated all follicles were held at −60 mV. In many following records voltage steps (+10 mV) were applied periodically to monitor follicle membrane resistance. Scale bars: horizontal; 240 sec (A, C–F), 120 sec (B); vertical; 100 nA (A, B, D–F), 50 nA (C). [(D–F) from Woodward and Miledi, 1987b].

see Dascal, 1987; Miledi *et al.*, 1989a). What is important to stress here is that all these receptor-stimulated currents are a consequence of IP$_3$ production, increase of intracellular calcium, and subsequent activation of Ca^{2+}-gated Cl$^-$ channels located in the oocyte membrane (Fig. 1 and Table I). Their persistence following defolliculation (manual or enzymatic) readily distinguishes these oocyte-based currents from those that originate in the follicular cells.

3.2. Follicular Cell–Based Muscarinic Responses (ACh-K$^+$, F_{in}, S_{in}, and S_{out} Currents)

Although the oscillatory $I_{Cl(Ca)}$ and the other Ca^{2+}-dependent current components elicited by ACh are preserved through defolliculation, some phases of the follicular current responses disappear completely (Fig. 3). This suggests that part of the ACh response arises in the membrane of follicular cells. In early studies (Kusano *et al.*, 1982; Dascal and Cohen, 1987; Miledi and Woodward, 1989a), at least two components of the follicular current responses elicited by ACh were consistently eliminated by defollicula-

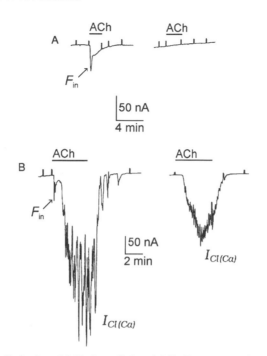

FIGURE 3. Selective elimination of follicular cell–based ACh-F_{in} responses after manual defolliculation. The first record in each panel shows responses to ACh (100 μM) in intact follicles from different frogs; second traces are responses in the same oocytes after manual defolliculation. In (A) the follicular cell–based F_{in} was completely abolished. In (B) F_{in} was similarly eliminated while the oscillatory $I_{Cl(Ca)}$ was largely preserved. [Panel (A) from Miledi and Woodward, 1989a].

tion: 1) a slow smooth outward current, associated with an increase in membrane conductance and carried by K^+ (referred to here as ACh-K^+) (e.g., Figs. 2C and 4B) and 2) a *Fast* transient inward current carried by Cl^-, which has a characteristic short latency of activation (referred to here as F_{in} current) (e.g., Figs. 2C, 3, and 4A,B).

In more recent studies (Arellano and Miledi, 1993, 1994) and recordings under conditions where the osmolarity of NR solution was reduced by decreasing NaCl content by approximately 20%, the follicular responses evoked by ACh included at least two additional components (Fig. 4A,B) that were also completely dependent on the coupling between follicular cells and the oocyte. One is a slow and *Smooth* inward current, associated with an increase in membrane conductance, mainly to Cl^- ions [denoted as S_{in} current]. This osmodependent S_{in} current is Ca^{2+}-independent. The second is an outward slow current, denoted *Smooth* outward (S_{out}) current, associated with a *decrease* in membrane conductance, presumably due to closing of Cl^- channels which are active under hypoosmotic conditions (e.g., Figs. 4A and 2B). The characterization of this current is still not complete and it will be mentioned only where necessary. The four components of the follicular responses to ACh are all mediated by muscarinic receptors. The subtype(s) of receptors involved still remains to be determined. Like other follicular currents, the amplitude of all these components varies between individual oocytes and

TABLE I

Main Types of Ion Membrane Currents in *Xenopus* Follicles

Localization	Ion current	Main ion carrier	Receptor-channel coupling pathway or activation mechanism
Oocyte	$I_{Cl(Ca)}$-oscillatory	Cl^-	IP_3, Ca^{2+} release (1–4)[a]
	$I_{Cl(Ca)}$-T_{in}	Cl^-	IP_3, Ca^{2+} influx (5–7)
	$I_{Cl(Ca)}$-T_{out}	Cl^-	Ca^{2+} influx, via voltage-dependent Ca^{2+} channels (8, 9)
Follicular cells	cAMP-K^+	K^+	cAMP (1, 10–13)
	ACh/ATP-K^+	K^+	Unknown[b] (1, 14)
	S_{in}	Cl^-	cAMP, unknown[b] (15)
	S_{out}	Cl^-	cAMP, unknown[b] (15)
	F_{in}	Cl^-	Unknown[b] (13, 15)
	$I_{K(osm)}$	K^+	External osmolarity decrease (15)
	$I_{Cl(osm)}$	Cl^-	External osmolarity decrease (15, 16)

[a]References: 1) Kusano *et al.*, 1977, 1982; 2) Miledi and Parker, 1984; 3) Oron *et al.*, 1985; 4) Parker and Miledi, 1986; 5) Parker *et al.*, 1985a; 6) Parker and Miledi, 1987; 7) Yao and Parker, 1993; 8) Miledi, 1982; 9) Barish, 1983; 10) Van Renterghem *et al.*, 1984, 1985; 11) Lotan *et al.*, 1982; 12) Woodward and Miledi, 1987a,b; 13) Miledi and Woodward, 1989a,b; 14) Dascal and Cohen, 1987; 15) Arellano and Miledi, 1993, 1994; 16) Arellano and Miledi, 1995.
[b]Pathway coupled to muscarinic and P_{2U} receptors.

between cells from different frogs, causing pronounced variability in the overall character of the electrical response (e.g., Figs. 2 and 4).

3.3. Follicular Cell–Based Responses Mediated by cAMP (cAMP-K^+, S_{in}, and S_{out} Currents)

Catecholaminergic agents (Kusano *et al.*, 1977, 1982; Van Renterghem *et al.*, 1984) and gonadotropins (Woodward and Miledi, 1987b) produce follicular outward currents carried by K^+ (e.g., Figs. 2D,E,F; 4C,D). Further characterization showed that these responses are mediated by a receptor-stimulated increase in the intracellular concentration of cyclic nucleotides, mainly cAMP (responses denoted here as cAMP-K^+ currents). The cAMP-K^+ current is abolished by defolliculation, indicating that it is of follicular origin (Kusano *et al.*, 1982; Woodward and Miledi, 1987b; Miledi and Woodward, 1989a). These and other studies have shown that the follicular cell membranes can contain a great variety of receptors to neurotransmitters, hormones, and peptides (see Table II), all of which are positively coupled to the cAMP pathway and cause activation of cAMP-K^+ currents (e.g., Fig. 2D,E,F).

In addition, neurotransmitters, hormones, and other drugs that increase the synthesis of follicular cAMP or direct intra-oocyte injection of cAMP also elicit a smooth inward current (cAMP-S_{in} current) which shares several characteristics with the ACh-S_{in} current (Arellano and Miledi, 1993, 1994) (Fig. 4C,D). This follicular current is carried mainly by Cl^- ions, is Ca^{2+}-independent, and shows a strong osmodependence. Activation of the cAMP pathway may also generate S_{out} currents associated with a *decrease* of membrane conductance (Fig. 4E). Defolliculation experiments show that all these

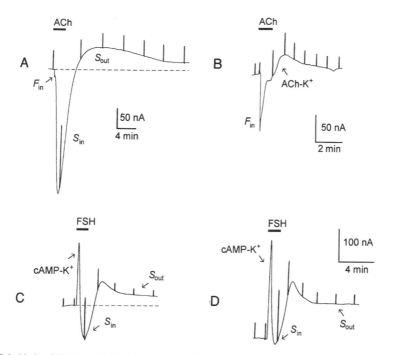

FIGURE 4. Native follicular cell–based responses to ACh and FSH in HR solution. (A,B) Responses to ACh (10 μM) in follicles from different frogs showing examples of total currents with different proportions of F_{in}, S_{in}, S_{out}, and ACh-K$^+$ currents as indicated. See text for details. (C,D) Responses to FSH (1 μg/ml) in two follicles from different frogs showing S_{in}, S_{out}, and cAMP-K$^+$ currents as indicated. Broken lines in (A) and (C) indicate the initial current level. (From Arellano and Miledi, 1993).

responses originate in the follicular cells. Again, the amplitude of these currents and the assortment of receptors that activate them vary greatly between follicles from different frogs.

3.4. Follicular Cell–Based Currents Elicited by Changes in Osmolarity

Another group of follicular ion currents are generated by decreasing the osmolarity of the external solution (i.e., hypoosmotic shock) or by increasing the follicle internal osmolarity by injecting substances such as sucrose or PEGs. These currents include components carried mainly by K$^+$ and Cl$^-$ ions, denoted here as $I_{K(osm)}$ and $I_{Cl(osm)}$ (Arellano and Miledi, 1993, 1995; Ackerman *et al.*, 1994). In our studies, follicular currents generated by changes in osmolarity were strictly dependent on the coupling of follicular cells to the oocyte, apparently as a consequence of the localization of both the osmosensor and the ion channels in the follicular cells (Arellano and Miledi, 1995) (Section 6).

The nomenclature of follicular currents used in this review is necessarily complex and descriptive. We hope a better terminology will arise when more is known about the

TABLE II
Follicular Cell Membrane Receptors
that Activate Ion Channels in *Xenopus*

	References[a]
Follicle receptors coupled to cAMP synthesis	
Neurotransmitters	
Noradrenaline	1, 2, 19
Dopamine	1, 20
Adenosine	3, 4
5-hydroxytryptamine	1, 5, 21
Gonadotropins (FSH[b], LH, cGH)	6–8, 22, 23
Prostaglandins	9, 10, 24–26
Peptides	
VIP	11, 27–29
Oxytocin	9, 30
ANF[c]	9, 31–33
CRF	12
GRH	13, 28
CGRP	14
Other	
Zn^{2+}	15, 18
Follicle receptors coupled to other pathways	
IP_3 synthesis	
Angiotensin II	16, 17, 26, 34
Unknown	
ACh (muscarinic)	1, 7, 35
ATP/UTP (P_{2U})	3, 18, 36

[a]References: *Xenopus*: 1) Kusano *et al.*, 1977, 1982; 2) Van Renterghem *et al.*, 1984; 3) Lotan *et al.*, 1982; 4) Stinnakre and Van Renterghem, 1986; 5) R. O. Arellano, unpublished results; 6) Woodward and Miledi, 1987a; 7) Arellano and Miledi, 1993, 1994; 8) Greenfield *et al.*, 1990b; 9) Miledi and Woodward, 1989b; 10) Mori *et al.*, 1989; 11) Woodward and Miledi, 1987b; 12) Moriarty *et al.*, 1988; 13) Yoshida and Plant, 1991; 14) Gillemare *et al.*, 1994; 15) Miledi *et al.*, 1989b; 16) Sandberg *et al.*, 1990, 1992; 17) Lacy *et al.*, 1992; 18) In this review. Mammals: 19) Adashi and Hsueh, 1981; 20) Bodis *et al.*, 1993a; 21) Bodis *et al.*, 1993b; 22) Hsueh *et al.*, 1983; 23) Nimrod *et al.*, 1976; 24) Goff and Armstrong, 1983; 25) McArdle, 1990; 26) Currie *et al.*, 1992; 27) Davoren and Hsueh, 1985; 28) Bagnato *et al.*, 1991; 29) Zhong and Kasson, 1994; 30) Nitray and Sirotkin, 1992; 31) Budnike *et al.*, 1987; 32) Pandey *et al.*, 1987; 33) Jonhson *et al.*, 1994; 34) Pucell *et al.*, 1991; 35) Luck, 1990; 36) Kamada *et al.*, 1994.
[b]Abbreviations: FSH, follicle stimulating hormone; LH, luteinizing hormone; cGH, chorionic gonadotropin hormone; VIP, vasoactive intestinal peptide; ANF, atrial natriuretic factor; CRF, corticotropin-releasing factor; GRH, growth hormone–releasing hormone; CGRP, calcitonin gene–related peptide.
[c]Follicular receptor coupled to cGMP synthesis (31, 32).

functional and molecular characteristics of the channels involved and the mechanisms that lead to their activation.

4.   FOLLICULAR CELL K$^+$ CURRENTS

As already mentioned, numerous neurotransmitters, peptides, and hormones applied to *Xenopus* follicles elicit membrane currents carried by K$^+$. Two groups of receptors show clear differences in their mechanisms of action, suggesting that the responses are due to activation of at least two pathways. We will distinguish these two groups by referring to them as cAMP-K$^+$ currents and K$^+$ currents activated by ACh or ATP (ACh-K$^+$, ATP-K$^+$).

4.1.   cAMP-K$^+$ Currents

cAMP-K$^+$ currents have been the most widely studied group of follicular responses and were first described in the context of activation by catecholamines such as adrenaline, dopamine, and serotonin (Kusano *et al.*, 1977, 1982). Typical responses consist of a smooth hyperpolarization from resting membrane potentials around -50 mV to values of -90 to -100 mV. This hyperpolarizing response is associated with a large increase in membrane conductance and desensitizes upon extended agonist application. Under voltage clamp, applications of adrenaline or dopamine elicit membrane currents which reverse polarity at a membrane potential of approximately -100 mV ($E_{rev} = -100$ mV). In Ringer's solution where the concentration of K$^+$ is increased either to 5 or 20 mM the current responses reverse polarity at approximately -80 and -50 mV, respectively, close to the value predicted by the Nernst equation for a potassium current. Substitutions of Na$^+$ and Cl$^-$ in the external solutions do not appreciably alter E_{rev}. All this indicates that the currents are carried mainly by K$^+$ ions and is in agreement with estimates for the equilibrium potential for K$^+$ ions of around -100 mV in NR (Kusano *et al.*, 1982; Van Renterghem *et al.*, 1984, 1985; Stinnakre and Van Renterghem, 1986; Lotan *et al.*, 1982; Woodward and Miledi, 1987a,b; Miledi and Woodward, 1989a,b) (Fig. 5).

The follicular cAMP-K$^+$ current is potently reduced by nonspecific potassium channel blockers such as Ba^{2+} applied externally (1–2 mM for complete effect) or by TEA$^+$ applied externally (20 mM) or internally (0.5–2 nmol). Furthermore, cAMP-K$^+$ currents are also blocked, at least partially, by antidiabetic sulfonylureas such as glibenclamide (Honoré and Lazdunski, 1991). These substances act on ATP-sensitive potassium (K$_{ATP}$) channels of several cellular systems (Davis *et al.*, 1991).

4.1.1.   Effects of Defolliculation

cAMP-K$^+$ currents are eliminated by oocyte defolliculation, independent of the method of defolliculation. For example, in one study we compared the efficacy of different defolliculation procedures and their effect on the cAMP-K$^+$ or oscillatory Cl$^-$ current responses elicited by various drugs (Miledi and Woodward, 1989a). Scanning electron microscopy showed that defolliculation by collagenase treatment removed or

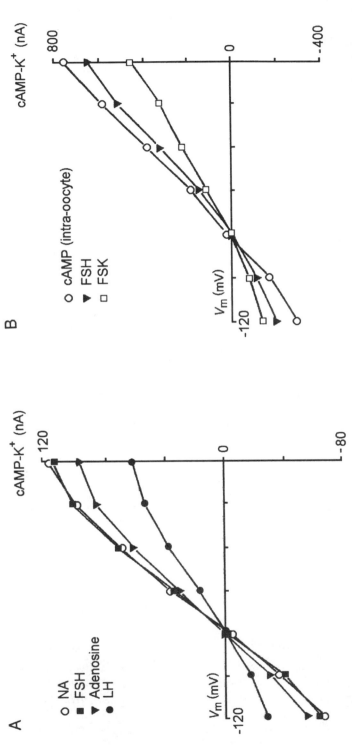

FIGURE 5. Current/voltage (I-V) relationships of cAMP-K⁺ currents. (A) Responses to NA (100 μM), FSH (5 μg/ml), adenosine (100 μM), and luteinizing hormone (LH, 5μg/ml), all in the same follicle. (B) Responses to FSH (1 μg/ml), FSK (1 μM), and intra-oocyte injection of cyclic AMP (~2 pmol) in the same follicle (different from A). All experiments in NR solution containing 5 mM KCl. [Panel (B) from Woodward and Miledi, 1987a].

destroyed between 40 to 99% of the enveloping follicular cells. On the other hand, manual defolliculation was a more reliable procedure for removing follicular cells. The microscopy results paralleled estimates of defolliculation obtained indirectly by electro-physiological recording of cAMP-K$^+$ current (Fig. 6A,C). In general, collagenase-treated oocytes retained some cAMP-K$^+$ current ($<$10%), while manually defolliculated oocytes did not show detectable responses (Fig. 6C). In addition, K$^+$ currents elicited by direct intra-oocyte cAMP or cGMP injection in follicles were also completely eliminated by the manual procedure (Fig. 7). As in other studies (Kusano *et al.*, 1982; Dascal and Landau, 1980), oscillatory Cl$^-$ currents arising in the oocyte membrane were not greatly affected by defolliculation (see Fig. 3B). Removal of the inner epithelium and thecal layers ("unzipped follicle") left most of the follicular cells intact, as assayed by scanning electron microscopy, and the follicles were still able to respond with K$^+$ currents up to 70% of the mean current elicited in whole follicles (Fig. 6A,B).

4.1.2.   Single-Channel K$^+$ Currents

We have obtained direct evidence that the current elicited by follicle stimulating hormone (FSH) originates in the follicular cells by recording of single-channel currents from the membrane of these cells. The patch-clamp recording pipette was brought to the follicular cell membrane through a small "window" made in the basement membrane (Section 2.2). A gigaseal was then obtained and the control activity of channels in the patch was monitored at different pipette potentials. The follicle was then superfused with FSH (1–2 μg/ml), a hormone which activates cAMP-K$^+$ currents (Woodward and Miledi, 1987b). After a variable delay (*ca.* 1–3 min), the FSH produced openings of single-channel outward currents close to the resting membrane potential of −40 to −60 mV (Fig. 8). The single-channel current activity was evoked, or increased if already present, in nine of 13 membrane patches studied under these conditions (four frogs).

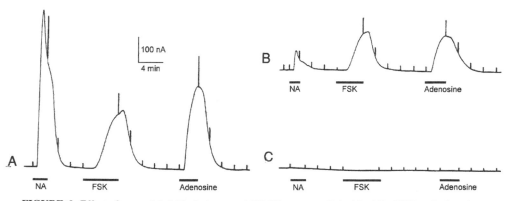

FIGURE 6. Effect of manual defolliculation on cAMP-K$^+$ currents elicited by NA, FSH, and adenosine. Examples of: (A) responses in an intact follicle-enclosed oocyte; (B) a follicle where the inner ovarian epithelium was removed manually together with the basement membrane, "unzipped follicle"; (C) an oocyte defolliculated manually (rolled oocyte). In all records, NA (100 μM), FSK (1 μM), and adenosine (100 μM) were applied in NR solution. Different follicles taken from the same frog.

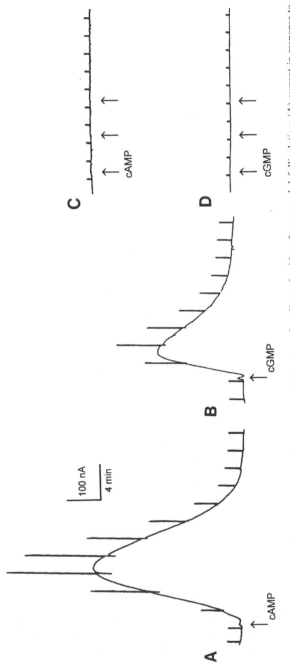

FIGURE 7. Abolition of K^+ currents produced by intra-oocyte injection of cyclic nucleotides after manual defolliculation: (A) current in response to injection of cAMP (1 pmol) in an intact follicle; (C) manually defolliculated oocyte from the same ovary; repeated injections of cAMP (1 pmol each time) failed to generate K^+ currents; (B) and (D) experiments as in (A) and (C) but injecting cGMP (10 pmol). Follicles from different frogs. All records in NR solution. (From Miledi and Woodward, 1989a).

Occasionally, single-channel activity of the follicular cell membrane stimulated by FSH included inward currents (Fig. 8A, arrow), and often the patches contained more than one active channel (e.g., Fig. 8B). Similar stimulation with FSH did not evoke activity in patches of the oocyte membrane (seven patches, same frogs as for follicle experiments).

In this preliminary analysis, we could detect at least two types of outward currents that differed markedly in their single-channel conductances, as estimated close to their reversal potential (Fig. 8B,C): conductances were 10–15 pS for one and *ca.* 40 pS for the other (Fig. 8C). Both channels showed an increase in activity (opening probability and open time) with membrane depolarization. This effect was stronger on the big conductance (BK^+) channel which, in addition, increased its conductance to ~110 pS over a depolarizing potential range of +20 to +70 mV. In contrast, the small (SK^+) channel had a more linear I-V relation. The FSH single-channel responses of the follicular cell membrane showed shifts in their reversal potential as expected for a K^+-selective channel. For example, follicles were stimulated by FSH superfusion for 2–4 min before obtaining a gigaseal. With the patch pipette filled with NR (2 mM K^+) the estimated reversal pipette potential for the high-conductance channels was about -20 mV (three patches), indicating that the E_{rev} for the unitary currents was -90 to -110 mV. Increase in the K^+ concentration in the patch pipette to 40 mM and 115 mM produced an enhancement of the single-channel conductance (120 pS and 165 pS, respectively) and shifted the reversal potential to more positive values (+44 mV and +75 mV, respectively) (not shown).

An independent study monitoring single-channel currents in follicular cells also showed two types of potassium channels (Honoré and Lazdunski, 1993). One type was opened by P1060 a pinacidil-derived K^+ channel opener which generally acts on K_{ATP} channels (e.g., McPherson, 1993; Davis *et al.*, 1991). These channels had a unitary conductance of 19 pS, were sensitive to intracellular cAMP and ATP, and were blocked by glibenclamide. The other type of follicular K^+ channel had a unitary conductance of 150 pS, showed Ca^{2+} dependence, was not sensitive to glibenclamide, but was blocked by charybdotoxin. At present we do not know the relation of these channels to those activated by follicular-cell stimulation by FSH. It is possible that the K^+ channel activity elicited by FSH in our experiments corresponds to the cAMP-dependent channel (SK^+ channel) and Ca^{2+}-dependent K^+ channels (BK^+ channel) described by Honoré and Lazdunski (1993). If this is correct, the follicular increase in cAMP induced by FSH would be activating Ca^{2+}-dependent K^+ channels in addition to the cAMP-dependent channels. This could explain why glibenclamide, which potently blocks K_{ATP} channels, fails to block completely whole-follicle cAMP-K^+ currents. For example 10 μM glibenclamide blocked *ca.* 70% of the K^+ current activated by superfusion of 8-Br-cAMP (Honoré and Lazdunski, 1991). More information about the electrical characteristics and physiological regulation of the channels activated by FSH are needed to clarify this point.

4.1.3. Membrane Receptors Involved in cAMP-K^+ Current Generation

The follicular cAMP-K^+ currents of *Xenopus* follicles can be activated by a bewildering number of substances (Table II). For agonists, such as noradrenaline,

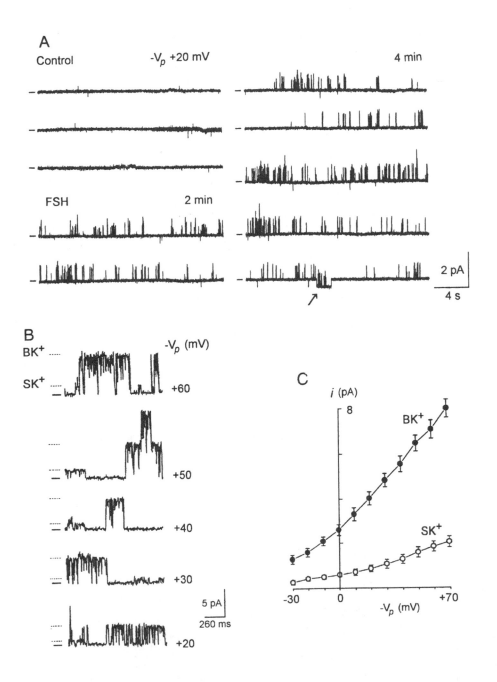

dopamine, adenosine, vasoactive intestinal peptide (VIP), atrial natriuretic factor (ANF), gonadotropins, prostaglandins, oxytocin, and others, there is evidence that the activation of responses is dependent on the stimulation of specific membrane receptors. For example, responses elicited by noradrenaline are not mimicked by agonists for α-adrenergic receptors such as phenylephrine but are mimicked by β-adrenergic agonists such as isoproterenol (Kusano *et al.*, 1982; Van Renterghem *et al.*, 1984). Also, structurally related transmitters such as adenosine and ATP act on distinct receptors. For example, theophylline, a blocker of P_1 purinergic receptors, specifically blocks hyperpolarizing follicular responses elicited by adenosine, but has little effect on the Cl^- currents evoked by ATP (Lotan *et al.*, 1982), which are probably mediated by P_{2U} purinergic/pyrimidergic follicular cell receptors (Section 5.5).

It is also clear that most, if not all, of the peptides and hormones act by stimulating the follicular membrane directly, as opposed to releasing a common transmitter from a source within the follicle (i.e., by a paracrine mechanism), a possibility that arises given the complexity of the follicle's structure. There is much evidence to support this: 1) Dissection of the internal epithelium and part of the thecal tissue of the follicle (i.e., epithelium removed; Section 2.2) eliminates an important source of possible cellular interactions but does not affect responses (Kusano *et al.*, 1982; R. O. Arellano, R. M. Woodward, and R. Miledi, unpublished results). 2) The follicular responses evoked by VIP, FSH (Woodward and Miledi, 1987a,b), prostaglandins, ANF, oxytocin (Miledi and Woodward, 1989b), or growth hormone–releasing hormone (GRH) (Yoshida and Plant, 1991) are not affected by a "cocktail" of antagonists to some of the other receptors that can be present in follicles, for example, atropine (muscarinic); propanolol, timolol (β-adrenergic); SCH23390 (dopaminergic); methysergide (serotoninergic); and theophylline (purinergic) (Fig. 9). 3) Sensitivity to the different agonists varies independently between follicles taken from different frogs (Fig. 10).

4.1.4. K^+ Currents Activated by Extracellular Zn^{2+}

One unexpected finding was the activation of K^+ currents by extracellular Zn^{2+} ions (Miledi *et al.*, 1989b). Like the follicular cell–based currents, Zn^{2+}-activated K^+ current responses are abolished by defolliculation. More than 90% of the follicles in all the frogs tested carry this type of response, which is not activated by other commonly used divalent cations (*e.g.*, Cd^{2+}, Co^{2+}, Ni^{2+}, Mn^{2+}, or Cr^{2+}) (Fig. 11A). Zn^{2+}-activated potassium currents are generated by low concentrations of Zn^{2+} (10–50 μM) and do not

---

FIGURE 8. Single-channel currents elicited by FSH in follicular cell membrane patches. (A) Record from a cell-attached membrane patch in follicles perfused continuously with NR solution (control) show only few and brief single-channel events. After ~2 min superfusion of FSH (2 μg/ml) in NR, outward currents were activated. After ~4 min the activity showed at least two channels with small conductances (SK^+), and occasionally prolonged opening of small inward currents (*arrow*). The line at the left of each trace shows zero current level. Pipette solution was NR, currents filtered at 500 Hz. (B) A different cell attached patch with at least two types of K^+ channels activated by FSH. The patch shows activity of channels with small (SK^+) and big (BK^+) conductances recorded at different pipette potentials (V_p). Pipette solution was NR, currents filtered at 1 kHz. (C) I-V relationships for the SK^+ and BK^+ channels elicited by FSH as in (B). Each point represents the mean ($\pm$ SD) current amplitude at different potentials in three different patches.

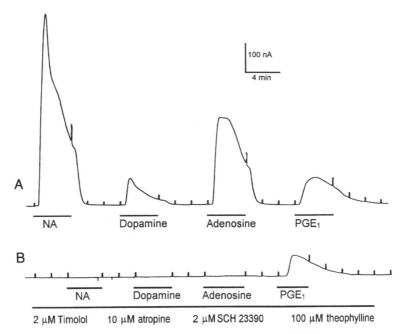

FIGURE 9. Selective preservation of the responses to prostaglandin (PGE₁) in a mixture of receptor antagonists. (A) Intact follicle exposed to NA (10 μM), dopamine (10 μm), adenosine (10 μM), and PGE₁ (1 μM) in NR solution. (B) The same follicle was then superfused with NR containing a "cocktail" of receptor antagonists, as indicated, and reexposed to the same agonists. cAMP-K⁺ currents generated by NA, dopamine, and adenosine were abolished, whereas the response to PGE₁ was preserved. (From Miledi and Woodward, 1989b).

correlate with the oscillatory Cl⁻ currents generated in the oocyte membrane by other divalent cations, or by Zn^{2+} itself. The I-V relationship is linear and appears to be similar to that for cAMP-K⁺ responses. Zn^{2+} responses are also blocked by Ba^{2+} (1–2 mM) (Fig. 11B) or TEA⁺ (200–500 pmol) injected intracellularly (see Fig. 19C). In addition, the follicular Zn^{2+}-activated K⁺ currents can be potentiated by 3-isobutyl-1-methyl-xanthine (IBMX) and forskolin (FSK), though the effect of these drugs varies. Applications of Zn^{2+} can also activate an inward current component which shows dependence on osmolarity and is carried mainly by Cl⁻ ions (Section 5.5). In some follicles the presence of this chloride current places the E_{rev} of the total currents at potentials more positive than the K⁺ equilibrium potential. Apparently, these responses are due to extracellular activation of a specific "receptor" located in the membrane of the follicular cells. The mechanisms activated by Zn^{2+} have not been studied in detail but probably warrant further attention because of evidence suggesting that this cation, and not others, increases the synthesis of cAMP induced by gonadotropins in rat gonadal tissue (Nishi *et al.*, 1984); and because Zn^{2+} elicits oocyte maturation (Wallace and Misulovin, 1980).

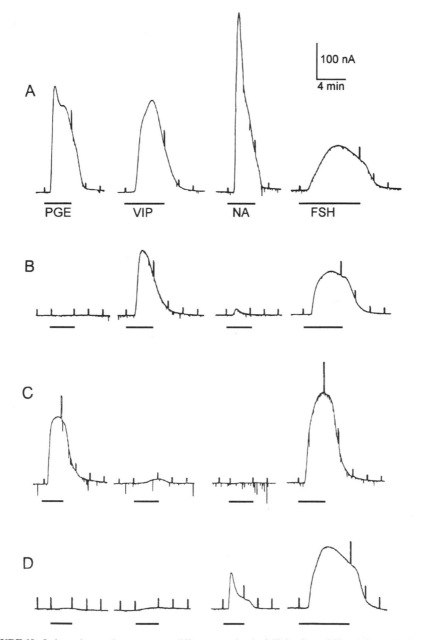

FIGURE 10. Independence of responses to different agonists in follicles from different frogs. (A–D) Four follicles from different frogs were exposed to the agonists [labeled only in (A)] at interval of 8–12 min. Concentrations: 1 μM PGE; 30 nM vasoactive intestinal peptide (VIP): 10 μM NA; and 2.5 μg/ml FSH. Note that the sensitivities to PGE, VIP, and NA varied independently between different frogs. Responses to FSH were less variable and tended to desensitize more slowly. (From Miledi and Woodward, 1989b).

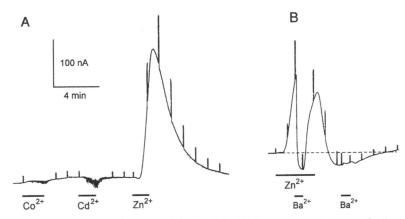

FIGURE 11. Smooth follicular K$^+$ currents elicited by Zn^{2+}. (A) Outward currents are specifically evoked by Zn^{2+} ions. Co^{2+} (1 mM) and Cd^{2+} (1 mM) ions elicited only small oscillatory responses, whereas in the same follicle Zn^{2+} (1 mM) generated a large K$^+$ current associated with an increase in membrane conductance. (B) Ba^{2+} (2 mM) selectively blocks the K$^+$ current revealing an additional inward current associated with an increase in membrane conductance. All records from "unzipped follicles" bathed in NR. (From Miledi *et al.*, 1989b).

4.1.5. Evidence for Involvement of cAMP in Follicular K$^+$ Current Generation

Due to the long latencies (several seconds to minutes) that characterize the oocyte and follicle responses (with the exception of F_{in}), it was originally suggested that agonists such as ACh and noradrenaline were activating the channels indirectly *via* a receptor-channel coupling mechanism (Kusano *et al.*, 1977, 1982).

The receptor-channel coupling mechanism for muscarinic receptors in the oocyte turned out to be the IP$_3$/Ca^{2+} system, wherein receptor-stimulated generation of IP$_3$ causes the release of Ca^{2+} from intracellular pools and the subsequent activation of Ca^{2+}-gated Cl$^-$ channels (Miledi, 1982; Miledi and Parker, 1984; Oron *et al.*, 1985; Parker and Miledi, 1986; Takahashi *et al.*, 1987). In contrast, follicular K$^+$ responses elicited by catecholaminergic and gonadotropin receptors showed no strict dependence on extracellular or intracellular Ca^{2+} (e.g., Stinnakre and Van Renterghem, 1986; Woodward and Miledi, 1987b). A more likely second messenger candidate for these responses was cAMP (Van Renterghem *et al.*, 1984, 1985) because 1) cellular responses stimulated by β-adrenergic agonists in numerous systems is *via* activation of adenylate cyclase (Aurbach, 1982; Cerione *et al.*, 1983), and 2) several aspects of follicular physiology are regulated via cAMP, and its synthesis as well as its degradation can be activated by membrane receptors (Kolena and Channing, 1972; Adashi and Hsueh, 1981; Kliachko and Zor, 1981).

There are now several studies implicating cAMP as an intermediary in the generation of follicular K$^+$ currents. Hyperpolarizing responses are mimicked by application of FSK (Woodward and Miledi, 1987a,b; Greenfield *et al.*, 1990a), a direct activator of adenylate cyclase (Seamon and Daly, 1981). Phosphodiesterase inhibitors such as IBMX and theophylline have potent amplifying effects on K$^+$ responses (Van Renterghem *et al.*, 1984; Lotan *et al.*, 1985; Woodward and Miledi, 1987a,b; Miledi and Woodward, 1989a,b). Both superfusion of membrane-permeable cAMP analogs (e.g., 8-Br-cAMP, db-cAMP) and direct injection of cAMP into follicle-enclosed oocytes mimic the K^+ currents activated by neurotransmitters and hormones (Van Renterghem *et al.*, 1984, 1985; Dascal *et al.*, 1985; Miledi and Woodward, 1989a,b). Activation of a membrane preparation from *Xenopus* follicular cells by either adenosine analogs or FSH increased synthesis of cAMP, and this correlated directly with the increase in the membrane conductance to potassium ions (Greenfield *et al.*, 1990b).

Intra-oocyte injections of cGMP into follicles mimic the K^+ responses elicited by cAMP (e.g., Fig. 7), but, in general, a higher concentration of cGMP is required to elicit currents comparable to those generated by cAMP in follicles from the same frog. Receptors like those for ANF (Miledi and Woodward, 1989b), which are known to be coupled to cGMP synthesis in mammalian granulosa cells (Budnik *et al.*, 1987; Pandey *et al.*, 1987), may be eliciting their effects in *Xenopus* follicles through a cGMP pathway.

Collectively, these studies indicate that the plasma membrane of the follicular cells contains specific receptors to a great diversity of neurotransmitters and hormones, most of which are coupled via G proteins to the activation of adenylate cyclase. The increase in cAMP generated by stimulation of these receptors activates K^+-selective channels also located in the membrane of the follicular cells. The mechanism by which cAMP activates the channels is not known in detail, but it may require stimulation of a kinase and a phosphorylation step on a regulatory molecule or on the K^+ channels themselves; alternatively, the K^+ channels may be opened directly by cAMP. Since the follicular responses are blocked, at least partially, by glibenclamide and the pinacidil-sensitive channels in follicular cells are also cAMP-dependent (Honoré and Lazdunski, 1993), we presume that the K^+ current response involves activation of channels related to the K_{ATP} family.

4.2. K^+ Currents Activated by ACh, ATP, or AII

K^+ responses elicited by ACh (Kusano *et al.*, 1982) or ATP (R. O. Arellano, R. M. Woodward, and R. Miledi, unpublished results) are generally less prominent than cAMP-K^+ currents. Although these responses differ in several respects, it is not yet clear if they use different types of channels or if it is only the receptor-channel coupling mechanisms which differ. The study of ACh-K^+ and ATP-K^+ currents has been hampered by the fact that these responses are generally small. If the potassium channels activated by ACh or ATP are the same as those involved in the generation of catecholaminergic and hormonal responses, then it will be necessary to explain how ACh (or ATP, Section 4.3) strongly attenuates cAMP-K^+ currents (Van Renterghem *et al.*, 1985; Dascal *et al.*, 1985; Woodward and Miledi, 1987b; Miledi and Woodward, 1989b). The balance of evidence tends to suggest that two types of K^+ channels are involved. There are two major differences in the mechanism of regulation between the two groups of responses. First, the ACh-K^+ current is either inhibited by IBMX (Dascal and Cohen, 1987) or does not seem to be affected by drugs that increase intracellular levels of cAMP (R. O. Arellano, R. M. Woodward, and R. Miledi, unpublished results). Second, the ACh-K^+ current is stimulated in the first minutes of incubation with phorbol esters (Dascal and Cohen, 1987), while these protein kinase C (PKC) activators have strong inhibitory effects on cAMP-K^+ currents (Dascal *et al.*, 1985; Woodward and Miledi, 1991; Miledi and Woodward, 1989b). The direct recording of at least two different types of K^+ channels in the membrane of the follicular cells is consistent with these results (Honoré and Lazdunski, 1993; see also Section 4.1.2). For example, it may be that the activity of one type of channel is inhibited specifically by muscarinic or purinergic activation, while the other type may underlie the small $ACh-K^+$ and $ATP-K^+$ currents. In addition, a divergence in the mechanism of activation is also suggested by the selective generation of F_{in} currents by ACh and ATP (Section 5). These currents are not seen when responses are activated through cAMP synthesis. Below we propose that ACh and ATP are coupled to an intracellular pathway that is distinct from those already described in follicles, that is, one in which the messenger molecules are neither cAMP nor IP_3/diacyglycerol. Experimental results supporting this hypothesis originate principally from studies of the inward currents elicited by ACh or ATP (Section 5) and from the attenuation of $cAMP-K^+$ current by ACh and ATP (following section). Although additional experiments are necessary to establish the relation of this hypothetical pathway to the mechanisms involved in the generation of $ACh-K^+$ and $ATP-K^+$ currents, we suggest that the activation of all these responses may have a direct mechanistic relation.

K^+ currents can also be elicited by stimulation of AII receptors. In general these responses were small (up to *ca.* 60 nA), and they may be regulated through a complex mechanism because in follicles maintained *in vitro* under conditions where the other responses are well preserved (Section 2.2), the K^+ responses elicited by AII cease or are strongly attenuated a few hours after isolating the follicles and incubating them in Barth's solution. K^+ responses elicited by AII were usually accompanied by the oscillatory $I_{Cl(Ca)}$, raising the possibility that the same IP_3 mechanism that generates the oscillatory current is involved in the generation of the K^+ current. This was indirectly supported by the finding that IBMX failed to potentiate the AII-elicited K^+ responses.

4.3. Modulation of $cAMP-K^+$ Current by Muscarinic and Purinergic Receptors

ACh receptor activation in follicles leads not only to the opening of membrane channels but also to a potent inhibition of $cAMP-K^+$ currents. This effect was initially detected while studying the effect of muscarinic receptor activation on responses evoked by adenosine and β-adrenergic agents (Dascal *et al.*, 1985; Van Renterghem *et al.*, 1985; R. O. Arellano, R. M. Woodward, and R. Miledi, unpublished results) and has also been shown for $cAMP-K^+$ currents evoked by VIP, gonadotropins, prostaglandins, ANF, and oxytocin, among others (Woodward and Miledi, 1987a,b; Miledi and Woodward, 1989a,b). The inhibitory effect is still evident on K^+ current responses generated by direct injection of cAMP into the follicle (Dascal *et al.*, 1985; R. O. Arellano, R. M. Woodward, and R. Miledi, unpublished results) by receptor-independent stimulation of adenylate cyclase using FSK (Stinnakre and Van Renterghem, 1986; Miledi and Woodward, 1989b; Woodward and Miledi, 1991) and on a follicular cAMP-dependent K^+ channel activated by P1060 (Honoré and Lazdunski, 1993). These results clearly show that the $cAMP-K^+$ current modulation by ACh is independent of cAMP synthesis

(Dascal *et al.*, 1985; Miledi and Woodward, 1989b). Also, it seems that the attenuation is on the cAMP-K$^+$ channels, as opposed to follicular cell–oocyte coupling, because in the same follicles voltage-activated currents which are probably located in the follicular cells are not inhibited by ACh (Miledi and Woodward, 1989b). The muscarinic modulation of follicular K$^+$ currents is of additional interest because it resembles the muscarinic modulation of potassium channels in other cells (e.g., Robbins *et al.*, 1993; Chen *et al.*, 1994).

Given the similarity of the follicular muscarinic responses to currents generated by ATP (Kusano *et al.*, 1982; Lotan *et al.*, 1982; see also Section 5), we hypothesized that ATP would also attenuate the cAMP-K$^+$ current. This was confirmed in follicles where the cAMP-K$^+$ channel was activated by FSH (1–2 μg/ml). Like muscarinic agonists, ATP (1–100 μM) also caused a rapid attenuation of the FSH-elicited K$^+$ responses (Fig. 12A,B). This inhibition was washed out in ~30 min (Fig. 12D,E). In the same follicles, ATP itself activated Cl$^-$ currents (Fig. 12C) that were completely dependent on the presence of follicular cells and their coupling to the oocyte (Section 5).

4.3.1. Possible Mechanism of cAMP-K$^+$ Current Modulation

An initial clue to the intracellular mechanism mediating the attenuation of cAMP-K$^+$ by ACh/ATP was the observation that activators of protein kinase C, such as phorbol esters and diacylglycerol analogs, seem to mimic the effect of ACh (Dascal *et al.*, 1985; Woodward and Miledi, 1991). The exact relationship between activation of PKC and the modulatory effects of ACh or ATP, however, remains uncertain. The muscarinic and purinergic receptors involved in this effect are almost certainly located in the membrane of the follicular cells (see below), but these receptors seem to activate an intracellular signaling pathway that is distinct from those involving the cAMP or phosphoinositide systems (see also Section 5.6).

There are a variety of indirect observations that argue against involvement of oocyte-based PKC in the attenuation of cAMP-K$^+$ currents. For example, follicles often contain muscarinic receptors in the membrane of the oocyte itself (Kusano *et al.*, 1977, 1982) and it was suggested that the oocyte's muscarinic receptors are involved in attenuation of follicular cAMP-K$^+$ currents (Dascal *et al.*, 1985). These oocyte-based ACh receptors are coupled to phospholipase C (PLC) stimulation, which catalyzes the hydrolysis of phosphatidylinositol 4,5-bisphosphate and produces IP$_3$ and diacylglycerol (Berridge, 1987). Stimulation of this ACh receptor–coupled mechanism can be monitored electrically due to activation of $I_{Cl(Ca)}$ channels located in the oocyte membrane (Miledi, 1982; Miledi and Parker, 1984). Therefore, we reasoned that if activation of the PLC pathway was responsible for the attenuation of the cAMP-K$^+$ currents, then other agonists coupled to that pathway should produce a similar attenuation. This, however, was not the case. Stimulation of the PLC pathway by a serum factor, AII (Fig. 13), or divalent cations did not cause attenuation of cAMP-K$^+$ currents even when these agents activated the characteristic Ca^{2+}-dependent oscillatory Cl$^-$ currents in the same follicles (Miledi and Woodward, 1989b; Woodward and Miledi, 1991). This suggests that the attenuation of follicular cAMP-K$^+$ current is a specific property of follicular cell ACh or ATP receptors and is not directly related to agonists that act *via* activation of PLC in the

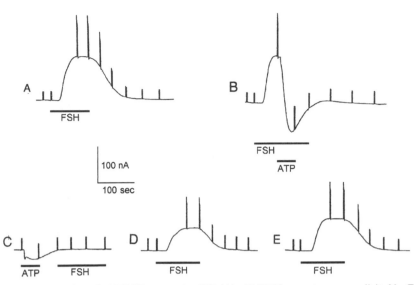

FIGURE 12. Modulation of cAMP-K$^+$ current by ATP. (A) cAMP-K$^+$ current response elicited by FSH (1 µg/ml) in an epithelium-removed follicle. (B) After a 15-min washing interval the follicle was reexposed to FSH and then, at the peak of the cAMP-K$^+$ current, ATP (1 µM) was applied. ATP application abolished the outward current; the effect was associated with a decrease in net membrane conductance and generation of follicular chloride currents. (C–E) Subsequent agonist applications on the same follicle at intervals of 15–20 min. (C) A second ATP application elicited F_{in} and S_{in} currents and immediate application of FSH did not elicit cAMP-K$^+$ current. The response to FSH largely recovered after a prolonged washing period (D–E). All records in HR solution.

oocyte membrane. Moreover, in cells where ACh or ATP inhibits the cAMP-K$^+$ currents, the follicular responses to these neurotransmitters, and especially to ATP (Fig. 12), often does not include activation of Ca^{2+}-dependent Cl$^-$ currents (Section 5.6).

In addition to cAMP-K$^+$ current modulation, the signaling pathways activated by ACh or ATP in follicular cells also generate follicular ACh-K$^+$, ATP-K$^+$ outward currents, as well as follicular inward currents F_{in} and S_{in} (see below). If a common mechanism is involved in all these effects one would expect that direct activation of PKC would be able to mimic, at least partially, these other components. This is not the case. Application of phorbol esters to follicles superfused with NR or HR solutions fails to activate any membrane currents similar to those elicited by the neurotransmitters in the same follicles (Section 5.6.2). Indeed, the only neurotransmitter effect mimicked by activators of PKC is attenuation of cAMP-K$^+$ currents.

Taken together, these results suggest that although the cAMP-K$^+$ currents are potently attenuated by PKC activators, this enzyme is not necessarily involved in the pathway used by ACh or ATP. Other characteristics of the inhibitory effect, for example, its independence of intracellular calcium (Miledi and Woodward, 1989b), suggest likewise that it is necessary to reexamine the proposed role of PKC activation in the inhibitory effects of ACh and ATP on the follicular cAMP-K$^+$ currents.

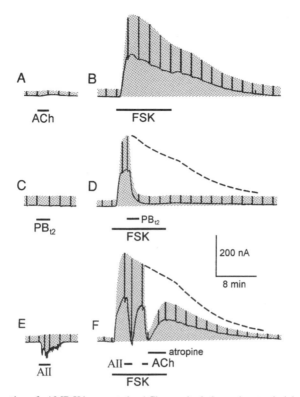

FIGURE 13. Attenuation of cAMP-K$^+$ currents by ACh or a phorbol ester is not mimicked by AII. Follicles held at -40 mV with periodic pulses to -20 mV. The shaded area indicates the changes in membrane current response occurring at -20 mV, the approximate reversal potential for Cl$^-$ ions in these follicles. Dotted line indicates presumed time course of FSK response under control conditions. Control follicle responses to (A) ACh (1 μM); (B) FSK (5 μM); (C) phorbol 12,13-dibutyrate (PB$_{t2}$, 100 nM); or (E) AII (1 μM). In (B) cAMP-K$^+$ current elicited by FSK desensitized slowly upon extended exposure. (D) FSK-elicited K$^+$ current response was rapidly attenuated by PB$_{t2}$; the inhibition was essentially irreversible. (F) FSK-elicited response was not affected by AII in a follicle where ACh rapidly attenuated the K$^+$ current, an effect that was partially reversed by washing out ACh in the presence of atropine (10 μM). Note that during the AII oscillatory response in (E) there was little change in membrane current at -20 mV and that during the cAMP-K$^+$ current in (F), AII elicited oscillatory current, but the step to -20 mV revealed little or no attenuation of K$^+$ current. Different follicles from same frog. All records in NR. (From Woodward and Miledi, 1991).

5.   FOLLICULAR CELL Cl$^-$ CURRENTS

From the earliest studies on intact follicles, the complexity and variability of membrane currents elicited by cholinergic agents implied the involvement of more than one response mechanism (Kusano *et al.*, 1977, 1982; Dascal and Landau, 1980). We have suggested that attentuation of cAMP-K$^+$ currents by ACh (Miledi and Woodward, 1989b) or ATP is mediated by receptors located in the follicular cells (Section 4.3). These observations, as well as the selective elimination of K$^+$ and F_{in} currents elicited by ACh or ATP after defolliculation (Miledi and Woodward, 1989a) (Figs. 3 and 17B,C), indicate that these responses also originate in the follicular cell compartment.

5.1. Osmodependent S_{in} Currents

Further evidence supporting this contention was obtained from an analysis of the inward membrane current responses elicited by FSH and ACh (Arellano and Miledi, 1993, 1994) while the follicles were superfused with an HR solution (Section 2.2). This simple change in the external solution has very pronounced effects on the responses evoked by agonists such as ACh and FSH (cf. Figs. 3 and 4). At first sight the changed pattern of the responses was somewhat puzzling, but further studies revealed that HR increases certain current components that are also observed but much smaller in NR (cf. Kusano *et al.*, 1982; Miledi and Woodward, 1989a,b) (Fig. 14). Making recordings in HR helps to "dissect" the responses and facilitates their electrophysiological analysis. Among the several components of the currents elicited by ACh and FSH in HR, a smooth and slowly developing inward current (S_{in}) is the most prominent (Fig. 14). S_{in} currents are generated by FSH, ACh, or ATP (Fig. 17A) and cannot be distinguished based on their general electrophysiological characteristics. Evidence suggests, however, that they are generated through different receptor-channel coupling mechanisms (Arellano and Miledi, 1994; Section 5.6), one of which involves cAMP synthesis; the other remains unknown. The idea of a dual pathway is indirectly supported by the observation that ACh

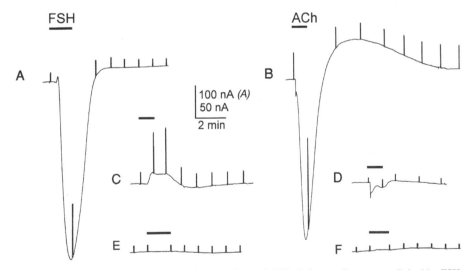

FIGURE 14. Effects of increasing external osmolarity or defolliculation on S_{in} currents elicited by FSH or ACh. In (A) and (B) responses to FSH (1 µg/ml) and ACh (10 µM) in HR solution. In (C) and (D), after equilibrating for 20–30 min in NR the follicles were reexposed to FSH or ACh. (E, F) The oocytes were manually defolliculated and reexposed to the agonists in HR (and in NR, not shown). S_{in} currents were reduced to < 5% in NR, and all the currents were abolished by defolliculation. F_{in} currents were not affected by the increase in osmolarity. The follicles (oocytes) were held at −95 mV (FSH), or −60 mV (ACh). (From Arellano and Miledi, 1993).

and ATP activate a specific fast Cl^- current (F_{in}) that is not an intrinsic feature of responses activated *via* the cAMP pathway.

In the presence of 1–2 mM Ba^{2+}, to block K^+ currents, E_{rev} of S_{in} currents elicited by ACh and FSH was between -20 to -35 mV, which agrees with the equilibrium potential for Cl^- ions (Kusano *et al.*, 1982). Moreover, a 66% decrease in Cl^- in the external medium (substituted by SO_4) causes a 10–15 mV positive shift of E_{rev} and a decrease in the total current, an effect that is more evident when the currents are outward at positive potentials. Substitutions of Na^+ and K^+ ions in the external solution have little effect on amplitude and time course of the S_{in} currents, as well as on their E_{rev}. These experiments suggest that S_{in} currents are carried mainly by Cl^- ions. The shift in E_{rev} of the S_{in} currents in low-Cl^- external solutions is generally smaller than predicted by the Nernst equation (*ca.* 23 mV in this case) (see also Section 5.6.1). This discrepancy could be due to at least three factors: 1) S_{in} channels may be permeable to other intracellular anions in addition to Cl^-. 2) The same mechanisms which activate S_{in} channels may also open K^+ channels that are not fully blocked by Ba^{2+}, driving the E_{rev} to more negative potentials. 3) Superfusion with low-Cl^- hypoosmotic solution may cause significant changes in the intracellular Cl^- concentration compared to that measured in NR.

Differences between S_{in} chloride currents and the $I_{Cl(Ca)}$ of the oocyte membrane (Arellano and Miledi, 1993) can be summarized as follows:

1. S_{in} currents are not Ca^{2+}-dependent.
2. They are eliminated by defolliculation.
3. I-V relationships of S_{in} currents show only weak rectification at membrane potentials more negative than -60 mV, whereas the oocyte's membrane Ca^{2+}-gated Cl^- currents show pronounced rectification at these potentials (Miledi and Parker, 1984).
4. S_{in} currents are strongly osmodependent; the current elicited by ACh increases eight to ten times with a decrease of 50 mosM, i.e., a 20% decrease of NR osmolarity.
5. S_{in} channels show similar permeabilities to Cl^-, I^-, and Br^-, while the oocyte's membrane Ca^{2+}-gated Cl^- channel is more permeable to I^- or Br^- than to Cl^-. All these characteristics indicate that S_{in} channels are intrinsically different from the Cl^- channels that are located in the membrane of the oocyte and are responsible for Ca^{2+}-dependent Cl^- currents. In the following sections we will discuss further the experiments supporting this statement, focusing on the characteristic properties of S_{in} and F_{in} currents.

5.2. S_{in} and F_{in} Currents Are Not Ca^{2+}-Dependent

Chelation of Ca^{2+} by intra-oocyte injections of EGTA or BAPTA has a rapid (3–5 min) and slowly reversible inhibitory effect on oocyte-based Ca^{2+}-gated Cl^- currents elicited by agonists (Parker *et al.*, 1985a,b; Woodward and Miledi, 1991) or by membrane depolarization (T_{out} currents; Miledi and Parker, 1984). When similar injections of chelators are made in follicles the S_{in} and F_{in} responses are not eliminated, even when the intracellular concentration of chelator is greater (*ca.* fivefold) than that required to abolish T_{out} currents (Arellano and Miledi, 1993). This failure to block the S_{in} and F_{in} currents is independent of the chelator injected (EGTA or BAPTA) or of the external solution (HR or NR).

The possibility remains that the oocyte–follicular cell gap junctions are weakly permeable to EGTA and BAPTA, which are thus prevented from attaining a sufficient concentration for effective chelation of Ca^{2+} in the follicular cell compartment. However, this seems unlikely, because the chelators have a molecular weight (EGTA, ~380 Da; BAPTA, ~476 Da) significantly lower than the average limit for permeation through gap junctions (~1000 Da; Lowenstein, 1981). Moreover, there is evidence that the oocyte–follicular cell gap junctions are permeable to substances with molecular weights similar to those of EGTA and BAPTA, *e.g.*, carboxyfluorescein (Browne *et al.*, 1979), cAMP and cGMP (Dascal *et al.*, 1985; Miledi and Woodward, 1989a), IP_3 (Sandberg *et al.*, 1992), sucrose, and PEGs with molecular weights up to ~600 Da (Arellano and Miledi, 1995; Section 6). Furthermore, experiments using the cell membrane–permeable acetoxymethyl (AM) esters of EGTA or BAPTA gave results similar to those obtained with intra-oocyte injection. For example, follicles were separated in two groups, one of which was incubated overnight in Barth's medium containing 10 μM BAPTA-AM (0.1% DMSO), the other in control Barth's medium with 0.1% DMSO (seven follicles in each group). The follicular responses elicited by ACh application were monitored in follicles bathed in HR solution, and T_{out} currents were recorded in Ringer's solution containing 10 mM $CaCl_2$ to measure indirectly the influx of calcium into the oocyte (Miledi, 1982). Under these conditions, the F_{in} and S_{in} currents were well preserved in both groups of follicles, while the T_{out} currents were abolished. In the control group the amplitudes of F_{in}, S_{in}, and T_{out} currents were 44 ± 21 nA, 74 ± 27 nA, and 262 ± 67 nA, respectively, while in the BAPTA-AM–treated group the currents were 46 ±20 nA, 93 ± 40 nA, and 12 ± 6 nA, respectively. Similar results were obtained in follicles from two other frogs. In one of these (four follicles per group), the currents activated by ACh included, in addition to follicular currents, the oocyte-based oscillatory $I_{Cl(Ca)}$. Like the T_{out} current, the oscillatory current was also eliminated by incubation with 10 μM EGTA-AM, while the F_{in} and S_{in} currents were preserved in follicles tested with ACh or ATP (Fig. 15). These experiments reinforce the idea that F_{in} and S_{in} currents activated by ACh or ATP are Ca^{2+}-independent. Similar results were obtained for the S_{in} elicited by FSH.

5.3. Cellular Localization of S_{in} and F_{in} Channels

Follicular S_{in} and F_{in} are strictly dependent on the electrical coupling between follicular cells and the oocyte. S_{in} currents elicited by FSH or by other hormones and neurotransmitters, together with the F_{in} currents generated specifically by ACh and ATP are all completely eliminated by defolliculation, be it enzymatic or manual (Figs. 15 and 18). Indeed, we have never observed F_{in} currents elicited by ACh or ATP in any type of defolliculated oocyte, even those expressing mRNA extracted from brain. Also, it is worth noting that currents activated by ATP in native follicles only rarely include oocyte-based components. For example, the spike component of the oscillatory current (Fig. 2A), which is elicited in native oocytes by ACh or AII (Miledi and Woodward, 1989a; Woodward and Miledi, 1991), has never been observed in follicular ATP responses (> 50

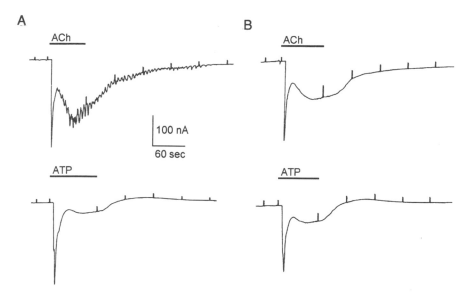

FIGURE 15. Effects produced by a membrane-permeable Ca^{2+} chelator on follicular Cl^- currents. (A) Examples of current responses of epithelium-removed follicles incubated overnight in medium containing 0.1% DMSO but not EGTA-AM. ACh (50 μM) and ATP (50 μM) elicited follicular cell–based F_{in} and S_{in} responses. ACh, in addition, evoked the oscillatory $I_{Cl(Ca)}$. (B) Follicles from the same frog incubated in the presence of EGTA-AM (10 μM, 0.1% DMSO). The agonist-elicited follicular cell–based inward currents were preserved, while oscillatory currents were eliminated. All follicles held at −60 mV in NR. Incubations with drugs in modified Barth's medium.

frogs, > 200 follicles). This strongly suggests that, contrary to what has been suggested earlier (Lotan *et al.*, 1982, 1986), the F_{in} and S_{in} currents evoked by ATP originate in the follicular cells. Similar to the follicular ACh responses (Arellano and Miledi, 1994), ATP responses only occasionally bear any relation to generation of oocyte-based Ca^{2+}-activated Cl^- currents. The one situation where there appears to be some interplay is in follicles which have substantial spontaneous oscillatory Cl^- current activity (Kusano *et al.*, 1982). Under this circumstance ATP (10–100 μM) can amplify the oscillations (Fig. 16). The effect may reflect a low-level production of IP_3 in the follicular cells, which then diffuses into the oocyte, adds to an already high basal level, and thus leads to potentiation of the signaling pathway (cf. Miledi *et al.*, 1989a). Whatever the mechanism, in "oscillating" follicles that were "silenced" by activation of a strong IP_3-mediated oscillatory response, for example, by AII, subsequent application of ATP activated only the follicular F_{in} and S_{in}, with no evident differences between the follicular ATP-elicited currents in the presence or absence of spontaneous oscillations. These experiments suggest that any ATP-induced increase in IP_3 in the follicular cells must be at a low level and that this has no great effect on the generation of F_{in} and S_{in} currents. This is in clear contrast with the interactions between different native or expressed receptors in the oocyte membrane acting through a common pathway of IP_3 production (Parker *et al.*, 1987; Miledi *et al.*, 1989a), where strong activation of one receptor system can be suf-

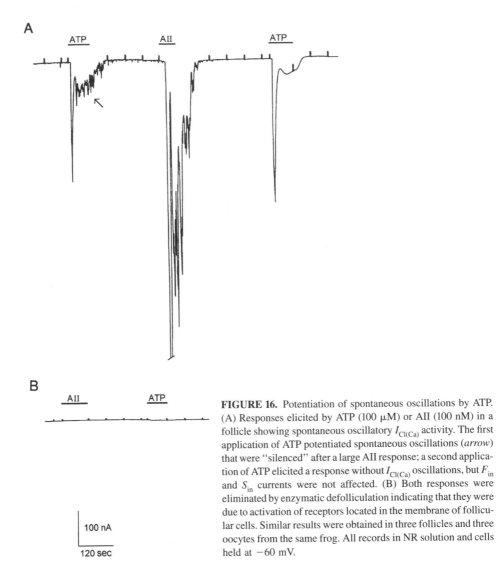

FIGURE 16. Potentiation of spontaneous oscillations by ATP. (A) Responses elicited by ATP (100 μM) or AII (100 nM) in a follicle showing spontaneous oscillatory $I_{Cl(Ca)}$ activity. The first application of ATP potentiated spontaneous oscillations (*arrow*) that were "silenced" after a large AII response; a second application of ATP elicited a response without $I_{Cl(Ca)}$ oscillations, but F_{in} and S_{in} currents were not affected. (B) Both responses were eliminated by enzymatic defolliculation indicating that they were due to activation of receptors located in the membrane of follicular cells. Similar results were obtained in three follicles and three oocytes from the same frog. All records in NR solution and cells held at −60 mV.

ficient to "exhaust" the messenger pathway for other receptors, or weak activation by one receptor greatly potentiates the action of another. Likewise, in follicles from some frogs where the muscarinic receptors seem to be located only in the follicular cell membrane (i.e., responses are completely eliminated by defolliculation) we have observed that ACh has a similar amplifying effect on spontaneous oscillations. We speculate that the elimination of this ancillary effect of follicular cell receptors may be one of the reasons why defolliculation can, in some frogs, attenuate Ca^{2+}-dependent Cl^- responses (e.g., Dascal *et al.*, 1984; Sandberg *et al.*, 1992).

5.4. Intrinsic Differences between S_{in} and F_{in} Currents

The most characteristic property of S_{in} currents is their dependence on the osmo-larity of the external medium (Arellano and Miledi, 1993, 1994). S_{in} currents are remarkably sensitive to changes in external osmolarity, and we have frequently observed current amplitude changes even with osmotic variations as small as 5%. F_{in} currents appear to be carried through different channels from those involved in S_{in} currents because they do not show osmodependence (Figs. 14 and 17) and because they have different permeability to I^- or Br^- ions (Arellano and Miledi, 1993). This has also been observed for the follicular Cl^- currents elicited by ATP. In these studies I-V relations were measured by applying voltage steps to different membrane potentials during the F_{in} currents generated by ATP or ACh in follicles bathed in NR. The pulse current was then subtracted from that obtained in unstimulated conditions, and similar experiments were made in Ringer's solutions where either Cl^- was reduced by 66% (substituted by SO_4) or NaCl was substituted by NaI (NaI-NR). For example, the F_{in} generated by ATP in follicles bathed in NR showed an E_{rev} of -22 ± 3 mV (n = 4), while in a low-Cl^- NR solution the E_{rev} was -7 ± 2 mV (n = 3); in NaI-NR solution the amplitude of the current increased and its E_{rev} was -46 ± 4 mV (n = 3). The experiments indicate that ACh- and ATP-elicited F_{in} are carried mainly by Cl^- and that the channels involved have a higher permeability to I^- ions, as do $I_{Cl(Ca)}$ channels (Arellano and Miledi, 1993). In contrast, analysis of the I-V relations of S_{in} currents evoked by ACh (Arellano and Miledi, 1993) or ATP indicates that their E_{rev} values are comparable in HR or NaI-HR (not shown) and suggests that Cl^- and I^- are similarly permeable through the S_{in} channels.

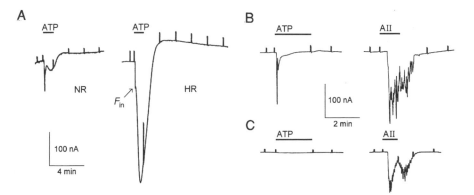

FIGURE 17. Effects of increasing external osmolarity or defolliculation on responses elicited by ATP. (A) F_{in} and S_{in} responses elicited by ATP (50 μM) in a follicle equilibrated in NR solution and after equilibrating for ~20 min in HR. S_{in} increased eightfold in HR while F_{in} (now partially masked) remained without much change. (B) Example of responses elicited by ATP (50 μM) and AII (500 nM) in the same follicle bathed in NR. (C) Oocytes from the same frog as in (B) were manually defolliculated and exposed to ATP and AII in NR. In all the oocytes tested the ATP responses were completely abolished while the AII responses were largely preserved (same result was obtained in four follicles and six oocytes from the same frog). All follicles and oocytes held at -60 mV.

Due to the superficial similarities between the Cl^- currents arising in the follicular cells and those generated in the oocyte membrane (e.g., Fig. 22) these currents have sometimes been confused. Most commonly, follicular F_{in} currents are confused with the initial spike component of the oscillatory $I_{Cl(Ca)}$ elicited in the oocyte itself. The absolute requirement of F_{in} currents on the enveloping follicular cells and their Ca^{2+} independence clearly establish them as different responses. In addition, we have shown that the F_{in} currents generated by ACh (Miledi and Woodward, 1989a; Arellano and Miledi, 1993) or ATP consistently have very short latency, even when generated by low concentrations of agonists (Section 5.6.2) (also see Fig. 22B). This contrasts strongly with the pronounced latency and dose dependence on the onset of Ca^{2+}-dependent Cl^- currents (Kusano *et al.*, 1977, 1982; Miledi and Parker, 1989; cf. Woodward and Miledi, 1991).

F_{in} and S_{in} currents have not yet been characterized at the single-channel level. We expect that, in addition to details about their gating mechanisms and modulation, such studies will provide further evidence of their distinct identity. In our patch-clamp studies on the membrane of follicular cells stimulated by FSH (Section 4.1.1), we occasionally observed single-channel currents with a reversal potential similar to that of currents carried by chloride (e.g., Fig. 8A). These single-channel currents had a linear I-V relationship and a unitary conductance of 9–12 pS at membrane potentials close to the resting potential (-40 to -60 mV). However, our results remain preliminary and more studies will be necessary to determine the characteristics of the channels and to ascertain their involvement in the Cl^- currents studied in the whole follicle.

5.5.   Membrane Receptors Involved in Cl^- Current Generation

S_{in} currents are activated by other substances besides ACh, ATP, and FSH. It now appears that several of the neurotransmitters, peptides, and hormones that activate K^+ currents can, to some degree, also activate S_{in} currents (see Table II; R. O. Arellano, R. M. Woodward, and R. Miledi, unpublished results). F_{in} currents, on the other hand, are highly specific to ACh and ATP. We have only occasionally observed very small (< 20 nA) F_{in}-like currents upon extracellular application of other agonists, for example, Cd^{2+}, Cu^{2+}, Zn^{2+} (Miledi *et al.*, 1989b), or AII (Section 5.6.2) (Fig. 23), and even more rarely by FSH. The activation of F_{in}-like currents seems to coincide with highly responsive follicles, that is, where the F_{in} currents generated by ACh or ATP were > 500 nA.

ACh acts on the follicular cell membrane through muscarinic receptors (Arellano and Miledi, 1993). We do not yet know if only one subtype of muscarinic receptor is responsible for the diverse responses to ACh in follicles or if different subtypes (see e.g., Caulfield, 1993) are involved. The concentration of ACh required to activate 50% of the maximal S_{in} response (EC_{50}) is 1–2.5 μM, and a similar value is observed for F_{in} currents in high to middle (100–500 nA) responsive follicles. S_{in} and F_{in} currents generated by 10–50 μM ATP are not blocked by 10–100 nM atropine or 1 μM pirenzipine (ten follicles, three frogs), which indicates that ATP does not act through muscarinic receptors. The responses generated by ATP are well mimicked by UTP, suggesting involvement of the purinergic/pyrimidergic receptor subtype P_{2U} (Zimmermann, 1994). S_{in} currents generated by ATP have an EC_{50} of 700 nM and, in follicles from the same frog, UTP-elicited

S_{in} currents have a lower EC_{50} of 350 nM. The F_{in} currents show values similar to those for each agonist. Several agonists for ATP/UTP receptors are able to elicit S_{in} currents with a potency sequence which that corresponds with that for the P_{2U} subtype (Fig. 18), that is, UTP $\geqslant$ ATP > 2-methylthio ATP > α,β-methylene ATP.

Follicle stimulating hormone has an approximate EC_{50} of 10 nM and its dose-response relationship has a high slope, indicating nonlinearities in the receptor-channel coupling mechanism (Arellano and Miledi, 1993). Other hormones which are able to activate S_{in} currents include luteinizing hormone (LH) and chorionic gonadotropin hormone (hCG). Furthermore, neurotransmitters such as adenosine, noradrenaline, and dopamine can activate S_{in} (R. O. Arellano, R. M. Woodward, and R. Miledi, unpublished results). However, responses to some of these agonists (e.g., noradrenaline) seem to be cut short because current generation coincides with desensitization of the receptors.

External applications of Zn^{2+} (1–100 μM) also generate large inward currents in follicles equilibrated in hypoosmotic medium, and this response develops simultaneously with the current carried by K^+ (Section 4.1.2). The Zn^{2+}-activated currents are similar to the S_{in} currents. In particular, they are osmodependent Cl^- currents (Fig. 19) that are eliminated by defolliculation and are Ca^{2+}-independent. More than 85% of follicles tested showed some sensitivity to Zn^{2+} (> 65 follicles from ~ seven frogs). For instance, in follicles from one frog where K^+ currents were blocked by 1 mM Ba^{2+}, the inward current responses to 10 μM Zn^{2+} had an average amplitude of 740 ± 95 nA (six follicles). The effect of Zn^{2+} was not mimicked by other divalent cations tested, i.e., Cu^{2+}, Cd^{2+}, Ni^{2+}, Sr^{2+}, or Mn^{2+}. In some follicles, the currents elicited by Zn^{2+} appeared to differ from the S_{in} activated by other agonists. In particular, during Zn^{2+} application depolarizing steps opened voltage-dependent channels which had slowly inactivating tail currents. Furthermore, Zn^{2+} frequently generated oscillatory currents that were substantially inhibited by Zn^{2+} itself, probably due to a direct blocking effect of this cation on the $I_{Cl(Ca)}$ channels (Miledi *et al.*, 1989b). It seems that Zn^{2+} acts at specific sites located extracellularly, because intra-oocyte injection of Zn^{2+} (40–100 pmol) did not

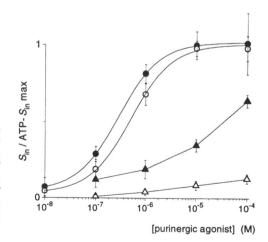

FIGURE 18. Dose-response relations of S_{in} currents elicited by ATP analogs. Normalized S_{in} (with respect to the ATP peak response) versus concentration of agonist: ATP (○), UTP (●), 2-methylthio ATP (▲), or α,β-methylene ATP (△). Each point represents the normalized mean current (± SD) elicited by the first application of the agonist in two to four epithelium-removed follicles. All records in HR (1 mM Ba^{2+}) and follicles held at −60 mV.

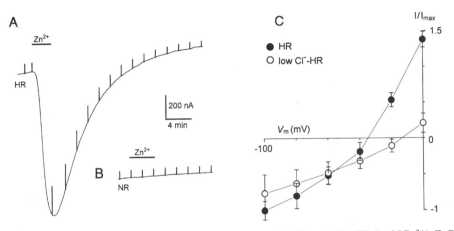

FIGURE 19. S_{in} currents elicited by Zn^{2+} in epithelium-removed follicles. (A) In HR (1 mM Ba^{2+}), $ZnCl_2$ (100 μM) generated large inward currents (S_{in}) that were potently inhibited in NR. (B) 1 mM Ba^{2+} added. (C) I-V relationships of S_{in} elicited by Zn^{2+} (100 μM) in follicles injected with TEA^+ (~400 pmol) and tested in HR (●) or low Cl^--HR (○). Each point represents the normalized mean S_{in} current at different potentials with respect to that at −100 mV in HR (three to four follicles, same frog).

elicit appreciable currents (n = 4), and because injections of the Zn^{2+} chelator N,N,N′,N′-tetrakis(2-pyridylmethyl)-ethylenediamine (*ca.* 100 pmol) had no effects on S_{in} currents generated by 10 μM Zn^{2+} (n = 3).

5.6. Receptor-Channel Coupling Mechanisms

With respect to the mechanisms underlying the generation of S_{in} and F_{in} currents, we have explored the possible involvement of second messenger systems. cAMP is an obvious candidate because its synthesis is stimulated by several receptors found in follicular cell membranes and because it is already known to be involved in the generation of cAMP-K^+ currents (Section 4.1.3). Diacylglycerol, another second messenger, activates PKC and has been proposed to mediate attenuation of the follicular K^+-cAMP current (Section 4.3.1). We tested for the involvement of these messenger systems using drugs which affect cAMP levels (Arellano and Miledi, 1994) or activate PKC directly.

5.6.1. cAMP Pathway

A growing body of evidence indicates that FSH stimulates the synthesis of cAMP in ovarian follicles (*e.g.*, Pierce and Parsons, 1981; Hsueh *et al.*, 1983; Greenfield *et al.*, 1990b) and that this second messenger is involved in the generation of cAMP-K^+ currents (Woodward and Miledi, 1987a,b; Miledi and Woodward, 1989a,b). Recently we have shown that in HR solution direct activation of adenylate cyclase by FSK or inhibition of phosphodiesterases by IBMX activates inward currents by themselves and strongly enhances S_{in} generated by FSH (Arellano and Miledi, 1994). The inward currents generated by FSK have an ionic basis and osmodependence similar to S_{in} currents elicited by either FSH or ACh, indicating that they are carried by the same type of channels. Nevertheless, it is also clear that the follicular S_{in} and F_{in} responses evoked by ACh application are not facilitated by either FSK or IBMX, suggesting that cAMP is not the second messenger used by the receptors activated by ACh. Similar results lead to this conclusion for S_{in} and F_{in} currents evoked by ATP (R. O. Arellano, R. M. Woodward, and R. Miledi, unpublished results).

Slow, smooth inward currents can also be elicited by intra-oocyte injection of cAMP (5–50 pmol) in follicles bathed in HR solution (using 1 mM Ba^{2+} to block K^+ currents) (Fig. 20A). The inward currents elicited by cAMP injection are osmodependent (Fig. 21A), are carried mainly by Cl^- ions, and have a linear I-V relationship over the -120 to $+20$ mV range (Fig. 20B). Similar or larger injections of cAMP into defollicu-lated oocytes do not elicit these currents, indicating that the Cl^- channels, and the rest of the machinery necessary for their activation, are all located in the follicular cell compart-ment. Furthermore, in follicles that respond well to cAMP, similar intra-oocyte injections of cGMP generate only weak responses (Fig. 21B) (Arellano and Miledi, 1994). Two other characteristics of these responses warrant comment: 1) S_{in} currents generated by

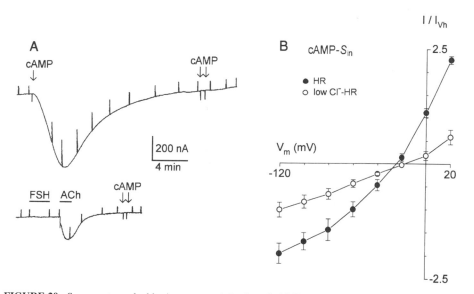

FIGURE 20. S_{in} currents evoked by intra-oocyte injection of cAMP, desensitization, and I-V relationships. (A) Upper trace: membrane current from an epithelium-removed follicle held at -60 mV in HR (1 mM Ba^{2+}). S_{in} current was generated by a single injection of cAMP. After recovery of the basal current, a second cAMP injection failed to elicit a response. Lower trace: (continuous with upper trace) After the cAMP-S_{in} had desensitized, external application of FSH (1 μg/ml) failed to elicit follicular currents, while currents elicited by ACh (100 μM) were preserved. (B) I-V relationships of the cAMP-S_{in} currents in follicles from one frog. Follicles were bathed in HR (○) containing TEA^+ (28 mM, Na^+ substitution) to block K^+ currents, or in low Cl^--HR (●) (same TEA^+ concentration). S_{in} currents were normalized with respect to the peak currents obtained at the holding potential in HR. Each point represents the normalization mean current ($\pm$ SD) from five follicles. (From Arellano and Miledi, 1994).

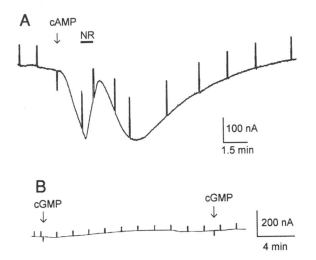

FIGURE 21. Osmodependence of S_{in} currents induced by intra-oocyte injection of cAMP and lack of effect by cGMP injection. (A) Example of S_{in} evoked by a single injection of cAMP (~10 pmol) and decrease of S_{in} by a brief increase in external osmolarity (NR) (Arellano and Miledi, 1994). (B) Follicles from the same frog failed to elicit S_{in} currents to injections of cGMP (~15 pmol). All epithelium-removed follicles from the same frog, superfused with HR (1 mM Ba^{2+}).

cAMP become refractory upon repeated stimulation; also, there is cross-desensitization with S_{in} currents generated by FSH, again suggesting that the activation pathways share some elements of the receptor-channel coupling mechanism. Interestingly, follicular inward currents evoked by ACh (Fig. 20A) or ATP are not affected by the generation and desensitization of S_{in} currents elicited by cAMP, all of which lends further support to the notion that FSH and ACh work *via* different receptor-channel coupling mechanisms. 2) S_{in} currents generated by cAMP injection can be reversibly downregulated by a transient increase in external osmolarity (superfusion with NR) (Fig. 21A).

The pathway used by cAMP to activate S_{in} channels has not been fully characterized (see Section 6.2 and 7). Although cAMP-dependent Cl^- channels are well known in several cell types (Welsh *et al.*, 1992), osmodependence has not yet been reported for these channels. In addition, although osmodependent Cl^- channels have been found in a variety of tissues, it appears that these channels are not activated by an increase in cAMP (*e.g.*, Lewis *et al.*, 1993; McCann *et al.*, 1989). More studies are needed in order to elucidate the processes involved in the cAMP-dependent activation of the follicular S_{in} channels.

5.6.2. Diacylglycerol Pathway

Two other important questions remain unresolved: 1) What are the receptor-channel coupling mechanisms involved in S_{in} and F_{in} currents activated by ACh or ATP, and 2) What are the mechanisms underlying S_{in} current osmodependence? (see also Section 7).

For the first problem, one possibility is diacylglycerol synthesis and consequent activation of PKC, particularly as this has been proposed in the inhibition of follicular cAMP-K^+ currents (Section 4.3.1). As in the latter case, there are, however, a variety of indirect results that argue against the idea that generation of F_{in} and S_{in} currents by ACh or ATP in follicular cells is mediated by activation of PLC. In our studies, follicular ACh responses from a large proportion of frogs ($> 50\%$) did not include any appreciable Ca^{2+}-dependent Cl^- current (Arellano and Miledi, 1993, 1994) (Fig. 22). This observation is even more striking in the case of follicular ATP responses which practically never have any Ca^{2+}-dependent currents, even in cells which show large follicular cell–based Cl^- currents (Figs. 16 and 23). Thus, activation of S_{in} and F_{in} currents is not correlated with the generation of Ca^{2+}-dependent currents subsequent to IP_3 synthesis. Another observation strongly supports this contention. Similar to muscarinic ACh receptors, AII receptors can be located either in the oocyte (Woodward and Miledi, 1991) or in follicular cell membranes (Sandberg *et al.*, 1990, 1992). As is the case for ACh receptors (Arellano and Miledi, 1993), the proportion of AII receptors in each compartment varies between

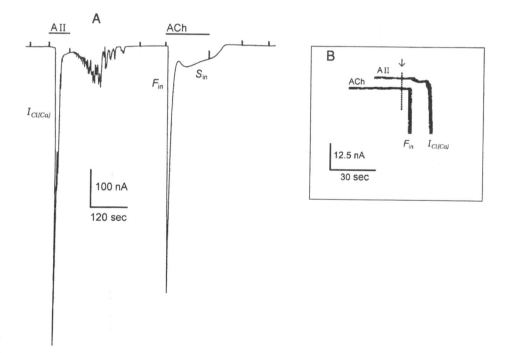

FIGURE 22. Cl^- currents elicited by activation of AII or ACh receptors located in follicular cell membranes. (A) Example of currents in follicles where the receptors to AII (500 nM) and ACh (50 μM) were located in follicular cells. The AII-elicited response was mostly oscillatory $I_{Cl(Ca)}$, while ACh elicited exclusively large F_{in} and S_{in} currents (cf. Fig. 17B,C). (B) The traces show the time of agonist application (*arrow*) in (A) on expanded time base, and the records were aligned in order to show difference in latencies (including dead time of the superfusion apparatus, in this case ~5 sec). Follicle in NR and held at −60 mV. Similar results were obtained in four follicles from the same frog.

follicles from different frogs. In follicles where AII receptors seem to be located exclusively in the follicular cells (i.e., responses are wholly eliminated by defollicula-tion), activation of these receptors does stimulate the PLC pathway, causing increases in synthesis of IP_3/diacylglycerol (Sandberg *et al.*, 1990, 1992). The synthesized IP_3 diffuses through gap junction channels into the oocyte causing activation of strong oscil-latory Ca^{2+}-gated Cl^- currents (Sandberg *et al.*, 1992) (Section 3.1, see also Fig. 1). There is, however, seldom any AII activation of S_{in} or F_{in} currents which are specifically evoked in the same follicles by ACh or ATP (Arellano and Miledi, 1994) (Figs. 22 and 23).

Occasionally, in follicles with AII receptors located in the follicular cells and which develop large (> 500 nA) F_{in} currents in response to ACh or ATP, AII can elicit, in addition to oscillatory Ca^{2+}-dependent Cl^- currents, a small (< 20 nA) inward current that resembles F_{in} (Fig. 23). Several properties of this F_{in}-like response are similar to those of the F_{in} current elicited by ACh or ATP: 1) The F_{in}-like currents evoked by AII show a short delay in their onset. 2) F_{in}-like currents are not Ca^{2+}-dependent (Fig. 23B). 3) F_{in}-like currents are eliminated by defolliculation. Overall, this result reaffirms the ineffectiveness of follicular AII receptors to elicit F_{in} or S_{in} currents, and taken together these observations strongly suggest that follicular ACh and ATP receptors do not act through a system coupled to IP_3/diacylglycerol synthesis to generate F_{in} and S_{in}.

We still cannot rule out the possibility that ACh or ATP induces the synthesis of diacylglycerol *via* stimulation of a different phospholipase (*e.g.*, PLD; Sandmann *et al.*, 1991; Murrin and Boarder, 1992). Experiments designed to activate follicular cell PKC directly, however, fail to mimic the ACh or ATP responses. For example, application of phorbol 12,13-dibutyrate (PB_{12} 0.1–1 μM) or sn-1,2-dioctanoylglycerol (DOG 0.1–1 μM), both of which activate PKC directly, do not elicit electrical responses in follicles bathed in NR (*e.g.*, Fig. 13C). In separate experiments performed in HR solution (1mM Ba^{2+}) these drugs do activate some inward and outward ion currents, but these responses have neither the amplitude nor the characteristics of the F_{in} or S_{in} activated by ACh or ATP (not shown). Follicular membrane currents evoked by PB_{12} (0.1–1 μM) showed a slower time course and the outward component was long-lasting and associated with a decrease in membrane conductance. Moreover, F_{in} and S_{in} responses elicited by 0.5 μM ACh were not enhanced by either extended or short pretreatment with PB_{12} or DOG (0.1–1 μM). The F_{in} currents elicited by ACh were, in fact, eliminated after a 10 min incuba-tion with 1 μM PB_{12}, and the associated S_{in} currents were inhibited. For example, in four follicles of a frog superfused with 0.5 μM ACh in HR, F_{in} and S_{in} currents of 50 ± 16 nA and 207 ± 14 nA, respectively, were evoked. In five follicles from the same frog incubated for 30–40 sec with 0.1 μM PB_{12} the responses evoked by ACh were 34 ± 9 nA and 237 ± 44 nA. Incubation of five follicles for 10 min with PB_{12} caused complete elimination of the F_{in} and reduced S_{in} to 56 ± 6 nA. Similar results were obtained in follicles from two more frogs and were not altered if PKC was activated using DOG or if ATP was used to evoke F_{in} and S_{in}. In occasional experiments, PKC activators do generate small, slowly developing inward currents. Since these responses are not observed in follicles bathed in NR it remains possible that PKC activation may cause some activation of the S_{in} channels. Nevertheless, the inconsistency of that response, together with the simultaneous development of the outward current linked to a decrease in conductance, complicates further evaluation. The very slow time course, low ampli-

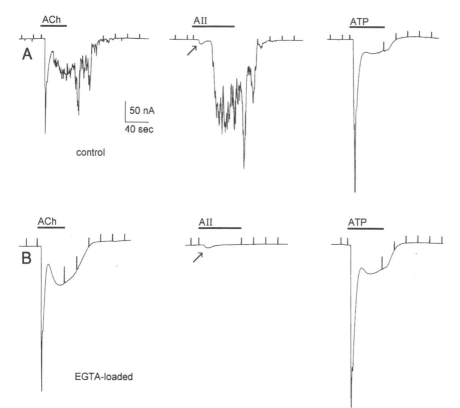

FIGURE 23. Comparison of currents elicited by ACh, AII, or ATP in the same epithelium-removed follicle. (A) Records from a follicle responding strongly to: 1) ACh (50 μM) with a mixture of oscillatory $I_{Cl(Ca)}$ (apparently due to potentiation of spontaneous oscillations; cf. Fig. 16A), along with typical F_{in} and S_{in}; 2) AII (600 nM) with strong oscillatory $I_{Cl(Ca)}$ and a small "F_{in}-like" current (*arrow*); and 3) ATP (50 μM) exclusively with F_{in} and S_{in} responses. (B) A different follicle from the same frog after loading with EGTA (~100 pmol). Oscillatory $I_{Cl(Ca)}$ were selectively abolished, while F_{in} and S_{in} were unaffected. Note that the small F_{in}-like current elicited by AII (*arrow*) was also preserved after EGTA loading. All records in NR solution.

tude (< 50 nA), and lack of potentiation of agonist-elicited currents all suggest that PKC activators do not open S_{in} channels in the same way as neurotransmitters and, therefore, that the PKC pathway is not the primary mechanism mediating transmitter-activated currents.

At present we have little or no direct insights into the mechanisms by which ACh and ATP activate follicular currents, though the latency of the F_{in} current response does begin to suggest some possibilities. For comparison, responses generated through the IP_3/Ca^{2+} pathway in oocytes have a long latency (several seconds to minutes), which is highly dependent on both temperature and agonist concentration (Kusano *et al.*, 1982; Miledi and Parker, 1989). This behavior is mainly a consequence of the process of IP_3 production, as opposed to Ca^{2+} release or Cl^- channel activation. The limiting step is an enzymatic reaction that is slowed by decreasing the temperature (the Q_{10} is ~5 at temperatures below 18°C) (Miledi and Parker, 1989). In contrast F_{in} responses have a brief latency, indiscernible from the dead time of our superfusion apparatus, and this latency is not greatly altered by decreasing agonist concentration (Fig. 24) or by decreasing the temperature (R. O. Arellano, R. M. Woodward, and R. Miledi, unpublished results).

In short, our results suggest that F_{in} currents generated by ACh or ATP are perhaps mediated by a "direct" mechanism of receptor-channel coupling, one that does not involve a cytoplasmic second messenger. The gating process may involve a common G protein, which would explain activation of the same channels by two different types of receptors (see Fig. 27). Alternatively, there may be a direct activation of follicular cell Cl^- channels by ATP, as described for epithelial airway cells (Stutts *et al.*, 1992).

6. FOLLICULAR CELL ION CURRENTS ELICITED BY CHANGES IN OSMOLARITY

While studying the osmodependence of S_{in} we noticed that even in the absence of agonists follicles can respond to a change in the external osmolarity (Arellano and Miledi, 1993). When switching the superfusion from NR to HR solutions, follicles generate membrane currents that may be up to 1 μA in amplitude and cease completely upon returning to NR. The responses evoked by decreasing the external osmolarity are

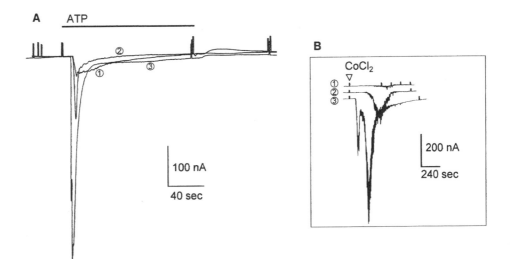

FIGURE 24. Effect of agonist concentration on latencies of the F_{in} currents elicited by ATP. (A) Superimposed traces of the F_{in} currents elicited by three different concentrations of ATP applied to one epithelium-removed follicle in NR: 1) 200 nM, 2) 250 nM, and 3) 5 μM. (B) For comparison, the insert shows the oocyte-based $I_{Cl(Ca)}$ elicited by Co^{2+}: 1) 30 μM, 2) 50 μM, and 3) 3 mM. Note different scales than in (A). (Records in (B) from Miledi *et al.*, 1989b).

associated with an increase in membrane conductance to both K^+ and Cl^- ions. Consequently, at a holding potential of -60 mV, this results in slow and smooth currents ($I_{(osm)}$) composed of a mixture of inward and outward components (Figs. 25 and 26). Like other follicular cell responses, the amplitude of currents generated by external hypo-osmolarity vary appreciably between follicles from different frogs. The inward currents, carried by Cl^- ions ($I_{Cl(osm)}$), are usually the more prominent and in most cases mask the outward current carried by K^+ ($I_{K(osm)}$). Even when masked, the K^+ current elicited by hypoosmotic medium can still be detected by a 3–6 mV positive shift in the E_{rev} of the total current, after blocking K^+ channels with Ba^{2+} (1–2 mM extracellular) or TEA^+ (~500 pmol intra-oocyte injection) (Arellano and Miledi, 1995). Occasionally, the follicles develop only the K^+ component (Fig. 25C). The follicular ionic currents generated by HR decay slowly if the follicles are maintained in this solution, and in approximately 5 hours the membrane conductance returns to a value close to that observed under control (NR) conditions.

It should be noted that follicles (and oocytes) are routinely stored in media that are hypoosmotic with respect to NR. The decay in osmo-induced currents suggests that after some hours in these media (e.g., Barth's, Merriam's, O-R media, *etc.*), the conductances originally elicited by hypoosmolarity have deactivated. The responses to hypoosmotic conditions can be reactivated either by incubating the follicles for 0.5–1 hr in NR and then reapplying HR or, more effectively, by a further decrease in osmolarity, for example, to 50% NR. These observations imply that down to a certain level of hypo-osmolarity (*ca.* 170 mosM), follicles have the capability of completely reestablishing the basal membrane conductance seen in NR, or at least to reach values very close to that level. This behavior is similar to that seen in cellular volume regulation in several other systems exposed to hypoosmotic stress (e.g., Cahalan and Lewis, 1988; Hoffmann and Simonsen, 1989; Hässinger and Lang, 1991). Typically, cells undergo an initial swelling followed by a slow recovery to basal volume. The mechanisms of adjustment appear to involve swelling-induced activation of ion currents, and in some cases the volume recovery is incomplete. All this leads us to suspect that the currents generated by hypoosmotic shock in *Xenopus* follicles, and probably the osmodependent S_{in} currents elicited by hormones and neurotransmitters, play a role in the regulation of cellular volumes in the follicular cell–oocyte complex. It is important, however, to note that we have no direct evidence for involvement of cellular "swelling" in the activation or modulation of $I_{Cl(osm)}$ and S_{in} currents.

6.1.   Localization of Elements Involved in $I_{Cl(osm)}$ Generation

The $I_{Cl(osm)}$ current generated by hypoosmolarity is similar to the S_{in} current (Section 6.2), not only in its ion selectivity and dependence on osmolarity, but also in that both are Ca^{2+}-independent (Arellano and Miledi, 1993; Ackerman *et al.*, 1994; R. O. Arellano, R. M. Woodward, and R. Miledi, unpublished results) and completely elimi-nated by defolliculation (Arellano and Miledi, 1993, 1995). This strongly suggests that the channels are located in the follicular cells and not in the oocyte membrane itself. Nevertheless, the response involves several elements, such as an osmo-sensor and a sensor-channel coupling mechanism, and these could be located in separate compart-

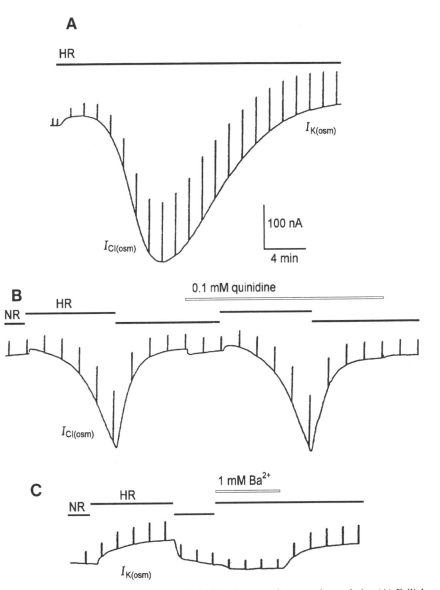

FIGURE 25. Follicular cell–based ion currents elicited by decreases in external osmolarity. (A) Follicles equilibrated in NR solution and clamped at $-60\,\mathrm{mV}$ were exposed to HR evoking outward and inward currents associated with increases in membrane conductance ($I_{(osm)}$). The currents were carried mainly by K^+ and Cl^- ions ($I_{K(osm)}$ and $I_{Cl(osm)}$, respectively). (B) Quinidine did not block $I_{(osm)}$ elicited by HR. (C) An example of a follicle in which HR generated mainly $I_{K(osm)}$; the current was blocked by Ba^{2+} added to HR. Follicles from different frogs. (From Arellano and Miledi, 1993).

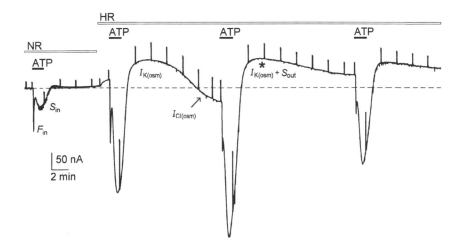

FIGURE 26. Time course of $I_{(osm)}$ and osmosensitivity of S_{in} currents elicited by ATP. Record shows a follicle initially equilibrated in NR solution and the F_{in} and small S_{in} currents elicited by ATP (0.5 μM) (bath solutions: empty *bars*; drug applications: solid *bars*). The follicle was then superfused with HR which generated $I_{(osm)}$, and ATP was reapplied three times: at the beginning of HR perfusion, on the peak of the $I_{Cl(osm)}$, and after its deactivation, where residual $I_{K(osm)}$ was still present as judged by the blocking effect of Ba^{2+} (1 mM). Note that the S_{in} current is almost fully generated in the first seconds in HR, and its amplitude remains large even at the end of the record where $I_{Cl(osm)}$ seems to have deactivated. During the peak of $I_{Cl(osm)}$ the ATP response included an outward current associated with decrease in membrane conductance (*), presumably S_{out} current which is less prominent or absent in the other ATP responses. The broken line indicates the initial level of current in NR. Similar results were obtained in five follicles from the same frog.

ments. As mentioned previously (Section 5.6.2) this situation resembles that of AII receptors which are located in the membrane of the follicular cells and are thought to activate Ca^{2+}-dependent Cl^- channels located in the oocyte membrane (Sandberg *et al.*, 1990). In an independent study of the follicular $I_{Cl(osm)}$ (Ackerman *et al.*, 1994; where the current was denoted $I_{Cl.swell}$), it was proposed that the channels are actually located in the oocyte membrane. That conclusion was based on experiments where $I_{Cl(osm)}$ was not completely eliminated by an unspecified procedure of "manual" defolliculation. The current was, however, strongly attenuated, and this was presumed to be due to removal of a positive modulatory effect of follicular cells on $I_{Cl(osm)}$ (Ackerman *et al.*, 1994). At present, there seems to be a discrepancy between that report and our results which clearly indicate that defolliculation, independent of the method used, completely abolishes $I_{Cl(osm)}$ (Arellano and Miledi, 1993, 1995). Resolution of the role played by the different follicle compartments in generation of $I_{Cl(osm)}$ and their relationship with S_{in} is important to understand follicular physiology and for the analysis of several osmosensitive molecules that have been expressed in oocytes following injection of mRNA or cRNA (e.g., Krapivinsky *et al.*, 1994; Gründer *et al.*, 1992; Preston *et al.*, 1992). For example, $I_{Cl(osm)}$ channel localization within the follicle may have direct implications for the elucidation of the role played by pI_{Cln}, a cytoplasmic protein that has been proposed as a swelling-induced Cl^- conductance regulator when expressed in oocytes (Krapivinsky et al., 1994).

To further investigate the localization of the different elements involved in the generation of $I_{Cl(osm)}$ we first examined the possibility that the channels are indeed located in the oocyte membrane itself and tried to open them directly. None of the common second messengers (e.g., cAMP, cGMP, diacylglycerol, Ca^{2+}) were able to mimic or enhance $I_{Cl(osm)}$ (Ackerman et al., 1994; R. O. Arellano, R. M. Woodward, and R. Miledi, unpublished results). We therefore tried to open $I_{Cl(osm)}$ channels in defollicu-lated oocytes using strategies designed to provoke an increase in oocyte volume; such strategies have been successful in activating swell-sensitive Cl^- channels in other cells (e.g., Doroshenko and Neher, 1992; Hagiwara et al., 1992; Lewis et al., 1993). Neither a severe reduction in external osmolarity (50–60%) nor a direct increase in oocyte volume produced by injection of substances such as mineral oil (~200 nl) were able to activate $I_{Cl(osm)}$ (or $I_{K(osm)}$, > 20 oocytes from several frogs).

We then considered the possibility that the osmo-sensor is located in the follicular cells, independent of whether it is directly associated with the $I_{Cl(osm)}$ channel or not. For these studies we increased the follicular internal osmolarity by injecting sucrose or polyethylene glycols into the oocyte and found that we were able to generate robust $I_{Cl(osm)}$ (Arellano and Miledi, 1995). For example, Cl^- currents evoked by intra-oocyte follicular injection of sucrose (4–25 nmol) had electrical characteristics that were indistinguishable from those of $I_{Cl(osm)}$: e.g., they had linear I-V relationships and the same ion selectivity, were dependent on the presence of the follicular cells, were Ca^{2+}- and cAMP-independent, were strongly osmosensitive, and were blocked by external La^{3+} (Arellano and Miledi, 1995).

We reasoned that if the osmo-sensor were located in the follicular cells, then our results could be explained by diffusion of sucrose from the oocyte into the follicular cell compartment via gap junctions. The consequent increase in tonicity within the follicular cells would activate $I_{Cl(osm)}$, perhaps by inducing water influx through the follicular membrane and swelling of the cells. If this were the case, the generation of $I_{Cl(osm)}$ would be dependent on the gap junction permeability of the substance injected and therefore strongly dependent on its molecular weight (Lowenstein, 1981). Conversely, if the osmo-sensor is located in the oocyte itself then $I_{Cl(osm)}$ should be generated independent of the molecular weight of the substance injected.

Taking into account that gap junction channels are permeable to substances with molecular weights up to ~1000 Da, we injected PEGs with mean molecular weights of 200, 300, 600, or 1000 Da and measured their ability to activate $I_{Cl(osm)}$ (Arellano and Miledi, 1995). The PEGs seemed to have few nonspecific effects when injected into the oocytes, at least not until they were injected at high concentrations at which they sometimes produced a sharp increase in oocyte membrane conductance accompanied by activation of oscillatory Cl^- currents. Fortunately, such effects occurred at concentra-tions well in excess of those necessary to activate $I_{Cl(osm)}$. Follicular oocyte injections of PEG 200, PEG 300, or PEG 600 in amounts similar to those necessary to generate $I_{Cl(osm)}$ using sucrose (i.e., 4–25 nmol) generated inward currents with the same electro-physiological characteristics as $I_{Cl(osm)}$. The potency of the different PEGs for generating

$I_{Cl(osm)}$ varied inversely to their molecular weight. Thus, PEG 200 and PEG 300 elicited strong $I_{Cl(osm)}$ in 95% of the follicles injected. PEG 600 was a weak activator, effective in only 20% of the follicles injected, while PEG 1000 (5–18 nmol) failed to active $I_{Cl(osm)}$ in any of the follicles tested. Moreover, in all cases where PEG 600 or PEG 1000 failed to elicit $I_{Cl(osm)}$, a reduction in the external osmolarity was still able to activate this current, indicating that PEGs with these high molecular weights do not in themselves impair the development of $I_{Cl(osm)}$.

These results show that generation of $I_{Cl(osm)}$ is strongly dependent on the molecular weight of the injected substances and are thus consistent with the notion that the osmo-sensor involved in $I_{Cl(osm)}$ channel activation is located in the follicular cells and not in the oocyte. Another piece of evidence supporting this argument is that generation of $I_{Cl(osm)}$ by injection of sucrose or PEG 300 was inhibited (50–86%) by superfusion of 1-octanol (1–1.5 mM), which is a gap junction blocker (Johnston *et al.*, 1980; Arellano and Miledi, 1995). Since there is no evidence for involvement of second messengers in the generation of $I_{Cl(osm)}$, and it has not been possible to activate $I_{Cl(osm)}$ channels in defolliculated oocytes, we consider it likely that the Cl⁻ channels involved are also located in the follicular cell compartment (Arellano and Miledi, 1995).

The specific mechanism for generation of $I_{Cl(osm)}$ remains unknown. In particular, it is unclear whether the follicular cell osmo-sensor and $I_{Cl(osm)}$ channels are part of the same molecule. A variety of possibilities have been proposed for other cells, for example, stretch-sensitive chloride channels (for a review, see Morris, 1990), a regulatory cyto-plasmic molecule associated with the cytoskeleton (e.g., Krapivinsky *et al.*, 1994), and a G protein–dependent process (e.g., Doroshenko *et al.*, 1991; Nilius *et al.*, 1994). We anticipate that the advantages that *Xenopus* follicles provide for electrophysiological and molecular biological studies will help elucidate the mechanisms of $I_{Cl(osm)}$ activation and their relation to osmo-sensor systems in other cells. For example, a type of osmodepen-dent Cl⁻ current flowing through a low conductance (~1.5–3 pS) channel has been described in several cellular types including T lymphocytes (Lewis *et al.*, 1993; Cahalan and Lewis, 1994), adrenal chromaffin cells (Doroshenko and Neher, 1992), and human endothelial cells (Nilius *et al.*, 1994). These channels are voltage-independent and, like those involved in the follicular $I_{Cl(osm)}$, are Ca^{2+}- and cAMP-independent. The relation-ship of these channels and the channels involved in generation of follicular $I_{Cl(osm)}$ are currently under investigation. Two other types of Cl⁻ channels activated by hypoosmotic treatments have been cloned and characterized: 1) the putative Cl⁻ channel formed by P-glycoprotein (Valverde *et al.*, 1992), a molecule which also operates as a transporter and belongs to the ABC protein family, and 2) the Cl⁻ channel ClC-2 (Gründer *et al.*, 1992), which is present in a great variety of cells and which, when expressed in *Xenopus* oocytes, can be activated by hyperpolarization and hypoosmotic shock. In both these cases the Cl⁻ channels have been proposed to play a role in regulation of cellular volume. For comparison, follicular $I_{Cl(osm)}$ channels are not voltage-dependent and unlike ClC-2 channels cannot be opened in isosmotic conditions. Furthermore, drugs such as quinidine (Fig. 26B), verapamil, vincristine, doxorubicin, and daunomycin (Valverde *et al.*, 1992; Gill *et al.*, 1992), which block the channel formed by P-glycoprotein, do not have appreciable effects on $I_{Cl(osm)}$ channels (Arellano and Miledi, 1993; R. O. Arellano, R. M.

Woodward, and R. Miledi, unpublished results). Thus it appears that although the latter two channels share with $I_{Cl(osm)}$ channels the characteristic of osmosensitivity they still have important differences in intrinsic properties.

6.2.   Relationship between $I_{Cl(osm)}$ and S_{in} Currents

Given their general properties, particularly with respect to osmodependence, it is tempting to suggest that S_{in} currents and $I_{Cl(osm)}$ are carried through the same type of channel. Their common features can be summarized as follows: 1) Both are Ca^{2+}-independent Cl^- currents. 2) Their I-V relations are linear over the voltage range -120 to $+40$ mV. 3) Both channels show lower permeability to I^- and Br^- than the Ca^{2+}-gated Cl^- channels in the oocyte. 4) Both are eliminated by oocyte defolliculation (Arellano and Miledi, 1993, 1995). Nevertheless, they present some important differences, especially in their mode of activation, and this may indicate that the two responses employ different channels: 1) S_{in} currents can be activated by an increase of intracellular cAMP (Arellano and Miledi, 1994), whereas $I_{Cl(osm)}$ is insensitive to this procedure (R. O. Arellano, R. M. Woodward, and R. Miledi, unpublished results). 2) Even when $I_{Cl(osm)}$ is deactivated following extended exposure to HR, S_{in} currents remain largely unaffected (Fig. 26). 3) The time course of $I_{Cl(osm)}$ generation and that of the S_{in} currents elicited by agonists (ATP, ACh, or FSH) are different. In these experiments follicles were first equilibrated in HR and tested using low (0.5 μM) concentrations of ATP (Fig. 26). Applications of the agonist were then repeated at the beginning (~1 min), at the peak (~10 min), and during the decay phase of $I_{Cl(osm)}$ (20–30 min). In such experiments the $I_{Cl(osm)}$ increases slowly in *ca.* 5–15 min (cf. Fig. 25), while ATP-S_{in} current is already near its maximum (75–90%) during the first minute in HR (Fig. 26). Furthermore, after $I_{Cl(osm)}$ is deactivated, it is still possible to evoke ATP-S_{in} currents with similar amplitude to those elicited at the beginning of HR superfusion (Fig. 26). Similar results were obtained for S_{in} generated by ACh or FSH. Another finding consistent with the faster kinetics of the S_{in} current's osmodependence is the rapid time course of the inhibition produced by NR on S_{in} elicited by cAMP injections (see Fig. 21A).

In spite of the different properties of $I_{Cl(osm)}$ and S_{in} currents it is still possible that these differences simply reflect multiple mechanisms of activation and modulation of a common population of Cl^- channels. Considering their many similarities, it would not be surprising if both $I_{Cl(osm)}$ and S_{in} are carried by the same type of channel.

7.   POTENTIAL PHYSIOLOGICAL ROLES

To state it simply, the physiological functions of the ionic currents elicited by the different agonists in *Xenopus* follicles, and the significance of the modulation of some conductances by osmolarity, remain entirely uncertain. Here we will restrict ourselves to mentioning two of the many possible functions: regulation of oocyte maturation and control of cell volumes.

7.1. Regulation of Oocyte Maturation

During growth of the follicle, meiotic division of the oocyte is halted at the first prophase. When the oocyte is fully developed (stage V-VI) meiosis can be reactivated through a complex series of reactions triggered by progesterone (Merriam, 1971; for a review, see Smith, 1989), which is synthesized and released by the follicular cells in response to hormonal stimulation by pituitary gonadotropins (Mulner *et al.*, 1978; Hsueh *et al.*, 1983; Shuetz and Glad, 1985). Progesterone acts through specific receptors in the oocyte membrane itself. A variety of evidence raises the possibility that follicular cell membrane receptors and channels influence the kinetics of this step in the maturation process. For example, stimulation of AII receptors in follicular cells seems to promote the progesterone-induced germinal vesicle breakdown (Sandberg *et al.*, 1990). Also, K^+ channel openers, which mimic some aspects of follicular cAMP-K^+ currents, have been shown to increase the percentage of germinal vesicle breakdown in oocytes treated with either gonadotropins or progesterone (Wibrand *et al.*, 1993).

7.2. Regulation of Follicle Volume

Hypoosmotic shock is commonly used in studies of cellular volume regulation. As discussed above (Section 6), most cells possess mechanisms to control their volumes and respond to hypoosmotic conditions with a initial swelling, which is followed by volume recovery. Defolliculated *Xenopus* oocytes, on the other hand, do not have an efficient system of volume regulation (Fry and Shaw, 1988; cf. Preston *et al.*, 1992; Chrispeels and Agre, 1994) and do not generate osmodependent membrane currents, which are thought to underlie, at least partly, volume recovery (Preston *et al.*, 1992; Arellano and Miledi, 1993, 1995; Section 6). In contrast, we have shown that external hypoosmotic solutions or a direct increase in the internal osmolarity of the follicular cells elicits substantial changes in follicle membrane conductance (Arellano and Miledi, 1993, 1995). These conductances may play an important role in the control of cellular volumes within the follicle, though the relation between the two phenomena has not been demonstrated directly. In this context, it should be recalled that follicular S_{in} elicited by different neurotransmitters and hormones exhibit an exquisite sensitivity to external osmolarity. We speculate that S_{in} currents are generated by neurotransmitters and hormones *via* a mechanism that in the presence of an osmotic gradient provokes swelling of the follicular cells and hence activation of osmosensitive Cl^- channels (perhaps those involved in $I_{Cl(osm)}$). Indeed, water flux and the associated changes in cellular volume are common processes induced by hormones and neurotransmitters in kidney, liver, and epithelia (for reviews, see Brown *et al.*, 1990; Hässinger and Lang, 1991). The hormone-induced changes in cellular volume have been proposed to be part of the physiological process whereby substances regulate cellular metabolism (Hässinger and Lang, 1992).

The ion current responses summarized herein are detectable in follicles from approximately stage IV onward (Woodward and Miledi, 1987b; Miledi and Woodward, 1989b; R. O. Arellano, R. M. Woodward, and R. Miledi, unpublished results). Detection of follicular currents from within the oocyte will be dependent on the establishment of the electrical coupling between oocyte and follicular cells, and the beginning of stage IV is characterized by an increase in this communication. Thus, it would not be surprising if follicular cell receptors and channels are present at earlier stages of development, even though they cannot be detected by electrical recording from within the oocyte.

Lastly, with respect to the physiological role of the receptors and conductances, it is worth noting that there appear to be similarities between follicular cell receptor systems in *Xenopus* and the receptor systems in mammalian granulosa cells (see Table II). Therefore, the study of *Xenopus* follicular responses may help toward a comprehensive understanding of the roles played by these receptors in the physiology of ovarian follicles in other species.

8. CONCLUSIONS

Xenopus follicles contain a great diversity of ion channels and receptors for neurotransmitters, hormones, and peptides, making them a very useful system for studies in biology and medicine. It is now clear that many of these receptor-channel systems are located in the membranes of follicular cells, which maintain electrical and metabolic communication with the oocyte *via* gap junctions. But, despite the advances in understanding the response mechanisms involved, little is known about the physiological roles played by the many receptors and channels. The variety of electrical responses and the large number of different membrane receptors in the *Xenopus* oocyte–follicular cell system is somewhat daunting and the picture is consequently very complicated (Fig. 27). However, we are really only beginning to understand follicular cells at the molecular level. The situation will become clearer as we learn more about the molecular structure and function of the receptors and channels that mediate the follicular cell responses. In

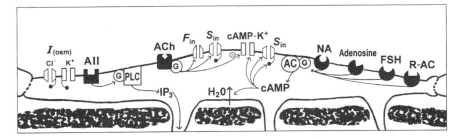

FIGURE 27. A summarizing diagram for some of the ion channels and receptors in the follicular cell membrane. Also illustrated are the signaling pathways used by the different receptors during ion channel activation, i.e., adenylate cyclase, PLC, and an undetermined signaling pathway activated by ACh (or ATP/UTP). S_{in} and $I_{(osm)}$ osmo-sensors have been represented with an hypothetical "chain and ball" model (cf. Gründer *et al.*, 1992). S_{in} channels activation either *via* cAMP or through muscarinic stimulation may be achieved by action on one or several putative sites (and *via* intermediary enzymes, e.g., PKA); here we have illustrated only some possibilities, namely, direct action on S_{in} channel molecule, b) the osmo-sensor molecule, and c) the rate of transmembrane water influx. Likewise, the activation mechanism of $I_{(osm)}$ channels remains unknown. R-AC, other receptors coupled to adenylate cyclase (see Table II); all other abbreviations as in Fig. 1 and text.

coming years, there will be more studies on the molecular biology of follicular responses and isolation and direct characterization of single follicular cells. We believe that continuing studies in this system will contribute to our understanding of ovarian physiology and provide insights into fundamental processes in other cellular systems. These studies should also promote a more rigorous use of the *Xenopus* follicle and oocytes when studying molecules expressed by injections of foreign mRNA or cloned DNA and RNA.

ACKNOWLEDGMENTS. We are grateful to Sandra Page and Mousa Shamonki for their help in the preparation of the figures, and to Rico Miledi for computer programming. R.O.A. acknowledges support from The Pew Charitable Trusts. This work was supported by a grant (NS-23284) from the National Institute of Neurological Disorders and Stroke.

9. REFERENCES

Ackerman, M. J., Wickman, K. D., and Clapham, D. E., 1994, Hypotonicity activates a native chloride current in *Xenopus* oocytes, *J. Gen. Physiol.* **103:**153–179.

Adashi, E., and Hsueh, A. J. W., 1981, Stimulation of β-adrenergic responsiveness by follicle-stimulating hormone in rat granulosa cells *in vitro* and *in vivo, Endocrinology* **108:**2170–2178.

Anderson, E., and Albertini, D. F., 1976, Gap junctions between the oocyte and companion follicle cells in the mammalian ovary, *J. Cell Biol.* **71:**680–686.

Arellano, R. O., and Miledi, R., 1993, Novel Cl⁻ currents elicited by follicle stimulating hormone and acetylcholine in follicle-enclosed *Xenopus* oocytes, *J. Gen. Physiol.* **102:**833–857.

Arellano, R. O., and Miledi, R., 1994, Osmo-dependent Cl⁻ currents activated by cyclic AMP in follicle-enclosed *Xenopus* oocytes, *Proc. R. Soc. London Ser. B* **258:**229–235.

Arellano, R. O., and Miledi, R., 1995, Functional role of follicular cells in the generation of osmolarity-dependent Cl⁻ currents in *Xenopus* follicles, *J. Physiol. (London)* **488:**351–357.

Aurbach, G. D., 1982, Polypeptide and amine hormone regulation of adenylate cyclase, *Annu. Rev. Physiol.* **44:**653–668.

Bagnato, A., Moretti, C., Frajese, G., and Catt, K. J., 1991, Gonadotropin-induced expression of receptors for growth hormone releasing factor in cultured granulosa cells, *Endocrinology* **128:**2889–2894.

Barish, M. R., 1983, A transient calcium-dependent chloride current in the immature *Xenopus* oocyte, *J. Physiol. (London.)* **342:**309–325.

Barnard, E. A., Miledi, R., and Sumikawa, R., 1982, Translation of exogenous messenger RNA coding for nicotinic acetylcholine receptors produces functional receptors in *Xenopus* oocytes, *Proc. R. Soc. London Ser. B.* **215:**241–248.

Barth, L. G., and Barth, L. J., 1959, Differentiation of cells of the *Rana pipiens* gastrula in unconditioned medium, *J. Embr. Exp. Morphol.* **71:**210–222.

Berridge, M. J., 1987, Inositol trisphosphate and diacylglycerol: Two interacting second messengers, *Annu. Rev. Biochem.* **56:**159–193.

Bodis, J., Tinneberg, H. R., Torok, A., Cledon, P., Hanf, V., and Papenfuss, F., 1993a, Effect of noradrenaline and dopamine on progesterone and estradiol secretion of human granulosa cells, *Acta Endocrinol.* **129:**165–168.

Bodis, J., Torok, A., Tinneberg, H. R., Hanf, V., Papenfuss, F., and Schuwarz, H., 1993b, Serotonin induces progesterone release from human granulosa cells in a superfused granulosa cell system, *Arch. Gynecol. Obstet.* **253:**59–64.

Brissette, J. L., Kumar, N. M., Gilula, N. B., Hall, J. E., and Dotto, G. P., 1994, Switch in gap junction protein expression is associated with selective changes in junctional permeability during keratinocyte differentiation, *Proc. Natl. Acad. Sci. USA* **91:**6453–6457.

Brown, D., Grosso, A., and De Sousa, R. C., 1990, Membrane architecture and water transport in epithelial cell membranes, in: *Advances in Membrane Fluidity*, Vol. 4 (R. C. Aloia, C. C. Curtain, and L. M. Gordon, eds.), Alan R. Liss, Inc., New York, pp. 103–132.

Browne, C. L., and Werner, W., 1984, Intercellular junctions between the follicle cells and the oocytes of *Xenopus laevis*, *J. Exp. Zool.* **230:**105–113.

Browne, C. L., Wiley, H. S., and Dumont, J. N., 1979, Oocyte-follicle cell gap junctions in *Xenopus laevis* and the effects of gonadotropin on their permeability, *Science* **203:**182–183.

Buccione, R., Schroeder, A. C., and Eppig, J. J., 1990a, Interactions between somatic cells and germ cells throughout mammalian oogenesis, *Biol. Reprod.* **43:**543–547.

Buccione, R., Vanderhyden, B. C., Caron, P. J., and Eppig, J. J., 1990b, FSH-induced expansion of the mouse cumulus oophorus *in vitro* is dependent upon specific factor(s) secreted by the oocyte, *Dev. Biol.* **138:** 16–25.

Budnik, L. T., Brunswig, B., and Mukhopadhyay, A. K., 1987, Atrial natriuretic factor stimulates luteal guanylate cyclase, *Regulat. Pept.* **19:**23–34.

Cahalan, M. D., and Lewis, R. S., 1988, Role of potassium and chloride channels in volume regulation by T lymphocytes, in: *Cell Physiology of Blood* (R. Gunn and J. Parker, eds.), Rockefeller University Press, New York, pp. 281–301.

Cahalan, M. D., and Lewis, R. S., 1994, Regulation of chloride channels in lymphocytes, in: *Current Topics in Membranes*, Vol. 42, Academic Press, pp. 103–129.

Caulfield, M. P., 1993, Muscarinic receptors—characterization, coupling and function, *Pharmacol. Ther.* **58:**319–379.

Cerione, R. A., Strulovici, B., Benovic, J. L., Lefkowitz, R. J., and Caron, M. G., 1983, Pure β-adrenergic receptor: The single polypeptide confers catecholamine responsiveness to adenylate cyclase, *Nature* **306:**562–566.

Chen, H., Jassar, B. S., Kurenny, D. E., and Smith, P. A., 1994, Phorbol ester–induced M-current suppression in bull-frog sympathetic ganglion cells: Insensitivity to kinase inhibitors, *Br. J. Pharmacol.* **113:** 55–62.

Chrispeels, M. J., and Agre, P., 1994, Aquaporins: Water channel proteins of plant and animal cells, *Trends Biol. Sci.* **19:**421–425.

Colman, A., 1984a, Expression of exogenous DNA in *Xenopus oocytes*, in: *Transcription and Translation: A Practical Approach* (B. D. Hames and S. J. Higgins, eds.), IRL Press, Oxford, pp. 49–69.

Colman, A., 1984b, Translation of eukaryotic messenger RNA in *Xenopus* oocytes, in: *Transcription and Translation: A Practical Approach* (B. D. Hames and S. J. Higgins, eds.), IRL Press, Oxford, pp. 271–302.

Currie, W. D., Li, W., Baimbridge, K. G., Yuen, B. H., and Leung, P. C., 1992, Cytosolic free calcium increased by prostaglandin $F_{2\alpha}$ ($PGF_{2\alpha}$), gonadotropin-releasing hormone, and angiotensin II in rat granulosa cells, *Endocrinology* **130:**1837–1843.

Dascal, N., 1987, The use of *Xenopus* oocytes for the study of ion channels, *CRC Crit. Rev. Biochem.* **22:** 317–387.

Dascal, N., and Cohen, S., 1987, Further characterization of the slow muscarinic responses in *Xenopus* oocytes, *Pflügers Arch.* **409:**512–520.

Dascal, N., and Landau, E. M., 1980, Types of muscarinic response in *Xenopus* oocytes, *Life Sci.* **27:**143–147.

Dascal, N., Landau, E. M., and Lass, Y., 1984, *Xenopus* oocyte resting potential, muscarinic responses and the role of calcium and guanosine 3′,5′-cyclic monophosphate, *J. Physiol. (London)* **352:**551–574.

Dascal, N., Lotan, I., Gillo, B., Lester, H. A., and Lass, Y., 1985, Acetylcholine and phorbol esters inhibit potassium currents evoked by adenosine and cyclic adenosine monophosphate in *Xenopus* oocytes, *Proc. Natl. Acad. Sci. USA* **82:**6001–6005.

Davis, N. W., Standen, N. B., and Stanfield, P. R., 1991, ATP-dependent potassium channels of muscle cells: Their properties, regulation, and possible function, *J. Bioenerg. Biomembr.* **23:**509–535.

Davoren, J. B., and Hsueh, A. J. W., 1985, Vasoactive intestinal peptide: A novel stimulator of steroidogenesis by cultured rat granulosa cells, *Biol. Reprod.* **33:**37–52.

Doroshenko, P., and Neher, E., 1992, Volume-sensitive chloride conductance in bovine chromaffin cell membrane, *J. Physiol. (London)* **449:**197–218.

Doroshenko, P., Penner, R., and Neher, E., 1991, Novel chloride conductance in the membrane of bovine chromaffin cells activated by intracellular GTPγS, *J. Physiol. (London)* **436:**711–724.

Dorrington, J. H., McKeracher, H. I., Chan, A. K., and Gore-Langton, R. E., 1983, Hormonal interactions in the control of granulosa cell differentiation, *J. Steroid. Biochem.* **19**:17–32.

Dumont, J. N., 1972, Oogenesis in *Xenopus laevis* (Daudin). I. Stages of oocyte development in laboratory maintained animals, *J. Morphol.* **136**:153–180.

Dumont, J. N., and Brummett, A. R., 1978, Oogenesis in *Xenopus laevis* (Daudin). Relationships between developing oocytes and their investing follicular cells, *J. Morphol.* **155**:73–98.

Eppig, J. J., 1979, A comparison between oocyte growth in co-culture with granulosa cells and oocytes with granulosa cells–oocyte junctional contact maintained *in vitro*, *J. Exp. Zool.* **209**:345–353.

Fry, D. J., and Shaw, S. R., 1988, Effects of anisotonic media on cell volume in oocytes isolated from the amphibians *Bufo bufo* and *Xenopus laevis*, *J. Physiol. (London)* **396**:174P.

Gill, D. R., Hyde, S. C., Higgins, C. F., Valverde, M. A., Mintenig, G. M., and Sepulveda, F. V., 1992, Separation of drug transport and chloride channel functions of the human multidrug resistance P-glyco-protein, *Cell* **71**:23–32.

Gillemare, E., Lazdunski, M., and Honoré, E., 1994, CGRP-induced activation of K_{ATP} channels in follicular *Xenopus* oocytes, *Plügers Arch.* **428**:604–609.

Goff, A. K., and Armstrong, D. T., 1983, Changes in responsiveness of rat granulosa cells to prostaglandin E_2 and follicle-stimulating hormone during culture, *Can. J. Physiol. Pharmacol.* **61**:608–613.

Goldin, A. L., 1991, Expression of ion channels by injection of mRNA into *Xenopus* oocytes, in: *Methods in Cell Biology*, Vol. 36 (B. K. Kay and H. B. Peng, eds.), Academic Press, Inc., New York, pp. 487–509.

Greenfield, L. J., Hackett, J. T., and Linden, J., 1990a, *Xenopus* oocyte K^+ current. I. FSH and adenosine stimulate follicle cell–dependent currents, *Am. J. Physiol.* **259**:C775–C783.

Greenfield, L. J., Hackett, J. T., and Linden, J., 1990b, *Xenopus* oocyte K^+ current. II. Adenylyl cyclase–linked receptors on follicle cells, *Am. J. Physiol.* **259**:C784–C791.

Gründer, S., Thiemann, A., Pusch, M., and Jentsch, T. J., 1992, Regions involved in the opening of ClC-2 chloride channel by voltage and cell volume, *Nature* **360**:759–762.

Gundersen, C. B., Miledi, R., and Parker, I., 1983, Serotonin receptors induced by exogenous messenger RNA in *Xenopus* oocytes, *Proc. R. Soc. London Ser. B* **219**:103–109.

Gundersen, C. B., Miledi, R., and Parker, I., 1984, Messenger RNA from human brain induces drug- and voltage-operated channels in *Xenopus* oocytes, *Nature* **308**:421–424.

Gurdon, J. B., Lane, C. D., Woodland, H. R., and Marbaix, G., 1971, Use of frog eggs and oocytes for the study of messenger RNA and its translation in living cells, *Nature* **233**:177–182.

Hagiwara, N., Masuda, H., Shoda, M., and Irisawa, H., 1992, Stretch-activated anion channels of rabbit cardiac myocytes, *J. Physiol. (London)* **456**:285–302.

Hamill, O. P., Marty, A., Neher, E., Sakmann, B., and Sigworth, P., 1981, Improved patch-clamp techniques for high-resolution current recording from cells and cell-free membrane patches, *Pflügers Arch.* **391**:85–100.

Häussinger, D., and Lang, F., 1991, Cell volume in the regulation of hepatic function: A mechanism for metabolic control, *Biochim. Biophys. Acta* **1071**:331–350.

Häussinger, D., and Lang, F., 1992, Cell volume and hormone action, *Trends Pharmacol. Sci.* **13**:371–373.

Heller, D. T., and Shultz, R. M., 1980, Ribonucleoside metabolism by mouse oocytes: Metabolic cooperativity between the fully grown oocyte and cumulus cells, *J. Exp. Zool.* **214**:355–364.

Hoffmann, E. K., and Simonsen, L. O., 1989, Membrane mechanisms in volume and pH regulation in vertebrate cells, *Physiol. Rev.* **69**:315–382.

Honoré, E., and Lazdunski, M., 1991, Hormone-regulated K^+ channels in follicle-enclosed oocytes are activated by vasorelaxing K^+ channel openers and blocked by antidiabetic sulfonylureas, *Proc. Natl. Acad. Sci. USA* **88**:5438–5442.

Honoré, E., and Lazdunski, M., 1993, Single-channel properties and regulation of pinacidil/glibenclamide-sensitive K^+ channels in follicular cells from *Xenopus* oocyte, *Pflügers Arch.* **424**:113–121.

Hsueh, A. J. W., Jones, P. B. C., Adashi, E. Y., Wang, C., Zhuang, L. Z., and Welsh, T. H., 1983, Intraovarian mechanisms in the hormonal control granulosa cell differentiation in rats, *J. Reprod. Fertil.* **69**:325–342.

Johnston, M. F., Simon, S., and Ramon, F., 1980, Interactions of anesthetics with electrical synapses, *Nature* **286**:498–500.

Johnson, K. M., Hughes, F. M., Fong, Y. Y., Mathur, R. S., Williamson, H. O., and Gorospe, W. C., 1994, Effects of atrial natriuretic peptide on rat ovarian granulosa cell steroidogenesis *in vitro*, *Am. J. Reprod. Immunol.* **31**:163–168.

Kado, R. T., Marcher, K. S., and Ozon, R., 1981, Electrical membrane properties of the *Xenopus laevis* oocyte during progesterone-induced meiotic maturation, *Dev. Biol.* **84:**471–476.

Kamada, S., Blackmore, P. F., Oehninger, S., Gordon, K., and Hodgen, G. D., 1994, Existence of P2-purinoceptors on human and porcine granulosa cells, *J. Clin. Endocrinol. Metab.* **78:**650–656.

Kay, B. K., and Peng, H. B. (eds.), 1991, *Methods in Cell Biology*, Vol. 36, Academic Press, Inc., New York.

Kliachko, S., and Zor, U., 1981, Increase in catecholamine-stimulated cyclic AMP and progesterone synthesis in rat granulosa cells during culture, *Mol. Cell. Endocrinol.* **23:**23–32.

Kolena, J., and Channing, C. P., 1972, Stimulatory effects of LH, FSH and prostaglandins upon cyclic 3',5'-AMP levels in porcine granulosa cells, *Endocrinology* **90:**1543–1550.

Krapivinsky, G., Ackerman, M. J., Gordon, E., Krapivinsky, L. D., and Clapham, D. E., 1994, Molecular characterization of a swelling-induced chloride conductance regulatory protein, pI_{Cln}, *Cell* **76:**439–448.

Kusano, K., Miledi, R., and Stinnakre, J., 1977, Acetylcholine receptors in the oocyte membrane, *Nature* **270:**739–741.

Kusano, K., Miledi, R., and Stinnakre, J., 1982, Cholinergic and catecholaminergic receptors in the *Xenopus* oocyte membrane, *J. Physiol. (London)* **328:**143–170.

Lacy, P., Murray-McIntosh, R. P., and McIntosh, J. E. A., 1992, Angiotensin II and acetylcholine differentially activate mobilization of inositol phosphates in *Xenopus laevis* ovarian follicles, *Pflügers Arch.* **420:**127–135.

Lasater, E. M., and Dowling, J. E., 1985, Dopamine decreases conductance of the electrical junctions between cultured retinal horizontal cells, *Proc. Natl. Acad. Sci. USA* **82:**3025–3029.

Lewis, R. S., Ross, P. E., and Cahalan, M. D., 1993, Chloride channels activated by osmotic stress in T lymphocytes, *J. Gen. Physiol.* **101:**801–826.

Lotan, I., Dascal, N., Cohen, S., and Lass, Y., 1982, Adenosine induced slow ionic currents in the *Xenopus* oocyte, *Nature* **298:**572–574.

Lotan, I., Dascal, N., Oron, Y., Cohen, S., and Lass, Y., 1985, Adenosine-induced K^+ current in *Xenopus* oocyte and the role of adenosine 3',5'-monophosphate, *Mol. Pharmacol.* **28:**170–177.

Lotan, Y., Dascal, N., Cohen, S., and Lass, Y., 1986, ATP-evoked membrane responses in *Xenopus* oocytes, *Pflügers Arch.* **406:**158–162.

Lowenstein, W. R., 1981, Junctional intercellular communication. The cell-to-cell membrane channel, *Physiol. Rev.* **61:**829–913.

Luck, M. R., 1990, Cholinergic stimulation, through muscarinic receptors, of oxytocin and progesterone secretion from bovine granulosa cells undergoing spontaneous luteinization in serum-free culture, *Endocrinology* **126:**1256–1263.

Lupu-Meiri, M., Shapira, H., Matus-Leibovitch, N., and Oron, Y., 1990, Two types of muscarinic responses in *Xenopus* oocytes: I. Differences in latencies and ^{45}Ca efflux kinetics, *Pflügers Arch.* **417:**391–397.

Masui, Y., and Markert, C. L., 1971, Cytoplasmic control of nuclear behavior during meiotic maturation of frog oocytes, *J. Exp. Zool.* **177:**129–146.

McArdle, C. A., 1990, Chronic regulation of ovarian oxytocin and progesterone release by prostaglandins: Opposite effects in bovine granulosa and early luteal cells, *J. Endocrinol.* **126:**245–253.

McCann, J. D., Li, M., and Welsh, M. J., 1989, Identification and regulation of whole-cell chloride currents in airway epithelium, *J. Gen. Physiol.* **94:**1015–1036.

McPherson, G. A., 1993, Current trends in the study of potassium channel openers, *Gen. Pharmacol.* **24:**275–281.

Merriam, R. W., 1971, Progesterone-induced maturational events in oocytes of *Xenopus laevis*. I. Continuous necessity for diffusible calcium and magnesium, *Exp. Cell Res.* **68:**75–80.

Miledi, R., 1982, A calcium-dependent transient outward current in *Xenopus laevis* oocytes, *Proc. R. Soc. London Ser. B* **215:**491–497.

Miledi, R., and Parker, I., 1984, Chloride current induced by injection of calcium into *Xenopus* oocytes, *J. Physiol. (London)* **357:**173–183.

Miledi, R., and Parker, I., 1989, Latencies of membrane currents evoked in *Xenopus* oocytes by receptor activation, inositol trisphosphate and calcium, *J. Physiol. (London)* **415:**189–210.

Miledi, R., and Woodward, R. M., 1989a, The effect of defolliculation on the membrane current responses of *Xenopus* oocytes, *J. Physiol. (London)* **416:**601–621.

Miledi, R., and Woodward, R. M., 1989b, Membrane currents elicited by prostaglandins, oxytocin and atrial natriuretic factor in follicle enclosed *Xenopus* oocytes, *J. Physiol. (London)* **416:**623–643.

Miledi, R., Parker, I., and Sumikawa, K., 1982, Synthesis of chick brain GABA receptors by frog oocytes, *Proc. R. Soc. London B Ser.* **216:**509–515.

Miledi, R., Parker, I., and Sumikawa, K., 1989a, Transplanting receptors from brains into oocytes, in: *Fidia Research Foundation Neuroscience Award Lectures*, Raven Press, New York, pp. 57–90.

Miledi, R., Parker, I., and Woodward, R. M., 1989b, Membrane currents elicited by divalent cations in *Xenopus* oocytes, *J. Physiol. (London)* **417:**173–195.

Mori, K., Oka, S., Tani, A., Ito, S., and Watanabe, Y., 1989, E-series prostaglandins activate cAMP-mediated potassium currents in follicle-enclosed *Xenopus* oocytes, *Biochem. Biophys. Res. Commun.* **162:**1535–1540.

Moriarty, T. M., Gillo, B., Sealfon, S., and Landau, E. M., 1988, Activation of ionic currents in *Xenopus* oocytes by corticotropin-releasing peptides, *Mol. Brain Res.* **4:**201–206.

Morris, C. E., 1990, Mechanosensitive ion channels, *J. Membr. Biol.* **113:**93–107.

Mulner, O., Thibier, C., and Ozon, R., 1978, Steroid biosynthesis by ovarian follicles of *Xenopus laevis in vitro* during oogenesis, *Gen. Comp. Endocrinol.* **34:**287–295.

Murrin, R. J. A., and Boarder, M. R., 1992, Neuronal "nucleotide" receptor linked to phospholipase C and phospholipase D? Stimulation of PC12 cells by ATP analogues and UTP, *Mol. Pharmacol.* **41:**561–568.

Neyton, J., and Trautmann, A., 1986, Acetylcholine modulation of the conductance of intercellular junctions between rat lacrimal cells, *J. Physiol. (London)* **377:**283–295.

Nilius, B., Oike, M., Zahradnik, I., and Droogmans, G., 1994, Activation of a Cl^- current by hypotonic volume increase in human endothelial cells, *J. Gen. Physiol.* **103:**787–805.

Nimrod, A., Erickson, G. F., and Ryan, K. J., 1976, A specific FSH receptor in rat granulosa cells: Properties of binding *in vitro*, *Endocrinology* **98:**54–64.

Nishi, Y., Hatano, S., Aihara, K., Okahata, H., Kawamura, H., Tanaka, K., Miyachi, Y., and Usui, T., 1984, Effect of zinc ion on human chorionic gonadotropin–stimulated *in vitro* production of cyclic-AMP and testosterone by rat testis, *Pediatric Res.* **18:**232–235.

Nitray, J., and Sirotkin, A. V., 1992, Reciprocal control of oxytocin and cAMP production by porcine granulosa cells *in vitro*, *Ann. Endocrinol.* **53:**28–31.

Oron, Y., Dascal, N., Nadler, E., and Lupu, M., 1985, Inositol 1,4,5-trisphosphate mimics muscarinic response in *Xenopus* oocytes, *Nature* **313:**141–143.

Pandey, K. N., Osteen, K. G., and Inagami, T., 1987, Specific receptor mediated stimulation of progesterone secretion and cGMP accumulation by rat atrial natriuretic factor in cultured human granulosa-lutein cells, *Endocrinology* **121:**1195–1197.

Parker, I., and Miledi, R., 1986, Changes in intracellular calcium and in membrane currents evoked by injection of inositol trisphosphate into *Xenopus* oocytes, *Proc. R. Soc. London Ser. B* **228:**307–315.

Parker, I., and Miledi, R., 1987, Inositol trisphosphate activates a voltage-dependent calcium influx in *Xenopus* oocytes, *Proc. R. Soc. London Ser. B* **232:**27–36.

Parker, I., Gundersen, C. B., and Miledi, R., 1985a, A transient inward current elicited by hyperpolarization during serotonin activation in *Xenopus* oocytes, *Proc. R. Soc. London Ser. B.* **223:**279–292.

Parker, I., Gundersen, C. B., and Miledi, R., 1985b, Intracellular Ca^{2+}-dependent and Ca^{2+}-independent responses of rat brain serotonin receptor transplanted to *Xenopus* oocytes, *Neurosci. Res.* **2:**491–496.

Parker, I., Sumikawa, K., and Miledi, R., 1987, Activation of a common effector system by different brain neurotransmitter receptors in *Xenopus* oocytes, *Proc. R. Soc. London Ser. B* **231:**37–45.

Piccolino, M., Neyton, J., and Gerschenfeld, H. M., 1984, Decrease of gap junction permeability induced by dopamine and cyclic adenosine 3′,5′-monophosphate in horizontal cells of turtle retina, *J. Neurosci.* **4:**2477–2488.

Pierce, J. C., and Parsons, T. F., 1981, Glycoprotein hormones: Structure and function, *Annu. Rev. Biochem.* **50:**465–495.

Preston, G. M., Carroll, T. P., Guggino, W. B., and Agre, P., 1992, Appearance of water channels in *Xenopus* oocytes expressing red cell CHIP28 protein, *Science* **256:**385–387.

Pucell, A. G., Hodges, J. C., Sen, I., Bumpus, F. M., and Husain, A., 1991, Biochemical properties of the ovarian granulosa cell type 2–angiotensin II receptor, *Endocrinology* **128:**1947–1959.

Robbins, J., Marsh, S. J., and Brown, D. A., 1993, On the mechanism of M-current inhibition by muscarinic m1 receptors in DNA-transfected rodent neuroblastoma × glioma cells, *J. Physiol. (London)* **469:**153–178.

Robinson, K. R., 1979, Electrical currents through full-grown and maturing *Xenopus* oocytes, *Proc. Natl. Acad. Sci. USA* **76:**837–841.

Robinson, S. R., Hampson, E. C. G. M., Munro, M. N., and Vaney, D. I., 1993, Unidirectional coupling of gap junctions between neuroglia, *Science* **262:**1072–1074.

Rudy, B., and Iverson, L. E. (eds.), 1992, *Methods in Enzymology*, Vol. 207, Academic Press, Inc., San Diego, pp. 225–390.

Sandberg, K., Bor, M., Ji, H., Markwick, A. J., Millan, M. A., and Catt, K. J., 1990, Angiotensin II-induced calcium mobilization in oocytes by signal transfer through gap junctions, *Science* **249:**298–301.

Sandberg, K., Ji, H., Iida, T., and Catt, K. J., 1992, Intercellular communication between follicular angiotensin receptors and *Xenopus laevis* oocytes: Mediation by an inositol 1,4,5-trisphosphate-dependent mechanism, *J. Cell Biol.* **117:**157–167.

Sandmann, J. S., Peralta, E. G., and Wurtman, R. J., 1991, Coupling of transfected muscarinic acetylcholine receptor subtypes to phospholipase D, *J. Biol. Chem.* **266:**6031–6034.

Schuetz, A. W., and Glad, R., 1985, *In vitro* production of meiosis inducing substance (MIS) by isolated amphibian (*Rana pipiens*) follicle cells, *Dev. Growth Differ.* **27:**201–212.

Seamon, K. B., and Daly, D. W., 1981, Forskolin: A unique diterpene activator of cyclic AMP-generating systems, *J. Cyclic Nucleic Res.* **14:**201–224.

Shultz, R. M., 1985, Roles of cell-to-cell communication in development, *Biol. Reprod.* **32:**27–42.

Smith, L. D., 1989, The induction of oocyte maturation: Transmembrane signaling events and regulation of the cell cycle, *Development* **107:**685–699.

Stinnakre, J., and Van Renterghem, C., 1986, Cyclic adenosine monophosphate, calcium, acetylcholine and the current induced by adenosine in the *Xenopus* oocyte, *J. Physiol. (London)* **374:**551–569.

Stutts, M. J., Chinet, T. C., Mason, S. J., Fullton, J. M., Clarke, L. L., and Boucher, R. C., 1992, Regulation of Cl^- channels in normal and cystic fibrosis airway epithelial cells by extracellular ATP, *Proc. Natl. Acad. Sci. USA* **89:**1621–1625.

Sumikawa, K., Parker, I., and Miledi, R., 1989, Expression of neurotransmitter receptors and voltage-activated channels from brain mRNA in Xenopus oocytes, in: *Methods in Neurosciences*, Vol. 1, Academic Press, Inc., San Diego, pp. 30–45.

Takahashi, T., Neher, E., and Sakman, B., 1987, Rat brain serotonin receptors in *Xenopus* oocytes are coupled by intracellular calcium to endogenous channels, *Proc. Natl. Acad. Sci. USA* **84:**5063–5067.

Tigyi, G., Dyer, D., Matute, C., and Miledi, R., 1990, A serum factor that activates the phosphatidylinositol phosphate signaling system in *Xenopus* oocytes, *Proc. Natl. Acad. Sci. USA* **87:**1521–1525.

Valverde, M. A., Diaz, M., Sepulveda, F. V., Gill, D. R., Hyde, S. C., and Higgins, C. F., 1992, Volume-regulated chloride channels associated with the human multidrug-resistance P-glycoprotein, *Nature* **355:**830–833.

van den Hoef, M. H. F., Dictus, W. J. A. G., Hage, W. J., and Bluemink, J. G., 1984, The ultrastructural organization of gap junctions between follicle cells and the oocyte in *Xenopus laevis*, *Eur. J. Cell Biol.* **33:**242–247.

Van Renterghem, C., Penit-Soria, J., and Stinnakre, J., 1984, β-adrenergic induced potassium current in *Xenopus* oocyte: Involvement of cyclic AMP, *Biochimie* **66:**135–138.

Van Renterghem, C., Penit-Soria, J., and Stinnakre, J., 1985, β-adrenergic induced potassium current in *Xenopus* oocytes: Role of cyclic-AMP, inhibition by muscarinic agents, *Proc. R. Soc. London Ser. B* **223:**389–402.

Wallace, R. A., and Misulovin, Z., 1980, The role of zinc and follicle cells in insulin-initiated meiotic maturation of *Xenopus laevis* oocytes, *Science* **210:**928–929.

Webb, D. J., and Nuccitelli, R., 1985, Fertilization potential and electrical properties of the *Xenopus laevis* egg, *Dev. Biol.* **107:**395–406.

Welsh, M. J., Anderson, M. P., Rich, D. P., Berger, H. A., Denning, G. M., Ostedgaard, L. S., Sheppard, D. M., Cheng, S. H., Gregory, R. J., and Smith, A. E., 1992, Cystic fibrosis transmembrane conductance regulator: A chloride channel with novel regulation, *Neuron* **8:**821–829.

Wibrand, F., Honoré, E., and Lazdunski, M., 1993, Opening of glibenclamide-sensitive K^+ channel in follicular cells promotes *Xenopus* oocyte maturation, *Proc. Natl. Acad. Sci. USA* **89:**5133–5137.

Woodward, R. M., and Miledi, R., 1987a, Hormonal activation of membrane currents in follicle enclosed *Xenopus* oocytes, *Proc. Natl. Acad. Sci. USA* **84:**4135–4139.

Woodward, R. M., and Miledi, R., 1987b, Membrane currents elicited by porcine vasoactive intestinal peptide (VIP) in follicle enclosed *Xenopus* oocytes, *Proc. R. Soc. London Ser. B* **231:**489–497.

Woodward, R. M., and Miledi, R., 1991, Angiotensin II receptors in *Xenopus* oocytes, *Proc. R. Soc. London Ser. B.* **244:**11–19.

Yao, Y., and Parker, I., 1993, Inositol trisphosphate-mediated Ca^{2+} influx into *Xenopus* oocytes triggers Ca^{2+} liberation from intracellular stores, *J. Physiol. (London)* **468:**275–296.

Yoshida, S., and Plant, S., 1991, A potassium current evoked by growth hormone-releasing hormone in follicular oocytes of *Xenopus laevis*, *J. Physiol. (London)* **443:**651–667.

Zhong, Y., and Kasson, B. G., 1994, Pituitary adenylate cyclase–activating polypeptide stimulates steroidogenesis and adenosine 3′,5′-monophosphate accumulation in cultured rat granulosa cells, *Endocrinology* **135:**207–213.

Zimmerman, H., 1994, Signalling via ATP in the nervous system, *Trends Neurosci.* **17:**420–426.

CHAPTER 7

CALCIUM-ACTIVATED POTASSIUM CHANNELS IN ADRENAL CHROMAFFIN CELLS

CHRISTOPHER J. LINGLE, CHRISTOPHER R. SOLARO, MURALI PRAKRIYA, and JIU PING DING

1. INTRODUCTION

The rich diversity of voltage-dependent K^+-selective ion channels emphasizes the central role that the K^+ channel superfamily plays in defining and modifying the electrical behavior of virtually all cells that express such channels (Rudy, 1988; Covarrubias *et al.*, 1991). As a consequence of differences in the rates and voltage dependence of activation, deactivation, and inactivation, various K^+-selective channels determine a variety of cellular electrical properties, including the rates of action potential repolarization, the duration of afterhyperpolarizations, the frequency of repetitive firing, and aspects of the resting potential. Furthermore, K^+ channels appear to be important targets of biochemical cascades that modulate ion channel function, since subtle alterations in K^+ channel function can markedly change the underlying electrical behavior of a cell.

Many cells, whether excitable or inexcitable, express K^+-selective channels whose activation requires the elevation of the submembrane Ca^{2+} concentration ($[Ca^{2+}]_i$). Such channels are of interest because they provide a means by which changes in concentration of an important intracellular signaling molecule can be rapidly coupled to regulation of cellular excitability. Since $[Ca^{2+}]_i$ will fluctuate both in response to influx of Ca^{2+} through Ca^{2+}-permeable ion channels and to biochemical cascades that mediate release of Ca^{2+} from intracellular stores, Ca^{2+}-dependent K^+ channels permit the status of the

CHRISTOPHER J. LINGLE, CHRISTOPHER R. SOLARO, MURALI PRAKRIYA, and JIU PING DING • Department of Anesthesiology, Washington University School of Medicine, St. Louis, Missouri 63110.

Ion Channels, Volume 4, edited by Toshio Narahashi, Plenum Press, New York, 1996.

internal biochemical milieu to influence the electrical activity of a cell. As with other K^+ channels, the specific properties of Ca^{2+}-dependent K^+ channels define their particular physiological roles.

Two major groups of Ca^{2+}-dependent K^+ channels with distinct pharmacological and physiological properties have been identified and are often found in the same cells (Romey and Lazdunski, 1984; Pennefather *et al.*, 1985; Ritchie, 1987a; Lancaster *et al.*, 1991). A third, less well-defined group also exists. One major group consists of voltage- and Ca^{2+}-dependent channels which have been termed maxi-K or BK (for big K^+) channels (Marty, 1981; Barrett *et al.*, 1982; Moczydlowski and Latorre, 1983). The large single-channel conductance of BK channels has made them a favorite subject for studies in artificial membrane bilayers and for detailed examination of gating behavior (Moczydlowski and Latorre, 1983; Vergara *et al.*, 1984; Barrett *et al.*, 1982; Magleby and Pallotta, 1983a,b; McManus and Magleby, 1988, 1991). Much is now known about permeation, blockade, and kinetic properties of these channels (reviews by Latorre, 1986; Blatz and Magleby, 1987; Latorre *et al.*, 1989; McManus, 1991), and these topics will not be addressed here. The second major group contains smaller conductance channels, often termed SK (for small K^+) channels (Blatz and Magleby, 1986; Lang and Ritchie, 1987; Lancaster *et al.*, 1991). SK channels are voltage-independent and, in some cases, are blocked by the bee-venom toxin apamin. The third group includes most Ca^{2+}-dependent K^+ channels which do not obviously fall into either of the other two groups. Because their single-channel conductances are intermediate to those of BK and SK channels, they have sometimes been termed IK channels. Representatives have been described in GH_3 cells (Lang and Ritchie, 1990b), T lymphocytes (Leonard *et al.*, 1992; Grissmer *et al.*, 1993), smooth muscle cells (Van Renterghem and Lazdunski, 1992), and granulocytes (Varnai *et al.*, 1993). This group will not be considered further in this chapter.

For several years we have been investigating the functional properties and physiological roles of Ca^{2+}-dependent K^+ channels in rat chromaffin cells. These cells express both the SK and BK types of Ca^{2+}-dependent currents (Neely and Lingle, 1992a,b). Furthermore, an unusual feature of rat chromaffin cells is that they express either of two BK channel variants: a non-inactivating form or an inactivating form (Solaro and Lingle, 1992; Herrington *et al.*, 1995; Solaro *et al.*, 1995a). Because inactivation is the most novel feature of this Ca^{2+}-dependent current, much of this article will focus on what we now know about inactivation of BK channels in chromaffin cells. In addition, we summarize work which describes the relative expression of both inactivating and non-inactivating variants within these cells, and we consider the functional significance of the two variants.

1.1. General Physiology and Pharmacology of BK Channels

The properties of BK channels have been the topic of several extensive reviews (Latorre, 1986; Latorre *et al.*, 1989; McManus, 1991) and will only be briefly summarized here. BK channels require both membrane depolarization and elevation of submembrane $[Ca^{2+}]$ for activation. Activation of BK channels appears to involve the binding of multiple Ca^{2+} ions (Moczydlowski and Latorre, 1983; Magleby and Pallotta, 1983a,b;

Golowasch et al., 1986; McManus and Magleby, 1991; Markwardt and Isenberg, 1992). The relationship of Ca^{2+}-dependent steps in channel activation to voltage-dependent steps remains an area of active investigation. One early activation model based on the properties of single BK channels in lipid bilayers proposed that the Ca^{2+}-binding steps involved in channel activation were intrinsically voltage-dependent (Moczydlowski and Latorre, 1983). Coupled with its simplicity, the strength of this general model has been its success in accounting for the voltage-dependent activation of BK channels over Ca^{2+} concentrations spanning at least four orders of magnitude. Any alternative proposal will have to be equally successful.

More recently, several studies have provided indirect evidence suggesting that the voltage-dependent and Ca^{2+}-dependent steps involved in BK channel activation are separable processes (Pallotta, 1985b; Blair and Dionne, 1985; Solaro and Lingle, 1992; Bulan et al., 1994). This assertion has been further supported by cloning of BK channels, which has shown that the core of the BK channel subunit is highly homologous to that of voltage-dependent K^+ channel subunits (Adelman et al., 1992; Butler et al., 1993). This suggests that the origin of voltage dependence in BK channels may be similar to that in other voltage-dependent K^+ channels. Furthermore, some of the apparent Ca^{2+} dependence of BK gating is influenced by a separable part of the channel subunit that is distinct from the presumably voltage-dependent core (Wei et al., 1994; Schreiber et al., 1994; Solaro et al., 1995b). Together, the above data have raised the possibility that Ca^{2+}-dependent steps in BK channel activation lack intrinsic voltage dependence and may be distinct from voltage-dependent transitions in channel activation.

Toxins which block BK channels have been useful tools for assessing the contribution of BK currents to normal cellular electrical activity. BK channels, in most cases, are blocked quite effectively by the scorpion toxins, iberiotoxin (IBX; Candia et al., 1992; Giangiacomo et al., 1992) and charybdotoxin (CTX; Miller et al., 1985). However, CTX can block other voltage-dependent K^+ channels (Sands et al., 1989; Leonard et al., 1992) or non-BK Ca^{2+}-dependent K^+ current (Leonard et al., 1992; Grissmer et al., 1993). Furthermore, CTX sensitivity may depend on posttranslational processing (Zagotta et al., 1989). As yet, the action of IBX appears specific to BK channels. Although a CTX-resistant BK channel variant has been isolated from brain and studied in lipid bilayers (Reinhart et al., 1989), CTX-resistant BK channels have not been widely observed in normal cells, with the possible exception of BK channels in posterior pituitary nerve terminals (Bielefeldt and Jackson, 1993).

1.2. Molecular Biology of Ca^{2+}- and Voltage-Dependent K^+ Channels

The recent cloning and expression of a BK channel gene, first from *Drosophila* (*dSlo*: Atkinson et al., 1991; Adelman et al., 1992) and subsequently from mouse (*mSlo*: Butler et al., 1993), have provided fresh perspective to issues concerning the diversity and functional properties of this channel class. BK channel variants have subsequently been isolated from specific tissues, including human vascular smooth muscle (McCobb et al., 1995) and human brain (Dworetsky et al., 1994; Pallanck and Ganetsky, 1994; Tseng-Crank et al., 1994). All BK channels cloned to date arise from a single *Slo* gene product which encodes a protein of approximately 1150–1200 amino acids. The first approximately 400 amino acids contain six hydrophobic segments which exhibit extensive homology with the six putative transmembrane-spanning segments of other voltage-dependent K^+ channels. In particular, both the H5 or P region, thought to form a portion of the ion channel pore, and the S4 region, thought to contribute to voltage-dependent gating, retain features characteristic of analogous domains in other voltage-dependent K^+ channels. The *Slo* channel subunit contains four additional hydrophobic segments downstream from the channel core. These have been designated S7–S10 (Butler *et al.*, 1993), although the topology of such segments is unknown.

   Slo channels from different species appear to differ in the number of consensus alternative splice sites. *Slo* channels from *Drosophila* contain at least five alternative splice sites (Adelman *et al.*, 1992; Lagrutta *et al.*, 1994), while *Slo* channels from mouse (Butler *et al.*, 1993), rat (Saito *et al.*, 1994), and human brain (Tseng-Crank *et al.*, 1994) contain two, two, and four alternative splice sites, respectively. The two splice sites in mouse and rat are shared with similar sites in the human sequence. For mammalian *Slo* genes, all known splice sites are located between the S7 and S10 hydrophobic domains. Multiple exons are available for insertion at each splice site, and in some cases, different exons have been reported to alter *Slo* channel gating behavior or shift the apparent Ca^{2+} dependence of activation (Adelman *et al.*, 1992; Lagrutta *et al.*, 1994; Saito *et al.*, 1994; Tseng-Crank *et al.*, 1994; McCobb *et al.*, 1995). Thus, variation at different alternative splice sites may be one mechanism by which the functional diversity of BK channels in normal cells arises.

1.3.   Diversity of BK Channels

   A remarkable feature of BK channels among different cell types is the rather widely variable apparent Ca^{2+} dependence of activation (reviewed by McManus, 1991). BK channels in chromaffin cells, for example, are open about half the time with 10 μM Ca^{2+} at about 0 mV. This is close to levels of activation seen for BK channels recorded in skeletal muscle (Barrett *et al.*, 1982) and rat sympathetic neurons (Smart, 1987). About 4 μM Ca^{2+} is sufficient for half-activation at 0 mV [$K_{50}(0)$] in cultured rat hippocampal neurons (Franciolini, 1988), turtle hair cells (Art *et al.*, 1995), and clonal pituitary GH_3 cells (Lang and Ritchie, 1987). In some smooth muscle tissues, the $K_{50}(0)$ is about 1 μM Ca^{2+} (e.g., Benham *et al.*, 1986; Singer and Walsh, 1987; Hu *et al.*, 1989). However, even more sensitive BK channels have been described (olfactory bulb neurons: $K_{50}(0)$ <50 nM, Egan *et al.*, 1993; pancreatic acinar cells: $K_{50}(0)$ < 100 nM, Maruyama *et al.*, 1983; mouse lacrimal acinar cells: $K_{50}(0)$ ~ 10 nM, Findlay, 1984). It has been suggested that some of the variability in reported Ca^{2+} sensitivities may arise because of differences in how Ca^{2+} concentrations were determined or other experimental factors (McManus, 1991). However, it is known that individual BK channels can exhibit marked differences in apparent Ca^{2+} sensitivity, even within a single preparation or membrane patch.

   As yet, molecular biological information about BK channels is insufficient to account for the diversity of BK channel function in mammalian cells. For the *dSlo* channel, alternatively spliced variants have substantial functional differences (Lagrutta *et al.*, 1994). However, for the mammalian splice variants that have been examined, the differences in gating behavior and/or apparent Ca^{2+} sensitivity are rather minor, involving at most about a 20-mV shift in gating at a given $[Ca^{2+}]$ (Tseng-Crank *et al.*, 1994; Saito *et al.*, 1994; McCobb *et al.*, 1995). It remains possible that additional variants exist which could account for further functional diversity.

Diversity among BK channels may also arise from association of the primary α subunit with an accessory β subunit originally purified from bovine tracheal smooth muscle (Knaus *et al.*, 1994a,b). This β subunit produces a marked shift in activation to more negative potentials at a particular Ca^{2+} concentration (McManus *et al.*, 1995; McCobb *et al.*, 1995) and may account for the higher sensitivity to $[Ca^{2+}]_i$ for BK channels in some smooth muscle cells. In other cells, however, the apparent Ca^{2+} sensitivity of cloned α subunits expressed in *Xenopus* oocytes is comparable to that of the BK channel in its native environment, e.g., BK channels in chromaffin cells (Saito *et al.*, 1994). This raises the possibility that a β subunit is not an obligatory component of the BK channel. As yet, there is no information as to whether there exists a family of β subunits whose members might affect BK gating in distinct ways.

A final factor which might contribute to BK channel diversity in native cellular environments is the existence of endogenous modulatory processes that might influence the apparent Ca^{2+} dependence and kinetic behavior of BK channels (Lechleiter *et al.*, 1988; Kume *et al.*, 1989; Reinhart *et al.*, 1991; White *et al.*, 1991; Chung *et al.*, 1991; Twitchell and Rane, 1993).

1.4. Properties of BK Channels that Are Important for Defining Potential Physiological Roles

Since most BK channels exhibit rapid and persistent activation in the continued presence of Ca^{2+} and depolarization, BK channels are thought to act as rapidly responding, passive sensors of membrane voltage and submembrane $[Ca^{2+}]$. Based on different apparent Ca^{2+} dependencies of BK channel activation among cells, two general categories of physiological roles for BK channels have been proposed. First, BK channels may participate in fast action potential repolarization. Second, persistent activation of BK channels near normal cell resting potentials by slight elevations of $[Ca^{2+}]_i$ may cause a reduction of cell excitability or, in smooth muscle, mediate muscle relaxation.

To assess possible physiological roles of BK channels, three properties of BK channels are important. First, the rate of channel activation at a given voltage and $[Ca^{2+}]$ is critical for determining whether BK channel activation that occurs during individual action potentials is sufficient to contribute to repolarization. Second, the rate of BK channel deactivation following repolarization defines the amplitude and duration of any afterhyperpolarization. Third, the steady-state probability that BK channels are open at a particular $[Ca^{2+}]_i$ and voltage defines the extent to which BK channels may contribute to conductance near the resting potential.

Evidence in support of a role for BK channels in action potential repolarization has been obtained in a number of systems. Generally, most evidence depends largely on the ability of BK channel blockers, e.g., tetraethylammonium (TEA) or CTX, to prolong action potentials. In sympathetic and hippocampal neurons, the kinetic properties of the tail current resulting from BK channel deactivation have been defined and, coupled with pharmacological evidence, are consistent with a role in rapid repolarization (Lancaster and Adams, 1986; Lancaster and Pennefather, 1987; Lancaster and Nicoll, 1987). Similarly, blockage of BK channels in GH_3 cells by CTX results in substantial action potential prolongation (Lang and Ritchie, 1990a), and single action potentials activate BK channels with high probability in cell-attached patches (Lang and Ritchie, 1987). This is direct evidence that BK channels are activated during the repolarization phase of an action potential. For each of these cell types, once the cell has repolarized below about -40 mV, BK channels deactivate rapidly, such that any afterhyperpolarization due to BK channel activity is quite brief (Adams *et al.*, 1982). Thus, in these cells, the Ca^{2+} and voltage dependence of BK channels is such that minimal BK channel activation is expected to occur near resting potentials, and the primary role of the current is simply to mediate fast action potential repolarization.

A different function of BK current has been described in particular molluscan neurons (Crest and Gola, 1993). In this system, BK current contributes to slow afterhyperpolarizations which influence repetitive firing rates by prolonging the interval between action potentials. Thus, the above examples make the point that depending on rates of deactivation at a particular voltage and $[Ca^{2+}]_i$, BK current can either solely mediate fast action potential repolarization or contribute to slow afterhyperpolarizations that influence cell firing rates. A BK channel has also been shown to mediate frequency dependent action potential failure in posterior pituitary nerve terminals (Bielefeldt and Jackson, 1993). This BK channel is half-activated at 0 mV with less than 250 nM $[Ca^{2+}]_i$. It is in the cells, in which BK currents have been reported to have higher apparent Ca^{2+} sensitivity (Findlay, 1984; Maruyama *et al.*, 1983; Singer and Walsh, 1987; Egan *et al.*, 1993), that one might expect a more substantial role for BK current in slow afterhyperpolarizations. Unfortunately, how BK channels participate in the normal voltage behavior of such cells is not yet directly known.

1.5. Physiology and Pharmacology of SK Channels

A number of cell types contain K^+ currents which require increases in submembrane Ca^{2+} for activation but which exhibit no obvious voltage dependence. Such currents give rise to slow afterhyperpolarizations which play an important role in spacing action potentials and in limiting the frequency of repetitive firing (Lang and Ritchie, 1990a). In other cases, slow activation of SK current may contribute to spike frequency adaptation (Goh and Pennefather, 1987; Lancaster and Nicoll, 1987). The activation of these currents is also prominent in many endocrine cells during the elevation of cytosolic Ca^{2+} triggered by release of Ca^{2+} from intracellular stores (Ritchie, 1987b; Horn and Marty, 1988; Tse and Hille, 1992; Neely and Lingle, 1992b). The purpose of SK current activation in such cases remains unclear.

Specific information about the properties of the single channels that produce the macroscopic Ca^{2+}-dependent SK currents is somewhat limited. Among different cells, such channels typically share a lack of voltage dependence and a similar small single-channel current amplitude (Blatz and Magleby, 1986; Lang and Ritchie, 1987; Lancaster *et al.*, 1991; Park, 1994). SK currents typically appear to be about half-activated in the range of 0.5–1.0 μM $[Ca^{2+}]_i$. Differences in pharmacological properties of SK channels occur among tissues. For example, the slow Ca^{2+}-dependent afterhyperpolarization in sympathetic neurons is apamin-sensitive (Goh and Pennefather, 1987), but a similar afterhyperpolarization in hippocampal neurons is resistant to apamin (Lancaster and Nicoll, 1987). In guinea pig vagal neurons, both apamin-sensitive and apamin-resistant components of the afterhyperpolarization are observed (Sah, 1992). In the absence of information about the genes encoding SK channels, it remains uncertain whether all such currents arise from a closely related family of ion channels.

2.   CALCIUM-DEPENDENT K$^+$ CURRENT IN CHROMAFFIN CELLS

Chromaffin cells, like pituitary and many other endocrine cells, exhibit Ca^{2+}-dependent exocytosis which can result from two distinct mechanisms of cytosolic $[Ca^{2+}]$ elevation. In one case, cytosolic $[Ca^{2+}]$ is elevated by changes in electrical activity and increased Ca^{2+} influx through voltage-dependent Ca^{2+} channels (Fenwick et al., 1982). In the other case, activation of particular cell surface receptors, e.g., muscarinic acetylcholine receptors (mAChRs) or bradykinin receptors, leads to activation of phospholipase C, elevation of inositol triphosphate (Eberhardt and Holz, 1987; Malhotra et al., 1988), and the release of Ca^{2+} stored in intracellular organelles (Stoehr et al., 1986). In our studies we have examined the activation and inactivation of Ca^{2+}-dependent current in response to elevation of cytosolic $[Ca^{2+}]$ initiated either by depolarization-stimulated Ca^{2+} influx or via secretagogue-induced release from intracellular stores (Neely and Lingle, 1992b; Herrington et al., 1995; Solaro et al., 1995a). We have attempted to define the types of Ca^{2+}-dependent current present in rat chromaffin cells, the properties of those currents, and the contributions of those currents to the electrical behavior of the cells (Neely and Lingle, 1992a; Solaro and Lingle, 1992; Solaro et al., 1995a).

2.1.   Rat Chromaffin Cells Express both BK and SK Ca^{2+}-Dependent Currents

The first description of a Ca^{2+}-dependent K$^+$ channel in any mammalian cell was of BK channels in excised patches from bovine chromaffin cells (Marty, 1981). Blockade of most of the Ca^{2+}-dependent outward current in bovine chromaffin cells by 1 mM TEA suggested that most, if not all, Ca^{2+}-dependent K$^+$ current arose from BK channels (Marty and Neher, 1985), although a more recent study indicates that bovine chromaffin cells do express some SK current (Artalejo et al., 1993).

Rat chromaffin cells also express the two major categories of Ca^{2+}-dependent K$^+$ current: a voltage-independent SK current sensitive to both apamin and curare and a voltage-dependent BK current. Both currents can be activated either by Ca^{2+} influx resulting from membrane depolarization (Neely and Lingle, 1992a) or by secretagogue-mediated elevation of $[Ca^{2+}]_i$ (Neely and Lingle, 1992b; Herrington et al., 1995). However, activation of BK current during secretagogue-induced elevation of Ca^{2+} requires concomitant membrane depolarization usually above -20 mV.

In rat chromaffin cells, current-voltage curves generated from peak outward currents activated by depolarizing voltage steps reveal a typical Ca^{2+}-dependent N shape

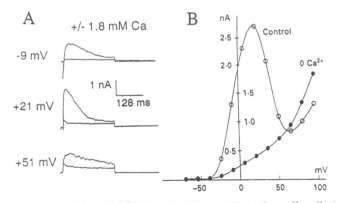

FIGURE 1. Current-voltage relations of Ca^{2+}-dependent K$^+$ current in rat chromaffin cells. (A) Traces show current elicited by the indicated voltage steps in the presence (1.8 mM) and absence of Ca^{2+}. Rapid inward current in traces at −9 and 21 mV is voltage-dependent Na$^+$ current. (B) Peak outward current elicited in another cell is plotted as a function of command potential, showing the Ca^{2+}-dependent N shape of the outward current.

relation (Fig. 1; Neely and Lingle, 1992a). This N shape was first described for bovine chromaffin cells (Marty and Neher, 1985) and reflects the dependence of Ca^{2+}-dependent outward current on activation of Ca^{2+} current. The outward current activated by Ca^{2+} influx contains two pharmacologically distinct components. One component is sensitive to 200 nM apamin, while the other is largely blocked by 1 mM TEA (Fig. 2). Similarly, when cytosolic [Ca^{2+}] is elevated by mAChR-induced release from cytosolic stores, Ca^{2+}-dependent K$^+$ current is also activated (Neely and Lingle, 1992b). This current can also be dissected into two pharmacologically distinct components (Fig. 3), each activated over a particular range of membrane potentials. At more negative membrane potentials (negative to −30 mV), current activated during the response to mAChR activation is largely voltage-independent and is almost completely blocked by 200 nM apamin or 200 μM curare. At potentials positive to −30 mV, a voltage-dependent current sensitive to TEA is also activated. To examine the ability of muscarine to affect Ca^{2+}- and voltage-dependent current, we have stimulated cells using repetitive voltage steps to about +80 mV. In the absence of muscarine, this protocol results in no activation of Ca^{2+}-dependent current. However, during mAChR-induced elevation of [Ca^{2+}]$_i$, steps to +80 mV result in the robust activation of voltage-dependent outward current, which is almost completely blocked by 10 nM IBX (Fig. 3). The voltage dependence and sensitivity of this current to 1 mM TEA also indicate that this current results from activation of BK channels (Neely and Lingle, 1992b). This work also demonstrated that both the voltage-dependent and voltage-independent components of Ca^{2+}-dependent current reflect activation of K$^+$-selective currents. A more recent study has defined the selectivity of rat chromaffin cell SK current for monovalent cations (Park, 1994).

During the mAChR-induced [Ca^{2+}]$_i$ elevation, many chromaffin cells exhibit a response that differs from that shown in Fig. 3. Specifically, following an initial

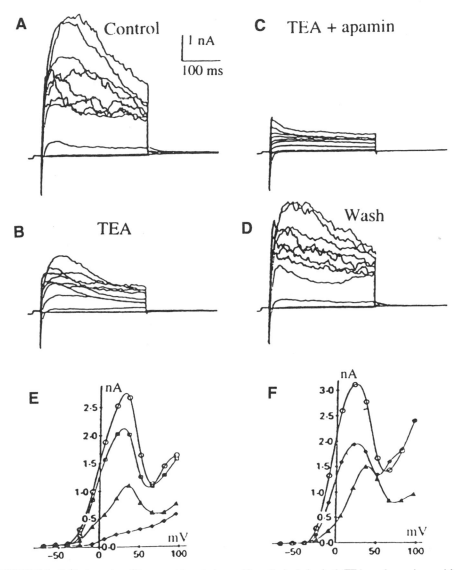

FIGURE 2. Ca^{2+}-dependent K^+ current in rat chromaffin cells includes both TEA- and apamin-sensitive components. (A–D) Panels show currents activated by voltage steps from -69 mV through $+96$ mV for control saline (A and D), 5 mM TEA (B), and 5 mM TEA plus 200 nM apamin (C). (E) The peak current activated during different command steps is shown for control saline (open circles), 5 mM TEA (filled triangles), 5 mM TEA plus 200 nM apamin (filled diamonds), and wash (open squares). (F) Currents were activated in control saline (open circles), 5 mM TEA (filled diamonds), or 200 nM apamin (filled triangles). (Modified from Neely and Lingle, 1992a.)

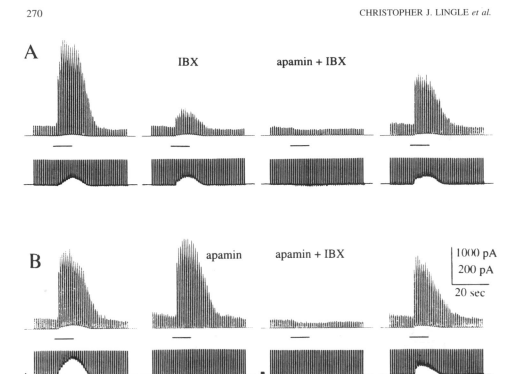

FIGURE 3. Apamin and iberiotoxin block separate components of current activated during mAChR-induced elevation of cytosolic $[Ca^{2+}]$. The cell was held at -49 mV and the membrane potential was stepped repetitively to $+81$ mV for 256 msec every 1 sec. Muscarine (50 μM) was applied every 4–5 min for the periods indicated by bars. Bottom traces show high gain records of current at the holding potential while upper traces emphasize current activated by voltage steps to $+81$ mV. (A) 10 nM IBX has no effect on current activated at -49 mV but blocks most current at $+81$ mV. (B) 200 nM apamin blocks virtually all outward current at the holding potential while having a small effect on current at $+81$ mV.

activation of voltage-dependent outward current, there is a marked suppression of outward current at the peak of the response to muscarine (Fig. 4A). In fact, the current-voltage relationship in the presence of muscarine becomes largely linear (Neely and Lingle, 1992b), reflecting almost exclusively the persistent activation of SK current. This suppression of voltage-dependent current results from the intrinsic inactivation behavior of the BK current found in most rat chromaffin cells (Herrington *et al.*, 1995). As shown in Fig. 5, single BK channels in patches from most rat chromaffin cells exhibit rapid inactivation (Solaro and Lingle, 1992). It is the inactivation of BK channels that is responsible for the marked suppression of whole-cell outward current seen in Fig. 4A (Herrington *et al.*, 1995). This phenomenon is also illustrated in more detail in Figs. 8 and 9.

In sum, rat chromaffin cells express both SK and BK Ca^{2+}-activated K^+ channels. Unlike BK channels in most other cells, most chromaffin cell BK channels inactivate. Because of this unusual aspect of Ca^{2+}-dependent K^+ current, the remainder of this

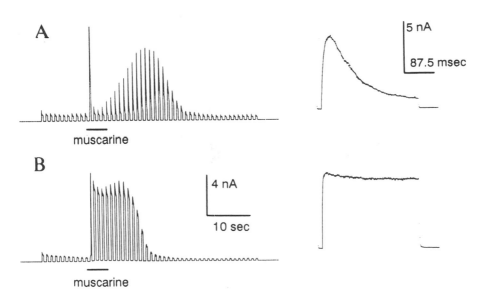

FIGURE 4. Inactivating and sustained voltage-dependent current activated during mAChR-induced eleva-tion of $[Ca^{2+}]_i$. A and B represent different cells. The holding potential was -49 mV and the membrane potential was stepped to $+81$ mV for 256 msec every 1.5 sec. Muscarine (50 μM) was applied as indicated resulting in elevation of cytosolic $[Ca^{2+}]$. In A, the first initial surge of current following the application of muscarine is truncated. On the right, current activated during a single voltage step (step 24 in A; step 16 in B) is shown on a faster time base to illustrate the pronounced inactivation of current for the cell in A and the relatively sustained current for the cell in B. Apamin (200 nM) was included in all salines. (Modified from Herrington *et al.*, 1995.)

article will focus on what is now known about BK channel inactivation and the possible physiological roles for this BK variant.

2.2.   Rat Chromaffin Cells Express Two Variants of BK Channel

Rat chromaffin cells abundantly express BK channels. In most patches (76%) from rat chromaffin cells, BK channels undergo relatively rapid and complete inactivation during depolarizing voltage steps in the presence of constant submembrane $[Ca^{2+}]$ (Solaro and Lingle, 1992; Solaro *et al.*, 1995a). We term these inactivating channels BK_i channels and the resulting current, BK_i current. A smaller percentage of patches (9%) contain BK channels which exhibit a more typical sustained activity during constant depolarization (Fig. 5). We term these BK_s channels. The remaining patches (15%) contain both BK_i and BK_s channels. Since most patches with exclusively BK_i channels contain three to ten channels, this distribution of BK_i and BK_s channels among patches cannot be explained by the random distribution in chromaffin cell membranes of two channel variants. Rather, the largely selective presence of either BK_i or BK_s channels among patches suggest either that similar types of channels are tightly associated in the

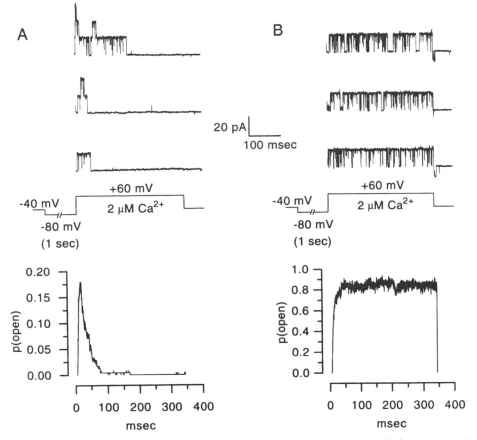

FIGURE 5. Either exclusively inactivating or non-inactivating BK channels are usually found in excised inside-out patches from rat chromaffin cells. In both A and B, an excised patch was stimulated every 3 sec with the protocol shown below the raw current traces. Ensemble currents are shown below. Subsequent treatment with trypsin revealed that the patch on the left contained four channels. The patch on the right contained one channel. (Modified from Solaro *et al.*, 1995a.)

membrane or that each channel type may be selectively associated with different cells. The latter possibility was shown to be the case using whole-cell recordings (Solaro *et al.*, 1995a).

Whole-cell recordings of Ca^{2+}-dependent K^+ current typically provide limited information about the properties of the underlying current because of uncertainty about the time course of Ca^{2+} current activation, inactivation, and cytosolic Ca^{2+} regulation. To assay the properties of whole-cell BK current in chromaffin cells, we have utilized three protocols, each of which robustly elevates $[Ca^{2+}]_i$ in distinct ways. In one case, muscarine is used to release Ca^{2+} from intracellular stores (Fig. 4); in the second, a prolonged depolarizing voltage step above the threshold for Ca^{2+} current activation is used to activate Ca^{2+} influx (Fig. 6); in the third, Ca^{2+} is directly introduced into the cell

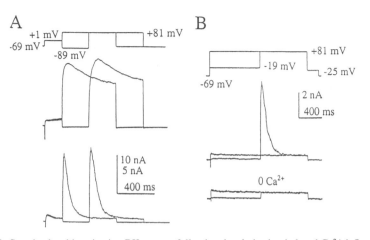

FIGURE 6. Sustained and inactivating BK current following depolarization-induced Ca^{2+} influx. (A) Ca^{2+} influx was activated by a 200 msec depolarizing loading step to +1 mV. Following the loading step, the membrane potential was stepped either directly to +81 mV or to −69 mV prior to a step to +81 mV. For the cell on the top, outward current activated at +81 mV is largely sustained, while on the bottom the BK current completely inactivates. The fact that activation of BK current occurs in both cases following 300 msec of recovery at −69 mV indicates that [Ca^{2+}]$_i$ remains sufficiently elevated to produce robust activation of current. This shows that the rapid decay of BK current on the bottom reflects inactivation and not decay of [Ca^{2+}]$_i$. (B) Currents were activated in a BK$_i$ cell by stepping to +81 mV with or without a loading step to −19 mV. Following the loading step, the step to +81 mV results in activation of a current that completely inactivates. The traces on the bottom were taken after removal of extracellular Ca^{2+} indicating that all of the inactivating voltage-dependent current is Ca^{2+}-dependent. (Modified from Solaro *et al.*, 1995a; Herrington *et al.*, 1995.)

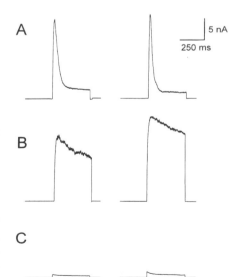

FIGURE 7. Inclusion of [Ca^{2+}]$_i$ in the recording pipette reveals either inactivating or non-inactivating outward current in rat chromaffin cells. Traces show examples of currents elicited by voltage steps to +87 mV following introduction of 4 µM Ca^{2+} into the cells from the recording pipette. Most cells exhibit inactivating current similar to that illustrated in A, but some cells show largely non-inactivating currents as seen in B. If the pipette saline contains 80 µM EGTA with no added Ca^{2+}, maximal current at +87 mV is typically less than 1 nA as shown in C. (Modified from Solaro *et al.*, 1995a.)

through the recording pipette (Fig. 7). Following the elevation of $[Ca^{2+}]_i$, when the cell membrane potential is stepped to $+80$ mV or above (near the Ca^{2+} equilibrium potential where Ca^{2+} influx is minimal) the properties of the BK current can be viewed with minimal contamination by other currents (Herrington *et al.*, 1995). Once cytosolic $[Ca^{2+}]$ is elevated, irrespective of the method, depolarizing steps applied to most rat chromaffin cells (~80%) result in activation of a Ca^{2+}- and voltage-dependent K^+ current which exhibits subsequent inactivation (Figs. 4, 6, and 7). Such current typically inactivates almost completely within about 300 msec. In contrast, a smaller set of cells (~20%) exhibits a more slowly inactivating or relatively non-inactivating BK current.

2.3. Properties of Macroscopic Inactivating BK Current Reflect Underlying Single-Channel Characteristics

That the macroscopic inactivating currents arise from the intrinsic properties of inactivating BK channels is supported by a number of arguments (Herrington *et al.*, 1995). First, the rates of inactivation onset are similar. In particular, during robust elevation of cytosolic $[Ca^{2+}]$ initiated by voltage steps, pipette dialysis, or mAChR activation, the outward current inactivates during a voltage step to $+90$ mV with a time constant of about 30–60 msec. This is similar to the 20–40 msec inactivation time constant measured from ensembles of BK_i single channels. Second, as illustrated in Fig. 8, the rates of recovery from inactivation of BK_i single channels are comparable to the rates of recovery of macroscopic inactivating currents (Herrington *et al.*, 1995). Third, the sensitivity to TEA and scorpion toxins is similar. Fourth, the steady-state inactivation properties of the single channels and the whole-cell currents appear similar under conditions in which channels are likely to be exposed to comparable concentrations of Ca^{2+} (Fig. 9). Thus, the phenotypic properties of whole-cell BK current appear to arise from properties intrinsic to the underlying BK channels.

Several observations suggest that inactivation is a stable property of the particular BK channels present in a patch or in a cell. First, upon patch excision we have never observed conversion of inactivating channels into non-inactivating channels or *vice versa*. Similarly, there are no slow changes in the properties of inactivation following excision. In perforated-patch recordings, when mAChR receptors are repeatedly activated to produce marked elevation of cytosolic $[Ca^{2+}]$, the properties of BK current, whether inactivating or non-inactivating, are unchanged even over several hours. Although it is provocative to suggest that mechanisms might exist for switching the behavior of a channel from inactivating to non-inactivating and back, our data at present would argue that if such a process does exist, it occurs very rarely under the conditions of our investigations.

In sum, rat chromaffin cells express two phenotypically distinct forms of BK current which correlate with the underlying properties of single BK channels (Solaro *et al.*, 1995a). However, it should be remembered that variation in the rate and extent of inactivation does exist to some extent among cells. A future question of interest is whether such variability may arise from heteromultimeric channels or other, as yet unknown regulatory processes.

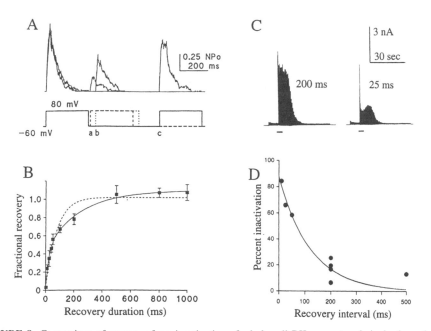

FIGURE 8. Comparison of recovery from inactivation of whole-cell BK_i current and single-channel ensemble BK_i current. (A) Ensemble currents were generated with the stimulation protocol shown below and plotted as $N*P_o$ for an inside-out patch containing BK_i channels. Following complete inactivation at $+80$ mV, channels were allowed to recover from inactivation for different intervals (a, 10 msec; b, 50 msec, c, 500 msec) at -60 mV in 2 μM Ca^{2+}. (B) Fractional recoveries of BK_i current at -60 mV and 2 μM Ca^{2+} are plotted for 19 patches. The dashed line represents a single exponential fit of 74.6 msec, while the solid line represents a two-exponential fit with a fast component of 22.2 msec (36%) and a slow component of 248.5 msec. (C) 250 msec voltage steps to $+81$ mV were separated by recovery steps to -49 mV of either 200 or 25 msec duration. Fifty μM muscarine was applied for the periods indicated by the bars, and the external saline included 200 nM apamin. Each vertical deflection of current results in activation and inactivation of BK current (e.g., Fig. 4A). With the longer recovery time (200 msec), most current that inactivates during a step to $+81$ mV recovers from inactivation during the interval at -49 mV. In contrast, 25 msec at -49 mV allows for little recovery from inactivation. (D) The percent inactivation during the maximal suppression of BK current relative to the initial amount of BK current activated during the response to muscarine is plotted versus the recovery time between voltage steps. A single exponential fit to the points yields a time constant of 124.3 ± 30.4 msec. Based on rates of inactivation of BK current at $+81$ mV during the muscarine-induced Ca^{2+} elevation, submembrane Ca^{2+} was estimated to be 2–5 μM. (Modified from Herrington et al., 1995.)

3. INACTIVATION OF BK_i CHANNELS: DOES A "BALL AND CHAIN" MODEL APPLY?

We will now address current information about the mechanism of inactivation of BK_i channels and how it compares to inactivation mechanisms of other voltage-dependent currents. We will begin with a brief summary of voltage-dependent K^+ channel inactivation.

Many voltage-dependent K^+ channels undergo inactivation that occurs on the time

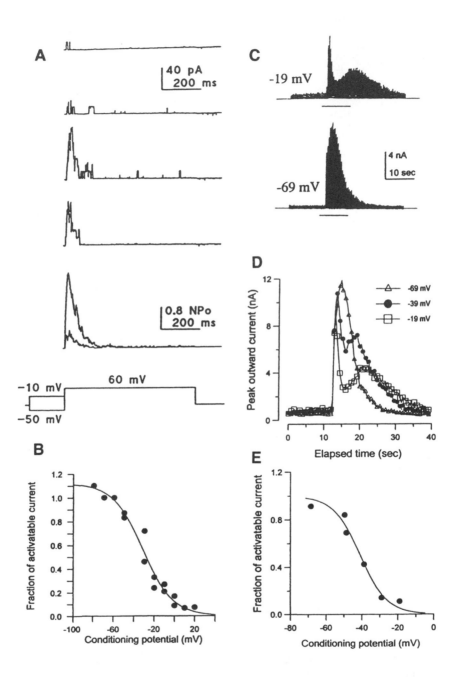

scale of 1–30 msec. The molecular basis of this rapid inactivation has been elegantly demonstrated through a combination of molecular biological and biophysical approaches and has been shown to arise from either of two mechanisms. The first studies evolved from the proposal of Armstrong and Bezanilla (1977) that Na^+ channel inactivation was caused by some physical component of the ion channel structure preventing ion permeation. Using the rapidly inactivating *Shaker*B (ShB) K^+ channel variant, Aldrich and associates were able to demonstrate that a particular cytosolic component of the channel located at the N-terminus was necessary and sufficient to produce fast inactivation (Hoshi *et al.*, 1990; Zagotta *et al.*, 1990). This "N-type" inactivation also occurs in some rapidly inactivating mammalian K^+ channels, including mammalian *Shaker* and *Shaw* homologues (Ruppersberg *et al.*, 1991). Interestingly, there is no amino acid homology between different K^+ channel inactivation–competent N-terminals, although both positive charge and hydrophobicity appear to be important (Murrell-Lagnado and Aldrich, 1993). The second form of inactivation has been shown to arise from association of *Shaker*-type (K_v1) subunits with an accessory β subunit ($K_v\beta1$) which contains an N-terminal domain that occludes the channel following channel activation (Rettig *et al.*, 1994). In both mechanisms, the inactivation or "ball" domain is thought to interact with sites close to the cytosolic mouth of the permeation pathway, thereby mediating blockade of the channel (Isacoff *et al.*, 1991).

Several arguments support the idea that N-terminal inactivation results from movement of the ball domain into a position occluding the mouth of the open ion permeation pathway. First, TEA and other quaternary amines block the *Shaker* channel from the cytosolic face of the channel by binding to a site which senses approximately 20% of the electrical field of the membrane (Choi *et al.*, 1993). Blockade by TEA in this site slows the rate of inactivation mediated by the ball domain (Choi *et al.*, 1991) and also competes for blockade of the channel by an exogenously applied peptide (ball peptide or BP) that corresponds to 22 amino acids at the *Shaker*B N-terminus (Murrell-Lagnado and Aldrich, 1993). Thus, when TEA occupies its binding site, it effectively prevents interaction

---

FIGURE 9. Comparison of steady-state inactivation of whole-cell BK_i current and single-channel ensemble BK_i current. (A) The top four sweeps show single traces of BK_i current activated in a single patch by steps to $+60$ mV with 1 μM Ca^{2+} following a conditioning step at either -10 mV (top two traces) or -50 mV (bottom two traces). Traces of average N^*P_o determined from many such traces are shown below. (B) The ensemble current amplitude normalized to current from -60 mV conditioning potential is plotted as a function of conditioning potential. The Boltzmann fit yielded a voltage of 50% inactivation of -34 mV and a slope factor of -13.4 mV. (C) Voltage-dependent current was activated by repetitive 40 msec steps to $+81$ mV and cytosolic Ca^{2+} was elevated by application of 50 μM muscarine as in Fig. 4. Following muscarine application, at the more positive holding potential, there is an initial activation of BK current followed by a marked suppression of BK current. (D) Peak current elicited by individual voltage steps to $+81$ mV (as in C) was plotted as a function of elapsed time for responses to muscarine. The amount of current during the maximal suppression of BK current was determined and compared to the peak BK current during the response to muscarine. This was used to generate the plot of fractional inactivation as a function of holding potential shown in E. The voltage of half-inactivation from the fit of a single Boltzmann was -39.3 ± 5.4 mV with a slope factor of -11.9 mV. As in Fig. 8, based on rates of BK_i current inactivation during voltage steps to $+81$ mV, submembrane $[Ca^{2+}]$ is expected to be in the range of 2–5 μM during the peak of the response to muscarine. (Modified from Herrington *et al.*, 1995.)

of the inactivation domain with its binding site. Second, N-type inactivation decreases with high concentrations of extracellular permeable ions (Demo and Yellen, 1991), suggesting that binding of ions in the permeation pathway decreases the stability of the interaction of the ball domain with its binding site. These two pieces of information are most easily explained if the ball domain interacts with the cytosolic mouth of the permeation pathway, close to the TEA and K^+ ion binding sites. A recent study has extended the above analysis to examine the effects of extracellular ions which can occupy but not fully traverse the ion permeation pathway (Gomez-Lagunas and Armstrong, 1994b). Ions that bind more than 50% of the way into the permeation pathway facilitate recovery from inactivation, while those that bind less deeply in the permeation pathway do not affect the inactivation process. These results again support the idea that either through electrostatic interactions or because of conformational consequences resulting from channel occupancy, the stability of the interaction of the inactivation domain with its binding site is reduced. A final argument that inactivation domains directly occlude the mouth of the ion permeation pathway is that residues which are thought to line the internal face of the ion permeation pathway have been shown to influence both the rate of inactivation and the binding of the "ball" peptides to the *Shaker* channel (Isacoff *et al.*, 1991).

The ability of the inactivation domain to bind to its blocking site reveals interesting information about conformational changes occurring during channel gating. For N-type inactivation, binding of the inactivation domain to its blocking site is strongly favored by opening of the ion channel. Thus, in the absence of channel activation there is little detectable inactivation. Similarly, binding of the inactivation domain effectively prevents the normal *Shaker* channel deactivation process (Demo and Yellen, 1991; Gomez-Lagunas and Armstrong, 1994a). As a consequence, during recovery from inactivation, a *Shaker* channel transiently reopens following dissociation of the inactivation domain from its binding site before the channel finally closes. This has also been demonstrated for both *Shaker* ($K_V 1.4$) and *Shaw* mammalian homologues (Ruppersberg *et al.*, 1991). However, a second component of fast inactivation in the *Shaker* channel, which does not exhibit sensitivity to external ions and does not require that the channel pass through a conducting state during recovery from inactivation, has also been observed (Gomez-Lagunas and Armstrong, 1994a).

Inactivation of *Shaker* channels results from the independent action of four inactivation domains (MacKinnon *et al.*, 1993; Gomez-Lagunas and Armstrong, 1994b), each associated with one subunit of the channel tetramer (MacKinnon, 1991). Occlusion by any one domain is sufficient for inactivation. By altering channels to contain less than four inactivation domains, it is possible to show that the rate of onset of inactivation depends on the number of inactivation domains associated with each ion channel tetramer, while the rate of recovery from inactivation is independent of the number of functional inactivation domains (MacKinnon *et al.*, 1993; Gomez-Lagunas and Armstrong, 1994a).

Work described in the next several sections addresses the extent to which inactivation of BK_i channels is similar to N-terminal inactivation of voltage-dependent K^+ channels.

3.1. Inactivation of BK_i Channels Is Rapid and Exhibits Apparent Ca^{2+} and Voltage Dependence

The basic properties of inactivation of BK_i channels have been studied using ensemble averages of single BK_i openings activated at multiple voltages and different submembrane $[Ca^{2+}]$ (Solaro and Lingle, 1992). As submembrane $[Ca^{2+}]$ is increased (Fig. 10A, B) or the command voltage made more positive (Fig. 10D, E), the rate of inactivation of the ensemble average current increases until it reaches a limiting value at

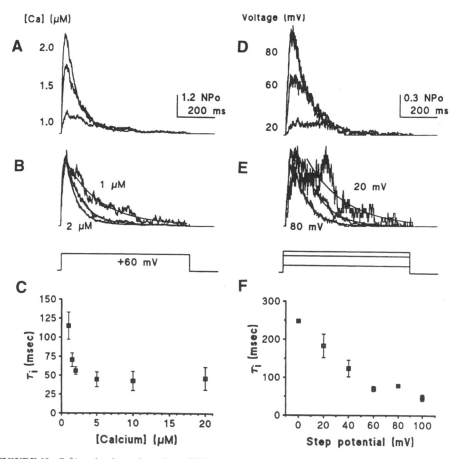

FIGURE 10. Ca^{2+} and voltage dependent of BK_i inactivation. Ensemble average currents of BK_i channels were generated as in Fig. 5. In all cases an inside-out patch was held at -60 mV. In A, the potential was stepped to $+60$ mV in the presence of 1, 1.5, or 2 μM Ca^{2+}. In B, the normalized currents with a single exponential fit to the current decay are shown. In C, the mean time constant of decay as a function of Ca^{2+} is plotted for estimates from eight patches. In D, E, and F equivalent information is plotted for the variation in inactivation time constant as a function of command potentials. In D, the patch was exposed to 2 μM Ca^{2+} and ensembles were obtained at $+20$, $+60$, and $+80$ mV. In E, the fitted normalized currents are shown. In F, the variation in the time constant of inactivation as a function of command potential is shown. (From Solaro and Lingle, 1992)

either elevated [Ca^{2+}] (Fig. 10C) or positive potentials (Fig. 10F). The fastest time constants of inactivation among patches range from about 20 to 40 msec at room temperature. These results suggest that the intrinsic rate of inactivation exhibits only weak voltage or Ca^{2+} dependence. Analogous to explanations for the voltage dependence of inactivation of voltage-dependent K^+ channels (Zagotta, Hoshi, and Aldrich, 1990), the apparent voltage and Ca^{2+} dependence of BK_i inactivation may derive from voltage- and Ca^{2+}-dependent activation steps.

3.2.   Inactivation of BK_i Channels Is Trypsin-Sensitive

Similar to N-terminal inactivation of the *Shaker*B K^+ channel (Hoshi *et al.*, 1990), inactivation of BK_i channels can be rapidly and completely removed by a brief application of trypsin to the cytosolic side of a patch (Solaro and Lingle, 1992; also Figs. 13 and 14). Similar applications of trypsin externally or to patches containing BK_s channels produce no obvious functional alterations in gating behavior. The sidedness of the trypsin effect and its rapid time course suggest that BK_i channels, or an associated subunit, contain readily accessible cytoplasmic residues suitable for digestion by trypsin that are important for maintaining the inactivation process.

3.3.   *Shaker*B Ball Peptides (BP) Block BK Channels

In addition to blocking voltage-dependent K^+ channels, the BP derived from the *Shaker*B channel is also able to block BK channels (Solaro and Lingle, 1992; Toro *et al.*, 1992; Foster *et al.*, 1992). In contrast, a peptide containing a single altered amino acid (L7E) does not block BK channels. This result suggests that BK channels may share recognition sites for BP molecules with voltage-dependent K^+ channels. In rat chromaffin cells, the effect of *Shaker*B BP has been examined on BK_i channels following removal of inactivation by trypsin (Solaro and Lingle, 1992). Application of 50 to 100 μM *Shaker*B BP to the cytoplasmic face of the channel reduces the amplitude of ensemble average BK_i single-channel currents by about 50%. In contrast to the effect of the BP on *Shaker* channels in which the ball produces a time-dependent block resulting in currents with a normal inactivation time course (Zagotta *et al.*, 1990), little time dependence was associated with the effect of BP on BK_i current time course. Although the lack of time dependence may imply that the peptide recognizes both open and closed conformations of the BK_i channel equally well, the currents were studied at relatively low concentrations of Ca^{2+}. As a consequence, any rapid, time-dependent blocking effect of the peptide may have been obscured by the slow activation rate of the channels.

The issue of whether the *Shaker* BP preferentially associates with the open conformation of BK channels has been addressed using cloned *mSlo* channels. As shown in Fig. 11, when *mSlo* channels are rapidly activated by a depolarizing voltage step to +90 mV in the presence of 60 μM Ca^{2+}, 200 μM *Shaker*B BP results in a clear time-dependent reduction of the macroscopic current. However, with 10 μM Ca^{2+}, 200 μM BP produces only a small time-dependent relaxation. A Hodgkin-Huxley activation model (Hodgkin and Huxley, 1952) was fit to the currents in the absence of BP to give an activation time constant (τ_m) and cooperativity factor (n). This model poorly describes many aspects of the *mSlo* current rising phase but is useful for the present purposes.

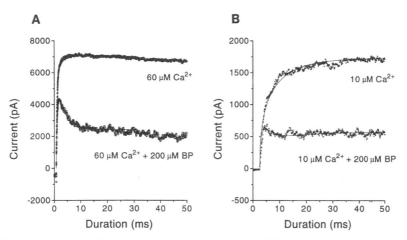

FIGURE 11. Blockade of *mSlo* currents by *Shaker*B ball peptide (BP). *mSlo* channels were expressed in *Xenopus* oocytes and currents were activated by voltage steps from −40 mV to +90 mV in inside-out patches. On the left, the currents were elicited with 60 μM Ca^{2+} with or without 200 μM *Shaker*B ball peptide. On the right for a different patch, currents were elicited with 10 μM Ca^{2+} with or without 200 μM BP. Both pipette and bath solution contained 160 mM K^+. Currents were fit with a Hodgkin-Huxley activation model. On the left, without BP, $\tau_m = 0.49$ msec, n = 1.13, $i_{max} = 6420$; with BP, τ_m and n were constrained to 0.49 msec and 1.13, respectively, resulting in $i_{max} = 5218$ with τ_i (an inactivation time constant) = 3.02 ± 0.09 msec with the fraction of unblocked channels = 0.35 at t = infinity. On the right, for 10 μM Ca^{2+}, without BP, $\tau_m = 9.59$, n = 0.38, and $i_{max} = 1745$. With BP, constraining $\tau_m = 9.59$, n = 0.38, and $i_{max} = 1745$, $\tau_i = 3.11 \pm 0.07$ msec and the unblocked fraction = 0.33. With 60 μM Ca^{2+}, 600 pA of ohmic current due to activation of channels at −40 mV was excluded from the fitted portion of the current. In 10 μM Ca^{2+}, there is little activation at −40 mV.

Currents in the presence of BP were then fit to a related model which included an inactivation time constant and a term for fractional block at steady state.

For both 10 and 60 μM Ca^{2+}, 200 μM BP results in a blocking time constant of about 3.0 msec with about 66% of the channels blocked at steady state. Thus, a similar rate of block describes both currents activated by two different $[Ca^{2+}]$, even though only a small time-dependent relaxation is observed with 10 μM Ca^{2+}. This indicates that with 10 μM Ca^{2+}, the blocking equilibrium is established during the rising phase of the slowly activating current. However, the results clearly show that when activation is faster (60 μM Ca^{2+}), time-dependent block by BP can be observed. This result has two possible interpretations. First, the peptide may require that the BK channel be open before it can bind and block. Second, the binding of the peptide to the channel, whether it is open or closed, may be strongly voltage-dependent.

3.4. Small Quaternary Ammonium Blockers Do Not Compete with the Endogenous Inactivation Mechanism in BK_i Channels

The previous result shows that the *Shaker*B peptide, similar to the native BK_i inactivation domain(s), can produce a time-dependent reduction of BK_i current following channel activation. If the native inactivation domain is acting at the same site as BP, molecules which compete with the ability of BP to bind to the channel mouth would be expected to compete similarly with the native inactivation domain. The ability of quaternary ammonium agents to impede the native inactivation process has therefore been examined. TEA is known to block BK channels at a site about 30% within the membrane electric field with a $K_d(0)$ of about 50 mM at 120 mM internal KCl (Villarroel *et al.*, 1988). In inside-out patches from rat chromaffin cells, 30 mM TEA reduced BK_i current by about 50% at +60 mV when applied to the cytoplasmic face of the membrane (Fig. 12A, B). However, no slowing of the inactivation process was observed. Similarly, when TEA was applied extracellularly no slowing of inactivation was noted. External TEA is known to compete with a slower *Shaker* inactivation process termed C-type inactivation (Choi *et al.*, 1991; Hoshi *et al.*, 1991).

QX-314 is another quaternary ammonium ion blocker of BK channels that has characteristics suitable for this type of experiment. In particular, internally applied QX-314 produces a rapid, flickery block of BK channels (Oda *et al.*, 1992) that is fast enough to be well separated from any effect on the inactivation rate. Similar to TEA, a molecule even as bulky as QX-314 does not retard the rate of inactivation of the chromaffin cell BK_i channels (Solaro and Lingle, unpublished observations).

These results suggest an interesting difference between rapid inactivation of the BK_i channels and N-terminal inactivation of voltage-dependent K^+ channels. Unlike the *Shaker* channel inactivation domain, the native BK_i inactivation domain associates with a site that is neither electrostatically nor sterically sensitive to the presence of TEA in either its intracellular or extracellular binding site. This difference raises the possibility that BK_i inactivation results from a structural and functional mechanism that is

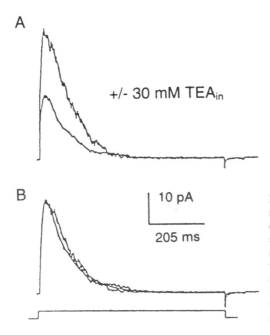

A

+/- 30 mM TEA$_{in}$

B

10 pA

205 ms

FIGURE 12. TEA does not compete with the native inactivation process. In A, ensemble BK_i currents were obtained in the presence and absence of 30 mM TEA applied to cytosolic side of an inside-out patch. The normalized currents with single exponentials fit to the decay phase are shown in B. There was no significant difference in the decay time constant at a TEA concentration which produced 50% block of the peak current.

quite different from the N-terminal pore occlusion of voltage-dependent K$^+$ channels. Alternatively, the mechanism of inactivation may be substantively similar, but the occluding structure may overlay access to the ion permeation pathway at some distance from the channel mouth.

3.5. Gradual Changes in Inactivation Rates Produced by Brief Trypsin Digestion Suggest that BK$_i$ Channels Contain Multiple Cytosolic Inactivation Domains

Fast inactivation of voltage-dependent K$^+$ channels, whether mediated by domains covalently attached to the channels or by accessory subunits, requires binding of only one of four independently acting inactivation domains (MacKinnon *et al.*, 1993; Gomez-Lagunas and Armstrong, 1994a). In contrast, inactivation of voltage-dependent Na$^+$ channels appears to involve a single site found in the cytoplasmic linker between the third and fourth cassette, corresponding to two linked K$^+$ channel monomers (Patton *et al.*, 1992). BK channels, like voltage-dependent K$^+$ channels, are thought to arise from the assembly of four separate, largely identical or homologous subunits (Shen *et al.*, 1994). If independent cytosolic domains are involved in BK$_i$ inactivation, up to four such domains would be predicted.

The effect of trypsin described above demonstrates that residues critical to the normal BK$_i$ inactivation process are readily accessible on the cytoplasmic face of the channel. The trypsin sensitivity of the inactivation process makes it possible to further test for similarities of BK$_i$ inactivation with voltage-dependent K$^+$ channel inactivation. If BK$_i$ inactivation results from multiple, independent inactivation domains, analogous to voltage-dependent K$^+$ channel inactivation, progressive removal of inactivation by brief trypsin treatments should result in a gradual slowing of the channel inactivation time course depending on the number of intact inactivation particles. In contrast, if inactivation is more like that of voltage-dependent Na$^+$ channels, involving a single inactivation domain (Vasillev *et al.*, 1988; Patton *et al.*, 1992, 1993; West *et al.*, 1992), or involves the concerted action of four functional domains, removal of inactivation by trypsin should be sudden and only a single inactivation rate should be observed.

The trypsin sensitivity of the inactivation time course was examined both in patches with a small number of channels and patches with a larger number of channels. In the first case, as shown in Fig. 13A, ensemble average BK currents were generated at +60 mV and 2 μM in a patch with two channels. Trypsin was applied for brief periods and ensemble average currents were generated between each application. Trypsin results in a gradual slowing of the inactivation rate and a substantial decrease in the resting inactivation level. Despite the slowing of the inactivation, these two channels continued to exhibit largely complete inactivation until nearly complete removal of inactivation. In Fig. 13B, a similar experiment from a patch with at least 12 channels is shown. This patch permitted 21 separate brief trypsin applications between generation of ensemble averages. As in Fig. 13A, trypsin resulted in a gradual slowing of the inactivation rate and, furthermore, resulted in an increasing amount of non-inactivating current. This result is expected if as inactivation domains are removed, some channels contain no functional inactivation domains, while other channels contain some residual number of domains. If

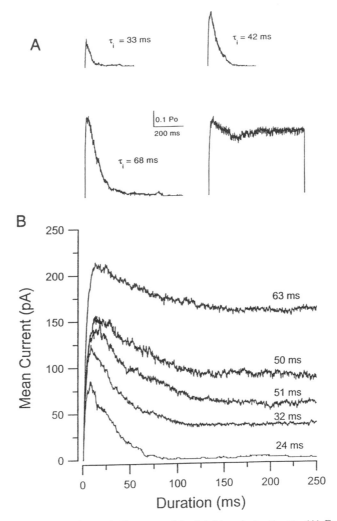

FIGURE 13. Inactivation can be gradually removed by brief trypsin treatments. (A) Ensemble average currents were generated from a patch with two channels. Brief application of trypsin between current acquisition resulted in a gradual slowing of inactivation and the removal of apparent resting inactivation. The time constant of decay for control ensemble currents was 33 msec. Following the first trypsin application, the time constant was 42 msec; following the second trypsin application, the time constant was 68 msec. A third application of trypsin resulted in almost complete removal of inactivation. Channels were activated with 2 μM Ca^{2+} by steps to +60 mV. (B) Ensemble currents in a patch with at least 12 channels are shown. Currents correspond to the 1st (24 msec), 5th, 10th, 15th, and 22nd (63 msec) ensemble with 21 separate trypsin applications. As above, trypsin results in a gradual slowing of the inactivation time constant as indicated on the figure. In addition, because of the large number of channels in the patch, some channels become totally non-inactivating at a time when other channels retain some inactivating behavior (C. R. Solaro and C. J. Lingle, unpublished).

channels are assumed to start with four inactivation domains, the simple prediction from this type of experiment is that the slowest inactivation time constant should be about four times the fastest time constant. We have never observed slowing in excess of fourfold and typically observe slowing on the order of two- to threefold. Given the limitation that there is some stochastic variation among each ensemble average, the minimal conclusion we draw is that BK_i channels in chromaffin cells appear to contain multiple, independently acting cytosolic inactivation domains. Gradual slowing of inactivation by trypsin does not support the possibility that inactivation results from the concerted action of multiple domains, all of which are required for inactivation.

The effect of trypsin on resting inactivation may seen surprising given that for the simple model involving multiple independent blocking domains, rates of recovery from inactivation are expected to be independent of the number of remaining inactivation domains (MacKinnon et al., 1993). However, in the presence of submembrane Ca^{2+}, BK_i channels exhibit significant resting inactivation (Herrington et al., 1995), which is defined by the relative rates of inactivation onset and recovery. As a consequence, even if microscopic rates of dissociation of the inactivation domain from its blocking site are unaffected by trypsin, slowing of the rate of inactivation onset by trypsin will diminish the level of resting inactivation.

3.6.   Resting Inactivation of BK_i Channels Is Ca^{2+}- and Voltage-Dependent

For voltage-dependent channels that exhibit inactivation, the fraction of channels available for activation depends on the cell membrane potential. This dependence is typically described by a Boltzmann function which represents the distribution of channels among inactivated states and states from which the channels can open rapidly (Hille, 1992). Although inactivation is typically coupled to channel opening, for most channels entry into inactivated states can also occur from closed states that precede open state(s) (Herrington and Lingle, 1992). However, the molecular rates that govern transitions between closed states that lead to channel opening are typically quite fast relative to rates of inactivation, so that for practical purposes, inactivation from closed states is minimal.

An interesting characteristic of BK_i channels is the large resting inactivation at potentials of -50 mV and more positive, with submembrane $[Ca^{2+}]$ of 1 μM or higher. Typically, from a holding potential of -40 mV, any individual BK_i channel has about a 25% to 50% chance of opening during a 400 msec step to $+60$ mV with 2–5 μM Ca^{2+}. As was shown in Fig. 9, steady-state inactivation of BK_i channels in inside-out patches with 1 μM Ca^{2+} exhibits a voltage dependence factor of -13.4 mV with half-inactivation at about -34 mV. The voltage dependence and Ca^{2+} dependence of steady-state inactivation is further illustrated in Fig. 14 in which BK_i channels were activated from different holding potentials at a fixed $[Ca^{2+}]_i$ (Fig. 14A) and with different $[Ca^{2+}]_i$ from a fixed holding potential (Fig. 14B). An interesting feature of steady-state inactivation at 1 μM Ca^{2+} is that an asymptote is reached at negative potentials that does not reflect full availability of channels for activation (Fig. 9). Thus, at potentials negative to -80 mV, a steady-state level of resting inactivation is reached, indicating that there is residual resting inactivation that is voltage-independent. Results at present do not distinguish

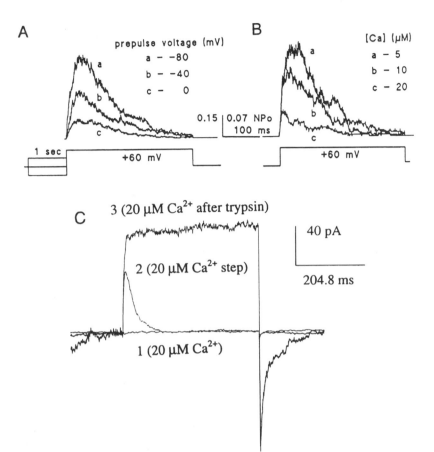

FIGURE 14. BK_i channels exhibit separable voltage- and Ca^{2+}-dependent resting inactivation. (A) Examples of ensemble BK_i currents activated at $+60$ mV with 2 μM Ca^{2+} after a 1 sec conditioning step at -80, -40, and 0 mV are shown. (B) Examples of ensemble BK_i currents activated at $+60$ mV at 5 (a), 10 (b), and 20 (c) μM Ca^{2+} from a holding potential of -40 mV are shown. Resting inactivation is increased by increases in cytosolic $[Ca^{2+}]$. (C) Resting inactivation at -40 mV is shown to be strictly dependent on Ca^{2+}. Trace 1 shows ensemble current in response to a depolarizing voltage step to $+60$ mV in the continuous presence of 20 μM Ca^{2+}. In trace 2, the cytosolic solution was switched from 0 Ca^{2+} to 20 μM Ca^{2+} just preceding the voltage step to $+60$ mV. Thus, in the absence of Ca^{2+}, there is little resting inactivation at -40 mV. Trace 3 shows the amount of BK_i current after trypsin removal of inactivation. (A and B from Solaro and Lingle, 1992; C from C. R. Solaro and C. J. Lingle, unpublished.)

whether this apparent resting inactivation results from true resting inactivation or from inactivation from closed states during the slow activation of current at $+60$ mV in the presence of 1 μM Ca^{2+}.

More recent experiments have tested whether the residual resting inactivation is Ca^{2+}-dependent. When a step in $[Ca^{2+}]$ from 0 to 20 μM is timed with a voltage step from -40 to $+60$ mV (Fig. 14), the nearly complete inactivation observed when 20 μM Ca^{2+} is continuously present is almost fully removed. Full removal of inactivation is defined by removal of inactivation by trypsin. Fits of a Hodgkin-Huxley (H-H) activation model to the current recorded after trypsin application define τ_m, n, and I_{max}. Fitting the inactivating current resulting from a simultaneous $[Ca^{2+}]_i$ and voltage step with an H-H activation-inactivation model gives an essentially identical estimate of I_{max}. Thus, the discrepancy between peak current following the coordinated step in $[Ca^{2+}]_i$ and voltage and that following trypsin can be largely explained by inactivation during the rising phase of current. Thus, at -40 mV in 0 Ca^{2+}, there is no resting inactivation of BK_i channels. This suggests that resting inactivation of BK_i channels arises from both Ca^{2+}-dependent and voltage-dependent steps and, furthermore, that these steps may be separate, although possibly coupled, processes.

3.7. Recovery of BK_i Channels from Inactivation Bypasses Open States

For some voltage-dependent K^+ channels (Ruppersberg *et al.*, 1991; Demo and Yellen, 1991; Gomez-Lagunas and Armstrong, 1994b), recovery from inactivation following repolarization is associated with channel reopening. This suggests that the native inactivation domain must dissociate from the channel, thereby unblocking the channel, before channel closure can occur. We have tested this possibility for BK_i channels using paired depolarizing voltage steps separated by a recovery period sufficient to allow for almost complete recovery from inactivation (Fig. 15A). For comparison, the amount of current flowing through deactivating BK_i channels following repolarization at the peak of the BK_i current was determined (Fig. 15B). If all inactivated BK_i channels recover through the same open states that must be traversed during deactivation, the total amount of current through BK_i channels at -60 mV should be identical for the two cases (Fig. 15C). This was not observed. The lack of openings observed during recovery from inactivation of BK_i channels strongly argues that deactivation of BK_i channels can proceed normally while channels remain inactivated. An alternative possibility is that recovery from inactivation involves passage through short-duration open states that differ from those occupied during deactivation.

3.8. Summary of Similarities between N-Terminal "Ball and Chain" Inactivation and Inactivation of BK_i Channels

Inactivation of BK_i channels described here appears to share some similarities with N-terminal inactivation of voltage-dependent K^+ channels:

1. Both involve multiple cytosolic, trypsin-sensitive domains.
2. The apparent rate of inactivation is dependent on the conditions favoring channel activation, i.e., voltage in both cases and $[Ca^{2+}]$ in the case of BK_i channels. At maximal rates of channel activation, the rates of inactivation become largely independent of voltage and, in the case of BK_i channels, independent of $[Ca^{2+}]$.
3. BK channels and some voltage-dependent K^+ channels can specifically bind peptides corresponding to known N-terminal inactivation domains; furthermore, these peptides appear to occlude the ion permeation pathway and are competitive with TEA binding.

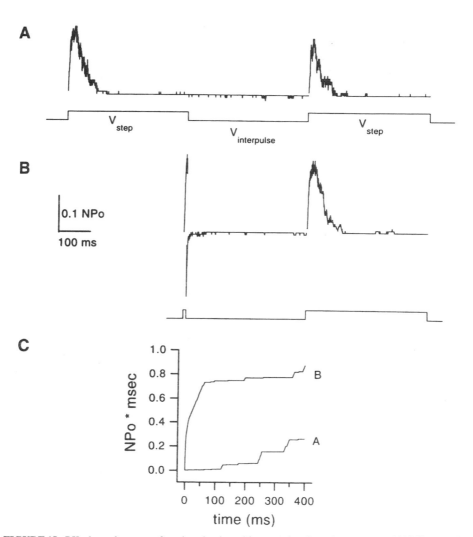

FIGURE 15. BK_i channels recover from inactivation without passing through open states. (A) BK_i ensemble currents were generated in 2 μM Ca^{2+} with a protocol in which BK_i channels were allowed to completely inactivate at +60 mV (V_{step}) before returning to −60 mV ($V_{interpulse}$) for 400 msec before a subsequent step to +60 mV. Note that ensemble currents during the recovery period are inward. (B) BK_i ensemble currents were generated with a brief activation step to define the amount of current expected at −60 mV from BK_i channels undergoing normal deactivation. (C) The cumulative current measured from detected openings during the recovery period either in A or B is plotted for the 400-msec recovery period. Although 400 msec is sufficient to result in almost complete recovery of BK_i channels from inactivation, total current during the recovery period in A is much less than that observed in B. Thus, most BK_i channels recover from inactivation through pathways distinct from the normal deactivation pathway. This indicates that recovery from inactivation does not occur through the same open states that occur during BK_i deactivation and suggests that BK_i deactivation may occur while the channel remains inactivated.

Although BK_i inactivation is reminiscent of N-terminal inactivation, there are several provocative differences:

1. Resting inactivation or inactivation from closed states preceding open states can be substantial in BK_i channels and is dependent on $[Ca^{2+}]_i$. This suggests that in the presence of $[Ca^{2+}]_i$ the channel may undergo conformational changes which permit blockade by an inactivation domain, even without channel opening.
2. Small blockers which occlude the cytosolic mouth of BK_i channels do not compete with the native inactivation domain. Therefore, at present, there is no evidence that inactivation of BK_i channels occurs by occlusion of the permeation pathway.
3. Recovery from inactivation occurs without detectable reopening of BK_i channels. This suggests that BK_i channels can deactivate while the inactivation domain is still bound.

Based on the above features of BK_i inactivation, we imagine the following: BK_i inactivation may occur by an occlusion mechanism analogous to N-terminal inactivation of voltage-dependent K^+ channels. However, we propose that the BK_i inactivation domain provides a lid over a small funnel that precedes or forms part of the permeation pathway. Because smaller quaternary blockers do not compete with the native inactivation domain, the native inactivation domain probably does not interact with a site immediately abutting the ion permeation pathway. Rather, if inactivation is occurring by physical blockade of access of ions to the open pore, this would require that a cavity of some currently undefined depth separates the bound inactivation domain from the beginning of the electrical field across the channel. Perhaps consistent with this picture, BK_i channels can deactivate without recovery from inactivation. This suggests that the positioning of the native inactivation domain in its blocking site does not impede the structural changes associated with activation and deactivation. In contrast to N-terminal inactivation of voltage-dependent K^+ channels, we would not expect native BK_i inactivation to exhibit a dependence on extracellular ions that occupy sites deep within the BK permeation pathway. This remains to be tested. It will be interesting to see whether any of these speculations about BK_i inactivation are borne out once information about the molecular components of BK channels in chromaffin cells becomes available (Saito *et al.*, 1994).

4. BK_i AND BK_s CURRENTS DIFFER IN OTHER WAYS

The most remarkable feature of BK current in rat chromaffin cells is the segregation among cells of inactivating and relatively non-inactivating currents. Differential expression of two phenotypic variants of BK current raises the possibility that the functional differences between these channels may have important consequences for the electrophysiological properties of chromaffin cells. It was therefore important to define whether there might be any other functional differences between BK_i and BK_s channels that could affect the physiological roles these channels play. We have focused on two functional properties of BK channels in chromaffin cells (Solaro *et al.*, 1995a). First, we have compared the activation rates of ensemble current resulting from BK_i and BK_s channels. Second, we have examined the deactivation rates of BK_i and BK_s channels.

4.1. Activation of BK_i and BK_s Current Is Similar, but Deactivation Differs

To compare the activation time course, ensemble currents from patches containing either BK_i or BK_s channels were generated at $+60$ mV and 2 μM Ca^{2+}. For BK_i currents, both the time to peak current and the apparent activation rate are influenced by the contemporaneous onset of inactivation. To facilitate comparison of activation alone, the time course of current activation and, in the case of BK_i channels, inactivation were fit by a Hodgkin-Huxley model (Solaro *et al.*, 1995a). Estimates for the time constant of activation and cooperativity were similar for both types of BK channels (Fig. 16A). This comparison was also made after briefly treating BK_i channels with cytosolic trypsin to remove inactivation. Fits of these ensemble currents yielded time constants of current activation which were indistinguishable from those obtained with BK_s channels. Thus, at comparable $[Ca^{2+}]$ and voltage, both BK_i and BK_s currents activate at similar rates.

Similarly, the deactivation behavior of BK_i and BK_s channels was compared using ensembles of the decay of current openings following repolarizing voltage steps to -40 mV (Fig. 16B). Such ensembles revealed a clear difference between BK_i and BK_s channels (Solaro *et al.*, 1995a). BK_s channel deactivation followed a single exponential time course with a time constant of about 2.4 msec at -40 mV and 2 μM Ca^{2+}, whereas BK_i channel deactivation required two exponential components to describe the current decay time course. The fast component of BK_i deactivation was similar to that of BK_s channel deactivation, while the second component, representing about 50% of the decaying current, was about 12.5 msec. Following trypsin removal of BK_i inactivation, BK_i channels still deactivate with a similar two-exponential time course, implying that the structure(s) affected by trypsin is(are) not responsible for the differences between BK_i and BK_s deactivation.

4.2. Are Inactivating BK Channels Found in Other Cells?

BK channels have been thought to play a stereotypical role, sensing rapid changes in membrane voltage and submembrane $[Ca^{2+}]$. Given the abundance of published records of steady-state BK channel activity during which channels are driven to well above 90% open probability, it is clear that the non-inactivating variants of these channels are common. However, it remains a possibility that inactivating BK channel variants may be more widespread than is currently appreciated. In particular, one must wonder, given the temptation to study BK channels under steady-state conditions of voltage and $[Ca^{2+}]$, is it possible that inactivating variants of BK channels may have been overlooked in some cell preparations? Precisely because of the tendency to study BK channels under steady-state conditions, it may be easy to overlook those patches with the occasional, isolated burst that would be characteristic of a channel exiting from inactivated states and then reentering inactivated states. The number of studies which have focused on the seductively appealing steady-state properties of BK_s variants sharply contrasts with the marked paucity of investigations which have studied BK channels using voltage steps.

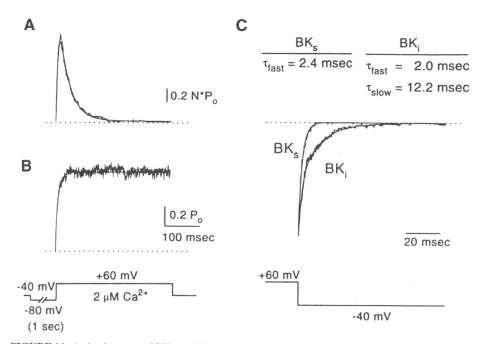

FIGURE 16. Activation rates of BK_i and BK_s currents are similar, but deactivation behavior differs. In A, ensemble average currents for a patch with BK_i channels and a patch with BK_s channels are shown. In each case, a fit of a Hodgkin-Huxley model of current activation is overlaid on the current time course. The fit to the BK_i channel ensemble included a term for inactivation time constant (τ_m) and fractional inactivation at t = infinity. At +60 mV and 2 μM Ca^{2+}, for two patches with BK_s current, τ_m was 8.7 msec and n = 0.70, while for seven patches with BK_i channels, τ_m was 10.2 ± 4.5 msec and n = 1.06 ± 0.3. In B, ensemble average deactivation currents for a patch with BK_s channels and a patch with BK_i channels are overlaid to compare the deactivation time course. In this example, at −40 mV and 2 μM Ca^{2+}, BK_s ensemble currents decay with a time constant of 2.4 msec, while BK_i currents decay with two components, one of 2.0 msec and one of 12.2 msec. For six patches with BK_i channels, 59.8% of the ensemble current decayed with a time constant of 2.36 ± 1.1 msec (mean ± SD) with the remainder of the current decaying at 11.95 ± 7.25 msec. For four patches with BK_s channels, the current decayed with a time constant of 2.01 ± 0.52 msec. (From Solaro *et al.*, 1995a.)

At the present time, there is very little direct evidence for the widespread occurrence of BK_i channels. Pallotta (1985a) presented data describing inactivation of BK channels in rat skeletal muscle. However, the steady-state open probability for the inactivating channels was inconsistent with previously reported steady-state properties of BK channels from the same cells (Barrett *et al.*, 1982), and inactivation does not appear to be a common feature of such channels (e.g., McManus and Magleby, 1988, 1991). BK channels which exhibit rapid inactivation following steps in submembrane Ca^{2+} have also been described in 15% of patches pulled from neonatal hippocampal neurons (Ikemoto *et al.*, 1989); in some cases, both inactivating and non-inactivating BK channels occurred in the same patches. Macroscopic currents and single channels recorded in cell-attached patches have also suggested the possible existence of inactivating BK channels in rodent pancreatic β cells (Findlay *et al.*, 1985; Smith *et al.*, 1990).

Using protocols applied to the study of BK_i channels in chromaffin cells (Solaro and Lingle, 1992), we have looked for BK_i channels in a few other cell types. We were able to find BK_i channels in PC12 cells and in some patches from bovine chromaffin cells (Solaro and Lingle, 1992). We have failed to find any evidence for BK_i channels in GH_3 cells, a small sample of hippocampal CA1 pyramidal cells, and rat sympathetic neurons. Prompted by earlier reports on β cells (Findlay *et al.*, 1985; Tabcharani and Misler, 1989; Smith *et al.*, 1990), we have also examined BK channels in rat pancreatic β cells. Of 14 patches which contained BK channels, 12 contained exclusively BK_i channels (Fig. 17) and two contained both BK_i and BK_s channels. Similar to chromaffin cell BK_i channels, inactivation of the β cell BK channel was removed by brief application of trypsin. Thus, most BK channels in primary cultures of rat β cells appear to be inactivating. However, BK channels in some β cells may be non-inactivating. For example, a study on BK

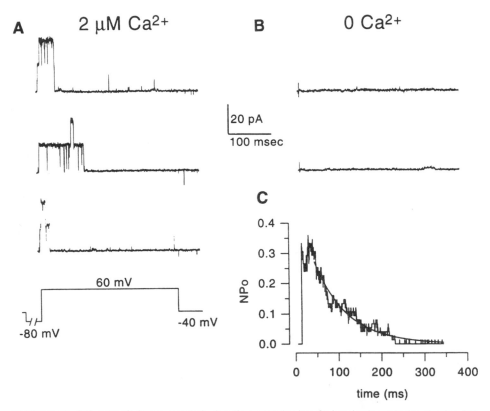

FIGURE 17. BK channels in rat pancreatic β cells are predominantly inactivating. (A) Inactivating BK channels were recorded in an inside-out patch using the voltage protocol and $[Ca^{2+}]_i$ indicated. Voltage steps were given every 3 sec. Three leak-subtracted current traces are shown. Adult rat β cells were provided by Dr. S. Misler. (B) Two leak-subtracted current traces activated by steps to $+60$ mV but with zero Ca^{2+} are shown. (C) An ensemble average plot of N^*P_o from the records shown in A is plotted. The solid line shows a single exponential of 76.1 ± 1.2 msec.

channels of neonatal rat β cells showed that such channels could be driven to near 100% open probability at high Ca^{2+} (Cook et al., 1984), indicating that BK channels in neonatal β cells may be largely non-inactivating.

4.3. Physiological Roles of BK_i and BK_s Channels in Chromaffin Cells

The existence of two functionally distinct BK channels, to a large extent selectively expressed in separate rat chromaffin cells, provides an interesting system for evaluating the role of each BK variant in the electrical firing behavior of the cells (Solaro et al., 1995a). This issue has been evaluated with two types of comparisons. First, action potentials were elicited by brief, just superthreshold current injections and the Ca^{2+} dependence and CTX dependence of the action potential waveform examined. Although CTX is less effective at reducing BK_i than BK_s current, 100 nM CTX typically blocks over 50% of the whole-cell BK_i current. Second, the ability of cells to fire during a 2 sec constant depolarizing current injection was examined. In all cases, standard voltage-clamp protocols were used to define whether a cell contained predominantly BK_i or BK_s currents, and apamin was used to block SK current.

Action potentials in rat chromaffin cells exhibit a Ca^{2+}-dependent afterhyperpolarization. Removal of external Ca^{2+} slows the latter half of action potential repolarization and abolishes any afterhyperpolarization in both BK_i and BK_s cells. Similarly, 100 nM CTX prolongs action potentials in both BK_i and BK_s cells with more marked effects on BK_i cell afterhyperpolarizations. Thus, both BK_i and BK_s channels contribute to action potential repolarization in chromaffin cells.

Inactivating Ca^{2+}-dependent K^+ current appears to be critical for the ability of chromaffin cells to repetitively fire action potentials when SK current is blocked by apamin. During constant depolarizing current injection, cells with BK_s current fire only one or a few action potentials over a broad range of injected currents, while cells with BK_i current are able to fire repetitively (Fig. 18A, B). Because other currents in the two cell types are similar (Solaro et al., 1995a), it appears that the presence of BK_i current permits chromaffin cells to fire repetitively. In contrast to rat chromaffin cells, it has been previously noted that most guinea pig chromaffin cells exhibit a phasic firing pattern during constant current injection (Holman et al., 1994).

Ironically, the ability of BK_i current to influence chromaffin cell excitability may stem not from inactivation, but rather from its slow deactivation. During individual action potentials, minimal inactivation of BK current is expected to occur (Herrington et al., 1995; J. P. Ding, unpublished). However, the slow deactivation kinetics of BK_i channels would be expected to produce, on average, a more prolonged and larger after-hyperpolarization than would be expected for cells with BK_s channels. This may ensure sufficient recovery of Na^+ channels from inactivation to support a subsequent action potential. Based on this interpretation, CTX should suppress tonic firing during depolarizing current injection in BK_i cells but have little effect on BK_s cells (Solaro et al., 1995a). This expectation is supported by the result in Fig. 18C indicating that the CTX-sensitive current in BK_i cells plays an important role in permitting repetitive firing, whereas CTX-sensitive current has no influence on the ability of BK_s cells to respond to depolarizing current injection. These conclusions about the possible roles of BK current

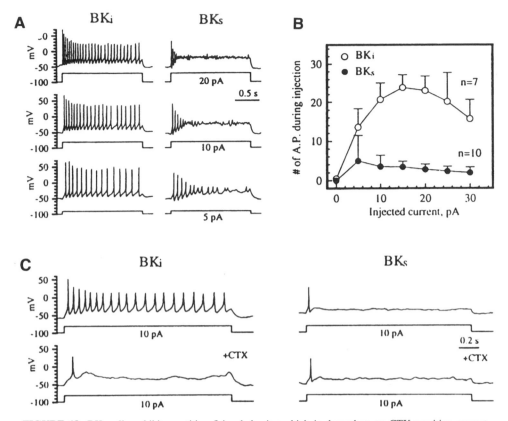

FIGURE 18. BK_i cells exhibit repetitive firing behavior which is dependent on CTX-sensitive current. (A) Depolarizing current was injected into cells with either BK_i (traces on left) or BK_s (traces on right) current. In BK_i cells, depolarizing current elicits repetitive firing. (B) The number of action potentials elicited during a 2-sec depolarizing pulse of different amplitudes is plotted for both BK_i cells and BK_s cells, showing that cells with BK_i current are more able to fire repetitively. (C) Action potentials were elicited as above for both a BK_i and BK_s cell, either in the absence or presence of 100 nM CTX. Although CTX only partially blocks BK_i current, blockade of BK_i channels by CTX results in suppression of the ability of a cell with BK_i current to fire repetitively. (Modified from Solaro *et al.*, 1995a.)

in chromaffin cell firing must be tempered by the fact that these experiments were done during blockade of SK current by apamin.

5.   SUMMARY

Rat chromaffin cells express an interesting diversity of Ca^{2+}-dependent K^+ channels, including a voltage-independent, small-conductance, apamin-sensitive SK channel and two variants of voltage-dependent, large-conductance BK channels. The two BK channel variants are differentially segregated among chromaffin cells, such that BK current is completely inactivating in about 75–80% of rat chromaffin cells, while the remainder express a mix of inactivating and non-inactivating current or mostly non-inactivating BK_s current.

The single-channel conductance of BK_i channels is identical to that of BK_s channels. Although rates of current activation are similar in the two variants, the deactivation kinetics of the two channels also differ. Furthermore, BK_i channels are somewhat less sensitive to scorpion toxins than BK_s channels.

The slow component of BK_i channel deactivation may be an important determinant of the functional role of these channels. During blockade of SK current, cells with BK_i current fire tonically during sustained depolarizing current injection, whereas cells with BK_s current tend to fire only a few action potentials before becoming quiescent. The ability to repetitively fire requires functional BK_i channels, since partial blockade of BK_i channels by CTX makes a BK_i cell behave much like a BK_s cell. In contrast, the physiological significance of BK_i inactivation may arise from the ability of secretagogue-induced $[Ca^{2+}]_i$ elevations to regulate the availability of BK_i channels during subsequent action potentials (Herrington et al., 1995). By reducing the number of BK channels available for repolarization, the time course of action potentials may be prolonged. This possibility remains to be tested directly.

These results raise a number of interesting questions pertinent to the control of secretion in rat adrenal chromaffin cells. An interesting hypothesis is that cells with a particular kind of BK current may reflect particular subpopulations of chromaffin cells. These subpopulations might differ either in the nature of the material secreted from the cell (e.g., Douglass and Poisner, 1965) or in the responsiveness to particular secretagogues. The differences in electrical behavior between cells with BK_i and BK_s current suggest that the pattern of secretion that might be elicited by a single type of stimulus could differ. For BK_i cells, secretion may occur in a tonic fashion during sustained depolarization, while secretion from cells with BK_s current may be more phasic.

In the absence of specific structural information about the domains responsible for inactivation of BK_i channels, our understanding of the mechanism of inactivation remains indirect. BK_i inactivation shares many features with N-terminal inactivation of voltage-dependent K^+ channels. However, there are provocative differences between the two types of inactivation which require us to propose that the native inactivation domain of BK_i channels may occlude access of permeant ions to the BK channel permeation pathway in a position at some distance from the actual mouth of the channel. Further understanding of the structural and mechanistic basis of inactivation of BK_i channels promises to provide new insights into both the cytoplasmic topology of BK channels and the Ca^{2+}- and voltage-dependent steps involved in channel activation.

ACKNOWLEDGMENTS. We wish to thank our colleagues, Dr. J. Herrington and Dr. A. Neely, for their key contributions to the work from this lab. We thank J. H. Steinbach for reading the manuscript. This work was supported by the National Institutes of Health (DK-46564).

6. REFERENCES

Adams, P. R., Constanti, A., Brown, D. A., and Clark, R. B., 1992, Intracellular calcium activates a fast, voltage-sensitive K$^+$ current in vertebrate sympathetic neurones, *Nature* **296**:746–749.

Adelman, J. P., Shen, K. Z., Kavanaugh, M. P., Warren, R. A., Wu, Y. N., Lagrutta, A., Bond, C. T., and North, R. A., 1992, Calcium-activated potassium channels expressed from cloned complementary DNAs, *Neuron* **9**:209–216.

Armstrong, C. M., and Bezanilla, F., 1977, Inactivation of the sodium channel. II. Gating current experiments, *J. Gen. Physiol.* **70**:567–590.

Art, J. J., Wu, Y.-C., and Fettiplace, R., 1995, The calcium-activated potassium channels of turtle hair cells, *J. Gen. Physiol.* **105**:49–72.

Artalejo, A. R., Garcia, A. G., and Neher, E., 1993, Small-conductance Ca^{2+}-activated K$^+$ channels in bovine chromaffin cells, *Pflügers Arch.* **423**:97–103.

Atkinson, N. S., Robertson, G. A., and Ganetsky, B., 1991, A component of calcium-activated potassium channels encoded by the *Drosophila Slo* locus, *Science* **253**:551–555.

Barrett, J. N., Magleby, K. L., and Pallotta, B. S., 1982, Properties of single calcium-activated potassium channels in cultured rat muscle, *J. Physiol. (London)* **331**:211–230.

Benham, C. D., Bolton, T. B., Lang, R. J., and Takewaki, T., 1986, Calcium-activated potassium channels in single smooth muscle cells of rabbit jejunum and guinea-pig mesenteric artery, *J. Physiol. (London)* **371**:45–67.

Bielefeldt, K., and Jackson, M. B., 1993, A calcium-activated potassium channel causes frequency-dependent action-potential failures in a mammalian nerve terminal, *J. Neurophysiol.* **70**:284–298.

Blair, L. A. C., and Dionne, V. E., 1985, Developmental acquisition of Ca^{2+}-sensitivity of K$^+$ channels in spinal neurones, *Nature* **315**:329–331.

Blatz, A. L., and Magleby, K. L., 1986, Single apamin-blocked Ca-activated K$^+$ channels of small conductance in cultured rat skeletal muscle, *Nature* **323**:718–720.

Blatz, A. L., and Magleby, K. L., 1987, Calcium-activated potassium channels, *Trends Neurosci.* **10**:463–467.

Bulan, E. J., Barker, J. L., and Mienville, J.-M., 1994, Immature maxi-K channels exhibit heterogeneous properties in the embryonic rat telencephalon, *Dev. Neurosci.* **16**:25–33.

Butler, A., Tsunoda, S., McCobb, D., Wei, A., and Salkoff, L., 1993, mSlo, a complex mouse gene encoding "maxi" calcium activated potassium channels, *Science* **261**:221–224.

Candia, S., Garcia, M. L., and Latorre, R., 1992, Iberiotoxin: A potent blocker of the large-conductance Ca^{2+}-activated K$^+$ channel, *Biophys. J.* **63**:583–590.

Choi, K. L., Aldrich, R. W., and Yellen, G., 1991, Tetraethylammonium blockade distinguishes two inactivation mechanisms in voltage-activated K$^+$ channels. *Proc. Natl. Acad. Sci. USA* **88**:5092–5095.

Choi, K. L., Mossman, C., Aube, J., and Yellen, G., 1993, The internal quaternary ammonium receptor site of *Shaker* potassium channels, *Neuron* **10**:533–541.

Chung, S., Reinhart, P. H., Martin, B. L., Brautigan, D., and Levitan, L. B., 1991, Protein kinase activity closely associated with a reconstituted calcium-activated potassium channel, *Science* **253**:560–562.

Cook, D. L., Ikeuchi, M., and Fujimoto, W. Y., 1984, Lowering pH$_i$ inhibits Ca^{2+}-activated K$^+$ channels in pancreatic B-cells, *Nature* **311**:269–271.

Covarrubias, M., Wei, A., and Salkoff, L., 1991, Shaker, Shal, Shab and Shaw express independent K$^+$ current systems, *Neuron* **7**:763–773.

Crest, M., and Gola, M., 1993, Large conductance Ca^{2+}-activated K$^+$ channels are involved in both spike shaping and firing regulation in *Helix* neurones, *J. Physiol. (London)* **465**: 265–287.

Demo, S. D., and Yellen, G., 1991, The inactivation gate of the *Shaker* K$^+$ channel behaves like an open-channel blocker, *Neuron* **7**:743–753.

Douglass, W. W., and Poisner, A. M., 1965, Preferential release of adrenaline from the adrenal medulla by muscarine and pilocarpine, *Nature* **208**:1102–1103.

Dworetzky, S. I., Trojnacki, J. T., and Gribkoff, V. K., 1994, Cloning and expression of a human large-conductance calcium-activated potassium channel, *Mol. Brain Res.* **27**:189–193.

Eberhardt, D. A., and Holz, R. W., 1987, Cholinergic stimulation of inositol phosphate formation in bovine adrenal chromaffin cells: Distinct nicotinic and muscarinic mechanisms, *J. Neurochem.* **49**:1634–1643.

Egan, T. M., Dagan, D., and Levitan, I. B., 1993, Properties and modulation of a calcium-activated potassium channel in rat olfactory bulb neurons, *J. Neurophysiol.* **69:**1433–1442.

Fenwick, E. M., Marty, A., and Neher, E., 1982, Sodium and calcium channels in bovine chromaffin cells, *J. Physiol. (London)* **331:**599–635.

Findlay, I., 1984, A patch-clamp study of potassium channels and whole-cell currents in acinar cells of the mouse lacrimal gland, *J. Physiol. (London)* **350:**179–195.

Findlay, I., Dunne, M. J., and Petersen, O. H., 1985, High-conductance K^+ channel in pancreatic islet cells can be activated by internal calcium, *J. Membr. Biol.* **83:**169–175.

Foster, C. D., Chung, S., Zagotta, W. N., Aldrich, R. W., and Levitan, I. B., 1992, A peptide derived from the *Shaker* B K^+ channel produces short and long blocks of reconstituted $Ca^{(2+)}$-dependent K^+ channels, *Neuron* **9:**229–236.

Franciolini, F., 1988, Calcium and voltage dependence of single Ca^{2+}-activated K^+ channels from cultured hippocampal neurons of rat, *Biochim. Biophys. Acta* **943:**419–427.

Giangiacomo, K. M., Garcia, M. L., and McManus, O. B., 1992, Mechanism of iberiotoxin block of the large-conductance calcium-activated potassium channel from bovine aortic smooth muscle, *Biochemistry* **31:**6719–6727.

Goh, J. W., and Pennefather, P. S., 1987, Pharmacological and physiological properties of the after-hyperpolarization current of bullfrog ganglion neurones, *J. Physiol. (London)* **394:**315–330.

Golowasch, J., Kirkwood, A., and Miller, C., 1986, Allosteric effects of Mg^{2+} on the gating of Ca^{2+}-activated K^+ channels from mammalian skeletal muscle, *J. Exp. Biol.* **124:**5–13.

Gomez-Lagunas, F., and Armstrong, C. M., 1994a, Inactivation in *Shaker*B K^+ channels: A test for the number of inactivating particles on each channel, *Biophys. J.* **68:**89–95.

Gomez-Lagunas, F., and Armstrong, C. M., 1994b, The relation between ion permeation and recovery from inactivation of *Shaker*B K^+ channels, *Biophys. J.* **67:**1806–1815.

Grissmer, S., Nguyen, A. N., and Cahalan, M. D., 1993, Calcium-activated potassium channels in resting and activated human T lymphocytes, expression levels, calcium dependence, ion selectivity, and pharmacology, *J. Gen. Physiol.* **102:**601–630.

Herrington, J., and Lingle, C. J., 1992, Kinetic and pharmacological properties of low-voltage–activated Ca^{2+} current in rat clonal (GH_3) pituitary cells, *J. Neurophysiol.* **68:**213–232.

Herrington, J., Solaro, C. R., Neely, A., and Lingle, C. J., 1995, Suppression of calcium- and voltage-activated current by muscarinic acetylcholine receptor activation in rat chromaffin cells, *J. Physiol. (London)* **485:**297–318.

Hille, B., 1992, *Ionic Channels of Excitable Membranes*, Sinauer Associates, Inc., Sunderland, Massachusetts.

Hodgkin, A. L., and Huxley, A. F., 1952, A quantitative description of membrane current and its application to conduction and excitation in nerve, *J. Physiol. (London)* **108:**37–77.

Holman, M. E., Coleman, H. A., Tonta, M. A., and Parkington, H. C., 1994, Synaptic transmission from splanchnic nerves to the adrenal medulla of guinea-pigs, *J. Physiol. (London)* **478:**115–124.

Horn, R., and Marty, A., 1988, Muscarinic activation of ionic currents using a new whole-cell recording method, *J. Gen. Physiol.* **92:**145–159.

Hoshi, T., Zagotta, W. N., and Aldrich, R. W., 1990, Biophysical and molecular mechanisms of *Shaker* potassium channel inactivation, *Science* **250:**533–538.

Hoshi, T., Zagotta, W. N., and Aldrich, R. W., 1991, Two types of inactivation in *Shaker* K^+ channels: Effects of alterations in the carboxy-terminal region, *Neuron* **7:**547–556.

Hu, S. L., Yamamoto, Y., and Kao, C. Y., 1989, Permeation, selectivity, and blockade of Ca^{2+}-activated K^+ channel of guinea pig taenia coli myocyte, *J. Gen. Physiol.* **94:**849–862.

Ikemoto, Y., Ono, K., Yoshida, A., Akaike, N., 1989, Delayed activation of large-conductance Ca^{2+}-activated K^+ channels in hippocampal neurons of the rat, *Biophys. J.* **56:**207–212.

Isacoff, E. Y., Jan, Y. N., and Jan, L. Y., 1991, Putative receptor for the cytoplasmic inactivation gate in the *Shaker* K^+ channel, *Nature* **353:**86–90.

Knaus, H. G., Folander, K., Garcia-Calvo, M., Garcia, M. L., Kaczorowski, G. J., Smith, M., and Swanson, R., 1994a, Primary sequence and immunological characterization of β-subunit of high conductance Ca^{2+}-activated K^+ channel from smooth muscle, *J. Biol. Chem.* **269:**17274–17278.

Knaus, H. G., Garcia-Calvo, M., Kaczorowski, G. J., and Garcia, M. L., 1994b, Subunit composition of the high conductance calcium-activated potassium channel from smooth muscle, a representative of the mSlo and slowpoke family of potassium channels, *J. Biol. Chem.* **269:**3921–3924.

Kume, H., Takai, A., Tokuno, H., and Tomita, T., 1989, Regulation of Ca^{2+}-dependent K^+-channel activity in tracheal myocytes by phosphorylation, *Nature* **341:**152–154.

Lagrutta, A., Shen, K. Z., North, R. A., and Adelman, J. P., 1994, Functional differences among alternatively spliced variants of Slowpoke, a *Drosophila* calcium-activated potassium channel, *J. Biol. Chem.* **269:**20347–20351.

Lancaster, B., and Adams, P. R., 1986, Calcium-dependent current generating the afterhyperpolarization of hippocampal neurones, *J. Neurophysiol.* **55:**1268–1292.

Lancaster, B., and Nicoll, R. A., 1987, Properties of two calcium-activated hyperpolarizations in rat hippocampal neurones, *J. Physiol. (London)* **389:**183–203.

Lancaster, B., and Pennefather, P., 1987, Potassium currents evoked by brief depolarizations in bull-frog sympathetic ganglion cells, *J. Physiol. (London)* **387:**519–548.

Lancaster, B., Nicoll, R. A., and Perkel, D. J., 1991, Calcium activates two types of potassium channels in rat hippocampal neurons in culture, *J. Neurosci.* **11:**23–30.

Lang, D. G., and Ritchie, A. K., 1987, Large and small conductance calcium-activated potassium channels in the GH_3 anterior pituitary cell line, *Pflügers Arch.* **410:**614–622.

Lang, D. G., and Ritchie, A. K., 1990a, Tetraethylammonium blockade of apamin-sensitive and insensitive Ca^{2+}-activated K^+ channels in a pituitary cell line, *J. Physiol. (London)* **425:**117–132.

Lang, D. G., and Ritchie, A. K., 1990b, TEA sensitivity of a 35 pS Ca^{2+}-activated K^+ channel in GH_3 cells that is activated by thyrotropin releasing hormone, *Pflügers Arch.* **416:**704–709.

Latorre, R., 1986, The large calcium-activated potassium channel, in: *Ion Channel Reconstitution* (C. Miller, ed.), Plenum Publishing Co., New York, pp. 431–467.

Latorre, R., Oberhauser, A., Labarca, P., and Alvarez, O., 1989, Varieties of calcium-activated potassium channels, *Annu. Rev. Physiol.* **51:**395–399.

Lechleiter, J. D., Dartt, D. A., and Brehm, P., 1988, Vasoactive intestinal peptide activates Ca^{2+}-dependent K^+ channels through a cAMP pathway in mouse lacrimal cells, *Neuron* **1:**227–235.

Leonard, R. J., Garcia, M. L., Slaughter, R. S., and Reuben, J. P., 1992, Selective blockers of voltage-gated K^+ channels depolarize human T lymphocytes: Mechanism of the antiproliferative effect of charybdotoxin, *Proc. Natl. Acad. Sci. USA* **89:**10094–10098.

MacKinnon, R., 1991, Determination of the subunit stoichiometry of a voltage-activated potassium channel, *Nature* **350:**232–235.

MacKinnon, R., Aldrich, R. W., and Lee, A., 1993, Functional stoichiometry of *Shaker* K channel inactivation, *Science* **262:**757–759.

Magleby, K. L., and Pallotta, B. S., 1983a, Calcium dependence of open and shut interval distributions from calcium-activated potassium channels in cultured rat muscle, *J. Physiol. (London)* **344:**585–604.

Magleby, K. L., and Pallotta, B. S., 1983b, Burst kinetics of single calcium-activated potassium channels in cultured rat muscle, *J. Physiol. (London)* **344:**605–623.

Malhotra, R. K., Wakade, T. D., and Wakade, A. R., 1988, Vasoactive intestinal polypeptide and muscarine mobilize intracellular Ca^{2+} through breakdown of phosphoinositides to induce catecholamine secretion. Role of IP_3 in exocytosis, *J. Biol. Chem.* **263:**2123–2126.

Markwardt, F., and Isenberg, G., 1992, Gating of maxi K^+ channels studied by Ca^{2+} concentration jumps in excised inside-out multi-channel patches (myocytes from guinea pig urinary bladder), *J. Gen. Physiol.* **99:**841–862.

Marty, A., 1981, Ca-dependent K channels with large unitary conductance in chromaffin cell membranes, *Nature* **291:**497–500.

Marty, A., and Neher, E., 1985, Potassium channels in cultured bovine adrenal chromaffin cells, *J. Physiol. (London)* **367:**117–141.

Maruyama, Y., Petersen, O. H., Flanagan, P., and Pearson, G. T., 1983, Quantification of Ca^{2+}-activated K^+ channels under hormonal control in pig pancreas acinar cells, *Nature* **305:**228–232.

McCobb, D. P., Fowler, N. L., Featherstone, T., Lingle, C. J., Saito, M., Krause, J. E., and Salkoff, L. B., 1995, A human calcium-activated potassium channel gene expressed in vascular smooth muscle, *Am. J. Physiol.* **269:**H767–H777.

McManus, O. B., 1991, Calcium-activated potassium channels, *J. Bioenerg. Biomembr.* **23:**537–560.

McManus, O. B., and Magleby, K. L., 1988, Kinetic states and modes of single large-conductance calcium-activated potassium channels in cultured rat muscles. *J. Physiol. (London)* **402:**79–120.

McManus, O. B., and Magleby, K. L., 1991, Accounting for the Ca^{2+}-dependent kinetics of single large-conductance Ca^{2+}-activated K^+ channels in rat skeletal muscle, *J. Physiol. (London)* **443:**739–777.

McManus, O. B., Helms, L. M. H., Pallanck, L., Ganetsky, B., Swanson, R., and Leonard, R. J., 1995, Functional role of the beta subunit of high-conductance calcium-activated potassium channels, *Neuron* **14:**645–650.

Miller, C., 1987, Trapping single ions inside single ion channels, *Biophys. J.* **52:**123–126.

Miller, C., Moczydlowski, E., Latorre, R., and Phillips, M., 1985, Charybdotoxin, a protein inhibitor of single Ca^{2+}-activated K^+ channels from mammalian skeletal muscle, *Nature* **313:**316–318.

Miller, C., Latorre, R., and Reisin, I., 1987, Coupling of voltage-dependent gating and Ba^{2+} block in the high conductance Ca^{2+}-activated K^+ channel, *J. Gen. Physiol.* **90:**427–449.

Moczydlowski, E., and Latorre, R., 1983, Gating kinetics of Ca^{2+}-activated K^+ channels from rat muscle incorporated into planar lipid bilayers. Evidence for two voltage-dependent Ca^{2+} binding reactions, *J. Gen. Physiol.* **82:**511–542.

Murrell-Lagnado, R. D., and Aldrich, R. W., 1993, Interactions of amino terminal domains of *Shaker* K channels with a pore blocking site studied with synthetic peptides, *J. Gen. Physiol.* **102:**949–975.

Neely, A., and Lingle, C. J., 1992a, Two components of calcium-activated potassium current in rat adrenal chromaffin cells, *J. Physiol. (London)* **453:**97–131.

Neely, A., and Lingle, C. J., 1992b, Effects of muscarine on single rat adrenal chromaffin cells, *J. Physiol. (London)* **453:**133–166.

Oda, M., Yoshida, A., and Ikemoto, Y., 1992, Blockade by local anesthetics of the single Ca^{2+}-activated K^+ channel in rat hippocampal neurones, *Br. J. Pharmacol.* **105:**63–70.

Pallanck, L., and Ganetsky, B., 1994, Cloning and characterization of human and mouse homologues of the Drosophila calcium-activated potassium channel gene, *slowpoke*, *Hum. Mol. Genet.* **3:**1239–1243.

Pallotta, B. S., 1985a, Calcium-activated potassium channels in rat muscle inactivate from a short-duration open state, *J. Physiol. (London)* **363:**501–516.

Pallotta, B. S., 1985b, N-Bromoacetamide removes a calcium-dependent component of channel opening from calcium-activated potassium channels in rat skeletal muscle, *J. Gen. Physiol.* **86:**601–611.

Park, Y. B., 1994, Ion selectivity and gating of small conductance Ca^{2+}-activated K^+ channels in cultured rat adrenal chromaffin cells, *J. Physiol. (London)* **481:**555–570.

Patton, D. E., West, J. W., Catterall, W. A., and Goldin, A. L., 1992, Amino acid residues required for fast Na^+-channel inactivation: Charge neutralizations and deletions in the III–IV linker, *Proc. Natl. Acad. Sci. USA* **89:**10905–10909.

Patton, D. E., West, J. W., Catterall, W. A., and Goldin, A. L., 1993, A peptide segment critical for sodium channel inactivation functions as an inactivation gate in a potassium channel, *Neuron* **11:**967–974.

Pennefather, P., Lancaster, B., Adams, P. R., and Nicoll, R. A., 1985, Two distinct Ca dependent K currents in bullfrog sympathetic ganglion cells, *Proc. Natl. Acad. Sci. USA* **82:**3040–3044.

Rae, J., Cooper, K., Gates, P., and Watsky, M., 1991, Low access resistance perforated patch recordings using amphotericin-B, *J. Neurosci. Methods* **37:**15–26.

Reinhart, P. H., Chung, S., and Levitan, I. B., 1989, A family of calcium-dependent potassium channels from brain, *Neuron* **2:**1031–1041.

Reinhart, P. H., Chung, S., Martin, B. L., Brautigan, D. L., and Levitan, I. B., 1991, Modulation of calcium-activated potassium channels from rat brain by protein kinase A and phosphatase 2A, *J. Neurosci.* **11:**1627–1635.

Rettig, J., Heinemann, S. H., Wunder, F., Lorra, C., Parcej, D. N., Dolly, J. O., and Pongs, O., 1994, Inactivation properties of voltage-gated K^+ channels altered by presence of b-subunit, *Nature* **369:**289–294.

Ritchie, A. K., 1987a, Two distinct calcium-activated potassium currents in a rat anterior pituitary cell line, *J. Physiol. (London)* **385:**591–609.

Ritchie, A. K., 1987b, Thyrotropin-releasing hormone stimulates a calcium-activated potassium current in a rat anterior pituitary cell line, *J. Physiol. (London)* **385:**611–625.

Romey, G., and Lazdunski, M., 1984, The coexistence in rat muscle cells of two distinct classes of Ca^{2+}-dependent K^+ channels with different pharmacological properties and different physiological functions, *Biochem. Biophys. Res. Commun.* **118:**669–674.

Rudy, B., 1988, Diversity and ubiquity of K channels, *Neuroscience* **25**:729–749.

Ruppersberg, J. P., Stocker, M., Pongs, O., Heinemann, S. H., Frank, R., and Koenen, M., 1991, Regulation of fast inactivation of cloned mammalian $I_K(A)$ channels by cysteine oxidation, *Nature* **352**:711–714.

Sah, P., 1992, Role of calcium influx and buffering in the kinetics of a Ca^{2+}-activated K^+ current in rat vagal motoneurons, *J. Neurophysiol.* **68**:2237–2247.

Saito, M., Nelson, C., McCobb, D., Salkoff, L., and Lingle, C. J., 1994, Molecular analysis of BK channels in adrenal chromaffin and PC12 cells, *Soc. Neurosci. Abstrs.* **20**:722.

Sands, S. B., Lewis, R. S., and Cahalan, M. D., 1989, Charybdotoxin blocks voltage-gated K^+ channels in human and murine T lymphocytes, *J. Gen. Physiol.* **93**:1061–1074.

Schreiber, M., Lingle, C., Wei, A. D., and Salkoff, L. B., 1994, A putative Ca^{2+}-sensing site in the mouse BK-type K^+ channel, *Soc. Neurosci. Abstrs.* **20**:721.

Shen, K.-Z., Lagrutta, A., Davies, N. W., Standen, N. B., Adelman, J. P., and North, R. A., 1994, Tetraethyl-ammonium block of *Slowpoke* calcium-activated potassium channels expressed in *Xenopus* oocytes: Evidence for tetrameric channel formation, *Pflügers Arch.* **426**:440–445.

Singer, J. J., and Walsh, Jr., J. V., 1987, Characterization of calcium-activated potassium channels in single smooth muscle cells using the patch-clamp technique, *Pflügers Arch.* **408**:98–111.

Smart, T. G., 1987, Single calcium-activated potassium channels recorded from cultured rat sympathetic neurones, *J. Physiol. (London)* **389**:337–360.

Smith, P. A., Bokvist, K., Arkhammar, P., Berggren, P.-O., and Rorsman, P., 1990, Delayed rectifying and calcium-activated K^+ channels and their significance for action potential repolarization in mouse pancreatic β-cells, *J. Gen. Physiol.* **95**:1041–1059.

Solaro, C. R., and Lingle, C. J., 1992, Trypsin-sensitive, rapid inactivation of a calcium-activated potassium channel, *Science* **257**:1694–1698.

Solaro, C. R., Prakriya, M., Ding, J. P., and Lingle, C. J., 1995a, Inactivating and non-inactivating Ca^{2+} and voltage-dependent K^+ current in rat adrenal chromaffin cells, *J. Neurosci.* **15**:6110–6123.

Solaro, C. R., Nelson, C., Wei, A., Salkoff, L., and Lingle, C. J., 1995b, Cytoplasmic Mg^{2+} modulates Ca^{2+}-dependent activation of *mSlo* by binding to a low affinity site on the channel core, *Biophys. J.* **68**:A30.

Stoehr, S. J., Smolen, J. E., Holz, R. W., and Agranoff, B. W., 1986, Inositol triphosphate mobilizes intracellular calcium in permeabilized adrenal chromaffin cells, *J. Neurochem.* **46**:637–640.

Tabcharani, J. A., and Misler, S., 1989, Ca^{2+}-activated K^+ channel in rat pancreatic islet B cells: Permeation, gating and blockade by cations, *Biochim. Biophys. Acta* **982**:62–72.

Toro, L., Stefani, E., and Latorre, R., 1992, Internal blockade of a Ca^{2+}-activated K^+ channel by Shaker B inactivating "ball" peptide, *Neuron* **9**:237–245.

Tse, A., and Hille, B., 1992, GnRH-induced Ca^{2+} oscillations and rhythmic hyperpolarizations of pituitary gonadotropes, *Science* **255**:462–464.

Tseng-Crank, J., Foster, C. D., Krause, J. D., Mertz, R., Godinot, N., DiChiara, T. J., and Reinhart, P. H., 1994, Cloning, expression, and distribution of functionally distinct Ca^{2+}-activated K^+ channel isoforms from human brain, *Neuron* **13**:1315–1330.

Twitchell, W. A., and Rane, S. G., 1993, Opioid modulations of Ca^{2+}-dependent K^+ and voltage-activated Ca^{2+} currents in bovine adrenal chromaffin cells, *Neuron* **10**:701–709.

Van Renterghem, C., and Lazdunski, M., 1992, A small-conductance charybdotoxin-sensitive, apamin-resistant Ca^{2+}-activated K^+ channel in aortic smooth muscle cells (A7r5 line and primary culture), *Pflügers Arch.* **420**:417–423.

Varnai, P., Demaurex, N., Jaconi, M., Schlegel, W., Lew, D. P., and Krause, K. H., 1993, Highly co-operative Ca^{2+} activation of intermediate-conductance K^+ channels in granulocytes from a human cell line, *J. Physiol. (London)* **472**:373–390.

Vasillev, P. M., Scheuer, T., and Catterall, W. A., 1988, Identification of an intracellular peptide segment involved in sodium channel inactivation, *Science* **241**:1658–1661.

Vergara, C., Moczydlowski, E., and Latorre, R., 1984, Conduction, blockade, and gating in a Ca^{2+}-activated K^+ channel incorporated into planar lipid bilayers, *Biophys. J.* **45**:73–76.

Villarroel, A., Alvarez, O., Oberhauser, A., and Latorre, R., 1988, Probing a Ca^{2+}-activated K^+ channel with quaternary ammonium ions, *Pflügers Arch.* **413**:118–126.

Wei, A., Solaro, C., Lingle, C., and Salkoff, L., 1994, Calcium sensitivity of BK-type K_{Ca} channels determined by a separable domain, *Neuron* **13**:671–681.

West, J. W., Patton, D. E., Scheuer, T., Wang, Y., Goldin, A. L., and Catterall, W. A., 1992, A cluster of hydrophobic amino acid residues required for fast Na^+-channel inactivation, *Proc. Natl. Acad. Sci. USA* **89:**10910–10914.

White, R. E., Schonbrunn, A., and Armstrong, D. L., 1991, Somatostatin stimulates Ca^{2+}-activated K^+ channels through protein dephosphorylations, *Nature* **351:**570–573.

Zagotta, W. N., Germeraad, S., Garber, S. S., Hoshi, T., and Aldrich, R. W., 1989, Properties of *Sh*B A-type potassium channels expressed in *Shaker* mutant *Drosophila* by germline transformation, *Neuron* **3:** 773–782.

Zagotta, W. N., Hoshi, T., and Aldrich, R. W., 1990, Restoration of inactivation in mutants of *Shaker* potassium channels by a peptide derived from ShB, *Science* **250:**568–571.

CHAPTER 8

REGULATION OF CALCIUM RELEASE CHANNEL IN SARCOPLASMIC RETICULUM

MICHIKI KASAI and TORU IDE

1. INTRODUCTION

The sarcoplasmic reticulum (SR) plays a key role in excitation-contraction coupling in muscle cells by regulating the intracellular Ca^{2+} concentration in response to an electrical signal from the motor nerve. However, the mechanism underlying the Ca^{2+} release from SR has remained unresolved (Ashley et al., 1991; Endo, 1977; Martonosi, 1984; Meissner, 1994; Williams, 1992). Recently, the Ca^{2+}-induced Ca^{2+} release channel was isolated as a ryanodine receptor (Fleischer et al., 1985; Imagawa et al., 1987; Inui et al., 1987; Lai et al., 1988), and its primary structure has been deduced by cDNA analysis (Takeshima et al., 1989). In skeletal muscle, only this channel is considered to participate in Ca^{2+} release from SR (Meissner, 1994). This channel is characteristically activated by Ca^{2+}, caffeine, and ATP and blocked by Mg^{2+}, procaine, and ruthenium red (Nagasaki and Kasai, 1983; Yamamoto and Kasai, 1982a,b; Tatsumi et al., 1988). To understand the mechanism of Ca^{2+} release from SR, investigation of the properties of the Ca^{2+} channel is important.

   The channel conceptually consists of gate, pore, and filter. To understand the properties of these elements, studies of the permeation of various ions and of the effect of various regulators are essential, since the former reflects the properties of the pore and the filter and the latter reflects mainly the properties of the gate. Especially important

MICHIKI KASAI and TORU IDE • Department of Biophysical Engineering, Faculty of Engineering Science, Osaka University, Toyonaka, Osaka 560, Japan.

Ion Channels, Volume 4, edited by Toshio Narahashi, Plenum Press, New York, 1996.

is the study of the effect of endogenous regulators because such regulators may participate in the excitation-contraction coupling mechanism.

To study the permeability and the regulation of the channel *in vitro*, two techniques
have been employed by various researchers (Meissner, 1994): the investigation of the
permeability properties of SR vesicles and single-channel recordings in lipid planar
bilayer. Although single-channel measurements provide more direct information, an
advantage of the vesicle flux technique is that it more readily yields representative
data by averaging the kinetic behavior of a large number of channels. Also, a possible
loss of alteration of function during channel purification and reconstitution into a foreign
lipid environment is less likely. Thus, the two techniques are used complementarily.

In this review, we summarize the results obtained by using various methods to study
the permeation of various ions and neutral molecules: the tracer method, the light
scattering method, the fluorescence quenching method and the lipid bilayer method
(Kasai and Nunogaki, 1988). The light scattering method was mainly used, since this
method is conventional and the time course of the permeation of slowly permeable
permeants could be followed easily (Kometani and Kasai, 1978; Kasai and Nunogaki,
1988). In this investigation, we used heavy fractions of sarcoplasmic reticulum (HSR)
vesicles, which contain a large number of ryanodine receptor Ca^{2+} channels, prepared
from rabbit skeletal muscle according to the method described previously (Kasai *et al.*,
1992).

2. PERMEABILITY OF Ca^{2+} CHANNEL

2.1. Choline and Monovalent Cation Permeation

The Ca^{2+} channel was identified as a Ca^{2+}-induced Ca^{2+} release channel *in vitro*
by flux measurement and was recently identified as a ryanodine receptor. For the first
time, we found this channel permeated choline when choline permeation was followed
by the light scattering method (Yamamoto and Kasai, 1981). First, we will explain the
principle of this method, since it is very convenient and was employed throughout this
report (Kasai and Nunogaki, 1988). As seen in the Fig. 1A, when the permeant concentration suspending membrane vesicles was increased suddenly by the stopped-flow apparatus, the scattered light intensity increased rapidly at the beginning and then decreased.
The fast increase is caused by the decrease of volume of the vesicles due to the outflow of
water and the subsequent decrease is caused by the increase of the volume due to the
inflow of the permeants and water. From the increasing phase, the permeation rate of
water can be estimated, and from the decreasing phase the permeation rate of the
permeants can be estimated. When choline Cl is used as a permeant, choline permeation
rate can be estimated since the permeation of Cl^- is much faster than that of choline
(Kometani and Kasai, 1978).

Figure 1A shows a typical example of choline influx to HSR vesicles. Choline influx
increased in the presence of micromolar concentrations of extravesicular Ca^{2+} (cytosolic
side) and inhibited in millimolar Ca^{2+}. The choline influx rate is shown in Fig. 1B as
a function of Ca^{2+} concentrations. Choline influx increased in the presence of micro-

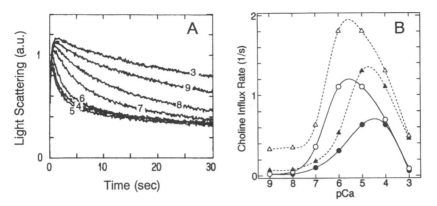

FIGURE 1. Choline permeation measured by light scattering method. The osmotic volume change of HSR vesicles was monitored by measuring the scattered light intensity (450 nm) at right angles to the incident beam using a stopped-flow spectrophotometer as described elsewhere (Kasai and Nunogaki, 1988; Kasai et al., 1992). HSR vesicles were diluted to 0.5 mg protein/ml in 5 mM tris maleate (pH 7.0). The diluted vesicles were mixed by a stopped-flow apparatus with a mixing solution containing 500 mM choline Cl, 5 mM tris maleate (pH 7.0), calcium buffer with various free calcium concentrations prepared with 2.5 mM EGTA and in the presence and absence of caffeine and/or ATP. (A) Change of the scattered light intensity observed at various Ca^{2+} concentrations without caffeine and ATP. Free Ca^{2+} concentrations, pCa values, are shown in the figure. (B) Choline permeation rate as a function of free Ca^{2+} concentration. Symbols: (●) without caffeine and ATP, (○) 5 mM caffeine, (▲) 0.5 mM ATP, (△) 5mM caffeine and 0.5 mM ATP. Note that the final concentrations of choline Cl and the calcium buffer were half of these values.

molar Ca^{2+} and decreased in millimolar Ca^{2+}. Such behavior of Ca^{2+} concentration dependence is a typical property of Ca^{2+}-induced Ca^{2+} release channel (ryanodine receptor). This result indicates that the Ca^{2+} channel has two kinds of Ca^{2+} binding sites: an activation site with high affinity for Ca^{2+} (micromolar) and an inhibition site with low affinity for Ca^{2+} (millimolar). It was also observed that this choline influx increased in the presence of millimolar caffeine and that the Ca^{2+} concentration required to open the Ca^{2+} channel shifted to the lower concentration of Ca^{2+} as shown in Fig. 1B. This result was interpreted as an increase in affinity of Ca^{2+} for the activation site by caffeine. ATP also increased the permeation rate without changing the apparent affinity of Ca^{2+} for activation and inhibition sites (Yamanouchi et al., 1984). These properties are typical of the Ca^{2+} channel in SR membrane. From these results, we have concluded that choline permeates the Ca^{2+} channel as Ca^{2+} does.

However, it was difficult to measure the permeation of small cations such as K^+ or Na^+ through the Ca^{2+} channel by this technique because HSR vesicles have another cation channel (Coronado et al., 1980). Later, permeation of monovalent cations was shown directly by single-channel recording using the lipid bilayer method (Lai et al., 1988; Smith et al., 1988). The Ca^{2+} channel has a large conductance, for example, 600 and 750 pS with symmetric 500 mM Na^+ and 250 mM K^+ as the current carrier, respectively (Meissner, 1994). Furthermore, the channel conducts large monovalent cations such as choline and tris, whose conductances are 26 and 22 pS, respectively (Smith et al., 1988).

2.2. Ca²⁺ and Divalent Cation Permeation

Measurement of fast flux of Ca^{2+} through the Ca^{2+} channel is difficult by the usual techniques. The Ca^{2+} efflux from HSR vesicles was measured by the fluorescence quenching method using chlortetracycline as a probe (Nagasaki and Kasai, 1984). Figure 2A shows a typical example of Ca^{2+} efflux from Ca^{2+} loaded vesicles. Ca^{2+} efflux increased in the presence of micromolar Ca^{2+} and decreased in millimolar Ca^{2+}. The Ca^{2+} efflux rate is shown in Fig. 2B as a function of Ca^{2+} concentrations. The result is similar to that of choline permeation shown in Fig. 1B, indicating the typical properties of the Ca^{2+} channel.

We found that the permeability of Mg^{2+} could be followed as easily as that of Ca^{2+} by this technique. The result of Mg^{2+} permeation is also shown in Fig. 2. The permeation rate of Mg^{2+} is exactly the same as that of Ca^{2+}.

Permeability of Ca^{2+}, Mg^{2+}, and Ba^{2+} was also studied by single-channel recording after fusing SR vesicles to lipid bilayer. The *trans* side (lumen side) was filled with these divalent cations and the *cis* side with monovalent cation such as tris or choline, and Ca^{2+} concentration of the *cis* side was changed. The Ca^{2+} channel was activated by micromolar Ca^{2+} and caffeine and ATP. Single-channel conductances were about 100 pS for Ca^{2+}, 40 pS for Mg^{2+} (Smith *et al.*, 1986), and 170 pS for Ba^{2+} (Smith *et al.*, 1985).

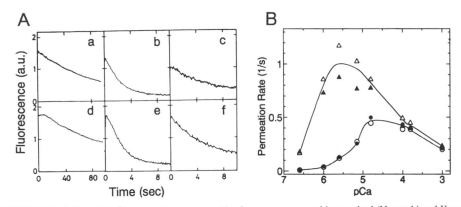

FIGURE 2. Ca^{2+} and Mg^{2+} permeation measured by fluorescence quenching method (Nagasaki and Kasai, 1984). HSR vesicles (10 mg protein/ml) were incubated in solution containing 150 mM KCl, 20 mM tris maleate (pH 6.7), 10 mM $CaCl_2$ or $MgCl_2$ overnight at 4° C. When Ca^{2+} efflux was to be determined, a 20 μl aliquot was diluted 100 times in a solution containing 150 mM KCl, 20 mM tris maleate (pH 6.7), 4.5 mM EGTA, 0.1 mM $MgCl_2$, 0.4 mM $CaCl_2$, and 20 μM chlortetracycline. When Mg^{2+} efflux was to be measured, the dilution medium was changed to a solution of 150 mM KCl, 20 mM tris maleate (pH 6.7), 4.5 mM EGTA, 0.5 mM $CaCl_2$, and 20 μM chlortetracycline. Then, 45 sec after the dilution, both solutions were mixed with an equal volume of a solution containing 150 mM KCl, 20 mM tris maleate (pH 6.7), 4.5 mM EGTA, 0.1 mM $MgCl_2$, 20 μM chlortetracycline, and a specified amount of Ca^{2+}. Fluorescence change was recorded at 470 nm by excitation at 390 nm at 23° C. (A) Fluorescence change: (a–c) Ca^{2+} efflux and (d–f) Mg^{2+} efflux. Extravesicular free Ca^{2+} concentrations; (a,d) 0.25 μM, (b,e) 16 μM, (c,f) 1 mM. (B) Efflux rates of Ca^{2+} and Mg^{2+} as a function of free Ca^{2+} concentration. Experiments similar to those in (A) were carried out in the presence and absence of caffeine (12.5 mM final). (●) Ca^{2+} flux in no caffeine, (○) Mg^{2+} flux in no caffeine, (▲) Ca^{2+} flux in caffeine, (△) Mg^{2+} flux in caffeine. Modified from Nagasaki and Kasai (1984).

2.3. Complementarity between Single-Channel and Flux Measurements

In the previous section, we found a discrepancy between two measurements. The flux rate of Mg^{2+} through the Ca^{2+} channel was exactly the same as that of Ca^{2+} from the fluorescence measurements in vesicles, but the conductance ratio, $\gamma_{Mg^{2+}}/\gamma_{Ca^{2+}}$, was about 0.4 from the single-channel conductance measurement. To explain this discrepancy we developed a theoretical model as described in the appendix.

If we assume that the single-channel conductance of Mg^{2+} is 40 pS, the diameter of the vesicles is 0.1 μm, and each vesicle has one channel, the permeation rate, k_f, is expected to be 7.5 msec (Kasai and Nunogaki, 1993). Since this value is larger than the gating rate of Ca^{2+} channel, we should apply Eq. (1) in the appendix to the permeation of Ca^{2+} and Mg^{2+} according to the theory. Thus, we could conclude that the result of the flux measurement reflected not the permeation rate but the open rate of the channel. The same apparent flux rate for Ca^{2+} and Mg^{2+} determined by fluorescence quenching shows the same open rate for Ca^{2+} and Mg^{2+} permeation. On the contrary, the real permeation rates, which are proportional to the conductance, are different for Ca^{2+} and Mg^{2+}. Accordingly, two kinds of measurement give us different and complementary information.

3. REQUIREMENT OF POTASSIUM CHLORIDE FOR NEUTRAL MOLECULE PERMEATION

3.1. Glucose Permeation

Since the Ca^{2+} channel incorporated into a lipid bilayer showed very large conductance (100 pS in 50 mM $CaCl_2$ and 1 nS in 1 M NaCl) (Smith *et al.*, 1985; Liu *et al.*, 1989), the channel is thought to have a large pore for solute permeation. Our observation that large cations such as choline or tris can permeate through the channel, and Meissner's finding that the channel passes glucose, as demonstrated by the tracer method (Meissner, 1986), also support the idea that the channel has a large pore.

At first, in order to confirm the result reported by Meissner (1986), glucose efflux from HSR vesicles was measured by the tracer method. As shown in Fig. 3A$_a$, glucose efflux was not accelerated by micromolar Ca^{2+} (pCa 5) or millimolar caffeine when KCl was not present. On the contrary, when submolar concentrations of KCl were present, glucose efflux was increased in the presence of micromolar Ca^{2+} (pCa 5) and this increment was enhanced by millimolar caffeine (Fig. 3A$_{b,c}$). The new finding is that submolar concentrations of KCl are required for glucose permeation. When the KCl concentration was further increased, the effects of Ca^{2+} and caffeine became more significant.

In the next step, the glucose influx was examined by the light scattering method, because the whole time course of the glucose permeation can be monitored on the subsecond time scale. Figure 4A shows a typical example of the light scattering change caused by glucose influx. Figure 4A$_a$ shows that an addition of micromolar Ca^{2+} (pCa 4) and millimolar caffeine and ATP did not affect glucose influx in the absence of KCl. Next, we followed the glucose influx in the presence of KCl by the same method. The

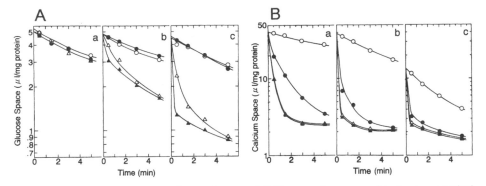

FIGURE 3. Effect of KCl concentration on glucose and Ca^{2+} permeation measured by the tracer method. HSR vesicles (10 mg protein/ml) were incubated overnight at $0°$ C in a solution containing various concentrations of KCl in addition to 1 mM $CaCl_2$, 1 mM glucose, 10 mM tris maleate (pH 7.0), and radioactive tracers. The amounts of radioactive tracers were 2 μCi/ml [^{14}C]glucose and 1 μCi/ml ^{45}Ca for glucose and Ca^{2+} release, respectively. An aliquot (100 μl) of each suspension was diluted 45-fold at room temperature with a dilution medium containing 1 mM glucose, 10 mM tris maleate (pH 7.0), calcium buffer prepared with 1 mM EGTA, and the same concentrations of KCl as for incubation in the presence and absence of 10 mM caffeine. Then, at each measurement time, a 1-ml aliquot was rapidly filtered and the amount of glucose or Ca^{2+} remaining in SR vesicles was determined. The radioactivities were expressed as isotope space. KCl concentrations are (a) 0, (b) 100 mM, and (c) 400 mM. Symbols: (○), pCa 9 and no caffeine, (●) pCa 9 and 10 mM caffeine, (△) pCa 5 and no caffeine, (▲) pCa 5 and 10 mM caffeine. (A) Glucose release. (B) Ca^{2+} release. Modified from Kasai et al. (1992).

same concentrations of KCl were added to the vesicular solutions and the mixing solutions. As discussed in our previous reports (Kometani and Kasai, 1978), the permeation of KCl does not interfere with the glucose permeation since KCl permeates very fast. In the presence of 100 mM KCl, a large increase in glucose influx was detected in the presence of caffeine and ATP at pCa 4 (Fig. $4A_b$). In the presence of 400 mM KCl, the presence of micromolar Ca^{2+} and millimolar caffeine and ATP increased the glucose influx (Fig. $4A_c$). These data are consistent with those in Fig. 3A and show that the light scattering method is useful in examining glucose permeation in the presence of KCl. When KCl concentrations were increased, the effect saturated at about 1 M when measured up to 3 M KCl (see Fig. 6).

To clarify whether the permeation of glucose observed in the presence of KCl is really mediated through the Ca^{2+}-induced Ca^{2+} release channel, the effect of Ca^{2+} concentration was studied in detail in the presence of 1 M KCl. As shown in Fig. 4B, the results were similar to those obtained for choline permeation shown in Fig. 1B. Thus, glucose permeation increased at micromolar Ca^{2+} and decreased at millimolar Ca^{2+}, and it was activated by millimolar caffeine and ATP. Furthermore, millimolar Mg^{2+} and micromolar ruthenium red blocked the glucose permeation (data not shown). The KCl-dependent glucose permeation was not observed in the light fraction of SR (LSR) vesicles. These properties are consistent with those of the Ca^{2+}-induced Ca^{2+} release channel.

The effect of KCl concentrations was studied more in detail. As shown in Fig. 5A,

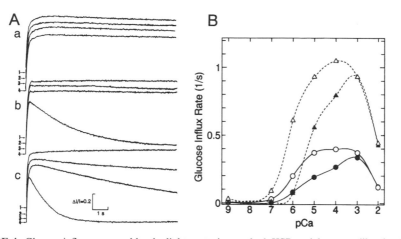

FIGURE 4. Glucose influx measured by the light scattering method. HSR vesicles were diluted to 0.4 mg protein/ml in 5 mM tris maleate (pH 7.0) containing various concentrations of KCl. The diluted vesicles were mixed by a stopped-flow apparatus with an equal volume of a mixing solution containing 200 mM glucose, 5 mM tris maleate (pH 7.0), calcium buffer prepared with 2.5 mM EGTA, and the same concentrations of KCl as for dilution in the presence and absence of caffeine and ATP. Then, the change of the scattered light intensity was recorded. (A) Effect of KCl concentration on glucose permeation. KCl concentrations: (a) 0, (b) 100 mM, and (c) 400 mM. In each trace, the conditions of the mixing solutions are: (1) pCa 9, without caffeine and ATP; (2) pCa 9, 10 mM caffeine and 1 mM ATP; (3) pCa 4, without caffeine and ATP; (4) pCa 4, 10 mM caffeine and 1 mM ATP. Note that the final concentrations of glucose, calcium buffer, caffeine, and ATP were half of these values. The arrows show the starting points of light scattering changes in each trace. (B) Calcium concentration dependence of glucose influx rate measured in 1 M KCl and in the absence and presence of caffeine and ATP. Symbols: ($\bullet$) without caffeine and ATP, ($\circ$) 10 mM caffeine, ($\blacktriangle$) 1 mM ATP, ($\triangle$) 10 mM caffeine and 1 mM ATP. Modified from Kasai *et al.* (1992).

the maximal rates of glucose influx increased with KCl concentrations and saturated at about 1 M KCl (Fig. 6A). In parallel, the Ca^{2+} concentration required to open the Ca^{2+} channel increased with increase in KCl concentrations. The shift of effective Ca^{2+} concentration can be explained by the competition between Ca^{2+} and K^+ at the activation site (see Section 4.2).

Next, the effect of molecules or salts other than KCl on the glucose permeation was studied. Sodium chloride and KNO_3 behaved like KCl qualitatively, whereas increase of the concentration of nonelectrolytes such as sucrose, glucose, and glycerol did not enhance the glucose permeation as measured by the tracer method. These results suggest that the effect of KCl on the Ca^{2+} channel, i.e., making it able to pass neutral molecules, is due to ionic strength and not to osmotic pressure of the solution.

3.2. Ca^{2+} Permeation

The above results showed that the Ca^{2+} channel did not pass glucose in the absence of KCl even when the Ca^{2+} channel is considered to be open in the presence of micromolar Ca^{2+} and millimolar caffeine. In order to know whether the gate of the Ca^{2+}

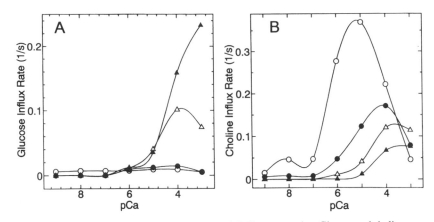

FIGURE 5. Effect of KCl concentrations on glucose and choline permeation. Glucose and choline permeation were measured as in Fig. 4 with changing KCl and Ca²⁺ concentrations. Glucose and choline concentrations of the mixing solutions were 200 mM. KCl concentrations: (○) 0, (●) 200 mM, (△) 500 mM, (▲) 1 M. (A) Glucose. (B) Choline permeations.

channel is really open under such conditions, Ca^{2+} release from HSR vesicles was studied under the same conditions used to study the glucose permeation by the tracer method. As shown in Fig. 3B, even in the absence of KCl, Ca^{2+} is released from HSR vesicles at pCa 5 both in the absence and in the presence of caffeine, and furthermore Ca^{2+} is released in the presence of caffeine even at pCa 9. These results indicate that the Ca^{2+} channel is open if micromolar Ca^{2+} and/or millimolar caffeine are present even in the absence of KCl. Furthermore, with increase of KCl concentration, an increase of Ca^{2+} release rate was observed. This effect of KCl may be essentially the same as that observed in the glucose permeation. The results in Fig. 3B include some interesting points in addition to the above findings. Ca^{2+} space extrapolated to $t = 0$ was very high. The high value is attributable to the effect of binding of Ca^{2+} to intravesicular proteins or membrane. Decrease of the calcium space with increasing KCl concentration can be explained by the screening effect of salt on the Ca^{2+} binding sites or competition of K^+ with Ca^{2+} for the site.

3.3.   Choline Permeation

Figure 5B shows the effect of KCl concentration on choline permeation. The Ca^{2+} concentration required to open the Ca^{2+} channel increased with increase in KCl concentrations as in the case of glucose. This effect is also explained by the competition between Ca^{2+} and K^+ at the activation site as in the case of glucose. However, the maximal rates of choline permeation decreased with KCl concentrations (Fig. 6B), which is opposite to the case of glucose. This effect may be due to the competition between K^+ and choline in the pore. Similar phenomena in Figs. 5B and 6B were observed in the presence of NaCl instead of KCl.

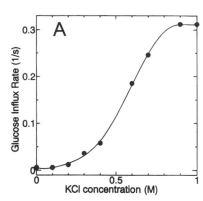

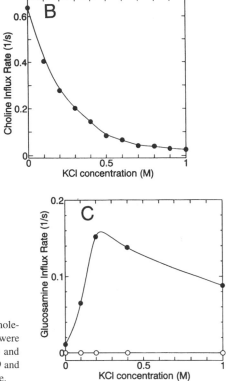

FIGURE 6. Effect of KCl concentrations on various molecules. Glucose, choline, and glucosamine permeations were measured as in Fig. 5. Permeation rates for glucose and choline were at pCa 4 and for glucosamine at (○) pCa 9 and (●) pCa 5. (A) Glucose. (B) Choline. (C) Glucosamine.

3.4. Permeation of Other Molecules

To test whether molecules other than glucose can permeate the Ca^{2+} channel, we studied permeations of other neutral molecules (Kasai *et al.*, 1992). It was found that xylose did not permeate through the Ca^{2+} channel in the absence of KCl, but did in the presence of KCl. Since xylose is a slightly smaller molecule than glucose, permeation rates were about twice those of glucose. Except for the absolute values of permeation rates, the overall features of xylose permeation were similar to those shown in Fig. 4. Next, in order to see whether amphoteric molecules can permeate through the Ca^{2+} channel like glucose, influx of glycine was studied. Except for the fact that the flux rates under all conditions are about four times larger than those of glucose, the overall features of glycine permeation are the same as those of glucose. These results indicate that xylose and glycine permeate through the Ca^{2+} channel in the same way as glucose. On the contrary, sucrose did not permeate under the same conditions, suggesting that the pore size of the Ca^{2+} channel is smaller than that of sucrose even in the presence of KCl.

The effect of KCl on the permeation of glucosamine was studied. At pH 7, glucosamine is considered to be a monovalent cation since its pK value is about 7.8. As shown in Fig. 6C, the flux rate increased at low KCl concentration and decreased at higher concentration. The behavior at low concentration is similar to that of glucose; at high concentration it is similar to that of choline.

Tetramethylammonium ion also permeated the Ca^{2+} channel. The effect of KCl was similar to the case of choline. Permeation of diaminopropane was also studied. This molecule may exist as a divalent cation. However, it did not permeate the channel even in the presence of KCl, perhaps because it is too large.

3.5. Model of Conformational Change in the Ca^{2+} Channel

We found that glucose permeates through HSR vesicles when the Ca^{2+} release channel is open in the presence of submolar concentrations of KCl, but does not do so in the absence of KCl. It has been reported that there exist at least four substates in the Ca^{2+} channel, based on measurements of Ca^{2+} and Na^+ conductances in a lipid bilayer system (Liu *et al.*, 1989). Occupation probability in each substate was dependent on ionic concentration and higher substates were occupied at higher ionic concentrations. As a result, the macroscopic conductance of Ca^{2+} (or Na^+) increases with ionic concentration. This effect may be related to our observation that flux rates of Ca^{2+} and neutral molecules increased with ionic strength. However, this effect cannot explain why glucose did not permeate in the absence of KCl. We propose a more plausible model, i.e., that the gating of the Ca^{2+} channel is little affected by ionic strength, but the conformation of the pore region of the channel is modified by ionic strength. A schematic representation of our model is shown in Fig. 7. In this model, the gate of the channel is opened by micromolar Ca^{2+} and millimolar caffeine and ATP, but the pore is too small to pass glucose in the absence of KCl. With an increase in KCl concentration, the pore in the channel molecule becomes large enough to allow both Ca^{2+} and neutral molecules to pass. If this is the case, we can expect the permeation of neutral molecules smaller than glucose even in the absence of KCl. Xylose did not display such an effect, but an even smaller molecule might do so. As far as the ion selectivity of the channel after the change of pore size is concerned, it may be maintained since ions are considered to permeate through binding to the selective filter.

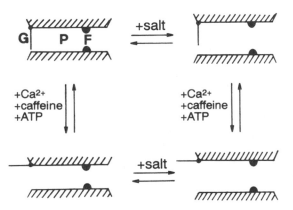

FIGURE 7. Schematic representation of conformational change of Ca^{2+} channel in the presence of salt. The channel consists of three parts: G, gate; P, pore (or path); F, ion-selective filter. The gate opens in the presence of Ca^{2+}, caffeine, and ATP, independent of the presence of salt. Ca^{2+} can permeate the channel even in the absence of salt, but glucose cannot. In the presence of salt, the channel becomes wider, so that it allows both Ca^{2+} and glucose to pass. Since ions permeate by binding to the selective filter, ion selectivity may be maintained when the pore size is changed. Modified from Kasai *et al.* (1992).

4. EFFECTS OF POTASSIUM CHLORIDE ON REGULATION OF CALCIUM CHANNEL

4.1. Activation and Inhibition Sites of Ca^{2+} Channel

As shown in Fig. 1B, the Ca^{2+} channel is open in micromolar Ca^{2+} and closed in millimolar Ca^{2+}. The activation constant (Ca^{2+} concentration required to open the channel) was decreased about tenfold by 5 mM caffeine, but the inhibition constant (Ca^{2+} concentration required to close the channel) was not affected by caffeine. ATP increased the permeation rate over the whole range of pCa in the presence and absence of caffeine but did not change the activation and inhibition constants for Ca^{2+}. The data indicate that ATP did not interact directly with the Ca^{2+} and caffeine binding sites. These results indicate that there exist at least three different sites of activation, for Ca^{2+}, caffeine, and ATP, and one site of inhibition for Ca^{2+}.

4.2. Effects of KCl on Activation and Inhibition Sites for Ca^{2+}

As shown in Fig. 5, activation and inhibition constants for Ca^{2+} observed in choline and glucose permeation increased with an increase in KCl concentration. The changes in the activation and inhibition constants for Ca^{2+} can be interpreted in terms of competition between K^+ and Ca^{2+} at the Ca^{2+} binding sites for activation and inhibition, respectively.

In Fig. 8, the apparent activation and inhibition constants for Ca^{2+} measured at various KCl concentrations are summarized (Kasai *et al.*, 1995). The apparent activation constants (pK_a) are the same for choline and glucose permeations if the KCl concentrations are the same (Fig. 8A). This result indicates that these properties reflect those of the gate. Caffeine shifted the affinity for Ca^{2+} toward higher pCa. Under the assumption of a single-site titration, the activation constants in the absence of KCl (pK_{a0}) were estimated to be 6.1 ± 0.2 and 6.9 ± 0.2 in the absence and presence of 5 mM caffeine, respectively, and the dissociation constants for K^+ were 120 ± 100 mM and 200 ± 150 mM, respectively. When choline concentrations were increased, the activation constant for Ca^{2+} did not change. Choline does not seem to interact with the Ca^{2+} binding site for activation.

Figure 8B shows the apparent inhibition constants (pK_i) at various KCl concentrations. As in the case of activation, the binding could be analyzed in terms of competition between Ca^{2+} and K^+. The inhibition constants for Ca^{2+} at zero K^+ (pK_{i0}) were 3.6 ± 0.2 and 4.0 ± 0.2 in the absence and presence of 5 mM caffeine, respectively, and dissociation constants for K^+ were 45 ± 40 mM and 30 ± 20 mM, respectively. The results show that caffeine has little effect on the inhibition site. Sodium ion also increased the activation and inhibition constants for Ca^{2+}, as did K^+.

The effect of KCl concentration on Ca^{2+} permeability was studied by fluorescence quenching using chlortetracycline (Nagasaki and Kasai, 1983). The result showed that the apparent Ca^{2+} dissociation constant of the activation site increased with KCl concentration. This effect is the same for glucose and choline permeation. However, the maximal rates of Ca^{2+} release did not change as much with KCl concentrations.

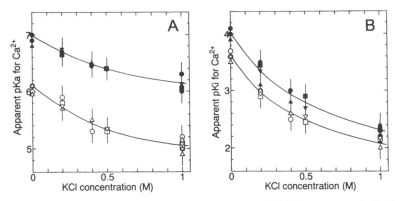

FIGURE 8. KCl concentration dependence of apparent activation and inhibition constants for Ca^{2+}. Choline and glucose permeation were determined in the presence and absence of caffeine and ATP by changing the KCl concentration at 100 mM choline chloride (final) and 100 mM glucose (final) as in Fig. 5. Caffeine and ATP concentrations were 5 mM and 0.5 mM (final), respectively. Apparent activation constant, K_a, was defined as the Ca^{2+} concentration which activated the channel to half of the maximal value and apparent inhibition constant, K_i, was that which inhibited the channel. Symbols: $(\bigcirc, \square)$ without caffeine and ATP, $(\bullet, \blacksquare)$ 5 mM caffeine, $(\triangle, \triangledown)$ 0.5 mM ATP, $(\blacktriangle, \blacktriangledown)$ 5 mM caffeine and 0.5 mM ATP. $(\bigcirc, \bullet, \triangle, \blacktriangle)$ for choline permeation and $(\square, \blacksquare, \triangledown, \blacktriangledown)$ for glucose permeation. The data are means of three to five experiments. The curves were fitted by a nonlinear least-squares method. The assumption was that K^+ and Ca^{2+} compete with each other at the same binding site and the channel opens when Ca^{2+} binds to the activation site and closes when Ca^{2+} binds to the inhibition site. The function used was $K_C/K_{CO} = 1 + ([K]/K_K)$, where K_C and K_{CO} are dissociation constants of Ca^{2+} in the presence and absence of K^+, respectively; K_K is the dissociation constant of K^+; and $[K]$ is the K^+ concentration. (A) Apparent activation constants, pK_a. (B) Apparent inhibition constants, pK_i. Modified from Kasai *et al.* (1995).

4.3.   Effects of KCl on Inhibition Sites for Ruthenium Red and Mg^{2+}

Figure 9 shows the effects of ruthenium red and Mg^{2+} on choline permeation at different concentrations of KCl. The inhibition constants for ruthenium red and Mg^{2+} increased with increases in KCl concentration. For example, K_i for ruthenium red changed from 1 ± 0.5 µM to 40 ± 20 µM and K_i for Mg^{2+} from 0.15 ± 1 mM to 3 ± 2 mM when KCl concentration was changed from 0.2 to 1 M. This result indicates that K^+ competes with ruthenium red or Mg^{2+} at the binding site for these inhibitors. We do not know whether the binding sites for ruthenium red and Mg^{2+} are the same or not. It is likely that ruthenium red and Mg^{2+} bind to the Ca^{2+} binding sites for activation and/or inhibition and block the channel opening (Yamamoto and Kasai, 1982c). Recently, it was suggested, by using a fusion protein of ryanodine receptor, trpE, that bound Ca^{2+} was displaced by ruthenium red and that this site is located in a 35–amino acid region from 4478 to 4512 (Chen *et al.*, 1992). This result suggests that ruthenium red binds to the activation site for Ca^{2+}, which is consistent with our finding, although more detailed analyses are required. When the choline concentration was changed, the apparent inhibition constants for ruthenium red and Mg^{2+} changed slightly (data not shown). Interaction of choline with the binding site for these inhibitors is weak.

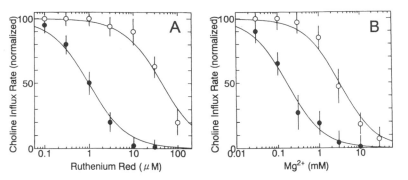

FIGURE 9. Effects of KCl on the action of inhibitors on choline permeation. Choline permeation was measured in the presence and absence of ruthenium red and Mg^{2+}. Choline concentration was 100 mM (final). Influx measurements were carried out in the presence of 5 mM caffeine and 0.5 mM ATP (final) at pCa 5. KCl concentrations: (●) 0.2 M KCl, (○) 1 M KCl. The points are means of three experiments carried out using the same preparation. Curves were fitted by single-site titration functions. (A) Effect of ruthenium red. (B) Effect of Mg^{2+}. Modified from Kasai *et al.* (1995).

5.   EFFECT OF RYANODINE

5.1.   Effect of Ryanodine on Pore Size

In Section 3, we showed that large neutral molecules such as glucose could permeate the Ca^{2+} channel in the presence of submolar concentrations of KCl. This result suggested that the Ca^{2+} channel is a large pore and that its conformation is modified depending on the ionic strength. In order to confirm such a conformational change induced by ionic strength, the effect of KCl on the permeability of choline and glucose through the Ca^{2+} channel opened by ryanodine was studied extensively. It is known that micromolar concentrations of ryanodine open the Ca^{2+} channel and lock it at an open state, and that higher concentrations close the channel (Meissner, 1986; Nagasaki and Fleischer, 1988; Lattanzio *et al.*, 1987).

5.1.1.   Open Locked State Revealed by Choline Permeation

At first, the effect of ryanodine on choline permeation was studied by the light scattering method (Kasai and Kawasaki, 1993). As shown in Fig. 10, before ryanodine treatment, choline does not permeate the Ca^{2+} channel in the absence of Ca^{2+} (pCa 9), but it permeates in the presence of Ca^{2+} (pCa 5). After treatment in 10 μM ryanodine at 37°C for 20 min, the channel allowed choline to pass even in the absence of Ca^{2+} (pCa 9), suggesting that the channel changed to a so-called open locked state. After treatment in 100 μM ryanodine, the channel did not allow choline to pass even in the presence of Ca^{2+} (pCa 5). These results are consistent with those obtained on Ca^{2+} permeation measurements with the tracer method (Meissner, 1986) or single-channel recording (Nagasaki and Fleischer, 1988; Lattanzio *et al.*, 1987; Kawasaki and Kasai, 1989). The results also show that the open locked state obtained on ryanodine treatment could be blocked by Mg^{2+} and ruthenium red (see also Fig. 13).

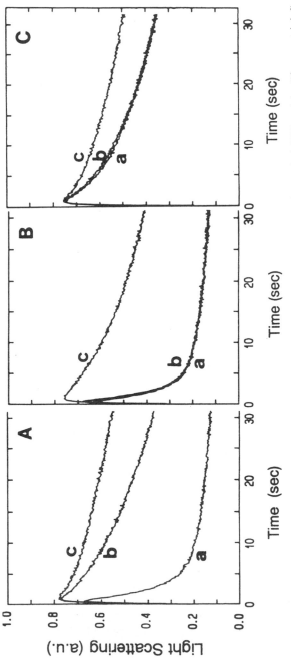

FIGURE 10. Effect of ryanodine on choline permeation. Choline influx was measured by the light scattering method. HSR vesicles (10 mg protein/ml) were treated in 0.1 M KCl, 5 mM tris maleate (pH 7.0), 100 μM free Ca^{2+} buffered with 2.5 mM EGTA at 37° C for 20 min in the presence and absence of ryanodine. The ryanodine concentrations were: (A) 0, (B) 10 μM, (C) 100 μM. The treated vesicles were diluted 20-fold in 5 mM tris maleate (pH 7.0), and choline influx was measured after mixing with a solution comprising 500 mM choline Cl, 5 mM tris maleate (pH 7.0), and calcium buffer prepared with 2.5 mM EGTA. The mixing solution conditions were: (a) pCa 5, (b) pCa 9, (c) pCa 9 plus 10 mM $MgCl_2$ and 10 μM ruthenium red. Modified from Kasai and Kawasaki (1993).

5.1.2.   Glucose Permeation in the Absence of KCl

In Section 3, we showed that glucose could permeate the Ca^{2+} channel in the presence of KCl, but could not permeate in the absence of KCl even in the open state. This result was explained by a model in which the pore size of the channel changed with ionic strength. In order to determine whether or not such a conformational change takes place in the channel locked in the open state by ryanodine, the effect of KCl on the glucose permeation of ryanodine-treated IISR vesicles was studied. Before the ryanodine treatment (Fig. 11A, D), glucose did not permeate even at pCa 4 in the absence of KCl, whereas it permeated in the presence of 1 M KCl as reported in Section 3. After 10 μM ryanodine treatment, glucose was able to permeate even at pCa 9 in the absence of KCl (Fig. 11B). Since glucose did not permeate even in the presence of activators such as caffeine and ATP, this result suggests that the conformation of the channel opened by ryanodine is different from that opened by activators such as Ca^{2+}, caffeine, and ATP. In the presence of 1 M KCl (Fig. 11E), the channel became more permeable and could not be blocked by 10 μM ruthenium red or 10 mM Mg^{2+}, corresponding to the open locked state. This result suggests that the pore of the channel in the open locked state can still be modified by ionic strength and the size of the pore can be increased with increasing KCl concentration. After treatment with 100 μM ryanodine, the Ca^{2+} channel changed to the closed state, where glucose could not permeate even at pCa 4 in the presence of KCl (Fig. 11C,F).

5.1.3.   Small Pore Size in Open Locked State

The effect of activator after ryanodine treatment was studied. As shown in Fig. 12, the permeabilities of choline became insensitive to caffeine or ATP in both the presence and absence of Ca^{2+} after 10 μM ryanodine treatment, but were sensitive to these reagents before the treatment. Interestingly, the maximal rates of choline flux of the ryanodine-treated channel were lower than those of the untreated one, which were attained in the presence of caffeine and/or ATP. Such behavior was also observed for glucose permeation in the presence of KCl after ryanodine treatment (Kasai and Kawasaki, 1993). This result may correspond to the observation on single-channel recording that the maximal conductance of the channel after the ryanodine treatment decreased to 40–50% of the value at the full open state of the untreated channel (Rousseau *et al.*, 1987; Nagasaki and Fleischer, 1988).

5.1.4.   Effect of Other Activators

It was reported that Ag^+, 4-acetoamido-4′-isothiocyanostilbene-2,2′-disulfonic acid (SITS), and polylysine open the Ca^{2+} channel (Tatsumi *et al.*, 1988; Kawasaki and Kasai, 1989). Glucose was found to become permeant in the absence of KCl after Ag^+ treatment. It was also shown that SITS and polylysine made the channel permeable for glucose even in the absence of KCl. These channel openers may act like ryanodine and not like physiological activators such as Ca^{2+}, caffeine, and ATP.

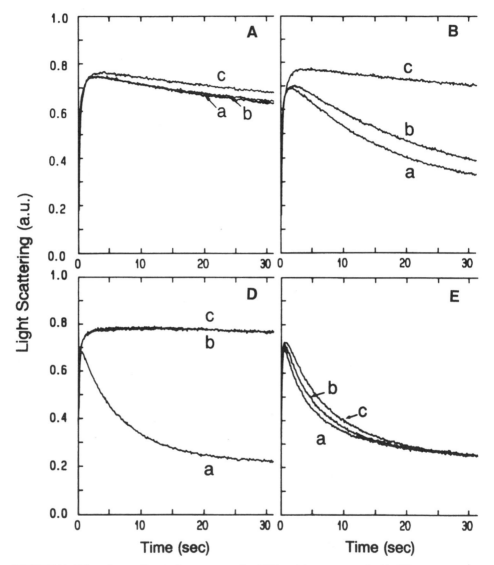

FIGURE 11. Effect of ryanodine on glucose permeation. HSR vesicles were treated with different concentrations of ryanodine as indicated in the legend to Fig. 10. The ryanodine concentrations were: (A,D) 0; (B,E) 10 μM; (C,F) 100 μM. The incubated vesicles were diluted 20-fold in solutions containing 5 mM tris maleate (pH

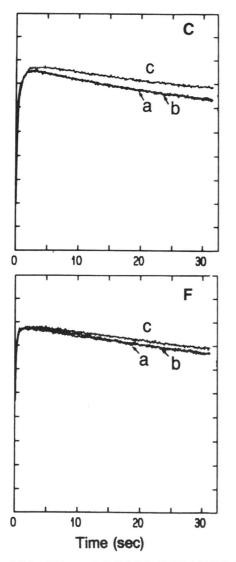

7.0) in addition to: (A,B,C) 0 KCl; (D,E,F) 1 M KCl. Glucose influx was measured after mixing with a solution comprising 200 mM glucose, 5 mM tris maleate (pH 7.0), and calcium buffer prepared with 2.5 mM EGTA in addition to: (A,B,C) 0 KCl; (D,E,F) 1 M KCl. The mixing solution conditions were: (a) pCa 4, (b) pCa 9, (c) pCa 9 plus 10 mM $MgCl_2$ and 10 μM ruthenium red. Modified from Kasai and Kawasaki (1993).

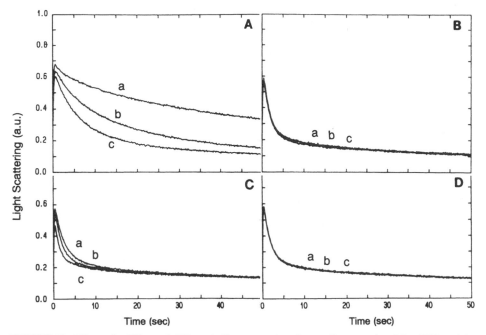

FIGURE 12. Effects of caffeine and ATP on choline permeation of ryanodine-treated vesicles. HSR vesicles were treated with different concentrations of ryanodine as indicated in the legend to Fig. 10. The ryanodine concentrations were: (A,C) 0; (B,D) 10 μM. The mixing solutions comprised 500 mM choline Cl, 5 mM tris maleate (pH 7.0), and calcium buffer prepared with 2.5 mM EGTA, in addition to activators. The calcium concentrations were: (A,B) pCa 9; (C,D) pCa 5. The activators were: (a) none; (b) 10 mM caffeine; (c) 1 mM ATP. Modified from Kasai and Kawasaki (1993).

5.2. Effect of Inhibitors on Ryanodine-Treated Channel

5.2.1. Effects of Ruthenium Red and Mg^{2+}

Figure 11 also demonstrates that the channel becomes less sensitive to ruthenium red and Mg^{2+} after ryanodine treatment. For this reason, the effects of inhibitors such as Mg^{2+} and ruthenium red were studied further. As shown in Fig. 13, the ryanodine-treated channel became less sensitive to these drugs for choline permeations than the nontreated one (compare with Fig. 9). For example, K_i for ruthenium red changed from 0.6 ± 0.3 to 20 ± 15 μM in the absence of KCl after 10 μM ryanodine treatment and from 15 ± 10 to > 1 mM in 1 M KCl. The K_i for Mg^{2+} also changed from 0.16 ± 0.10 to 2 ± 1 mM in the absence of KCl and from 0.9 ± 0.5 to 60 ± 30 mM in 1 M KCl. Thus, when KCl concentration was low, these inhibitors still blocked the channel, but they became practically ineffective in the presence of 1 M KCl. The glucose permeation showed similar behavior for inhibitors in the presence of KCl (Kasai *et al.*, 1995). The effects of inhibitors can be interpreted as the changes of the dissociation constants depending on the state of the channel and on ionic strength.

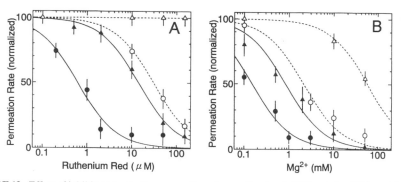

FIGURE 13. Effect of inhibitors on choline permeation of ryanodine-treated SR vesicles. HSR vesicles were treated with ryanodine as in Fig. 10. Choline permeation was measured with changing KCl concentrations using 200 mM choline Cl at pCa 5 as in Fig. 10. KCl concentrations: ($\bullet$, $\circ$) 0; ($\blacktriangle$, $\triangle$) 1 M. Ryanodine concentrations: ($\bullet$, $\blacktriangle$) 0; ($\circ$, $\triangle$) 10 μM. (A) Effect of ruthenium red. (B) Effect of Mg^{2+}.

We propose the following interpretation. The open states of the gate are more stabilized by the ryanodine treatment, and the affinity of the binding sites for inhibitors decreases in the order of stabilization: at pCa 5 before the ryanodine treatment, at pCa 9 after the treatment, and at pCa 5 after the treatment. Furthermore, K^+ competes with the inhibitors for binding sites. The inhibitors do not appear to block the channel directly by plugging the pore, but do so by interacting with the gate. On the contrary, the direct plugging model by ruthenium red was also proposed by single-channel analysis (Ma, 1993). Again, further analyses are required.

5.2.2. Effects of Procaine and Tetracaine

It is well known that procaine and tetracaine inhibit the Ca^{2+} channel (Yamamoto and Kasai, 1982b). The effect of KCl concentration and ryanodine treatment on the inhibitory actions of procaine were studied in terms of choline permeation. As shown in Fig. 14, procaine inhibited choline permeation in the presence and absence of 1 M KCl at similar concentrations. This is very different from the cases of ruthenium red and Mg^{2+} as seen in Fig. 13. This result indicates that procaine does not compete with K^+. After ryanodine treatment, the apparent inhibition constants increased as in the case of ruthenium red and Mg^{2+}, but this state did not depend on KCl concentration. Apparent dissociation constants for procaine obtained from similar experiments were 3.0 ± 2.0 mM, 1.5 ± 2.0 mM, 50 ± 20 mM, and 80 ± 30 mM, in the absence and presence of 1 M KCl before ryanodine treatment, and in the absence and presence of 1 M KCl after 10 μM ryanodine treatment, respectively. Tetracaine had similar effects. The corresponding values for tetracaine were 47 ± 25 μM, 46 ± 35 μM, 2 ± 1 mM, and 3 ± 2 mM, respectively. The affinity of tetracaine was 20 to 50 times stronger than that of procaine. We can conclude that procaine or tetracaine binds to a site different from those of ruthenium red or Mg^{2+}. Furthermore, the binding site of procaine or tetracaine is affected by ryanodine

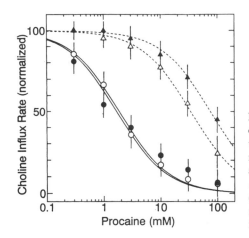

FIGURE 14. Effects of KCl and ryanodine treatment on inhibition of choline permeation by procaine. HSR vesicles were treated with ryanodine and choline permeation was measured as in Fig. 10 at pCa 5 in the absence and presence of KCl. KCl concentrations: (○, △) 0; (●, ▲) 1 M. Ryanodine concentrations: (○, ●) 0; (△, ▲) 10 μM. The points are means of three experiments carried out using the same preparation. Modified from Kasai et al. (1995).

binding, probably through allosteric interaction. Single-channel analysis of the effect of local anesthetics was also reported (Xu et al., 1993). Further analyses are required.

6. REGULATION OF Ca^{2+} CHANNEL BY ENDOGENOUS REGULATORS

6.1. Loss of Ca^{2+} Response Accompanied by Calsequestrin

In order to find intrinsic molecules which regulate Ca^{2+} release from HSR vesicles, permeation of choline through the channel was studied by the light scattering method (Kawasaki and Kasai, 1994). Opening of the Ca^{2+} channel was estimated from the difference between choline permeabilities at pCa 5 and pCa 9. Figures 15A and 15B show that Ca^{2+} response was lost when HSR vesicles were incubated overnight at 0°C in the presence of 1 mM EDTA or EGTA. In this state, caffeine and ATP could not activate the Ca^{2+} channel. It was shown that the loss of Ca^{2+} response was attributed to the effect of Ca^{2+}. In order to confirm the reversibility of the effect of EDTA, 1 mM Ca^{2+} was added after the overnight treatment by EDTA. The Ca^{2+} response recovered partially in a few hours (Fig. 15C).

It is suggested that some important factors might be lost during the incubation with EDTA. The supernatant solution was subjected to SDS-PAGE after centrifugation of the EDTA-treated HSR vesicles. Many proteins were found in the supernatant solution. In order to find the relation between the loss of Ca^{2+} response and released proteins, Ca^{2+} response and SDS-PAGE of the supernatant solution were compared after incubation under various Ca^{2+} concentrations (Fig. 16). At pCa higher than 7, Ca^{2+} response was practically lost (Fig. 16A). At pCa lower than 6, Ca^{2+} response increased with increases in Ca^{2+} concentration. Noticeably, at pCa 5 to 3, Ca^{2+} response was larger than that of control. When compared with the result of SDS-PAGE (Fig. 16B), a large amount of calsequestrin was found in the supernatant solution when the Ca^{2+} response was lost. It

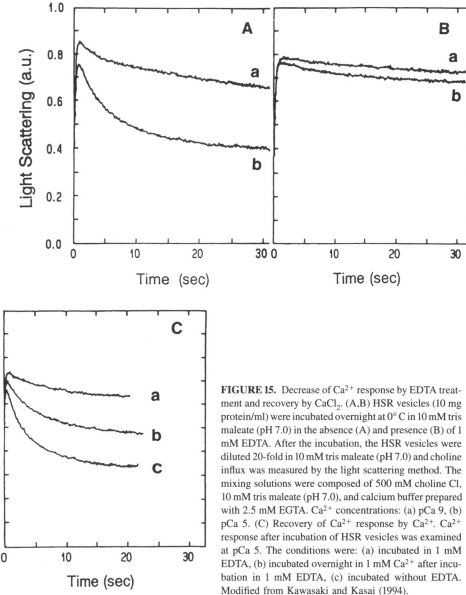

FIGURE 15. Decrease of Ca^{2+} response by EDTA treatment and recovery by $CaCl_2$. (A,B) HSR vesicles (10 mg protein/ml) were incubated overnight at $0°$ C in 10 mM tris maleate (pH 7.0) in the absence (A) and presence (B) of 1 mM EDTA. After the incubation, the HSR vesicles were diluted 20-fold in 10 mM tris maleate (pH 7.0) and choline influx was measured by the light scattering method. The mixing solutions were composed of 500 mM choline Cl, 10 mM tris maleate (pH 7.0), and calcium buffer prepared with 2.5 mM EGTA. Ca^{2+} concentrations: (a) pCa 9, (b) pCa 5. (C) Recovery of Ca^{2+} response by Ca^{2+}. Ca^{2+} response after incubation of HSR vesicles was examined at pCa 5. The conditions were: (a) incubated in 1 mM EDTA, (b) incubated overnight in 1 mM Ca^{2+} after incubation in 1 mM EDTA, (c) incubated without EDTA. Modified from Kawasaki and Kasai (1994).

should be noted that significant amounts of calsequestrin remained in the membrane fraction after the Ca^{2+} response was lost completely. The remaining calsequestrin might be attributed to the recovery of the Ca^{2+} response after the addition of Ca^{2+}.

HSR vesicles were incorporated into a lipid bilayer membrane and the effect of calsequestrin was studied (Fig. 17). When the vesicles were incorporated into the lipid

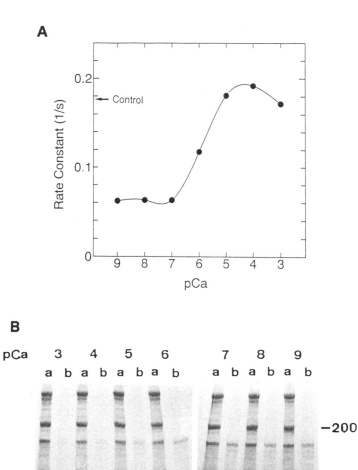

FIGURE 16. Ca^{2+} response and released proteins. HSR vesicles were incubated at various concentrations of free Ca^{2+}. The incubation conditions were 10 mg protein/ml, 1 mM EGTA, 10 mM tris maleate (pH 7.0), and various concentrations of CaCl$_2$. (A) Choline permeation was followed at pCa 5 as in Fig. 15 and initial rate of choline permeation is shown. An arrow shows control. (B) After incubation as in (A), the HSR vesicles were centrifuged at 15,000 × g for 30 min. The pellet (a) and the supernatant (b) were analyzed by SDS-PAGE. CSQ shows the band of calsequestrin. Molecular weights are indicated on the right side in kDa. Modified from Kawasaki and Kasai (1994).

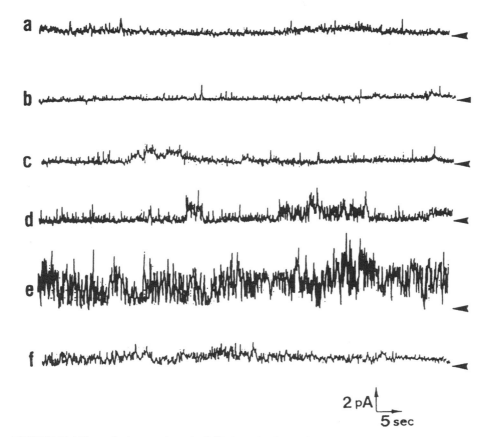

FIGURE 17. Effects of calsequestrin on the Ca^{2+} channel activity. HSR vesicles suspended in 1.5 M KCl were incorporated into the lipid bilayer and electric currents were recorded. (a) Just after the incorporation; (b) 1 mM EGTA was added to the *trans* side; (c) 2 m M $CaCl_2$ was added to the *trans* side of (b); (d) 500 μg calsequestrin was added to the *trans* side of (c); (e) stirring was started 5 min after the recording of (d); (f) 20 μM ruthenium red was added to the *cis* side. The holding potential was 20 mV in all cases. Arrowheads indicate the closed state of the channel. Modified from Kawasaki and Kasai (1994).

bilayer, slight opening of the Ca^{2+} channel was observed. In this experiment, open probability was measured as K^+ current and the *trans* side corresponds to the lumen side of SR. When 1 mM EGTA was added to the *trans* side, open probability did not change. After 2 mM $CaCl_2$ was added to the *trans* side, open probability increased slightly. When 500 μg calsequestrin was added, the channel began to open just after the initiation of stirring. Without addition of Ca^{2+} to the *trans* side, no activation was observed. The channel was blocked by ruthenium red, indicating that this current was through the Ca^{2+} channel. These results show that calsequestrin activates the Ca^{2+} release channel from the lumen side and that Ca^{2+} was required for this interaction.

Calsequestrin is considered to be a Ca^{2+} storing protein in the lumen since it has a large number of carboxyl groups and can bind many Ca^{2+} ions (MacLennan and Wong,

1971). The present results suggest a new function of calsequestrin as a regulator of the Ca^{2+} channel. Such an idea was also proposed by Ikemoto et al. (1989, 1991), who studied conformational change accompanied with the opening of the Ca^{2+} channel by using a fluorescent probe, and suggested the contribution of calsequestrin to the regulation of the channel. The present results shows a more direct contribution of calsequestrin. We showed the recovery of the Ca^{2+} response after incubation with Ca^{2+} in Fig. 15C. This phenomenon can be explained as follows. Some calsequestrin remained inside the vesicles after EDTA treatment, but calsequestrin lost Ca^{2+}. To express the function of calsequestrin, Ca^{2+} is required. Added Ca^{2+} might stimulate the remaining calsequestrin to the active state.

The Ca^{2+} release channel was purified as a ryanodine receptor and the channel protein showed many pharmacological properties, such as Ca^{2+} response. That is, the channel protein itself shows Ca^{2+} response without calsequestrin (Smith et al., 1988). However, in the present study, Ca^{2+} channel became insensitive to Ca^{2+} after loss of calsequestrin. To explain the contradiction, the difference in the experimental conditions should be considered. In the case of the purified channel protein, the channel is free from all regulators, but the channel in SR vesicles may interact with other proteins. Under such circumstances, it is plausible that the Ca^{2+} channel requires calsequestrin and that calsequestrin should bind Ca^{2+} to open the channel.

6.2. Proteins Bound to Calsequestrin and Ryanodine Receptor

In the previous section, we showed the possibility that calsequestrin regulates calcium release. Calsequestrin is known to bind to SR membrane from the lumen side, but the binding protein is not known. An affinity column of calsequestrin fixed to Sepharose 4B was constructed. Junctional face membrane (JFM) was solubilized in 1% Lubrol-PX and passed through the affinity column. As shown in Fig. 18, many proteins bound to the column as if the T-SR junction might be reconstituted on the column. From the elution pattern of the KCl gradient, it was difficult to distinguish which proteins bound to calsequestrin directly. We focused on the proteins that disappeared after passing the column. As shown in Fig. 18B, these proteins were near 200 kDa, 100 kDa, and 30 kDa. The protein around 210 kDa was slightly larger than myosin (200 kDa). The protein near 100 kDa was eluted at 150 mM KCl and 3 mM EGTA (Fig. 18A). Around 100 kDa, three bands could be distinguished clearly. These molecular weights were estimated to be 106, 100, and 96 kDa. The 106-kDa protein was close in weight to the Ca^{2+} release protein reported by Hilkert et al. (1992), the 100-kDa protein was similar to Ca^{2+}-Mg^{2+} ATPase, and the 96-kDa protein was similar to triadin reported by Brandt et al. (1990). The protein adsorbed to the calsequestrin column was 96 kDa. The protein near 30 kDa was around 25 kDa.

If calsequestrin regulates the Ca^{2+} release channel, there should exist a third protein which bridges calsequestrin and the ryanodine receptor, since these proteins do not interact directly. To find proteins that interact with both the ryanodine receptor and calsequestrin, an affinity column of ryanodine receptor was prepared. The solubilized protein of JFM with Lubrol-PX was passed through the column and was eluted with 1 M KCl. The eluted proteins were then passed through the calsequestrin column. The obtained proteins included two bands near 200 kDa, one band near 100 kDa, and few

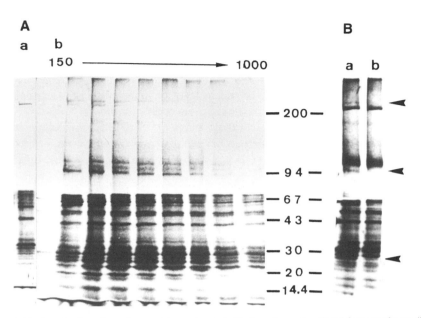

FIGURE 18. Proteins bound to calsequestrin-conjugated affinity column. Junctional face membrane (JFM) was prepared according to the method of Costello *et al.* (1988). JFM (1 mg/ml) was incubated for 1 hr at 4° C in 1 M KCl, 1% Lubrol-PX, 20 mM HEPES-Na (pH 7.4), and protease inhibitors (PI). Unsolubilized material was removed by centrifugation and the supernatant was diluted by 1% Lubrol-PX, 20 mM HEPES-Na (pH 7.4), and PI to 150 mM KCl. After adjusting $CaCl_2$ to 1.5 mM, the solution was passed for 5 hr at 4° C through the column conjugated with calsequestrin (Marer *et al.*, 1988). The proteins were eluted with a gradient of 150–1000 mM KCl in the presence of 1% Lubrol-PX and PI. (A) (a) Eluate at 150 mM KCl, (b) eluates from 150 to 1000 mM KCl. (B) (a) Sample before passing the column, (b) sample not bound to the column.

bands around 30–60 kDa. The band around 200 kDa seemed to be myosin and another was the 210-kDa protein again. The 96-kDa protein as well as the protein near 30 kDa and 25 kDa were also found.

We concluded that the 96-kDa protein is triadin, reported by Brandt *et al.* (1990), from the following facts: both have similar molecular weight, are abundant in the T-SR region, are solubilized by Lubrol-PX but not by CHAPS, and bind to a hydroxyapatite column. Furthermore, the 96-kDa protein may exist in polymer form (more than 10-mer) since it was recovered at a higher density position than ryanodine receptor (4-mer) by sucrose density gradient centrifugation. If the 96-kDa protein is triadin, it should be multifunctional since it interacts with calsequestrin in addition to DHP receptor and ryanodine receptor as reported (Brandt *et al.*, 1990; Kim *et al.*, 1990).

7. SUMMARY

In this review, we summarized the results obtained mainly by flux measurements through Ca^{2+} channel in HSR vesicles. The Ca^{2+} channel has a large pore which passes not only divalent cations such as Ca^{2+}, Mg^{2+}, and Ba^{2+} and monovalent cations such as Na^+, K^+, and Cs^+, but also large ions such as choline and tris. The permeation rates of choline and glucose through the Ca^{2+} channel were measured quantitatively by the light scattering method. The slow permeation of such molecules may reflect the structure of pores since the permeation process is the rate-limiting step for such large molecules. Neutral molecules such as glucose became permeable in the presence of submolar KCl, which suggests that pore size of the channel becomes larger in KCl.

The apparent permeation rates of Ca^{2+} and Mg^{2+} obtained from the flux measurement were the same, although their single-channel conductances were different. This discrepancy was explained by the fact that flux measurements reflects the open rate of the channel. Thus, complementarity between the flux measurement and single-channel recording was demonstrated.

From the effects of K^+ on the action of regulators on Ca^{2+} channel, it was suggested that the Ca^{2+} channel has many binding sites for activators and inhibitors. There are two kinds of Ca^{2+} binding sites for activation and inhibition. Activation sites for Ca^{2+}, caffeine, and ATP are different and inhibition sites for Ca^{2+} and procaine are different. The binding sites for ruthenium red and Mg^{2+} are the same as the activation and/or inhibition sites for Ca^{2+}.

Ryanodine-treated Ca^{2+} channel became permeable to glucose even in the absence of KCl. The conformational state of the channel opened by ryanodine is different from that opened by Ca^{2+}, caffeine, and ATP. The maximal flux rates of choline and glucose induced by ryanodine were smaller than those attained by caffeine and ATP. This result is consistent with the observation obtained by single-channel recording; the maximal value of single-channel conductance after ryanodine treatment becomes 40–50% of the value before the treatment. It is likely that the radius of the pore opened by ryanodine is smaller than that opened by Ca^{2+}, caffeine, or ATP. The flexibility of the channel may be decreased in the open locked state induced by ryanodine.

The Ca^{2+} response to open the channel by micromolar Ca^{2+} was lost when calsequestrin was released from the vesicles. It is possible that calsequestrin acts as an endogenous regulator of Ca^{2+} channel through triadin in excitation-contraction coupling.

8.   APPENDIX

8.1.   Channel Open Process Is Rate-Limiting for the Slow Component in the Flux Measurement of Membrane Vesicles

Here we explain briefly a theoretical model of ion permeation through ion channels in membrane vesicles (Kasai and Nunogaki, 1993). We consider the ion efflux from membrane vesicles. The size of vesicles is small enough (about 0.1 μm in diameter) to neglect the concentration gradient inside the vesicles in the relevant time scale (longer than 1 msec). Furthermore, for simplicity, we assume that all vesicles are the same size and each vesicle has one ion channel of the same property. The ion channel has two states: closed and open. The transition is expressed as

$$\text{Closed} \underset{\beta}{\overset{\alpha}{\rightleftharpoons}} \text{Open}$$

where α and β are the opening and the closing rate constants, respectively.

The channels are expected to be in equilibrium between the open and the closed states at time $t = 0$, when the flux measurement is started. The probability of the channel being in the closed state at $t = 0$ is expressed by $\beta/(\alpha + \beta)$ and in the open state by $\alpha/(\alpha + \beta)$.

In this model, if we assume that the ion flux rate, k_f, is much larger than the gating rates, i.e., $k_f \gg \alpha, \beta$, the content of permeants in the vesicles at time t, $F(t)$, is expressed by

$$F(t) = \frac{\beta}{\alpha + \beta} e^{-\alpha t} \tag{1}$$

On the contrary, if $k_f \ll \alpha, \beta$, i.e., ion flux rate is much lower than the gating rates, the content of the permeants is given by the usual flux formula:

$$F(t) = e^{-k_f' t}, \tag{2}$$

where k_f' is an averaged flux rate constant given by $k_f \alpha/(\alpha + \beta)$, since the channels are in fast dynamic equilibrium between the open and the closed states.

For the analysis of quickly permeating ions such as Ca^{2+} or Mg^{2+}, Eq. (1) should be applied, while for slowly permeating ions or molecules such as choline or glucose, Eq. (2) should be applied.

9. REFERENCES

Ashley, C. C., Mulligan, I. P., and Lea, T. J., 1991, Ca^{2+} and activation mechanisms in skeletal muscle, *Quart. Rev. Biophys.* **24**:1–73.

Brandt, N. R., Caswell, A. H., Wen, S.-R., and Talvenheimo, J. A., 1990, Molecular interactions of the junctional foot protein and dihydropyridine receptor in skeletal muscle triads, *J. Membr. Biol.* **113**:237–251.

Chen, S. R. W., Zhang, L., and MacLennan, D. H., 1992, Characterization of a Ca^{2+} binding and regulatory site in the Ca^{2+} release channel (ryanodine receptor) of rabbit skeletal muscle sarcoplasmic reticulum, *J. Biol. Chem.* **267**:23318–23326.

Coronado, R., Rosenberg, T. R. L., and Miller, C., 1980, Ionic selectivity, saturation, and block in a K^+-selective channel from sarcoplasmic reticulum, *J. Gen. Physiol.* **76**:425–446.

Costello, B., Chadwick, C., and Fleischer, S., 1988, Isolation of junctional face membrane of sarcoplasmic reticulum, in: *Methods in Enzymology*, Vol. 157 (S. Fleischer and B. Fleischer, eds.), Academic Press, New York, pp. 46–50.

Endo, M., 1977, Calcium release from the sarcoplasmic reticulum, *Physiol. Rev.* **57**:71–108.

Fleischer, S., Ogunbunmi, E. M., Dixon, M. K., and Fleer, E. A. M., 1985, Localization of Ca^{2+} release channels with ryanodine in junctional cisternae of sarcoplasmic reticulum of fast skeletal muscle, *Proc. Natl. Acad. Sci. USA* **82**:7256–7259.

Hilkert, R., Zaidi, N., Shome, K., Nigam, M., Lagenauer, C., and Salama, G., 1992, Properties of immuno-affinity purified 106–kDa Ca^{2+} release channels from the skeletal sarcoplasmic reticulum, *Arch. Biochem. Biophys.* **292**:1–15.

Ikemoto, N., Ronjat, M., Meszaros, L. G., and Koshita, M., 1989, Postulated role of calsequestrin in the regulation of calcium release from sarcoplasmic reticulum, *Biochemistry* **28**:6764–6771.

Ikemoto, N., Antoniu, B., Kang, J.-J., Meszaros, L. G., and Ronjat, M., 1991, Intravesicular calcium transient during calcium release from sarcoplasmic reticulum, *Biochemistry* **30:**5230–5237.

Imagawa, T., Smith, J. S., Coronado, R., and Campbell, K., 1987, Purified ryanodine receptor from skeletal muscle sarcoplasmic reticulum is the Ca^{2+}-permeable pore of the calcium release channel, *J. Biol. Chem.* **262:**16636–16643.

Inui, M., Saito, A., and Fleischer, S., 1987, Purification of the ryanodine receptor and identity with feet structures of junctional terminal cisternae of sarcoplasmic reticulum from fast skeletal muscle, *J. Biol. Chem.* **262:**1740–1747.

Kasai, M., and Kawasaki, T., 1993, Effects of ryanodine on permeability of choline and glucose through calcium channels in sarcoplasmic reticulum vesicles, *J. Biochem* **113:**327–333.

Kasai, M., and Nunogaki, K., 1988, Permeability of sarcoplasmic reticulum, in: *Methods in Enzymology*, Vol. 157 (S. Fleischer and B. Fleischer, eds.), Academic Press, New York, pp. 437–468.

Kasai, M., and Nunogaki, K., 1993, Channel opening process is responsible for the slow component in the flux measurement of membrane vesicles, *J. Theor. Biol.* **161:**461–480.

Kasai, M., Kawasaki, T., and Yamamoto, K., 1992, Permeation of neutral molecules through calcium channel in sarcoplasmic reticulum vesicles, *J. Biochem.* **112:**197–203.

Kasai, M., Yamaguchi, N., and Kawasaki, T., 1995, Effect of KCl concentration on gating properties of calcium release channels in sarcoplasmic reticulum vesicles, *J. Biochem.* **117:**251–256.

Kawasaki, T., and Kasai, M., 1989, Disulfonic stilbene derivatives open the Ca^{2+} release channel of sarcoplasmic reticulum, *J. Biochem.* **106:**401–405.

Kawasaki, T., and Kasai, M., 1994, Regulation of calcium channel in sarcoplasmic reticulum by calsequestrin, *Biochem. Biophys. Res. Commun.* **199:**1120–1127.

Kim, K. C., Caswell, A. H., Talvenheimo, J. A., and Brandt, N. R., 1990, Isolation of a terminal cisterna protein which may link the dihydropyridine receptor to the junctional foot protein in skeletal muscle, *Biochemistry* **29:**9281–9289.

Kometani, T., and Kasai, M., 1978, Ionic permeability of sarcoplasmic reticulum vesicles measured by light scattering method, *J. Membr. Biol.* **41:**295–308.

Lai, F. A., Erickson, H. P., Rousseau, E., Liu, Q.-Y., and Meissner, G., 1988, Purification and reconstitution of the calcium release channel from skeletal muscle, *Nature* **331:**315–319.

Lattanzio, F. A., Schlatterer, R. G., Nicar, M., Campbell, K. P., and Sutko, J. L., 1987, The effects of ryanodine on passive calcium fluxes across sarcoplasmic reticulum membrane, *J. Biol. Chem.* **262:**2711–2718.

Liu, Q.-Y., Lai, F. A., Rousseau, E., Jones, R. V., and Meissner, G., 1989, Multiple conductance states of the purified calcium release channel complex from skeletal sarcoplasmic reticulum, *Biophys. J.* **55:**415–424.

Ma, J., 1993, Block by ruthenium red of the ryanodine-activated calcium release channel of skeletal muscle, *J. Gen. Physiol.* **102:**1031–1056.

MacLennan, D. H., and Wong, P. T., 1971, Isolation of a calcium-sequestering protein from sarcoplasmic reticulum, *Proc. Natl. Acad. Sci. USA* **68:**1231–1235.

Martonosi, A., 1984, Mechanisms of Ca^{2+} release from sarcoplasmic reticulum of skeletal muscle, *Physiol. Rev.* **64:**1240–1320.

Maurer, T., Tanaka, M., Ozawa, T., and Fleischer, S., 1988, Purification and crystallization of calcium-binding protein from skeletal muscle sarcoplasmic reticulum, in: *Methods in Enzymology*, Vol. 157 (S. Fleischer and B. Fleischer, eds.), Academic Press, New York, pp. 321–328.

Meissner, G., 1986, Ryanodine activation and inhibition of the Ca^{2+} release channel of sarcoplasmic reticulum, *J. Biol. Chem.* **261:**6300–6306.

Meissner, G., 1994, Ryanodine receptor/Ca^{2+} release channels and their regulation by endogenous effectors, *Annu. Rev. Physiol.* **56:**485–508.

Nagasaki, K., and Fleischer, S., 1988, Ryanodine sensitivity of the calcium release channel of sarcoplasmic reticulum, *Cell Calcium* **9:**1–7.

Nagasaki, K., and Kasai, M., 1983, Fast release of calcium from sarcoplasmic reticulum vesicles monitored by chlortetracycline fluorescence, *J. Biochem.* **94:**1101–1109.

Nagasaki, K., and Kasai, M., 1984, Channel selectivity and gating specificity of calcium-induced calcium release channel in isolated sarcoplasmic reticulum, *J. Biochem.* **96:**1769–1775.

Rousseau, E., Smith, J. S., and Meissner, G., 1987, Ryanodine modifies conductance and gating behavior of single calcium release channel, *Am. J. Physiol.* **253:**C364–C368.

Smith. J. S., Coronado, R., and Meissner, G., 1985, Sarcoplasmic reticulum contains adenine nucleotide–activated calcium channels, *Nature* **316**:446–449.

Smith, J. S., Coronado, R., and Meissner, G., 1986, Single channel measurements of the calcium release channel from skeletal muscle sarcoplasmic reticulum. Activation by Ca^{2+} and ATP and modulation by Mg^{2+}, *J. Gen. Physiol.* **88**:573–588.

Smith, J. S., Imagawa, T., Ma, J., Fill, M., Campbell, K. P., and Coronado, R., 1988, Purified ryanodine receptor from rabbit skeletal muscle is the calcium-release channel of sarcoplasmic reticulum, *J. Gen. Physiol.* **92**:1–26.

Takeshima, H., Nishimura, S., Matsumoto, T., Ishida, H., Kangawa, K., Minamino, N., Matsuo, H., Ueda, M., Hanaoka, M., Hirose, T., and Numa, S., 1989, Primary structure and expression from complementary DNA of skeletal muscle ryanodine receptor, *Nature* **339**:439–445.

Tatsumi, S., Suzuno, M., Taguchi, T., and Kasai, M., 1988, Effects of silver ion on the calcium-induced calcium release channel in isolated sarcoplasmic reticulum, *J. Biochem.* **104**:279–284.

Williams, A. J., 1992, Ion conduction and discrimination in the sarcoplasmic reticulum ryanodine receptor/calcium-release channel, *J. Muscle Res. Cell Motil.* **13**:7–26.

Xu, L., Jones, R., and Meissner, G., 1993, Effects of local anesthetics on single channel behavior of skeletal muscle calcium release channel, *J. Gen. Physiol.* **101**:207–233.

Yamamoto, N., and Kasai, M., 1981, Studies on the cation channel in sarcoplasmic reticulum vesicles. I. Characterization of Ca^{2+}-dependent cation transport by using a light scattering method, *J. Biochem.* **90**:1351–1361.

Yamamoto, N., and Kasai, M., 1982a, Characterization of the Ca^{2+}-gated cation channel in sarcoplasmic reticulum vesicles, *J. Biochem.* **92**:465–475.

Yamamoto, N., and Kasai, M., 1982b, Kinetics of the actions of caffeine and procaine on the Ca^{2+}-gated cation channel in sarcoplasmic reticulum vesicles, *J. Biochem.* **92**:477–484.

Yamamoto, N., and Kasai, M., 1982c, Mechanism and function of the Ca^{2+}-gated cation channel in sarcoplasmic reticulum vesicles, *J. Biochem.* **92**:485–496.

Yamanouchi, Y., Kanemasa, T., and Kasai, M., 1984, Effects of adenine nucleotides on the Ca^{2+}-gated cation channel in sarcoplasmic reticulum vesicles, *J. Biochem.* **95**:161–166.

CHAPTER 9

SINGLE-CHANNEL STUDIES IN MOLLUSCAN NEURONS

MICHAEL FEJTL and DAVID O. CARPENTER

1. INTRODUCTION

Throughout the modern history of neuroscience, molluscs have served as a model system to disclose the neuronal basis of cell-to-cell communication and the underlying mechanisms of neuronal excitability. The ease with which individual neurons within a particular molluscan species can be recognized led to the establishment of *in vitro* preparations ranging from the isolated squid axon nerve to single-cell culture systems. Hence it became possible to study the effects of the various neurotransmitters, second messengers, and toxic substances on properties of single identified neurons.

Ever since Hodgkin and Huxley examined the ionic mechanisms governing action potential generation in the squid (Hodgkin and Huxley, 1952), molluscs have been widely used by neuroscientists to study ion channel behavior. Their relatively simple nervous system, which consists of only a few thousand neurons with diameters of several tens of micrometers up to 1 mm (R2 in the abdominal ganglion of the marine mollusc *Aplysia californica* is one of the biggest neurons known today), allows easy identification and access to a particular cell. Subsequent studies using two-electrode and single-electrode voltage-clamp techniques revealed the ionic currents at the whole-cell level. By applying statistical methods such as fluctuation analysis, information was also gathered about single-channel properties, conductance in particular. However, this

MICHAEL FEJTL • Wadsworth Center for Laboratories & Research, New York State Department of Health, Albany, New York 12201, and Institute of Neurophysiology, University of Vienna, A-1090 Vienna, Austria.   DAVID O. CARPENTER • Wadsworth Center for Laboratories & Research, New York State Department of Health, Albany, New York 12201, and School of Public Health, State University of New York at Albany, Albany, New York 12203-3727.

Ion Channels, Volume 4, edited by Toshio Narahashi, Plenum Press, New York, 1996.

methodology used to derive conductivity values was still an indirect approach, and it was not until 1976, when Neher and Sakmann (1976) and Hamill *et al.* (1981) introduced the patch-clamp technique to neuroscientists, that events at the single-channel level could be resolved directly. The discrete transitions from a closed state to an open state and *vice versa* are the basis for any descriptive and quantitative approach at the molecular level of ion channels. For instance, whole-cell currents could be compared with single-channel events simultaneously and revealed a nonuniform distribution of ion channels across the cell body (hot spots). Moreover, without the single-channel patch-clamp technique, information about stretch-activated channels could never have been gathered. Although various ionic currents in molluscan nervous tissue and their modulation by either transmitters or second messengers have been described at the single-channel level, a comprehensive overview has yet to be published. The goal of this chapter is to give the reader an understanding of the voltage- and ligand-gated ion channels in molluscan neurons at the single-channel level.

2. VOLTAGE-ACTIVATED CURRENTS

2.1. The Sodium Channel

The fast, voltage-dependent sodium current underlies action potential generation. Most of the single-channel data available in molluscs were derived almost exclusively from the squid. This comes as no surprise, since it was then possible to investigate the Hodgkin and Huxley model of sodium channel gating (Hodgkin and Huxley, 1952) at the single-channel level in the very same preparation. The original model postulated that three or four charged particles govern the activation m, and one charged particle mediates the inactivation process h. Moreover, activation and inactivation were thought to be independent processes. This view has been challenged by the observation that the gating current, i.e., charge movement upon sodium-channel activation prior to channel opening, was partially immobilized during inactivation (Armstrong and Bezanilla, 1977). To reconcile models of sodium channel gating at the single-channel level in the squid giant axon, a new technique had to be developed, since it is virtually impossible to patch the external membrane owing to the presence of Schwann cells. The development of the squid giant axon cut-open technique (Bezanilla, 1987a,b; Llano *et al.*, 1988) made it possible to record single sodium channels in the outside-out patch configuration. Briefly, a 1-cm piece of the squid giant axon was pinned down and cut open longitudinally so as to expose the internal face of the nerve. Thus a patch pipette could be positioned onto the internal side of the membrane; after gigaseal formation and patch excision, an outside-out patch was formed. Figure 1 shows single-channel currents of a single sodium channel (no double openings were observed after more than 3000 sweeps) after depolarizing voltage steps to -48 mV and -28 mV, respectively (Vandenberg and Bezanilla, 1991). The most obvious features are: 1) the channel reopened after closing during the voltage step, 2) the latencies to first openings decreased at the more depolarized level, and 3) no openings occurred during maintained depolarization. The best kinetic model that can account for the observed single-channel data incorporates several closed states, one open state, and a reversible inactivation pathway from either the open or closed state

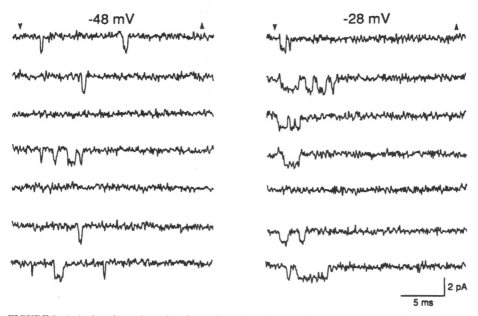

FIGURE 1. A single voltage-dependent fast sodium channel recorded in an outside-out patch configuration. Shown are two steps to −48 mV and −28 mV from a holding potential of −108 mV. The arrowheads indicate the onset and offset of the voltage step. (From Vandenberg and Bezanilla, 1991.)

(Fig. 2). At very positive potentials and under the conditions of an inverse sodium gradient, a second open state can be revealed, which is likely to occur from the inactivated state (Correa and Bezanilla, 1994). Several closed states are necessary to account for the latencies to first openings. The inactivation occurring from both open and closed states explains why there can be no channel openings despite depolarization and why reopenings take place after inactivation has been initialized (see Fig. 1). It also allows the removal of inactivation after a hyperpolarizing voltage step, which is essentially a return from the inactivated state to a closed state.

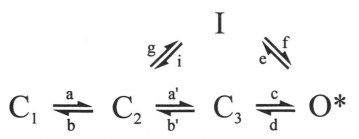

FIGURE 2. Proposed gating model for the Na^+ channel. C, O, and I stand for the closed, open, and inactivated state, while the lowercase letters represent the respective forward and backward rate constants (modified from Vandenberg and Bezanilla, 1991).

High concentrations of divalent cations such as Ca^{2+} and Mg^{2+} have been shown to block the sodium conductance in the squid axon (Taylor *et al.*, 1976). This is evident at the single-channel level, where a conductance of 12–15 pS has been reported in Ca- and Mg-free sea water (Bezanilla, 1987a,b), whereas a 3.5-pS conductance was revealed by noise analysis in the presence of divalent cations (Llano and Bezanilla, 1984). These properties of the Na^+ channel can be well studied when the inactivation of the channel is blocked by the steroidal alkaloid batrachotoxin (BTX), derived from Colombian frogs of the species *Phyllobates, P. aurotaenia*, in particular (see Khodorov, 1985). Application of this toxin renders the channel in an open steady state and allows one to study kinetic properties of the open and closed states without having to deal with the inactivation process. In Fig. 3 a BTX-modified sodium channel was recorded in the presence and absence of 10 mM Ca^{2+}, and it is clear that the current amplitude is reduced when calcium is present. The conductance was 20 pS with no calcium added and was reduced to about 2 pS in 10 mM Ca^{2+} (Behrens *et al.*, 1989). This study was conducted in symmetrical 200 mM NaCl, but under a normal sodium concentration gradient, i.e., high $[Na^+]_0$ and low $[Na^+]_i$, the conductance for the unmodified channel was 29.7 pS and 9.9 pS for the BTX-treated Na^+ channel (Correa *et al.*, 1991). In regular sea water, external Ca^{2+} reduces the conductance of unmodified Na^+ channels from 14–16 pS to about 4–6 pS at $4°$ C (Levis *et al.*, 1984; Llano and Bezanilla, 1986). This result has been confirmed in Na^+ channels taken from axoplasmic organelles (Wonderlin and French, 1991) and in sodium channels from the squid optic nerve incorporated into lipid bilayers (Latorre *et al.*, 1987). Although BTX-modified channels have a lower unit conductance compared with the normal Na^+ channel (Quandt and Narahashi, 1982) due to the larger energy barrier that Na^+ ions have to cross in the BTX-treated channel (Correa *et al.*, 1991), the overall higher conductance under normal sodium gradients can be explained by charge screening. Fixed negative charges at the mouth of the channel create an electrostatic negative potential which attracts Na^+ ions. Hence, the local Na^+ concentration close to the ion channel mouth is higher than in the bulk solution, which in turn may increase the single-channel conductance (Correa and Bezanilla, 1988; Correa *et al.*, 1989). Figure 4 shows the voltage dependence of sodium channel opening in an outside-out configura-

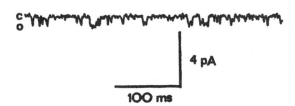

4 pA

100 ms

FIGURE 3. A single sodium channel was held open by applying BTX, which renders Na^+ channels in the open state due to removal of inactivation. The two traces show a BTX-modified Na^+ channel in the absence (top) and presence (bottom) of 10 mM Ca^{2+} (holding potential was -90 mV in 200 mM symmetrical Na^+; Behrens *et al.*, 1989).

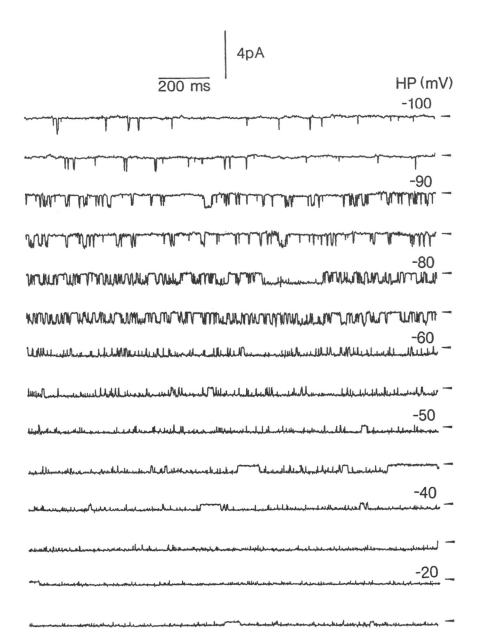

FIGURE 4. Voltage dependence of a single BTX-modified sodium channel. Half-maximal activation occurred around −80 mV. Note that at the more depolarized level the channel almost never leaves the open state. The arrowheads indicate the closed level. (From Correa *et al.*, 1992.)

tion. The voltage at which the open probability P_o was 50% was about $-80\,mV$ (Behrens et al., 1989; Correa et al., 1992). The P_o/V relation was fit by a double Boltzmann distribution, indicating the presence of more than one closed state. Increasing the temperature increases channel kinetics, and in the case of the Na^+ channel it decreases the open probability P_o. Thus, increased temperature stabilizes the closed state of the channel, thereby affecting the mean open time more than the mean closed time. This is why at temperatures $> 8°\,C$ the second open state of the BTX-modified Na^+ channel is hard to resolve (Correa et al., 1992). Hence, the preferred model for the BTX-modified sodium channel has two open states, three closed states, and no inactivation pathway (Fig. 5).

Purified sodium channels have an α subunit of ca. 260–290 kDa associated with smaller β subunits (Catterall, 1986). Four homologous domains line the channel, and the putative helix S4 has been proposed to participate in the voltage-sensing mechanism (Noda et al., 1984). If all the subunits acted independently, as the Hodgkin-Huxley model proposes, four forward and four backward rate constants would emerge. However, since m and h do not act independently (Horn and Vandenberg, 1984), and in view of the single-channel studies carried out in the squid giant axon, a different gating model for the voltage-dependent Na^+ channel has been suggested to account for the actual data. Activation and inactivation do not occur independently, since the gating charge is partially immobilized during inactivation. The channel can inactivate either from the open or the closed state and can reenter an open or closed configuration from the inactivated state. This explains the removal of inactivation after hyperpolarization, i.e., the return from an inactivated to a closed state, and can account for the late channel openings after the peak sodium current.

Another type of sodium current, which has been implicated in the bursting behavior in certain molluscan neurons, has been investigated in the marine mollusc *Pleurobranchea* (Green and Gillette, 1983). This channel displayed a higher conductance (ca. 40 pS) than the classic voltage-dependent Na^+ channel. It did not depend on membrane voltage, but its opening frequency increased after treatment with cAMP. Moreover, when kinase inhibitors were applied, they failed to block the cAMP-induced enhancement of this Na^+ current (Sudlow et al., 1993). Thus, it was concluded that the cyclic nucleotide acted directly at the channel site and that phosphorylation did not mediate the increase

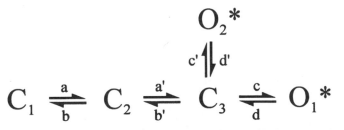

FIGURE 5. Gating model for the BTX-treated sodium channel. The inactivation gateway has been omitted (compare with the model for the native Na^+ channel in Fig. 2) and a second open state has been added, which can only be resolved at lower temperature (<8 °C; Correa et al., 1992).

of the cAMP-dependent sodium current. FMRFamide, an endogenous neuropeptide, can activate a 6.6-pS (symmetrical NaCl) Na$^+$ current in some *Helix* neurons without relying on a second messenger system (Cottrell *et al.*, 1990). No channel openings were observed when FMRFamide was present in the bath but not in the electrode of a cell-attached patch. Assuming the involvement of a second messenger cascade, the Na$^+$ channels should have been activated even when FMRFamide was not present in the pipette. Another indication that this is a fast ligand-activated Na$^+$ channel was that the current could be activated repeatedly in outside-out patches, a situation where second messenger–mediated currents normally undergo run-down due to wash-out of diffusible compounds. This action of FMRFamide clearly differs from its modulatory effects on other ion currents, such as the serotonin-sensitive potassium current, which rely on second messenger pathways.

In growth cones of identified *Helisoma* neurons in culture, Cohan *et al.* (1985) have found a current which is mainly, although not exclusively, permeable to Na$^+$ ions. Its conductance near 0 mV holding potential was 70 pS and its activity expressed itself in long open and closed states lasting for seconds, with occasional brief returns from the open to the closed state. Since actively growing growth cones differ morphologically, from stable nonelongating growth cones, the former having a complex filipodial structure extending from large flattened lamellipodia and the latter being smaller in size and club-shaped, a corresponding physiological difference could be expected. Inside-out patches taken from growing and stable growth cones displayed a similar Na$^+$-dependent current, whereas on-cell patches in stable growth cones were silent. A putative intracellular factor kept the channel in a quiet state, since after withdrawal of the patch pipette in the inside-out mode, the activity of the channel resumed. Hence, during active growth of a neuron, this Na$^+$ current might act as a physiological signal until the growth cone has reached its respective target. Then the cell would produce some yet unknown factor(s) and transform this channel into the inactive state.

2.2. The Potassium Channels

Among the various voltage-dependent potassium channels (Rudy, 1988), three types of potassium currents have been investigated in molluscan neurons at the single-channel level: 1) the classic delayed rectifier (I_K) responsible for the repolarization of the membrane after the upstroke of an action potential (Hodgkin and Huxley, 1952), 2) the fast A-current (I_A), which is activated at subthreshold levels upon a depolarizing voltage step and is thought to participate in the generation of the resting membrane potential (RMP) and to modulate the frequency of pacemaker neurons (Hagiwara *et al.*, 1961; Connor and Stevens, 1971), and 3) the anomalous (inward) rectifier (I_{anom}), which exhibits a higher conductance for K$^+$ ions in the hyperpolarized voltage range (Adrian and Freygang, 1962; Armstrong and Binstock, 1965).

2.2.1. The Delayed Rectifier I_K

The delayed rectifier potassium current is the main contributor to the outward current that governs the repolarization of the membrane after the peak of the action

potential (Hodgkin and Huxley, 1952). This current has three distinct characteristics:
1) the activation occurs a few milliseconds after a depolarizing voltage step, 2) the open
probability is highly voltage-dependent, and 3) the inactivation rate is very slow. Several
studies indicate that I_k may comprise at least two potassium channels with conductances
of 22 pS and 40 pS, respectively. However, single-channel recordings from reconstituted
potassium channel proteins from the squid axon (*Loligo vulgaris*) revealed three types of
channels, with conductances of 11 pS, 22 pS, and 32 pS, respectively. Since the 11-pS
channel was the predominant one, the authors concluded that it is the main contributor to
the delayed rectifier (Prestipino *et al.*, 1989). In outside-out patches from the giant axon
of the squid *Loligo pealei* and *Loligo opalescens*, Llano and coworkers (Llano and
Bezanilla, 1985; Llano *et al.*, 1988) recorded currents through single K$^+$ rectifier
channels. The most abundant channel was the 20 pS channel; Figure 6 shows individual
traces of the 20 pS and the 40 pS channel. The mean open time for the 20-pS channel
was 2.1 msec and the open-time histogram was best fit with two exponentials with time
constants τ_1 = 2.84 msec and τ_2 = 0.6 msec. The 40-pS channel had different kinetic
properties from the 20-pS channel. The open-time distribution could be best fit with two
exponentials (τ_1 = 4.84 msec; τ_2 = 11.2 msec), whereas the closed-time distribution
needed a third time constant for the best fit (τ_1 = 0.46 msec; τ_2 = 6.39 msec;

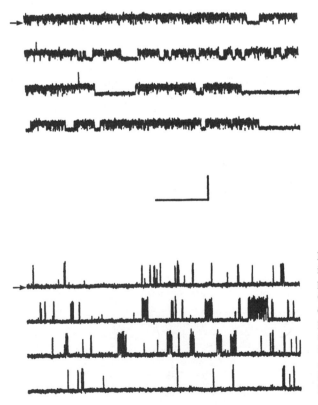

FIGURE 6. Individual traces showing single-channel activity of the 20
pS (top) inactivating and the 40 pS
(bottom) delayed rectifier potassium
current recorded in the cut-open axon
preparation. Holding potential was 42
mV in both cases, and the recording
was obtained in the presence of 2 mM
ATP. Arrows point to the closed state.
Scale bar is 5 pA; 500 msec. (From
Perozo *et al.*, 1991a.)

$\tau_3 = 41.3$ msec). The 40-pS channel was highly selective for K^+ ions. When the axon was preexposed to 0 K^+, a third type of channel was often observed with a conductance of about 10 pS. However, since the 10-pS channel and the 20-pS channel were never observed simultaneously in the same patch, it is conceivable that the former was a breakdown of the latter. The 20-pS channel resembles the current described by Conti and Neher (1980), although they reported a conductance of 9–11 pS. This discrepancy might be due to a block of the potassium channel by sodium ions, as has been reported by Bezanilla and Armstrong (1972). The sequential gating model for the delayed rectifier channel consists of several closed states and a single open state. Fast flickering of the channel, i.e., rapid transitions between the last closed and the open state, occurred at a much higher frequency than the transitions between the various closed states. Thus, the transition from the closed to the open state was not rate-limited (Llano et al., 1988). Ensemble fluctuation analysis of the I_k current in outside-out patches from isolated neurons of the giant fiber lobe (GFL) of the stellate ganglia in the squid gave a single-channel conductance of ca. 13 pS (Llano and Bookman, 1985, 1986). The difference in conductance between fluctuation analysis and actual single-channel currents was probably due to the lower $[K^+]$ used when ensemble analysis was performed (295 mM versus 530 mM). Fluctuation analysis of the delayed rectifier was also performed in neurons of the snail *Helix roseneri* and revealed a conductance of 2.4 pS, which was independent of voltage. This is consistent with the model that the I_k channel has only a single open state (Reuter and Stevens, 1980). An estimate of the channel density yielded 7.2 ± 1.6 channels per μm^2, assuming a capacitance of 1 $\mu F/cm^2$. When ion selectivity was tested, it could be shown that the I_k channel was selective for the following ions: $Tl^+ > K^+ > Rb^+ > Cs^+ > NH_4^+ \gg Li^+ > Na^+$, and this has also been shown in *Lymnaea* neurons. However, the single-channel conductance in the latter was ca. 60 pS and therefore it remains questionable whether this current contributes to the delayed rectifier (Kazachenko and Geletyuk, 1983). Activation time to half peak was about 3 msec; delivering a 50-msec depolarizing step, the current inactivated to 80% of its peak value with a time constant of 20–25 msec. When longer depolarizations were used (> 100 msec), inactivation was about 40% of the peak value. This was largely due to a decrease of the potassium conductance and not due to an accumulation of K^+ ions at the external side of the membrane, as revealed by tail-current analysis (Llano and Bookman, 1985, 1986). As one would expect for the delayed rectifier, the open probability P_o was highly voltage-dependent. The value of P_o was 0.2 at -20 mV, 0.3 at 0 mV, and 0.6 at $+20$ mV (Llano et al., 1986).

At least two compounds have been used to investigate modifying actions on the delayed rectifier. Trinitrobenzene sulfonic acid (TNBS), an agent which modifies amino groups, increased the magnitude of the steady-state outward current in squid axons by a factor of 1.3–1.5 and slowed the activation kinetics by a factor of 4. At pH 9 the time course of that action was faster than at pH 7.5, indicating a reactivity toward neutral amino groups (Spires et al., 1988). Since single-channel recordings did not show an increase in conductance in GFL neurons, the enhancing effect was most likely caused by an increase in open probability. Nealy et al. (1993) confirmed the enhancing action of TNBS on the potassium current in the squid axon, but found an opposing action in GFL neurons. The two main conductances found were a 10-pS and a 25-pS channel, but occasionally a channel of 40–50 pS could be observed. In contrast to the findings of Llano *et al.* (1988) the 10-pS and 40-pS channels were also present in the soma and not only in the axon. TNBS decreased the somatic current by increasing the latency to first opening after a voltage step. Untreated channels opened after 1.2 msec, but after application of TNBS the latencies increased by more than a magnitude for 30% of the channels observed. Since there was no effect on the unitary current (see Spires *et al.*, 1988), TNBS might decrease the somatic current by slowing the rate constants of the closed states, which would result in a delayed opening. Phosphorylating agents like ATP modulated the 20-pS channel and the 40-pS channel in different ways. While the activity of the 40-pS channel was enhanced, the 20-pS channel underwent a decrease (Perozo *et al.*, 1990, 1991a,b). The authors concluded that the 40 pS channel is the main contributor to I_k, since the 20 pS channel suffered from slow inactivation and decrease of the open probability after several depolarizing voltage steps (but see Llano *et al.*, 1988). Caged ATP and the catalytic subunit of protein kinase A were included in the patch electrode, and outside-out patches were taken from the quid axon using the cut-open technique. The 40-pS channel was transformed into a high-activity mode upon release of ATP, caused by a brief illumination of the electrode with UV light (Fig. 7). This was clearly due to ATP, since neither caged Ca^{2+} nor normal light had an effect on the 40-pS channel, although Ca^{2+} opened a 95-pS channel (see Vandenberg *et al.*, 1989). The unitary current was not affected by phosphorylation but the open probability P_o changed more than 100-fold. On average, P_o increased from 0.00022 ± 0.00015 (n = 7) to

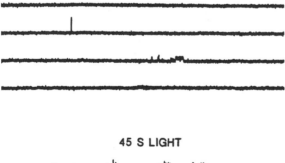

FIGURE 7. ATP increased the activity of the 40-pS K^+ channel. Caged ATP was included in the patch pipette and the single-channel activity was recorded in control (top trace) and after illumination of the electrode, which released ATP (bottom trace). Holding potential was 42 mV. (From Perozo *et al.*, 1991a.)

0.019 ± 0.00096 (n = 3) (Perozo et al., 1991a). The onset of the transformation of the 40-pS channel into a high-activity state ranged from 30 sec to 8 min and lasted for the remainder of the experiment. The mean open time decreased only slightly from 1.86 msec to 1.52 msec, but the mean closed time changed dramatically from 260.56 msec to 23.97 msec; hence the increase in P_o. Dwell-time distributions yielded a double exponential fit for the open state and three exponentials for the closed state. The two fast components of the closed time distribution remained unchanged after phosphorylation, but the third slow component decreased from 22.7 sec to 170 msec, representing a decrease in the interburst interval. In contrast to the enhancing effect of ATP on the 40-pS channel, the activity of the 20 pS channel was reduced (Perozo et al., 1990, 1991b). Figure 8 shows the decrease of channel activity 7 min after UV illumination. As in the 40-pS channel, caged Ca^{2+} had no effect, and the unitary current remained the same. The reduction of the channel activity was accompanied by a shift of the steady-state activation toward a more depolarized range. The mean open time and the P_o were voltage-independent in the range of -70 mV to $+20$ mV, and this indicates that the transition of an open state to an adjacent closed state is also voltage-independent. Thus, since the mean open time was not affected by phosphorylation, some of the rate constants might have been affected. This was, in fact, the case. The open time distribution was fit with a two-exponential function, and the slow component was moved toward more depolarized potentials. In addition, the inactivation curve was also shifted to more depolarized levels. Plots of latency distributions showed that after release of caged ATP, the latency to first opening increased for every potential tested (Fig. 9). Overall, phosphorylation shifted the voltage dependence of the rate constants by 9–17 mV in the depolarized direction, resulting in the observed slow-down of the open kinetics. Therefore, the I_K current seems to comprise at least two conductances, a 20 pS and a 40 pS channel, as has been reported for other preparations, including neuroblastoma cells (Quandt, 1988), amphibian spinal neurons (Harris et al., 1988), and T lymphocytes (Decoursey et al., 1987), and both types can be modulated by ATP.

2.2.2.   The A Current (I_A)

The fast outward current carried by potassium ions, termed the A current, is activated at subthreshold levels upon a depolarizing voltage step. It was first described in molluscan neurons of the species Onchidium (Hagiwara et al., 1961) and has been implicated in stabilizing the RMP and to play a role in the fast repolarization phase of the action potential. It also contributes to frequency modulation in pacemaker neurons (Connor and Stevens, 1971). The channel has been identified at the single-channel level according to the general features of the transient A current: 1) A hyperpolarizing prepulse to -100 mV fully removes steady-state inactivation. 2) Channel opening occurs within the first few msec (< 10 msec) after a depolarizing voltage step. 3) The open probability declines rapidly after the peak current. 4) A-channels are not blocked by external tetraethylammoniumion (TEA; Thompson, 1977).

In inside-out patches taken from neurons of the pond snail Lymnaea stagnalis, the single-channel conductance for the A channel was 40 pS (Kazachenko and Geletyuk, 1984; Geletyuk and Kazachenko, 1983). Individual traces show that the burst duration is

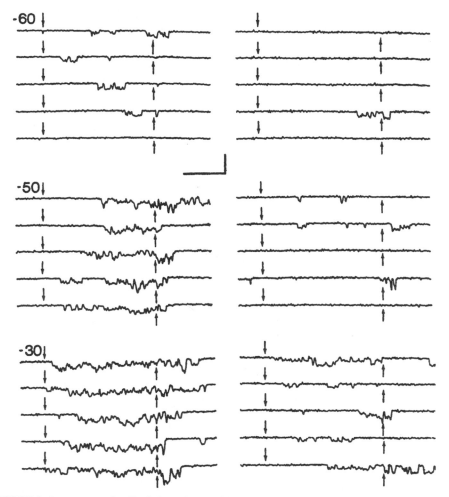

FIGURE 8. In contrast to the 40-pS channel, the activity of the 20-pS channel decreased in the presence of phosphorylating agents. The left panel shows the control situation with the negative numbers indicating three depolarizing voltage steps from a holding potential of -80 mV. The right panel shows the reduced activity (only three active channels versus four in control) of the 20-pS channel 5 min after release of caged ATP by illumination. The upward and downward arrows indicate the onset and offset of the voltage step. Note that the channel is still active after the voltage pulse, generating a "tail" current. Scale bar is 3 pA; 3 msec. (From Perozo *et al.*, 1991b.)

voltage-dependent, i.e., it decreases with depolarization, which is to be expected for the transient fast potassium current (Fig. 10). Interestingly, the unitary conductance was broken down into subconductance levels of equal spacing of about 2.5 pS, either spontaneously or after introduction of channel blockers such as Cd^{2+} (Fig. 11). Amplitude distributions revealed up to 16 sublevels and thus, the single-channel pore may consist of elementary subunits with a conductance of 2.5 pS. Such a breakdown of the unit conductance has also been detected in molluscan glial cells, where a 100-pS K^+

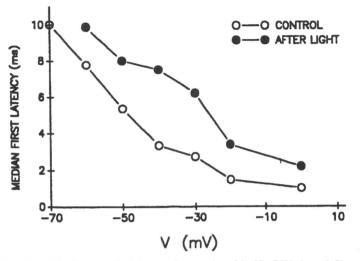

FIGURE 9. Phosphorylation increases the latency to first opening of the 20-pS K[+] channel. Shown is a plot of the latencies to first opening before and after the release of ATP. The increase in the latencies is apparent at all membrane potentials. (From Perozo *et al.*, 1991b.)

channel has been studied (Geletyuk and Kazachenko, 1989). When a very high K[+] gradient was used (100 mM [K[+]]$_0$ / 1 M [K[+]]$_i$) the 16 sublevels could be broken down even further. Each of the 16 substates had four inherent sublevels, giving a total of 64 quantized substates (Geletyuk and Kazachenko, 1987). It has been proposed that this quantization determines the kinetic parameters for the open and closed times, giving multiples of frequencies of 1, 2, 4, and up to 8 kHz (Geletyuk *et al.*, 1988).

In *Doriopsilla albopunctata* and *Anisoderis nobilis* Premack and coworkers estimated the number of A channels present in identified neurons and compared the current density in an entire axotomized cell body versus the axon hillock (Premark *et al.*, 1989). In whole-cell recordings, about 2.9×10^6 channels were activated by a voltage step from -90 mV to -30 mV, whereas 1.3×10^3 channels were found in cell-attached patches using a large patch pipette (20-μm orifice). The time course of activation and inactivation was the same for the whole-cell and the patch current, but when the current densities, i.e., the peak current normalized to the surface area, were compared, a significant difference was found. The overall ratio of patch current density to whole-cell current density was 0.64, indicating a higher current density (more channels) in the initial segment. This was further confirmed with patch recordings from the latter, where the ratio was found to be 1.3. Hence the potassium A channels are not distributed randomly across the cell body. In fact, there is considerable variability in patch current density recorded from different locations of the soma in the same neuron (Fig. 12). When the variances of the current densities recorded with patch electrodes of different sizes (ranging from 50–1000 μm²) were compared, the current densities recorded with smaller patch electrodes had a greater variance. This indicates a cluster organization of A channels. With a smaller patch electrode one is more likely to record from regions devoid of channels or from

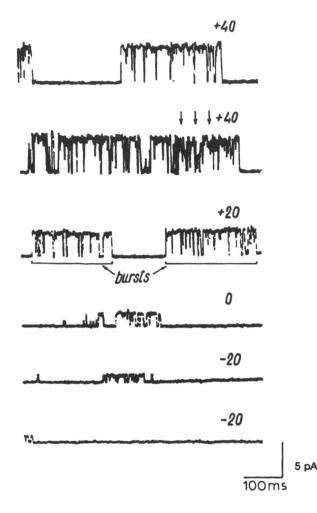

FIGURE 10. Inside-out patch showing the bursting activity of the A channel. The numbers above each trace indicate the respective voltage step from a holding potential of −50 mV. Note that the frequency of bursts decreased with depolarization. (From Kazachenko and Geletyuk, 1984.)

regions with "hot spots" which have a greater number of channels. This explains why the variance of the current density is larger when recordings are done with small patch electrodes. This result is in good agreement with the work by Johnson and Thompson (1989), in which they showed a cluster organization of channels for the inward Na$^+$ current, the potassium A current, and the delayed rectifier in *Aplysia* buccal ganglion B cells. It is noteworthy that one cluster of A channels may be composed entirely of A channels, since 1) there was no evidence of other types of potassium channel present in the same cluster, and 2) when unitary currents were recorded in a cell-attached patch, only voltage steps from −80 mV to 0 mV elicited single-channel currents, whereas a step from the RMP potential to 0 mV did not evoke a response (Fig. 13). The reversal potential of the A current was −64 mV and the derived slope conductance of the unitary current was about 9 pS. Although the activation and inactivation parameters resembled those of the whole-cell currents, there was a difference in the inactivation time constant in homologous neurons of different animals, ranging from 59 msec to 127 msec. Inactiva-

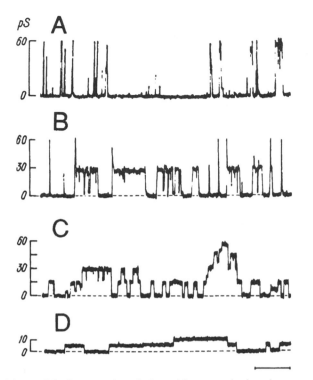

FIGURE 11. Breakdown of the fast potassium A channel into several subconductances of equal spacing. Initial conductance was 60 pS (A), which underwent degradation after 15 min into sublevels of 30 pS (B), 16 pS (C), and 8 pS after recording the channel for 1 hr (D). Inside-out patch; holding potential was −50 mV (A, B, C) and −100 mV (D). Scale bar is 100 msec. (From Kazachenko and Geletyuk, 1984.)

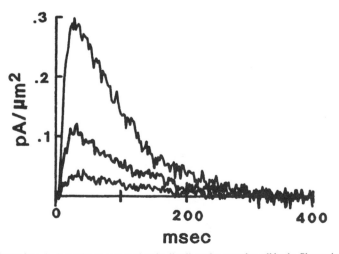

FIGURE 12. Potassium A channels are not randomly distributed across the cell body. Shown is the variability of the current density recorded from three patches in the same neuron. Currents were corrected for patch size by dividing the peak current by the patch capacitance. Voltage steps were delivered to −30 mV after a conditioning step to −90 mV to remove inactivation. (From Premack et al., 1989.)

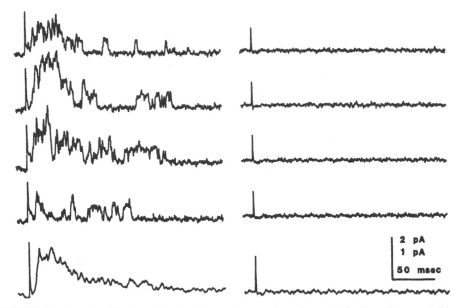

FIGURE 13. A cluster of A channels (note that more than one channel opened after the voltage step) did not contain the delayed rectifier K^+ channel. The left panel shows four individual sweeps with a voltage-step protocol to activate A channels, i.e., step from -80 mV to 0 mV. The fifth trace is the ensemble current composed of ten individual sweeps. The right panel shows voltage steps in the same patch delivered from the resting potential (-40 mV). There was no channel opening, indicating that there were no channels present other than the A-type potassium channel. These channels have been rendered inactive due to the different voltage protocol applied. (From Premack *et al.*, 1989.)

tion was about 30% at the RMP; above -40 mV the activation was linear and did not reach saturation up to $+60$ mV. A difference in inactivation kinetics between homologous neurons of the same species as well as a difference in current density has been shown in two nudibranch species by Serrano and Getting (1989). They compared the A current density in identified neurons of *Archidoris montereyensis* and *Anisoderis nobilis*. A higher current density was found for *Anisoderis* (14 ± 2.2 $\mu A/\mu F$) than for *Archidoris* (5.8 ± 1.1 $\mu A/\mu F$) when homologous neurons were compared. In *Helix aspersa* neurons the single-channel conductance of the A current was about 14 pS (Ram and Dagan, 1987; Taylor, 1987), which is in the range reported by Premack *et al.* (1989). This is only slightly smaller than the values reported for rat nodose ganglia cells (22 pS; Cooper and Shrier, 1985) or dorsal-root ganglia neurons (20 pS; Kasai, 1985), but differs considerably from the value in *Lymnaea* reported by Kazachenko and Geletyuk (1984). Permeability studies showed that the A channel is permeable to $Tl^+ > K^+ > Rb^+ > NH_4^+ > Cs^+$ and, to a much smaller extent, Na^+ and Li^+ (Taylor, 1987).

2.2.3. The Anomalous (Inward) Rectifier (I_{anom})

The anomalous rectifier (I_{anom}) was studied in inside-out patches from neurons isolated from the left pedal ganglia of the snail *Planorbarius corneus* (Kachman *et al.*,

1989). The density of K^+ channels appeared to be high, since only in the presence of blockers such as Cs^+, Ba^{2+}, or TEA (2–10 mM), could a discrete opening and closing of single inwardly rectifying currents be detected. In the voltage region more negative than the equilibrium potential for potassium ions, E_k, the single-channel conductance in the linear portion was about 81 pS. Moreover, as is characteristic for the anomalous rectifier, the I-V relation became nonlinear at potentials more positive than E_k. The conductance decreased and eventually fell to zero at holding potentials of more than +80 mV. Openings of about 1 msec were grouped into bursts of about 10-msec duration, whereas the closed state was either short-lasting (about 1 msec) or long-lasting (> 10 msec). The mean open time τ_o was 1.4 ± 0.5 msec, the short-lived closed time τ_{cl} was 0.74 ± 0.17 msec, and the burst duration τ_t was 9.8 ± 2.1 msec. Lifetime histograms could be best fit with one exponential for the open state and two exponentials for the closed state, indicating one and two kinetically different states, respectively. Neither of the two time constants of the closed state was voltage-dependent, whereas open time, burst duration, and open probability decreased with membrane depolarization. One major difference between *Planorbarius* and other nonmolluscan preparations is the conductance, which is 80 pS in *Planorbarius* and varies between 6 and 27 pS in sea urchin oocytes (Fukushima, 1982), rat myotubules (Ohmori *et al.*, 1981), and cardiomyocytes (Sakmann and Trube, 1984).

2.2.4. The Ca^{2+}-Dependent K^+ Current ($I_{K(Ca)}$)

The presence of potassium channels that are dependent on the presence of Ca^{2+} ions was reported in molluscan neurons some 20 years ago (Gorman and Thomas, 1980; Heyer and Lux, 1976; Meech and Standen, 1975). Single-channel analysis revealed the presence of two major conductances, a potassium channel with a big conductance (BK channel, also often referred to as C current) and a smaller conductive channel, the SK channel. Besides their differences in conductance, the BK and SK channels have unique properties with regard to voltage dependence and sensitivity to scorpion toxin. While the BK channel is highly voltage-dependent, selectively blocked by charybdotoxin (ChTX), and sensitive to TEA, the SK channel exhibits only a weak voltage dependence and is not sensitive to either ChTX or TEA. Hence, the BK channel has been proposed to play a role in spike repolarization, whereas the SK current may underlie the post-spike afterhyperpolarization (I_{AHP}).

To investigate $I_{K(Ca)}$ without contamination of other potassium channels contributing to the total outward current, several authors took advantage of U-cells in the right parietal ganglion of *Helix*, which are unique in the sense that they possess Ca^{2+}-dependent action potentials and an outward current exclusively carried by $I_{K(Ca)}$ (Lux and Hofmeier, 1982). Gola *et al.* (1990) and Crest and Gola (1993) studied the properties of the BK channel in U-neurons using cell-attached patches in conjunction with an intracellular electrode to elicit whole-cell spikes. The unitary conductance ranged from 24 pS in normal $[K]_o = 5$ mM to 65 pS in low $[K]_o = 0.5$ mM (mean = 48 ± 13 pS; n = 64); in contrast to the delayed rectifier K^+ current, the time course of activation was slow for the C current after a depolarizing step. However, when the depolarizing voltage pulse was preceded by either a small conditioning step or by a whole-cell spike, the activation rate increased dramatically (Fig. 14). This facilitation of the C current was likely to result

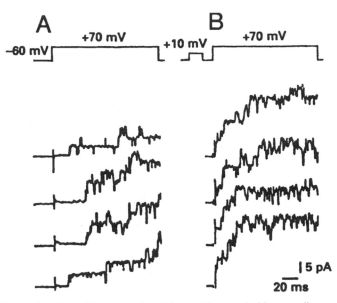

FIGURE 14. The activity of the calcium-dependent K$^+$ current increased with a preceding prepulse. (A) Individual sweeps from a holding potential of -60 mV to $+70$ mV. (B) Shown is the dramatic increase of the single-channel activity caused by a 10-mV prepulse for 20 msec just prior to the voltage step. This allowed additional Ca^{2+} ions to enter and resulted in the enhanced activity of I$_{K(Ca)}$. (From Gola *et al.*, 1990.)

from an increase in open probability rather than an increase in the number of active channels, since the steady-state level of the control and enhanced C current was the same. In *Aplysia* neurons, the open probability increased from 1–2% (nominal 0 [Ca^{2+}]$_i$) to 11–19% (0.3 μM Ca^{2+}) and to 79–93% when internal [Ca^{2+}] was raised to 1 μM (Hermann and Erxleben, 1987). Calcium ions not only play a permissive role in the activation process of I$_{K(Ca)}$ (the C current was completely blocked within 1–2 min when EGTA was ionophoresed with the intracellular electrode), but a persistent Ca^{2+} influx is necessary to keep the channel in the open state (Fig. 15). A depolarizing pulse evoked a burst-like single-channel activity which ceased after 2–4 sec. A whole-cell spike, i.e., additional Ca^{2+} entry, elicited while the cell was still depolarized caused a similar burst (compare the amplitude histograms in Fig. 15), indicating that the shutdown of the burst was actually due to inactivation of voltage-dependent Ca^{2+} channels, not to inactivation of C channels. No channel opening was observed upon depolarization when a Ca^{2+}-free solution was used in the patch pipette, but opening could be induced by introducing a whole-cell spike 50 msec before the depolarization. Then the C channel opened with a delay of about 140–420 msec, which was possibly the time necessary for Ca^{2+} ions to diffuse to the cytoplasmic channel mouth. Lux *et al.* (1981) compared the delay of opening of single channels (Ca^{2+} ions were present in the patch electrode) with the delay of the macroscopic I$_{K(Ca)}$ in *Helix pomatia* D-cluster neurons of the right parietal ganglion. They reported a unit conductance of 18.5 pS and found the delay to be in the range of 7–30 msec for the single channel as well as for the macroscopic current. Thus,

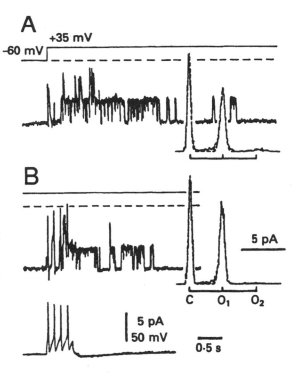

FIGURE 15. The presence of Ca^{2+} ions is necessary for a persistent activation of $I_{K(Ca)}$. In (A) the opening of single Ca^{2+}-dependent K^+ channels is shown after a voltage step. The activity ceased after about 4 sec, but with the patch still depolarized, a whole-cell spike induced by an intracellular electrode reopened the C channel (B). Note that the amplitude histograms revealed a similar distribution. (From Gola *et al.*, 1990.)

the rate-limiting step in the activation of $I_{K(Ca)}$ might be the binding of Ca^{2+} ions to the channel mouth. This could be shown by estimating the rise time of the ensemble C currents (average of sweeps of single-channel activity) recorded with an incremental delay between the whole-cell spike and the test depolarization. Since there was no Ca^{2+} in the patch electrode, the only way calcium could enter the cell was by the whole-cell spike *via* voltage-dependent calcium channels located outside the patch area. Figure 16 illustrates this experiment. It is apparent that the rise time decreased with increasing delay, even when, at delays longer than 400 msec, the peak current declined owing to diffusion of Ca^{2+} away from the channel mouth. However, a portion of available Ca^{2+} still remained bound to the channel, dominating the activation process of the $I_{K(Ca)}$ current. The decrease of the rise time of the ensemble current can be attributed to a decrease in latency to first openings, resulting in an increase of P_o. This may reflect an increase in the rate constant for binding calcium. Hence, Ca^{2+} plays a crucial role as a rate-limiting factor in the activation phase of the C current. The mean burst duration was, however, independent of $[Ca^{2+}]$, but nevertheless voltage-dependent. Mean burst duration was 11.5 msec at -30 mV, 14 msec at -20 mV, 24 msec at -10 mV, and 42 msec at $+20$ mV. Crest and Gola (1993) estimated that 25–30% of the C current participated in spike repolarization due to its voltage dependence. The mean closed time within a burst was 0.5 msec, and both the open and closed time distributions were fit with a double exponential. The time constants for the open time were $\tau_1 = 1.42$ msec and $\tau_2 = 15$ msec and for the closed time, $\tau_1 = 82$ msec and $\tau_2 = 0.62$ msec. Interestingly, during long-term

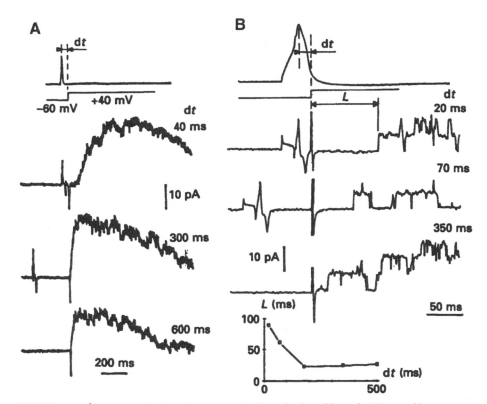

FIGURE 16. Ca^{2+} is the rate-limiting factor governing the activation of $I_{K(Ca)}$. In (A) ensemble currents are shown (patch electrode was Ca^{2+}-free; 20–25 sweeps) and the voltage steps were delivered with an increasing delay (40 msec, 300 msec, and 600 msec) following the initiation of a whole-cell spike. Note that although the peak current declined with increasing delay, the rising phase became steeper. In (B) individual openings are shown. The latency to first opening decreased as the delay increased. (See text; from Gola et al., 1990.)

recordings the number of channels present in the patch decreased. This was paralleled by a decrease in the patch's capacitance, indicating that this was due to a decrease in surface area and not channel inactivation. Two patches taken from the same neuron showed that only one of them contained C channels, supporting the view of a cluster organization. The number of channels in a particular patch varied considerably (between four to six), and on average a single patch contained 5.05 ± 0.4 (n = 75) channels. It is noteworthy that voltage-dependent Ca^{2+} channels seem to be normally distributed across the cell body while $I_{K(Ca)}$ channels appear in clusters. Thus, only those Ca^{2+} channels associated with C channel clusters may provide the calcium influx necessary for C channel activation.

At least two toxins, both derived from scorpion venom, have been shown to specifically block the BK-type Ca^{2+}-dependent potassium current. Kaliotoxin (KTX; Crest et al., 1992) and charybdotoxin (Miller et al., 1985; Hermann, 1986; Hermann and Erxleben, 1987), derived from the species *Androctonus mauretanicus mauretanicus* and *Leiurus quinquestriatus*, respectively, blocked the BK-type current in a reversible

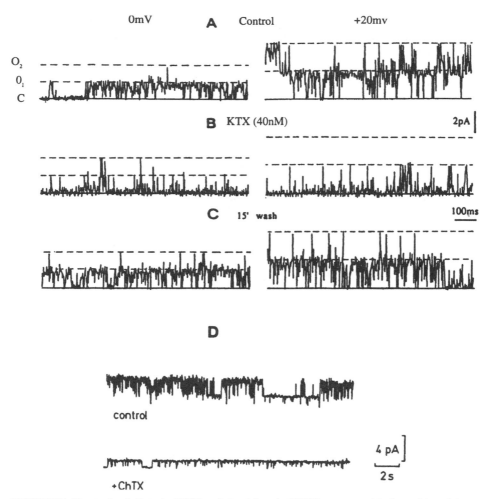

FIGURE 17. Two toxins, kaliotoxin (KTX) and charybdotoxin (ChTX), are potent blockers of the calcium-dependent potassium current. (A), (B), and (C) show the single-channel activity in an outside-out patch of a *Helix* neuron and the block of $I_{K(Ca)}$ by KTX, whereas (D) shows the block by ChTX (200 nM) in an *Aplysia* neuron. (From Crest *et al.*, 1992; Hermann and Erxleben, 1987.)

manner when applied to the external face of the membrane (Fig. 17). Neither one of the compounds blocked the BK channel when it was applied to the internal face of the membrane. KTX was slightly more potent than ChTX, ($K_{D(KTX)} = 20$ nM; $K_{D(ChTX)} = 30$ nM at 0 mV), but the two toxins differed in two ways. While the block by ChTX was voltage-dependent and immediate, KTX showed no voltage-dependent blockade but induced fast flickering before total shutdown of the channel. These toxins have a *ca.* 3000-fold potency over TEA in blocking $I_{K(Ca)}$ and are therefore useful pharmacological tools to study the calcium-dependent K^+ current.

It has been known that the macroscopic $I_{K(Ca)}$ current increases under phosphorylating conditions (DePeyer *et al.*, 1982). When the catalytic subunit of cAMP and Mg^{2+}-ATP were applied to the internal side of an inside-out patch taken from an F-cluster neuron of *Helix aspersa*, the activity of the BK channel increased dramatically over the whole voltage range (Fig. 18). The phosphorylation probably caused an increase in the affinity of the channel for calcium, since its activation curve shifted toward the hyperpolarized range. This provides evidence that the phosphorylation site is located in the channel itself, or at least near the channel mouth (Ewald *et al.*, 1985). Another second messenger system, the hydrolysis of phosphatidylinositol-4,5-biphosphate to diacylglycerol and inositol triphosphate (IP_3) has been shown to be involved in the activation of a calcium-dependent potassium current (Fink *et al.*, 1987, 1988). In cultured bag cell neurons of *Aplysia*, intracellular injection of IP_3 increased the activity of a 40-pS K^+ conductance recorded in on-cell patches. The increase in channel openings resulted from a shift of the activation threshold toward the hyperpolarized range. When the patch was withdrawn from the cell, increasing the Ca^{2+} concentration in the bath, and not the presence of IP_3, led to the same observation. Thus, the enhanced activity of the $I_{K(Ca)}$ current was due to the release of intracellular calcium by IP_3 and not to a direct action of this second messenger.

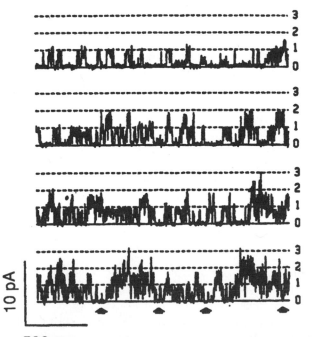

FIGURE 18. The calcium-dependent K^+ channel is modulated by cAMP. Application of the catalytic subunit (CS) of cAMP-dependent protein kinase to the internal side of an inside-out patch (held at 0 mV) increased the single-channel activity. The figure shows a continuous record 50 sec after introducing the CS into the bath. The arrows at the bottom record indicate the occasional shutdown of all active channels. (From Ewald *et al.*, 1985.)

In 1990, Kramer introduced the "patch cramming" technique, in which an inside-out patch taken from a donor cell was calibrated with various second messengers, such as Ca^{2+}, and subsequently inserted into a recipient neuron to study the presence of putative second messengers in the intact cell (Kramer, 1990). He recorded Ca^{2+}-dependent potassium channels in *Helix* F1 neurons and showed that the majority of the recipient neurons as well as the donor patch survived the cramming procedure. The activity of $I_{K(Ca)}$ channels depended highly on the position of the patch electrode within the recipient neuron. If the patch electrode was advanced too far into the cytosol, the activity ceased but was reintroduced upon moving the pipette toward the original position, closer to the cell membrane. This suggested that the channel actually sensed the local Ca^{2+} concentration rather than bulk concentration within the cell. This study also implied that the local calcium concentration near the cell membrane was actually higher than in the cytosol. Taken together, the $I_{K(Ca)}$ current participates in spike repolarization because of its voltage-dependent property (the falling phase of the action potential increased when the $I_{K(Ca)}$ current was blocked; Hermann and Erxleben, 1987; see also Chagneux *et al.*, 1989) and in regulating cell-firing frequency due to its Ca^{2+}-binding mechanism. The channel can be best described as a calcium-modulated voltage-dependent current (Gola *et al.*, 1990).

2.3. The Calcium Channel

By using the single-channel technique the activation and inactivation process of calcium channels was resolved at the molecular level in conjunction with simultaneous recording of the whole-cell calcium current. Lux and Nagy (1981) managed to isolate the calcium current from the contaminating late outward calcium-dependent potassium current in cell-attached patches in *Helix* neurons by using external and internal TEA. They reported a slope conductance of 5 to 15 pS and a mean open time of 3 msec. The open probability increased upon depolarization, but the mean open time was voltage-independent. When the Ca^{2+} current was studied in the cell-attached mode in combination with a two-electrode voltage clamp, it was shown that the activation and inactivation kinetics of the averaged ensemble currents were similar to the obtained whole-cell current (Fig. 19; Lux and Brown, 1984a; Brown *et al.*, 1986). No outwardly single-channel activity was observed after a depolarizing step from rest to 0 mV, providing evidence that inactivation was not due to contamination of a late outward current. Neither the mean open times nor the unitary current amplitude was altered during the voltage pulse. Thus, inactivation had to be mediated by a change of the closed time and that was, in fact, the case. When the closed times were measured during the first 50 msec of a voltage step and compared with closed times during the last 200 msec of a 300-msec test pulse, the closed times during the inactivation phase increased threefold. Thus, the inactivation of the calcium current was caused by a reduction of the open probability of single calcium channels. Accumulation of Ca^{2+} could also contribute to the inactivation process (see Eckert and Chad, 1984, for inactivation of the calcium current). To rule out the latter, Ca^{2+} was omitted from the external bath solution and 10 mM Ni^{2+} was added to block potential Ca^{2+} entry. With Ca^{2+} still present in the patch electrode, the same single-channel activity was observed, refuting the hypothesis that the inactivation was

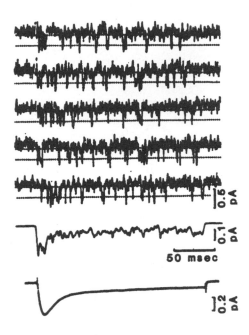

FIGURE 19. Simultaneous recording of voltage-dependent Ca^{2+} channels in the two-electrode voltage-clamp and cell-attached mode. Five individual single-channel traces are shown (voltage step from −50 mV to 0 mV). Below is the ensemble current of 42 individual sweeps and the Ca^{2+} current recorded in the whole-cell voltage-clamp mode. Note that the activation and inactivation kinetics of the ensemble and whole-cell currents are similar. (From Lux and Brown, 1984.)

caused by accumulation of Ca^{2+} near the channel. When barium ions were used as the charge carrier, the unitary currents were larger and the inactivation slower than in the Ca^{2+} solution (0.43 pA in Ca^{2+} and 0.52 pA in Ba^{2+}; Lux and Brown, 1984b). The slope conductance for the calcium current was *ca.* 7 pS, and assuming a capacitance of 1 $\mu F/cm^2$, the channel density was estimated at 31 μm^2. The barium currents are known to inactivate less than the calcium current owing to a shift of the activation curve toward hyperpolarized potentials (Wilson *et al.*, 1983). Thus, a higher open probability of the channel was expected in the Ba solution. Measuring P_o for one particular patch gave a value of 0.15 for the calcium current and 0.2 for the barium current. Lux and Brown (1984b) also investigated temperature effects on single calcium channel currents. Cooling the chamber from 29° C to 9° C resulted in a decrease of the unitary current (0.49 pA versus 0.33 pA) with the mean open times nearly unaffected (1.05 msec versus 1.2 msec). However, the slower time constant of the two-exponential distribution of the closed time histogram, which represents the interburst closed time, increased rather dramatically upon cooling ($\tau_1 = 1.88$ msec and 1.82 msec and $\tau_2 = 4.9$ msec and 11 msec at 29° C and 9° C, respectively). In addition, cooling increased the latency to first openings, also referred to as waiting time, resulting in a slowing down of the averaged ensemble current (Brown *et al.*, 1984). Thus, the decrease of the peak current at lower temperature can be accounted for by a reduction of the open probability during the first 25 msec of the voltage step.

When Ba currents were recorded in isolated neurons of the snail *Lymnaea*, a channel with a higher conductance in the range of about 20 pS and an open time of 1 msec was observed upon patch excision (Yazejian and Byerly, 1987). Moreover, the smaller voltage-dependent channel disappeared in inside-out patches and was replaced by a large

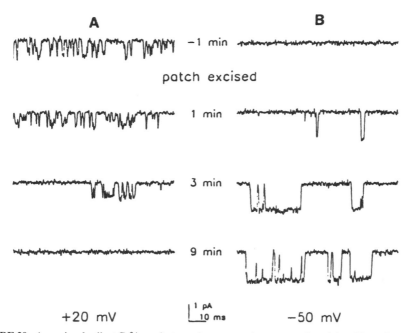

FIGURE 20. A previously silent Ca^{2+} conductance becomes active upon patch excision. Shown in (A) is the single-channel activity recorded with Ba^{2+} as the charge carrier. Note that after going into the inside-out mode, the small-conductance voltage-dependent Ca^{2+} current vanished. The same experiment was conducted in a different cell but at a different holding potential (B). Often there was no indication of the small conductance channel, but when the patch was excised, a large-conductance, non-voltage-dependent channel appeared. (From Yazejian and Byerly, 1989.)

non-voltage-dependent conductance, termed the HP channel, since it is active at the resting holding potential (Fig. 20; Yazejian and Byerly, 1989). The voltage-dependent current never reappeared, even when Mg-ATP or the catalytic subunit of cAMP-dependent protein kinase was added to the media. The two channels are probably distinct entities, since they often appeared simultaneously within the first 2 min after patch excision and since the larger conductive channel appeared even when the smaller voltage-dependent conductance was not observed in the cell-attached mode. The slope conductance of the HP channel was 27 pS in the hyperpolarized range (-110 mV to -40 mV), the mean open time was about 5 msec, and the channel exhibited inward rectification in the more depolarized region. The activity of the HP channel recorded in the outside-out configuration is reversibly blocked by replacing 30 mM Ba^{2+} of the 40 mM Ba solution with Mg^{2+}. Adding 3 mM Ba^{2+} or Ca^{2+} to the internal media, however, caused a reduction of the current in the inside-out mode. On the other hand, this channel was activated in on-cell patches by internally perfusing the cell with either a high-Ca^{2+} solution or prolonged perfusion with a low-Ca^{2+}/no ATP solution. It is likely that the HP channel underlies the macroscopic regenerative increase in internal Ca^{2+}, which is triggered by a transient rise in the internal calcium concentration (Byerly and Moody,

1984). The HP channel was also investigated in cultured *Lymnaea* neurons (Strong and Scott, 1992). It had a similar conductance (between 25 and 30 pS) and was rapidly activated in an excised patch. Open and closed times were fit with a two-component exponential function on the order of 5–10 msec. The current was reversibly blocked by adding 1 mM Mg^{2+} to the bath. Magnesium ions had no effect on the open time but increased the slower time constant of the closed time from 5.7 msec to 17.8 msec thereby reducing the open probability. The block differed from the known open channel block mechanism by Mg^{2+} ("flickering"), since additional bursts were not seen and the open time was not affected. It is therefore conceivable that the HP channel is blocked under physiological conditions by the presence of intracellular Mg^{2+} ions. The majority of the channels were located in the neurites and not in the cell soma. On average, 44% of the patches taken from the soma showed the HP channel, whereas 88% of the patches taken from the neurites had the HP channel.

A calcium channel of similar magnitude can be activated by the second messenger protein kinase C (PKC). Bag cell neurons of *Aplysia* were kept in culture and cell-attached patch recording was performed with Ba^{2+} as charge carrier (Strong *et al.*, 1987a). In control conditions, one class of Ca^{2+} channels had a conductance of 12 pS. Treatment of the cell with the phorbol ester TPA (activates PKC) revealed a new class of previously silent calcium channels with a conductance of about 24 pS (Fig. 21). The 12-pS channel was not affected by the TPA treatment and kinetic analysis showed no striking difference between these two channels (open time histogram was fit monoexpo-

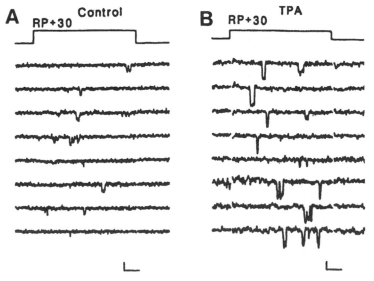

FIGURE 21. Treatment of a bag cell neuron with the PKC-activator TPA (12-O-tetradecanoyl-phorbol-13-acetate) uncovers a distinct Ca^{2+} channel. Cell attached patches are shown before (A) and after (B) treatment with 10 nM TPA. In control, only a 12-pS Ca^{2+} channel was observed, but after TPA treatment a higher conductance channel (24 pS) appeared (estimated resting potential was −61 mV). Scale bars: 1 pA, 10 ms (20 ms in B; from Strong *et al.*, 1987a.)

nentially with a time constant of < 1 msec). Hence, the enhancement of the macroscopic Ca^{2+} current may be due to a recruitment of a new class of Ca^{2+} channels.

Internal Cs^+ has been shown to cause run-down of the Ca^{2+} current in outside-out patches in *Aplysia* neurons, but when NaCl was used in the intracellular solution, the single Ca^{2+} channel currents could be observed for several hours (Chesnoy-Marchais, 1985a,b). Two types of conductances were found, but only the larger one was subjected to further analysis since studying the smaller conductive channel was hampered by the size of its unitary current. The elementary current of the larger conductance was about 1 pA at -40 mV, but owing to a strong rectification in the depolarized range, a slope conductance could not be estimated. High concentrations of standard calcium channel blockers such as Ni^{2+}, Co^{2+}, Mn^{2+}, and Cd^{2+} were necessary to block the channel completely (10 mM $NiCl_2$ or 30 mM $CoCl_2$). Most of the channels studied had a very long open time in the range of 100 msec, but occasionally a faster channel was seen. Open dwell time was fit with a monoexponential and sometimes with a biexponential, representing the contribution of the fast channel. The closed time histogram was fit with a double exponential function. Both the open and closed times were voltage-dependent, and the mean open time increased e-fold for a depolarizing voltage change of 37 mV. Replacing Ca^{2+} with Ba^{2+} increased the elementary current by a factor of 2.6 and reduced the mean open and closed time by a factor of 9 and 15, respectively. Other divalent cations such as Sr^{2+} and Mg^{2+} had a similar effect on the kinetic parameters but did not change the unitary current. The overall conductance sequence for the Ca^{2+} channel was: $Ba \geqslant Ca = Sr \approx Mg$. An "anomalous mole fraction behavior" was observed at membrane potentials more positive than -40 mV. Although a total substitution of Ca^{2+} with Ba^{2+} (60 mM) resulted in an increase of the elementary current, a 20 mM Ca^{2+} + 40 mM Ba^{2+} solution resulted in a smaller current in the depolarized range (0–10 mV) compared with the 60 mM Ca^{2+} solution (Fig. 22). The substitution experiments suggest that the open channel has binding sites for more than one specific ion species. Strong *et al.* (1987b) reported the activation of a similar Ca^{2+} channel in excised patches taken from cultured *Aplysia* bag cell neurons.

Reconstituted Ca^{2+} channel proteins in lipid bilayers at the tip of a patch electrode revealed the presence of three conductances of 10, 25, and 36 pS. The gating of all three types was similar and the Ca^{2+} channel agonist Bay K 8644 increased the duration of the open state, enhancing the channel open probability (Coyne *et al.*, 1987). A Ca^{2+} channel with a conductance of 24 pS has also been reported in axosomes prepared from the squid axon (Fishman and Tewari, 1990).

Two different types of Ca^{2+} channels were disclosed in a coculture system of *Aplysia* sensory and motor neurons (Edmonds *et al.*, 1990). A slow-inactivating current active at voltages below 0 mV was blocked by the L channel blocker nifedipine, whereas a fast-inactivating, nifedipine-insensitive current active above 0 mV was blocked by the neuropeptide FMRFamide. The fast current had a unitary amplitude too small to establish a slope conductance, but the nifedipine-sensitive Ca^{2+} current revealed two conductances of 15 and 25 pS. Nifedipine had no effect on the postsynaptic potential recorded in the motor neuron, while FMRFamide caused presynaptic inhibition. This result is consistent with the idea that the dihydropyridine-sensitive channel does not contribute to transmitter release (neither homosynaptic depression nor presynaptic

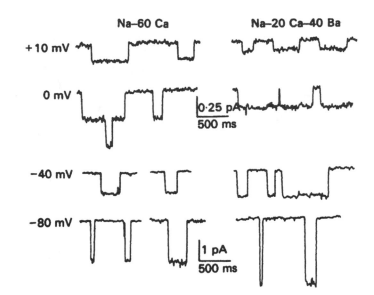

FIGURE 22. "Anomalous mole fraction behavior" of a Ca^{2+} channel in *Aplysia* neurons. A substitution of a 60 mM Ca^{2+} solution with a 20 mM Ca^{2+}–40 mM Ba^{2+} solution resulted in a decreased Ca^{2+} current at potentials more depolarized than -40 mV. At all other potentials, the Ca^{2+} current was higher in the Ba^{2+}-containing solution. (From Chesnoy-Marchais, 1985b.)

facilitation was affected), whereas the fast-inactivating, small-conductance channel may mediate synaptic release. However, since FMRFamide has multiple actions in the *Aplysia* nervous system (see above), these conclusions have to be taken with care. In the terminology of mammalian Ca^{2+} channels, the nifedipine-sensitive current is analogous to the L-type, and the fast-inactivating current analogous to the N-type Ca^{2+} channel. There is also evidence for the presence of the P-type Ca^{2+} channel (first described in Purkinje neurons of the cerebellum) in the molluscan nervous system, at least for the giant synapse of the squid (Llinàs *et al.*, 1989). A toxin (FTX) derived from the funnel-web spiders *Agelenopsis asperta*, *Hololina curta*, and *Calilena* blocked synaptic transmission in the squid giant synapse without affecting the presynaptic action potential. This blockade was also produced by Cd^{2+} and Co^{2+}, but not by ω-conotoxin or dihydropyridines. Thus, the Ca^{2+} channel, with a conductance of 20 pS and 8 pS in 80 mM and 100 mM Ca^{2+}, respectively, shared some of the properties of the P-type channel, at least with regard to its pharmacology. Similarities between molluscan and vertebrate Ca^{2+} channels have also been reported by others. Brown *et al.* (1982) recorded Ca^{2+} currents from pacemaker cells of the snail *Helix*, cultured dorsal-root ganglion (DRG) cells from the chick *Gallus gallus*, and PC12 cells from the rat adrenal medulla. Unitary currents in all three species were in the range of 0.5 pA and occurred in bursts upon a 30-mV depolarization. In *Helix* neurons, the slope conductance was *ca.* 7 pS, as has been reported by others.

2.4. The Chloride Channel

In *Aplysia* neurons a hyperpolarizing voltage step activates a slowly developing chloride conductance (Chesnoy-Marchais, 1983), which can also be observed in cell-free patches. Ion substitution experiments and estimation of the equilibrium potential established the presence of an ion channel permeable preferentially to Cl^- ions (Chesnoy-Marchais, 1984; Chesnoy-Marchais and Evans, 1986a). Long hyperpolarizing voltage steps (30 sec) below E_{Cl} elicited inwardly directed single-channel currents carried by chloride ions. The channel exhibited bursting behavior, and the mean burst duration increased with hyperpolarization from 230 msec at -2 mV to 950 msec at -42 mV. The closed time intervals separating these bursts were *ca.* 1 sec at -42 mV and > 10 sec at -2 mV, and this translated into an open probability of 0.45 at -42 mV and 0.15 at -2 mV. Upon further inspection of individual bursts a short-lived subconductance level was resolved with half the unitary conductance, which was in the range of 10–15 pS. Interestingly, this channel was seen only in a small number of patches. It is known, however, that this chloride conductance is abundant in *Aplysia* neurons. Hence, an intracellular factor might control the functioning of these channels, and a wash-out process due to patch excision might be the cause for the low occurrence in outside-out patches. Another type of chloride channel, which is clearly different from the Cl^- current activated by hyperpolarization, was also described in *Aplysia* neurons (Chesnoy-Marchais and Evans, 1986b). One major difference concerns its unitary conductance, which was 100 pS in the presence of external NaCl and internal CsCl. Upon replacing NaCl with mannitol the reversal potential changed as expected for the Cl^--selective channel. However, in the mannitol solution the reversal potential was slightly more negative than the calculated reversal potential for pure Cl^- ions. Thus the channel was also conductive for cations, Cs^+ in this case. When Cl^- ions were substituted with NO_3^-, the permeability was actually higher than in the chloride solution, and the channel was also permeable to isothionate $> SO_4^{2-} >$ methanesulfonate $>$ gluconate, in that order. Thus, this chloride channel was rather nonselective and differed from the hyperpolarization-activated conductance also in its weak voltage dependence. Three different conductance states were observed, and the three associated currents all reversed at different holding potentials, suggesting that the individual conductance levels differ in their selectivity for a particular ratio of small cations to large anions. In intact cells this current may be active at rest, thereby contributing to the resting membrane potential. A chloride-selective channel with a similar conductance (100–120 pS) has been reported in dissociated neurons from the squid giant fiber lobe of the stellate ganglion (Oberhauser, 1989). In neurons of the snail *Lymnaea stagnalis*, Geletyuk and Kazachenko (1985) described a chloride current sensitive to external K^+, Rb^+, Cs^+, and to internal Ca^{2+} ions. The channel was recorded in inside-out patches and had a conductance of 200 pS. However, its activity depended on the presence of both internal Ca^{2+} and external K^+. With nominal 0 mM calcium and potassium in the respective solutions, the channel opened rarely. In addition, a breakdown of the channel into multiple subconductance states was observed over a time period of *ca.* 10 min. Moreover, as additional substates appeared (as many as 16 equally spaced at 12–14 pS), the higher conductivity levels vanished, indicating that the 100-pS channel broke down into several subconductance levels. Thus, this channel might be active in a normal physiological solution, but it remains to be seen, by utilizing on-cell patches, whether this current contributes to either the resting membrane potential or to the afterhyperpolarization following an action potential.

2.5. The Proton Current

In snail neurons, a voltage-dependent outward current carried by H^+ ions was recognized by Thomas and Meech (1982). Owing to a voltage dependence similar to the Ca^{2+} current, inactivation of the latter was thought to be potentially obscured by an increased H^+ current flowing through calcium channels. One possible physiological role proposed for the proton current was the maintenance of the internal acidity of the cell, since H^+ may be extruded from the cytosol by activation of the Ca^{2+}/H^+ exchanger after calcium influx. However, Byerly and Suen (1989) have utilized the inside-out patch clamp technique in *Lymnaea* neurons and have shown that stable proton currents could be recorded even after run-down of the calcium current. Thus, it was concluded that the H^+ ions did not permeate Ca^{2+} channels. In addition, it has yet to be established whether the observed H^+ current (with a very small unit conductance) is, in fact, mediated *via* specific ion channels since no individual channel openings and closings were detected, even though a proton current was seen in inside-out patches.

3. LIGAND-ACTIVATED CURRENTS

3.1. The Serotonin (5-HT)-Activated K^+ Channel

Application of serotonin causes the closure of a specific serotonin-sensitive potassium current, I_s. This current and its modulation by second messenger compounds has been extensively studied in the nervous system of *Aplysia*, where I_s underlies a simple form of learning, the defensive gill withdrawal reflex (Klein *et al.*, 1982). In *Aplysia* sensory neurons of the abdominal ganglion, 5-HT leads to a small depolarization accompanied by an increase in input resistance and spike broadening. The latter enhances the influx of Ca^{2+} into the nerve terminal and ultimately increases transmitter release. This mechanism of presynaptic facilitation strengthens the motor response of the gill withdrawal reflex in *Aplysia* (see Belardetti and Siegelbaum, 1988, for review). The S current displays a single-channel conductance of 55 pS in sensory neurons in the abdominal ganglia, 73 pS in sensory neurons of the pleural ganglia, and 23 pS in heart excitatory neurons of the giant African snail, *Achatina fulica* Fèrrusac (Furukawa and Kobayashi, 1988). It contributes to the resting membrane potential, does not inactivate upon depolarization, is calcium-independent, and shows outward rectification in normal sea water due to the asymmetry of the K^+ concentration (Camardo *et al.*, 1983; Pollock and Camardo 1987; Shuster *et al.*, 1991). It is, however, specifically blocked by serotonin, and single-channel studies have shown that the reduction of the macroscopic S current by 5-HT is due to a reduction in the number of active channels. Cell-attached patches were taken from sensory neurons, and bath application of 30 μM 5-HT reduced the number of active channels in the patch from five to two. Increasing the concentration to 60 μM shut

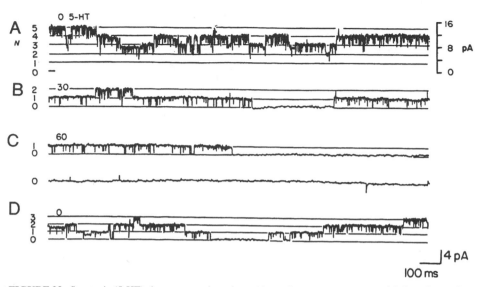

FIGURE 23. Serotonin (5-HT) closes a potassium channel in mechanosensory neurons of *Aplysia*. Increasing the concentration of 5-HT from 0 to 60 μM (A–C) successively reduced the number of active channels in the patch. At 60 μM, a total shutdown of the activity was observed, which was partially reversible upon wash-out (D). (From Siegelbaum *et al.*, 1982.)

down the channels completely, while this effect was reversible upon wash-out (Fig. 23; Siegelbaum *et al.*, 1982). The unitary current level was not affected by serotonin; only the number of active channels was reduced. Another implication of this experiment was that the action of 5-HT had to be mediated by a second messenger, since serotonin was only applied to the external solution and was not present in the patch pipette. Intracellular injection of cAMP mimicked the response of serotonin, suggesting that 5-HT led to an increase of cAMP *via* an elevation of adenylate cyclase. This corroborates previous results which have implicated a cAMP-dependent protein kinase to mediate the serotonin-induced shutdown of the S channel (Castelluci *et al.*, 1980). Further studies were undertaken to find out the site of the phosphorylating process. The catalytic subunit of cAMP (CS-cAMP) was applied to the internal side of inside-out patches in the presence and absence of Mg-ATP. Thirty-eight percent of the S channels in the patch closed in the presence of ATP, but only 23.5% closed in the absence of ATP (Shuster *et al.*, 1984). This showed that the effect of serotonin can be mimicked in excised membrane patches and that the protein phosphorylated was the channel itself, or at least a protein closely associated with the S channel. There was, however, a marked difference between the action of serotonin in the cell-attached mode and the effect of cAMP in inside-out patches. The channel closure caused by 5-HT actually outlasted the time of application, whereas the CS-cAMP only led to a transient blockade. While 5-HT blocked about a 0.46 fraction of the channels, CS-cAMP only blocked an average fraction of 0.34. Some of the closed channels even reopened in the continuous presence of CS-cAMP. An explanation for this behavior is the presence of putative diffusible proteins that act as blockers of a specific phosphatase in the intact cell. Hence, phosphorylation continues in the intact cell while it is disrupted in inside-out patches due to the lack of the putative phosphatase blocker. This has been examined further by applying fluoride ions, which are known to act as nonspecific phosphatase inhibitors, to the internal face of an inside-out patch. Potassium fluoride potentiated the effect of cAMP to about the same level as in the intact cell, providing evidence that a phosphatase was still present in the membrane patch while the inhibitor was washed out (Camardo et al., 1986; Shuster et al., 1985). Since the S current has been implicated in presynaptic facilitation, it was of interest to find out whether the channel is restricted to the cell body or is also present in nerve terminals. When Aplysia neurons are isolated and put in a culture dish, they begin to grow neurites. These neuronal elongations exhibit growth cones at their very end, and the growth cone can serve as a model for the nerve terminal. Action potentials recorded from either an intact growth cone or from a growth cone physically separated from the cell body showed the same response upon application of serotonin, that is, a broadening of the spike. In single-channel studies it was shown that the S channel present in growth cones had the same properties as the S channel in the soma. (Belardetti et al., 1986; Siegelbaum et al., 1986; Green et al., 1990 found an S-like potassium channel in growth cones in cultured neurons of Helix, but also reported a variety of voltage-dependent K$^+$ currents with conductances of 4, 8, 14, and 26 pS.)

Several compounds known to block either the fast transient or the delayed rectifier potassium current have been used to study their ability to block the S channel. Agents such as 5 mM 4-AP, 10 mM TEA, and 10 mM Ba^{2+} or Co^{2+} have very little effect on the macroscopic S current. However, in outside-out patches of Aplysia sensory neurons both external and internal TEA caused a dose-dependent reduction in the current amplitude (K_D = 85 mM and 42 mM for external and internal TEA, respectively). At higher concentrations (300 mM), TEA even increased the open probability of the S channel. This discrepancy with regard to the macroscopic current may result from the fact that the macroscopic current resembles the mean current amplitude, expressed as $I = N \times p \times i$ (N = the number of active channels, p = the open probability, i = single-channel current amplitude). Thus, the reduced single-channel amplitude might be counteracted by the increase in the open probability, resulting in an unaffected macroscopic current. The voltage dependence of the TEA blockade was very weak, changing e-fold with a 257-mV change in membrane potential for internal TEA. The blockade of the S channel by external TEA was virtually voltage-independent (Shuster and Siegelbaum, 1987). It seems therefore unlikely that internal and external TEA bind to the same site, since internal TEA probably acts closer to the voltage sensor of the channel than external TEA.

In addition to the closure of the S channel by serotonin, two endogenous neuropeptides, small cardioactive peptides A and B (SCP$_A$ and SCP$_B$) have been shown to have similar effects on the S channel. They produce presynaptic facilitation in Aplysia sensory neurons and increase the cAMP level when applied to the bath (Abrams et al., 1984). SCP$_B$ at 1–100 μM reduced the number of active channels in the patch without affecting the unit conductance. The two peptides and 5-HT did not cross-desensitize, suggesting that they acted on independent receptors, but nevertheless elevated the cAMP level. However, after a prolonged application of 5-HT, a decay of the peptide-evoked response was observed, and this process of heterotrophic desensitization of the adenylate cyclase system has been reported for other preparations (Harden, 1983). The two peptides did, however, cross-desensitize, indicating that they act *via* a common receptor-signaling pathway distinct from the 5-HT cascade. Another neuropeptide, FMRFamide, in contrast to the action of 5-HT, enhanced the S current by activating the metabolism of arachidonic acid (Belardetti *et al.*, 1987; Brezina *et al.*, 1987; Piomelli *et al.*, 1987). In contrast to 5-HT, FMRFamide did not merely increase the number of active channels in the patch, but also increased the time the channels spent in the open state, enhancing the open probability (Fig. 24). Hence, FMRFamide caused a slow hyperpolarization, a decrease in action potential duration, and a reduction in transmitter release (in some neurons in the snail *Helix aspersa*, FMRFamide also led to a small depolarization due to an increase in the Na^+ conductance; Cottrell *et al.*, 1984). And as in the case of 5-HT, a second messenger system had to be involved, since the peptide was present in the bath but not in the patch pipette. Arachidonic acid is a major constituent of phospholipids in the *Aplysia* nervous system. It is metabolized to bioactive compounds such as the eicosanoids *via* the 12- and 5-lipoxygenase pathway, and to prostaglandins through the cyclooxygenase pathway. When applied extracellularly, 5–50 μM arachidonic acid mimicked the FMRFamide response. One of the metabolites of arachidonic acid is 12-hydroxyperoxy-eicosatetraenoic acid (12-HPETE); this bioactive compound enhanced the S current in inside-out patches in the absence of any other second messengers like GTP or ATP (Buttner *et al.*, 1989). Thus, it acted directly at the ion channel without an intermediate step of phosphorylation. Moreover, its potency was higher when applied to outside-out patches (< 1 μM versus 20 μM in inside-out patches), suggesting that the 12-HPETE binding site is located near the extracellular face of the membrane. There is also evidence for a heme-containing enzyme mediating the enhancement of the S current by

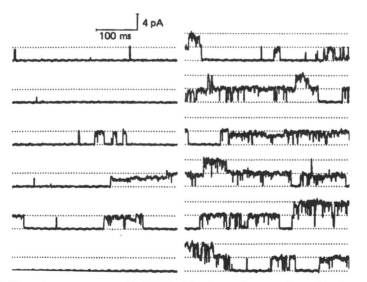

FIGURE 24. The endogenous neuropeptide FMRFamide enhances the serotonin-sensitive S current. Shown is the activity of a potassium S channel recorded in the cell-attached mode. In control (left panel) only occasional channel openings were seen but application of 20 μM FMRFamide into the bath enhanced the activity of the S current (right panel). (From Brezina *et al.*, 1987.)

FMRFamide. The channel activity increased in the presence of hematin, which catalyzes the metabolism of 12-HPETE (Belardetti *et al.*, 1989).

3.2. The Cholinergic Current

Acetylcholine (ACh) activates inwardly and outwardly directed currents in most molluscan neurons. In *Aplysia*, it was shown that the ACh-evoked depolarization is mainly carried by Na^+ ions, whereas the two inhibitory conductances are mainly Cl^-- and K^+-dependent (Kehoe, 1972). Several investigators have used either noise analysis or current-relaxation techniques to reveal single-channel characteristics of ACh-coupled channels in molluscan neurons, but actual patch-clamp data are only scarcely available (see Ascher *et al.*, 1978a,b; Ikemoto and Akaike, 1988). Since some cholinomimetics such as suberyldicholine were shown to selectively activate a specific conductance (Cl^-, in the case of suberyldicholine), it became feasible to study a particular ACh-coupled channel in isolation. In cell-attached and outside-out patches of *Helix* neurons, Ascher and Erulkar (1983) first described the ACh-chloride channel at the single-channel level. They reported a conductance of *ca.* 18 pS and a pronounced desensitization, even when suberyldicholine was used at very low concentrations (50 nM). Thus, the activity of the channel was very bursty in nature, and the burst duration was in the range of 70 to 100 msec. In the cell-attached configuration, an inwardly directed chloride current was not observed when the membrane was hyperpolarized below E_{Cl}, indicating some form of rectification. In the freshwater snail *Planorbarius corneus*, suberyldicholine activated a 10-pS chloride current, but even at high concentrations (mM range), no strong desensitization was observed (Kachmann *et al.*, 1990). An ACh-induced chloride current of similar magnitude (12 pS) was reported in outside-out patches of *Lymnaea* neurons, and the open time histogram was fit with two exponentials of 4 msec and 20 msec (Bregestovski and Redkozubov, 1986). Interestingly, Green *et al.* (1989) used a pressure application system and applied ACh onto outside-out patches from Cl neurons of *Helix aspersa*. They did not find any indication of a chloride conductance, but rather a Na^+-dependent inward current of 10 pS. Our own studies of the ACh-Cl current in cultured *Aplysia* neurons using cell-attached patches showed some similarities to the results in *Helix* neurons (Fejtl and Carpenter, 1991; Fejtl *et al.*, 1994b). Desensitization was evident with 0.5 µM ACh in the pipette, since channel activity usually occurred in bursts > 100 msec, separated by long closed periods on the order of seconds. On the other hand, when 1 µM carbachol was used, no such long bursts were observed, indicating much less desensitization of the ACh receptor (Fig. 25). As with the ACh-Cl current in *Helix*, we did not see an inward chloride current below E_{Cl}, suggesting a rectification process in this voltage region. The *Aplysia* ACh-Cl current, however, had a much higher slope conductance in the range of 50 pS (Fejtl *et al.*, 1994a), and this could reflect a difference between the two freshwater snails, *Helix* and *Lymnaea* and the sea slug *Aplysia*.

4. MECHANOSENSITIVE (STRETCH-SENSITIVE) ION CHANNELS

Since the first finding of nonselective cationic stretch-activated ion channels (SACat) in muscle tissue of the chick (Guharay and Sachs, 1984), mechanosensitive

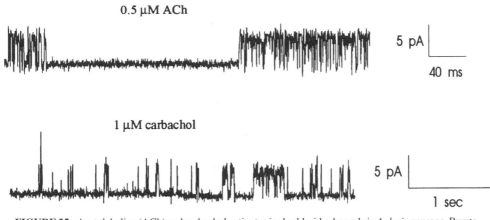

FIGURE 25. Acetylcholine (ACh) and carbachol activate single chloride channels in *Aplysia* neurons. Bursts, separated by long closed intervals, are indicative of a desensitized receptor and occurred with ACh at 0.5 μM. Fewer bursts and no long closed periods were seen at an even higher concentration of carbachol (1 μM).

channels have been reported in virtually all cell types, ranging from plant, bacteria, and yeast to molluscan and vertebrate cells (see Morris, 1990, 1992, for review). In molluscs, it was thought that such channels would be found exclusively in organs which naturally undergo periodic changes of tension, such as the heart (see Brezden and Gardner, 1986, for example). However, later studies revealed the presence of stretch-sensitive channels in the nervous system as well. Culture systems of isolated molluscan neurons have proven to be a useful model for neurite elongation, since during neurite growth the growth cones sprouting from the cell soma presumably undergo mechanical stress. Cell-attached and inside-out patches were taken from the growth cone and the soma of *Lymnaea* neurons in culture. Applying negative pressure to the patch pipette (-50 to -100 mm Hg) induced a current, selectively permeable to K^+ ions (Sigurdson and Morris, 1989). This stretch-activated potassium (SAK) channel had a conductance of 44 pS and the open probability increased with increasing pressure. It was hypothesized that a greater number of SAK channels would be present in the growth cone than in the soma, but no obvious channel clustering was found. The overall channel density was estimated to be *ca.* 1–2 channels/μm^2 in the soma and in the growth cone. Open and closed times were fit with a double- and three-exponential function, and the increase in open probability was due to a decrease of the long, third time constant of the closed time distribution and not to an increase in mean open time. Since the activity of the channel always expressed itself in bursts, the long closed state has been interpreted as the interburst interval. Quinidine blocked the SAK channel and introduced a flickering behavior, consistent with the scheme of an open-channel block. Interestingly, in the same preparation, Morris and Sigurdson (1989) found a 6-pS K^+-selective conductance that was inactivated by pressure. This SIK channel had not been recognized previously because, in all likelihood, residual pressure present in the patch pipette, even at nominal 0 mm Hg, led to an inactivation of this channel (Fig. 26). The SIK channel shared similar kinetics with the SAK channel, i.e., the decrease in the observed open probability was caused by an increase in the interburst interval rather than by a decrease of the mean open

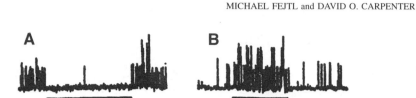

FIGURE 26. Stretch-activated (SAK) and stretch-inactivated (SIK) potassium channels coexist in the same neuron. In (A) the activity of a 6-pS potassium current was abolished by applying −40 mm Hg to the patch pipette (the SAK channel was blocked by 1 mM quinidine). In contrast, the SAK channel has a conductance of 44 pS and was activated by −40 mm Hg (B). Note the different calibration: 1 pA, 2 sec in A and 4 pA, 2 sec in B. (From Morris and Sigurdson, 1989.)

time. A plausible role for the SIK channel is to counteract the activity of the SAK channel. Considering the putative importance of stretch-sensitive channels in cell motility underlying neuronal growth and neurite elongation, a reduced adhesion to the substrate, i.e., less tension, would activate the SIK current, while stretching the membrane during active motion on the substrate would inactivate SIK but activate the SAK channel. This mechanism could play a role in stabilizing the resting membrane potential. In *Aplysia* sensory neurons a similar SAK-type channel was found (Vandorpe and Morris, 1992). Its conductance was 50 pS and, together with the pharmacological evidence that this SAK channel was suppressed by 5-HT, activated by FMRFamide, and not blocked by TEA, the conclusion was drawn that this SAK channel was, in fact, the serotonin-sensitive S channel. The open and closed kinetics revealed two time constants in the millisecond range and one third closed time parameter between ten and several hundreds of milliseconds (Vandorpe *et al.*, 1994). Unlike the S channel, where the increase in open probability is not paralleled by an increase in the number of active channels, stretch often induced a marked increase in active channels. It is questionable, however, whether the increase in the number of SAK channels was indeed a physiological response, or rather, caused by an increase of membrane area in the patch due to mechanical stress. There is evidence that the patch formation alters the properties of mechanosensitive channels. All the data of stretch-sensitive channels were gathered in the single-channel gigaohm-seal configuration. It is obvious that if the single-channel activity of SIK and SAK channels reflects any kind of mechanotransducing mechanism, a stretch-sensitive current was to be expected upon mechanical stimulation of an intact cell. When such an attempt was made to resolve this issue, however, no such current was observed (Morris and Horn, 1991). Growth cones and cell bodies of *Lymnaea* neurons in culture were studied in the whole-cell configuration. Under the assumption of 1–2 channels/μm^2, several hundreds of pA of current (or > 3000 channels in the cell body) were expected under increased mechanical stress. However, no matter what kind of stimuli were used (osmotic stimulation, negative and positive pressure, spritzing of bath solution, all of which resulted in a visible deformation of either the growth cone or the cell soma), a corresponding whole-cell current could not be detected. Although a small current of roughly 5 pA was seen, it was attributed to a nonspecific increase in leakage current, since it occurred at a very high pressure (> 100 mm HG) just before patch rupture. Hence, a discrepancy exists between whole-cell and single-channel recording in

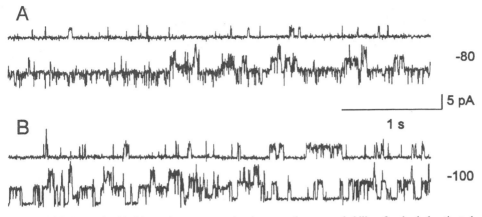

FIGURE 27. Mercuric chloride, at a low concentration, increases the open probability of carbachol-activated chloride channels in *Aplysia* neurons. Shown are two cell-attached patches taken from the same neuron at two different pipette potentials (A and B). The upper trace shows the single channel activity with 1 μM carbachol, and the lower trace depicts the enhanced opening frequency in the presence of 1 μM carbachol + 1 μM $HgCl_2$. (From Fejtl *et al.*, 1994b.)

that the stretch-sensitive channels were only observed in the tight gigaseal configuration. This was further exemplified by a report of a chloride-selective stretch-activated channel (SACl) in neurons of the land snail *Cepea nemoralis* (Bédard and Morris, 1992). This channel was seen only in excised patches, suggesting that patch excision might have disrupted cytoskeletal proteins, leading to its activation. Moreover, the delayed activation of SAK channels seen in *Lymnaea* neurons was reduced by cytochalasin D, which promotes the depolymerization of actin filaments (Small and Morris, 1994). This also indicates a crucial involvement of cytoskeletal proteins in the characterization of stretch-sensitive channels. Clearly, it remains to be seen whether SAK and SIK channels, which can be identified by the single-channel recording technique, participate in any kind of mechanotransducing physiologic processing in molluscan neurons *in situ*.

5. THE EFFECTS OF LEAD AND MERCURIC IONS

Molluscs have been used as a model system for a long time to study cellular mechanisms related to human metal neurotoxicity. However, with regard to single-channel patch-clamp analysis, studies are surprisingly very limited. Two studies are available thus far, one dealing with the effect of lead ions on sodium channels in *Helix* neurons (Osipenko *et al.*, 1992; Györi *et al.*, 1993) and one investigating the action of inorganic mercury on the ACh-induced chloride current in *Aplysia* neurons (Fejtl *et al.*, 1994a,b). In isolated single neurons of *Helix*, 50–100 μM Pb^{2+} closed a 14-pS sodium current recorded in the cell-attached configuration. This Na^+ conductance was active at rest, and application of Pb^{2+} led to a reversible hyperpolarization, reflecting the closure of the sodium channels in the presence of lead ions. Dwell-time analysis showed that both open and closed times were distributed monoexponentially, suggesting a single binding site for Pb^{2+} at the sodium channel. The blockade of the sodium current was mainly caused by a decrease of the mean open time and not by a reduction of the unitary current amplitude. This current is possibly a new type, since thus far only a Pb-induced inward current has been reported in snail neurons (Salánki *et al.*, 1991).

Carbachol-induced chloride currents were recorded in the cell-attached patch configuration in cultured *Aplysia* neurons. The effect of a low concentration of $HgCl_2$ was investigated, and since cultured molluscan neurons are big, it was possible to compare the effect of the control solution (1 μM carbachol) with the test solution (1 μM carbachol + 1 μM Hg^{2+}) in the same neuron by successively obtaining two patches. It was shown that the frequency of carbachol-induced channel openings was markedly increased by mercury, leading to an increase in open probability (Fig. 27). In both the control and the test solution the open and closed time histograms were fit with two exponentials, and Hg^{2+} caused a significant decrease of the mean closed time (10.37 ± 1.08 msec in control versus 3.32 ± 0.02 msec in the test solution), without affecting the unitary conductance. Kinetic analysis showed that Hg^{2+} caused a reduction of the slow time constant of the fit of the closed time histogram, suggesting a reduction of dwell-time in the longer closed state, which resulted in the observed increase in P_o (Fejtl *et al.*, 1994b). Since the unitary current was not increased by mercury, it is likely that Hg^{2+} ions affect the binding kinetics of the ACh-receptor and not the ion channel pore.

6. REFERENCES

Abrams, T. W., Castellucci, V. F., Camardo, J. S., Kandel E. R., and Lloyd, P. E., 1984, Two endogenous neuropeptides modulate the gill and siphon withdrawal reflex in *Aplysia* by presynaptic facilitation involving cAMP-dependent closure of a serotonin-sensitive potassium channel, *Proc. Natl. Acad. Sci. USA* **81:**7956–7960.

Adrian, R. H., and Freygang, W. H., 1962, Potassium and chloride permeability of frog muscle membrane, *J. Physiol.* **163:**61–103.

Armstrong, C. M., and Bezanilla, F., 1977, Inactivation of the sodium channel. II. Gating current experiments, *J. Gen. Physiol.* **70:**567–590.

Armstrong, C. M., and Binstock, L., 1965, Anomalous rectification in the squid axon injected with tetra-ethylammonium chloride, *J. Gen. Physiol.* **48:**859–872.

Ascher, P., and Erulkar, S., 1983, Cholinergic chloride channels in snail neurones, in: *Single-Channel Recording* (B. Sakmann and E. Neher, eds.), Plenum Press, New York, pp. 401–407.

Ascher, P., Marty, A., and Neild, T. O., 1978a, Life time and elementary conductance of the channels mediating the excitatory effects of acetylcholine in *Aplysia* neurones, *J. Physiol.* **278:**177–206.

Ascher, P., Marty, A., and Neild, T. O., 1978b, The mode of action of antagonists of the excitatory response to acetylcholine in *Aplysia* neurones, *J. Physiol.* **278:**207–235.

Bédard, E., and Morris, C. E., 1992, Channels activated by stretch in neurons of a *Helix* snail, *Can. J. Physiol. Pharmacol.* **70:**207–213.

Behrens, M. I., Oberhauser, A., Bezanilla, F., and Latorre, R., 1989, Batrachotoxin-modified sodium channels from the squid optic nerve in planar bilayers, *J. Gen. Physiol.* **93:**23–41.

Belardetti, F., and Siegelbaum, S. A., 1988, Up- and down-modulation of a single K^+ channel function by distinct second messengers, *Trends Neurosci.* **11:**232–238.

Belardetti, F., Schacher, S., and Siegelbaum, S. A., 1986, Action potentials, macroscopic and single channel currents recorded from growth cones of *Aplysia* neurones in culture, *J. Physiol.* **374:**289–313.

Belardetti, F., Kandel, E. R., and Siegelbaum, S. A., 1987, Neuronal inhibition by the peptide FMRFamide involves opening of S-K$^+$ channels, *Nature* **325**:153–156.

Belardetti, F., Schacher, S., Kandel, E. R., and Siegelbaum, S. A., 1986, The growth cones of *Aplysia* sensory neurons: Modulation by serotonin of action potential duration and single potassium channel currents, *Proc. Natl. Acad. Sci. USA* **83**:7094–7098.

Belardetti, F., Campbell, W. B., Falck, J. R., Demontis, G., and Rosolowsky, M., 1989, Products of heme-catalyzed transformation of the arachidonate derivative 12-HPETE open S-type K$^+$-channels in *Aplysia*, *Neuron* **3**:497–505.

Bezanilla, F., 1987a, Single sodium channels recorded from the cut-open squid giant axon, *Biophys. J.* **51**(2):195a.

Bezanilla, F., 1987b, Single sodium channels from the squid giant axon, *Biophys. J.* **52**:1087–1090.

Bezanilla, F., and Armstrong, C. M., 1972, Negative conductance caused by entry of sodium and cesium ions into the potassium channels of the squid axons, *J. Gen. Physiol.* **60**:588–608.

Bregestovski, P. D., and Redkozubov, A. E., 1986, Acetylcholine-activated single chloride channels in neurons of the mollusc *Lymnaea stagnalis, Biol. Membr.* **3**:960–969.

Brezden, B. L., and Gardner, D. R., 1986, A potassium-selective channel in isolated *Lymnaea stagnalis* heart muscle cells, *J. Exp. Biol.* **123**:175–189.

Brezina, V., Eckert, R., and Erxleben, C., 1987, Modulation of potassium conductances by an endogenous neuropeptide in neurones of *Aplysia californica, J. Physiol.* **382**:267–290.

Brown, A. M., Camerer, H., Kunze, D. L., and Lux, H. D., 1982, Similarity of unitary Ca^{2+} currents in three different species, *Nature* **299**:156–158.

Brown, A. M., Lux, H. D., and Wilson, D. L., 1984, Activation and inactivation of single calcium channels in snail neurons, *J. Gen. Physiol.* **83**:751–769.

Brown, A. M., Kunze, D. L., and Lux, H. D., 1986, Single calcium channels and their inactivation, *Membr. Biochem.* **6**:73–81.

Buttner, N., Siegelbaum, S. A., and Volterra, A., 1989, Direct modulation of *Aplysia* S-K$^+$ channels by a 12-lipoxygenase metabolite of arachidonic acid, *Nature* **342**:553–555.

Byerly, L., and Moody, W. J., 1984, Intracellular calcium ions and calcium currents in perfused neurones of the snail, *Lymnaea stagnalis, J. Physiol.* **352**:637–652.

Byerly, L., and Suen, Y., 1989, Characterization of proton currents in neurones of the snail, *Lymnaea stagnalis, J. Physiol.* **413**:75–89.

Camardo, J. S., Shuster, M. J., Siegelbaum, S. A., and Kandel, E. R., 1983, Modulation of a specific potassium channel in sensory neurons of *Aplysia* by serotonin and cAMP-dependent protein phosphorylation, *Cold Spring Harbor Symp. Quant. Biol.* **48**:213–220.

Camardo, J. S., Shuster, M. J., and Siegelbaum, S. A., 1986, Modulation of a potassium channel in *Aplysia* sensory neurons: Role of protein phosphorylation, in: *Molecular Aspects of Neurobiology* (R. L. Montalcini *et al.*, eds.), Springer-Verlag, Berlin, Heidelberg, pp. 106–112.

Castellucci, V. F., Kandel, E. R., Schwartz, J. H., Wilson, F. D., Nairn, A. C., and Greengard, P., 1980, Intracellular injection of the catalytic subunit of cyclic AMP-dependent protein kinase stimulates facilitation of transmitter release underlying behavioral sensitization in *Aplysia, Proc. Natl. Acad. Sci. USA* **77**:7492–7496.

Catterall, W. A., 1986, Molecular properties of voltage-sensitive sodium channels, *Annu. Rev. Biochem.* **55**:953–985.

Chagneux, H., Ducreux, C., and Gola, M., 1989, Voltage-dependent opening of single calcium-activated potassium channels in *Helix* neurons, *Brain Res.* **488**:336–340.

Chesnoy-Marchais, D., 1983, Characterization of a chloride conductance activated by hyperpolarization in *Aplysia* neurons, *J. Physiol.* **342**:277–308.

Chesnoy-Marchais, D., 1984, Two types of chloride channels in outside-out patches from *Aplysia* neurones, *J. Physiol.* **357**:64P.

Chesnoy-Marchais, D., 1985a, Single channels permeable to calcium in *Aplysia* neurons, *J. Physiol.* **365**:83P.

Chesnoy-Marchais, D., 1985b, Kinetic properties and selectivity of calcium-permeable single channels in *Aplysia* neurones, *J. Physiol.* **367**:457–488.

Chesnoy-Marchais, D., and Evans, M. G., 1986a, Chloride channels activated by hyperpolarization in *Aplysia* neurones, *Pflügers Arch.* **407**:694–696.

Chesnoy-Marchais, D., and Evans, M. G., 1986b, Nonselective ionic channels in *Aplysia* neurons, *J. Membr. Biol.* **93**:75–83.

Cohan, C. S., Haydon, P. G., and Kater, S. B., 1985, Single channel activity differs in growing and nongrowing growth cones of isolated identified neurons of *Helisoma*, *J. Neurosci. Res.* **13**:285–300.

Connor, J. A., and Stevens, C. F., 1971, Prediction of repetitive firing behavior from voltage clamp data on an isolated neurone soma, *J. Physiol.* **213**:31–53.

Conti, F., and Neher, E., 1980, Single channel recordings of K^+ currents in squid axons, *Nature* **285**:140–143.

Cooper, E., and Shrier, A., 1985, Single-channel analysis of fast transient potassium currents from rat nodose neurons, *J. Physiol.* **369**:199–208.

Correa, A. M., and Bezanilla, F., 1988, Properties of BTX-treated single Na channels in squid axon, *Biophys. J.* **53**:226a.

Correa, A. M., and Bezanilla, F., 1994, Gating of the squid sodium channel at positive potentials: II. Single channels reveal two open states, *Biophys. J.* **66**:1864–1878.

Correa, A. M., Latorre, R., and Bezanilla, F., 1989, Na-dependence and temperature effects on BTX-treated sodium channels in squid axon, *Biophys. J.* **55**:403a.

Correa, A. M., Latorre, R., and Bezanilla, F., 1991, Ion permeation in normal and batrachotoxin-modified Na^+ channels in the squid giant axon, *J. Gen. Physiol.* **97**:605–625.

Correa, A. M., Bezanilla, F., and Latorre, R., 1992, Gating kinetics of batrachotoxin-modified Na^+ channels in the squid giant axon, *Biophys. J.* **61**:1332–1352.

Cottrell, G. A., Davies, N. W., and Green, K. A., 1984, Multiple actions of a molluscan cardioexcitatory neuropeptide and related peptides on identified *Helix* neurones, *J. Physiol.* **356**:315–333.

Cottrell, G. A., Green, K. A., and Davies, N. W., 1990, The neuropeptide Phe-Met-Arg-Phe-NH_2 (FMRFamide) can activate a ligand-gated ion channel in *Helix* neurones, *Pflügers Arch.* **416**:612–614.

Coyne, M. D., Dagan, D., and Levitan, I. B., 1987, Calcium and Barium permeable channels from *Aplysia* nervous system reconstituted in lipid bilayers, *J. Membr. Biol.* **97**:205–213.

Crest, M., and Gola, M., 1993, Large conductance Ca^{2+}-activated K^+ channels are involved in both spike shaping and firing regulation in *Helix* neurones, *J. Physiol. (London)* **465**: 265–287.

Crest, M., Jacquet, G., Gola, M., Zerrouk, H., Benslimane, A., Rochat, H., Mansuelle, P., and Martin-Eauclaire, M.-F., 1992, Kalitoxin, a novel peptidyl inhibitor of neuronal BK-type Ca^{2+}-activated K^+ channels characterized from *Androctonus mauretanicus mauretanicus* venom, *J. Biol. Chem.* **267**:1640–1647.

Decoursey, T. E., Chandy, K. G., Gupta, S., and Cahalan, M. D., 1987, Two types of potassium channels in murine T lymphocytes, *J. Gen. Physiol.* **89**:379–404.

DePeyer, J., Cachelin, A., Levitan, I. B., and Reuter, H., 1982, Ca^{2+}-activated K^+ conductance in internally perfused snail neurons is enhanced by protein phosphorylations, *Proc. Natl. Acad. Sci. USA* **79**:4207–4211.

Eckert, R., and Chad, J. E., 1984, Inactivation of Ca channels, *Prog. Biophys. Mol. Biol.* **44**:215–267.

Edmonds, B., Klein, M., Dale, N., and Kandel, E. R., 1990, Contributions of two types of calcium channels to synaptic transmission and plasticity, *Science* **250**:1142–1147.

Ewald, D. A., Williams, A., and Levitan, I. B., 1985, Modulation of single Ca^{2+}-dependent K^+-channel activity by protein phosphorylation, *Nature* **315**:503–506.

Fejtl, M., and Carpenter, D. O., 1991, Acetylcholine-activated single channel currents in cultured *Aplysia* neurons, *Soc. Neurosci. Abstr.* **17**:583.

Fejtl, M., Györi, J., and Carpenter, D. O., 1994a, Hg^{2+} increases the open probability of carbachol-activated Cl^- channels in *Aplysia* neurons, *Neuroreport* **5**:2317–2320.

Fejtl, M., Györi, J., and Carpenter, D. O., 1994b, Mercuric (II) chloride modulates single channel properties of carbachol-activated Cl^- channels in cultured neurons of *Aplysia californica*, *Cell Mol. Neurobiol.* **14**: 665–674.

Fink, L., Connor, J. A., and Kaczmarek, L. K., 1987, Inositol triphosphate activates K^+ channels through elevation of intracellular calcium in peptidergic neurons of *Aplysia*, *Soc. Neurosci. Abstr.* **13**:152.

Fink, L. A., Connor, J. A., and Kaczmarek, L. K., 1988, Inositol triphosphate releases intracellularly stored calcium and modulates ion channels in molluscan neurons, *J. Neurosci.* **8**:2544–2555.

Fishman, H. M., and Tewari, K. P., 1990, Single Ca^{2+} channels in patches of axosomes from transfected squid axon, *Biophys. J.* **57**:523a.

Fukushima, Y., 1982, Blocking kinetics of the anomalous potassium rectifier of tunicate eggs studied by single channel recording, *J. Physiol.* **331**:311–331.

Furukawa, Y., and Kobayashi, M., 1988, Two serotonin-sensitive potassium channels in the identified heart excitatory neurone of the African giant snail, *Achatina fulica* Ferussac, *Experientia* **44:**738–740.

Geletyuk, V. I., and Kazachenko, V. N., 1983, Discrete character of potassium channel conductance in *Lymnaea stagnalis* neurons, *Biofizika* **28:**994–998.

Geletyuk, V. I., and Kazachenko, V. N., 1985, Single Cl⁻ channels in molluscan neurones: Multiplicity of the conductance states, *J. Membr. Biol.* **86:**9–15.

Geletyuk, V. I., and Kazachenko, V. N., 1987, Discreteness of the parameters of current oscillations in single ion channels, *Biophysics* **32:**290–294.

Geletyuk, V. I., and Kazachenko, V. N., 1989, Single potential-dependent K⁺ channels and their oligomers in molluscan glial cells, *Biochim. Biophys. Acta* **98:**343–350.

Geletyuk, V. I., Kazachenko, V. N., and Tseeb, B. E., 1988, Quantization of kinetic parameters of single ion channels, *Dokl. Akad, Nauk. SSSR* **301:**465–469.

Gola, M., Ducreux, C., and Chaux, H., 1990, Ca²⁺-activated K⁺ current involvement in neuronal function revealed by *in situ* single-channel analysis in *Helix* neurones, *J. Physiol.* **420:**73–109.

Gorman, A. L. F., and Thomas, M. V., 1980, Potassium conductance and internal calcium accumulation in a molluscan neurone, *J. Physiol.* **308:**287–313.

Green, D. J., and Gillette, R., 1983, Patch- and voltage-clamp analysis of cyclic AMP–stimulated inward current underlying neurone bursting, *Nature* **306:**784–785.

Green, K. A., Cadogan, A., and Cottrell, G. A., 1989, Nicotinic-type unitary currents in *Helix* neurons, *Comp. Biochem. Physiol.* **93A:**47–51.

Green, K. A., Powell, B., and Cottrell, G. A., 1990, Unitary K⁺ currents in growth cones and perikaryon of identified *Helix* neurones in culture, *J. Exp. Biol.* **149:**79–94.

Guharay, F., and Sachs, F., 1984, Stretch-activated single ion channel currents in tissue-cultured embryonic chick skeletal muscles, *J. Physiol.* **352:**685–701.

Györi, J., Osipenko, O. N., and Kiss, T., 1993, Voltage-clamp and single channel analysis of Pb²⁺-induced current in isolated snail neurons, *Acta Biol. Hung.* **44:**3–7.

Hagiwara, S., Kusano, K., and Saito, N., 1961, Membrane changes of *Onchidium* nerve cell in potassium-rich media, *J. Physiol.* **155:**470–489.

Hamill, O. P., Marty, A., Neher, E., Sakmann, B., and Sigworth, F. J., 1981, Improved patch-clamp techniques for high-resolution current recording from cells and cell-free membrane patches, *Pflügers Arch.* **391:**85–100.

Harden, T. K., 1983, Agonist-induced desensitization of the beta-adrenergic receptor-linked adenylate cyclase, *Pharmacol. Rev.* **35:**5–32.

Harris, G. L., Henderson, L. P., and Spitzer, N. C., 1988, Changes in densities and kinetics of delayed rectifier potassium channels during neuronal differentiation, *Neuron* **1:**739–750.

Hermann, A., 1986, Selective blockade of a Ca-activated K-current in *Aplysia* neurons by charybdotoxin, *Pflügers Arch.* **406**(Suppl):R54.

Hermann, A., and Erxleben, C., 1987, Charybdotoxin selectively blocks small Ca-activated K channels in *Aplysia* neurons, *J. Gen. Physiol.* **90:**27–47.

Heyer, C. B., and Lux, H. D., 1976, Control of the delayed outward potassium currents in bursting pacemaker neurones of the snail *Helix pomatia*, *J. Physiol.* **262:**349–382.

Hodgkin, A. L., and Huxley, A. F., 1952, A quantitative description of membrane current and its application to conduction and excitation in nerve, *J. Physiol.* **117:**500–544.

Horn, R., and Vandenberg, C. A., 1984, Statistical properties of single sodium channels, *J. Gen. Physiol.* **84:**505–534.

Ikemoto, Y., and Akaike, N., 1988, Kinetic analysis of acetylcholine-induced chloride current in isolated *Aplysia* neurones, *Pflügers Arch.* **412:**240–247.

Johnson, J. W., and Thompson, S., 1989, Measurement of nonuniform current densities and current kinetics in *Aplysia* neurons using a large patch method, *Biophys. J.* **55:**299–308.

Kachman, A. N., Samoilova, M. V., and Snetkov, V. A., 1989, A single potassium channel of anomalous (inward) rectification in molluscan neurons, *Neirofiziologiya* **21:**31–38.

Kachman, A. N., Frolova, E. V., and Gapon, S. A., 1990, Chloride channels activated by suberyldicholine in mollusc neurons, *Neirofiziologiya* **22:**697–700.

Kasai, H., 1985, Single transient potassium currents in mammalian sensory neurons studied using patch clamp techniques, *Soc. Neurosci. Abstr.* **11:**955.

Kazachenko, V. I., and Geletyuk, V. I., 1983, Single potential-dependent potassium channel in the neurons of mollusc *Lymnaea stagnalis, Biofizika* **28**:270–273.

Kazachenko, V. N., and Geletyuk, V. I., 1984, The potential-dependent K$^+$ channel in molluscan neurones is organized in a cluster of elementary channels, *Biochim. Biophys. Acta* **773**:132–142.

Kehoe, J. S., 1972, Three acetylcholine receptors in *Aplysia* neurons, *J. Physiol.* **225**:115–146.

Khodorov, B. I., 1985, Batrachotoxin as a tool to study voltage-sensitive sodium channels of excitable membranes, *Prog. Biophys. Mol. Biol.* **45**:57–148.

Klein, M., Camardo, J., and Kandel, E. R., 1982, Serotonin modulates a specific potassium current in the sensory neurons that show presynaptic facilitation in *Aplysia, Proc. Natl. Acad. Sci. USA* **79**:5713–5717.

Kramer, R. H., 1990, Patch cramming: Monitoring intracellular messengers in intact cells with membrane patches containing detector ion channels, *Neuron* **2**:335–341.

Latorre, R., Oberhauser, A., Condrescu, M., DiPolo, R., and Bezanilla, F., 1987, Incorporation of sodium channels from squid optic nerve into planar lipid bilayers, *Biophys. J.* **51**:195a.

Levis, R. A., Bezanilla, F., and Torres, R. M., 1984, Estimate of the squid axon sodium channel conductance with improved frequency resolution, *Biophys. J.* **45**:11a.

Llano, I., and Bezanilla, F., 1984, Analysis of sodium current fluctuations in the cut-open squid giant axon, *J. Gen. Physiol.* **83**:133–142.

Llano, I., and Bezanilla, F., 1985, Two types of potassium channels in the cut-open squid giant axon, *Biophys. J.* **47**:221a.

Llano, I., and Bezanilla, F., 1986, Batrachotoxin-modified single sodium channels in squid axon, *Biophys. J.* **49**:43a.

Llano, I., and Bookman, R. J., 1985, The K$^+$ conductance of squid giant fibre lobe neurons, *Biophys. J.* **47**:223a.

Llano, I., and Bookman, R. J., 1986, Ionic conductances of squid giant fiber lobe neurons, *J. Gen. Physiol.* **88**:543–569.

Llano, I., Bookman, R. J., and Armstrong, C. M., 1986, Single K channels recorded from squid GFL neurons, *Biophys. J.* **49**(2):216a.

Llano, I., Webb, C. K., and Bezanilla, F., 1988, Potassium conductance of the squid giant axon. Single-channel studies, *J. Gen. Physiol.* **92**:179–196.

Llinàs, R., Sugimori, M., Lin, J.-W., and Cherksey, B., 1989, Blocking and isolation of a calcium channel from neurons in mammals and cephalopods utilizing a toxin fraction (FTX) from funnel-web spider poison, *Proc. Natl. Acad. Sci. USA* **86**:1689–1693.

Lux, H. D., and Brown, A. M., 1984a, Single channel studies on inactivation of calcium currents, *Science* **225**:432–434.

Lux, H. D., and Brown, A. M., 1984b, Patch and whole cell calcium currents recorded simultaneously in snail neurons, *J. Gen. Physiol.* **83**:727–750.

Lux, H. D., and Hofmeier, G., 1982, Activation characteristics of the calcium-dependent outward potassium current in *Helix, Pflügers Arch.* **394**:70–77.

Lux, H. D., and Nagy, K., 1981, Single channel Ca^{2+} currents in *Helix pomatia* neurons, *Pflügers Arch.* **391**:252–254.

Lux, H. D., Neher, E., and Marty, A., 1981, Single channel activity associated with the calcium dependent outward current in *Helix pomatia, Pflügers Arch.* **389**:293–295.

Meech, R. W., and Standen, N. B., 1975, Potassium activation in *Helix aspersa* neurones under voltage clamp: A component mediated by calcium influx, *J. Physiol.* **249**:211–239.

Miller, C., Moczydlowski, E., Latorre, R., and Phillips, M., 1985, Charybdotoxin, a protein inhibitor of single Ca^{2+}-activated K$^+$ channels from mammalian skeletal muscle, *Nature* **313**:316–318.

Morris, C. E., 1990, Mechanosensitive ion channels, *J. Membr. Biol.* **113**:93–107.

Morris, C. E., 1992, Are stretch-sensitive ion channels in molluscan cells and elsewhere physiological mechanotransducers? *Experientia* **48**:852–858.

Morris, C. E., and Horn, R., 1991, Failure to elicit neuronal macroscopic mechanosensitive currents anticipated by single-channel studies, *Science* **251**:1246–1249.

Morris, C. E., and Sigurdson, W. J., 1989, Stretch-inactivated ion channels coexist with stretch-activated ion channels, *Science* **234**:807–809.

Nealey, T., Spires, S., Eatock, R. A., and Begenisich, T., 1993, Potassium channels in squid neuron cell bodies: Comparison to axonal channels, *J. Membr. Biol.* **132**:13–25.

Neher, E., and Sakmann, B., 1976, Single channel currents recorded from membrane of denervated frog muscle fibres, *Nature* **260**:799–802.

Noda, M., Shimizu, S., Tanabe, T., Takai, T., Kayano, T., Ikeda, T., Takahashi, H., Nakayama, H., Kanaoka, Y., Minamino, M., Kangawa, K., Matsuo, H., Raftery, M. A., Hirose, T., Inayama, S., Hayashida, H., Miyata, T., and Numa, S., 1984, Primary structure of *Electrophorus electricus* sodium channel deduced from cDNA sequence, *Nature* **312**:121–127.

Oberhauser, A., 1989, Single ion channels recorded from squid GFL neurons, *Biophys. J.* **55**:172a.

Ohmori, S., Yoshida, S., and Hagiwara, S., 1981, Single K⁺-channel currents of anomalous rectification in cultured rat myotube, *Proc. Natl. Acad. Sci. USA* **78**:4960–4964.

Osipenko, O. N., Györi, J., and Kiss, T., 1992, Lead ions close steady-state sodium channels in *Helix* neurons, *Neuroscience* **50**:483–489.

Perozo, E., Jong, D. S., and Bezanilla, F., 1990, Single channel analysis of the phosphorylation of the squid axon delayed rectifier, *Biophys. J.* **57**:22a.

Perozo, E., Vandenberg, C. A., Jong, D. S., and Bezanilla, F., 1991a, Single channel studies of the phosphorylation of K⁺ channels in the squid giant axon. I. Steady-state conditions, *J. Gen. Physiol.* **98**:1–17.

Perozo, E., Jong, D. S., and Bezanilla, F., 1991b, Single channel studies of the phosphorylation of K⁺ channels in the squid giant axon. II. Nonstationary conditions, *J. Gen. Physiol.* **98**:19–34.

Piomelli, D., Volterra, A., Dale, N., Siegelbaum, S. A., Kandel, E. R., Schwartz, J. H., and Belardetti, F., 1987, Lipoxygenase metabolites of arachidonic acid as second messengers for presynaptic inhibition of *Aplysia* sensory cells, *Nature* **328**:38–43.

Pollock, J. D., and Camardo, J. S., 1987, Regulation of single potassium channels by serotonin in the cell bodies of the tail mechanosensory neurons of *Aplysia californica*, *Brain Res.* **410**:367–370.

Premack, B. A., Thompson, S., and Coombs-Hahn, J., 1989, Clustered distribution and variability in kinetics of transient K channels in molluscan neuron cell bodies, *J. Neurosci.* **2**:4089–4099.

Prestipino, G., Valdivia, H. H., Liévano, A., Darszon, A., Ramirez, A. N., and Possani, L. D., 1989, Purification and reconstitution of potassium channel proteins from squid axon membranes, *FEBS Lett.* **250**:570–574.

Quandt, F., 1988, Three kinetically distinct potassium channels in mouse neuroblastoma cells, *J. Physiol.* **395**:401–418.

Quandt, F., and Narahashi, T., 1982, Modifications of single Na⁺ channels by batrachotoxin, *Proc. Natl. Acad. Sci. USA* **79**:6732–6736.

Ram, J. L., and Dagan, D., 1987, Inactivating and non-inactivating outward current channels in cell-attached patches of *Helix* neurons, *Brain Res.* **405**:16–25.

Reuter, H., and Stevens, C. F., 1980, Ion conductance and ion selectivity of potassium channels in snail neurones, *J. Membr. Biol.* **57**:103–118.

Rudy, B., 1988, Diversity and ubiquity of K channels, *Neuroscience* **25**:729–749.

Sakmann, B., and Trube, G., 1984, Conductance properties of single inwardly potassium channels in ventricular cells from guinea-pig heart, *J. Physiol.* **347**:659–683.

Salánki, J., Osipenko, O. N., Kiss, T., and Györi, J., 1991, Effect of Cu²⁺ and Pb²⁺ on membrane excitability of snail neurons, in: *Molluscan Neurobiology. Symposium on Molluscan Neurobiology* (K. S. Kits *et al.*, eds.), North-Holland, Amsterdam, pp. 214–220.

Serrano, E. E., and Getting, P. A., 1989, Diversity of the transient outward potassium current in somata of identified molluscan neurons, *J. Neurosci.* **9**:4021–4032.

Shuster, M. J., and Siegelbaum, S. A., 1987, Pharmacological characterization of the serotonin-sensitive potassium channel of *Aplysia* sensory neurons, *J. Gen. Physiol.* **90**:587–608.

Shuster, M. J., Camardo, J. S., Siegelbaum, S. A., Eppler, C. M., and Kandel, E. R., 1984, Modulation of the serotonin-sensitive potassium channel by cAMP-dependent protein kinase, *Soc. Neurosci. Abstr.* **10**:145.

Shuster, M. J., Camardo, J. S., Siegelbaum, S. A., and Kandel, E. R., 1985, Cyclic AMP–dependent protein kinase closes the serotonin-sensitive K⁺ channels of *Aplysia* sensory neurones in cell-free membrane patches, *Nature* **313**:392–395.

Shuster, M. J., Camardo, J. S., and Siegelbaum, S. A., 1991, Comparison of the serotonin-sensitive and Ca⁺⁺-activated K⁺ channels in *Aplysia* sensory neurons, *J. Physiol.* **440**:601–621.

Siegelbaum, S. A., Camardo, J. S., and Kandel, E. R., 1982, Serotonin and cyclic AMP close single K⁺ channels in *Aplysia* sensory neurones, *Nature* **299**:413–417.

Siegelbaum, S. A., Belardetti, F., Camardo, J. S., and Shuster, M. J., 1986, Modulation of the serotonin-sensitive potassium channel in *Aplysia* sensory neurone cell body and growth cone, *J. Exp. Biol.* **124:** 287–306.

Sigurdson, W. J., and Morris, C. E., 1989, Stretch-activated ion channels in growth cones of snail neurons, *J. Neurosci.* **9:**2801–2808.

Small, D. L., and Morris, C. E., 1994, Delayed activation of single mechanosensitive channels in *Lymnaea* neurons, *Am. J. Physiol.* **267:**C598–C606.

Spires, S., Eatock, R. A., Nealey, T., and Begenisich, T., 1988, Chemical modification of K channels: Macroscopic ionic, gating, and single channel currents, *Biophys. J.* **53:**261a.

Strong, J. A., and Scott, S. A., 1992, Divalent-selective voltage-independent calcium channels in *Lymnaea* neurons: Permeation properties and inhibition by intracellular magnesium, *J. Neurosci.* **12:**2993–3003.

Strong, J. A., Fox, A. P., Tsien, R. W., and Kaczmarek, L. K., 1987a, Stimulation of protein kinase C recruits covert calcium channels in *Aplysia* bag cell neurons, *Nature* **325:**714–717.

Strong, J. A., Fox, A. P., Tsien, R. W., and Kaczmarek, L. K., 1987b, Formation of cell-free patches unmasks a large divalent-permeable, voltage-independent channel in *Aplysia* neurons, *Soc. Neurosci. Abstr.* **13:** 1011.

Sudlow, L. C., Huang, R.-C., Green, D. J., and Gillette, R., 1993, cAMP-activated Na^+ current of molluscan neurons is resistant to kinase inhibitors and is gated by cAMP in the isolated patch, *J. Neurosci.* **13:**5188–5193.

Taylor, P. S., 1987, Selectivity and patch measurements of A-current channels in *Helix aspersa* neurones, *J. Physiol.* **388:**437–447.

Taylor, R. E., Armstrong, C. M., and Bezanilla, F., 1976, Block of sodium channels by external calcium ions, *Biophys. J.* **16:**27a.

Thomas, R. C., and Meech, R. W., 1982, Hydrogen ion currents and intracellular pH in depolarized voltage-clamped snail neurones, *Nature* **299:**826–828.

Thompson, S. H., 1977, Three pharmacologically distinct potassium channels in molluscan neurons, *J. Physiol.* **265:**465–488.

Vandenberg, C. A., and Bezanilla, F., 1991, A sodium channel gating model based on single channel, macroscopic ionic, and gating currents in the squid giant axon, *Biophys. J.* **60:**1511–1533.

Vandenberg, C. A., Perozo, E., and Bezanilla, F., 1989, ATP modulation of the K current in squid giant axon. A single channel study, *Biophys. J.* **49:**215a.

Vandorpe, D. H., and Morris, C. E., 1992, Stretch-activation of the *Aplysia* S-channel, *J. Membr. Biol.* **127:** 205–214.

Vandorpe, D. H., Small, D. L., Dabrowski, A. R., and Morris, C. E., 1994, FMRFamide and membrane stretch as activators of the *Aplysia* S-channel, *Biophys. J.* **66:**46–58.

Wilson, D. L., Morimoto, K., Tsuda, Y., and Brown, A. M., 1983, Interaction between calcium ions and surface charge as it relates to calcium currents, *J. Membr. Biol.* **72:**309–324.

Wonderlin, W. F., and French, R. J., 1991, Ion channel in transit: Voltage-gated Na^+ and K^+ channels in axoplasmic organelles of the squid *Loligo pealei*, *Proc. Natl. Acad. Sci. USA* **88:**4391–4395.

Yazejian, B., and Byerly, L., 1987, Single channel Ba currents in neurons of the snail *Lymnaea stagnalis*, *Biophys. J.* **51:**423a.

Yazejian, B., and Byerly, L., 1989, Voltage-independent barium-permeable channel activated in *Lymnaea* neurons by internal perfusion or patch excision, *J. Membr. Biol.* **107:**63–75.

CHAPTER 10

NEURONAL NICOTINIC ACETYLCHOLINE RECEPTORS

JON LINDSTROM

1. INTRODUCTION

Studies of the structure and function of neuronal nicotinic acetylcholine receptors (AChRs) evolved out of studies of muscle AChRs. This review will begin with a brief summary of muscle type AChRs because they are the archetype for studies of neuronal nicotinic AChRs in particular and ligand-gated ion channels in general.

Throughout, the perspective will be from the work done in our laboratory rather than a general survey. An extensive general survey has been done by Sargent (1993), and shorter reviews have been prepared by Role (1992), McGehee and Role (1995), Papke (1993), and Lindstrom (1995). A little history will be given to show how these studies developed and to deal with problems of nomenclature that changed with time.

It is now clear that AChRs are members of a gene superfamily of ligand-gated ion channels that includes the homologous $GABA_A$ receptors, glycine receptors, and $5-HT_3$ serotonin receptors (Betz, 1990; Barnard, 1992), but not the structurally dissimilar ligand-gated ion channels comprising glutamate (Seeburg, 1993) or ATP receptors (Brake *et al.*, 1994; Valera *et al.*, 1994). Probably all the receptors in the AChR superfamily are formed by five homologous subunits oriented around a central ion channel like barrel staves (Unwin, 1995; Anand *et al.*, 1991; Cooper *et al.*, 1991; Langosch *et al.*, 1988; Boess *et al.*, 1995; Green *et al.*, 1995).

The basic homologies in the structures of receptors in this superfamily are illustrated by two elegant experiments by Changeux and coworkers. One of these experiments showed that alteration of only three amino acids in the sequence lining the cation-

JON LINDSTROM • Department of Neuroscience, Medical School of the University of Pennsylvania, Philadelphia, Pennsylvania 19104-6074.

Ion Channels, Volume 4, edited by Toshio Narahashi, Plenum Press, New York, 1996.

specific channel of excitatory $\alpha7$ neuronal AChRs to amino acids typical of the anion-specific channels of inhibitory glycine or $GABA_A$ receptors altered the selectivity of the mutated $\alpha7$ AChR channels from cations to anions (Galzi et al., 1992). The second experiment showed that the extracellular domain of $\alpha7$ AChRs could replace the extracellular domain of $5\text{-}HT_3$ receptors to produce acetylcholine-gated cation channels with the ion selectivity of $5\text{-}HT_3$ receptors (Eisile et al., 1993). In fact, the precise structures of all of the subunits in the superfamily are not so similar that it is possible to casually make functional receptors from any mix-and-match mosaic among the subunits in the superfamily. However, these experiments elegantly support the belief based on the pattern of sequence similarities in the subunits of this superfamily that all of these subunits evolved by a process of gene duplication from an ancestral homomeric ligand-gated ion channel and that the subunits have basically similar shapes and functional domains, resulting in assembly into receptors of basically similar shapes and functional properties. The synthesis, structure, and function of AChRs are the best characterized among the receptors in the superfamily. Thus, they serve as a model for understanding the other receptors in this superfamily and to some extent other ligand-gated ion channels.

It is useful, and perhaps evolutionarily relevant, to think of the AChR gene family as having three branches: 1) AChRs of skeletal muscles and fish electric organs, 2) neuronal AChRs which, unlike muscle AChRs, do not bind the snake venom toxin α-bungarotoxin (αBgt), and 3) neuronal AChRs that do bind αBgt.

This grouping reflects the historical order in which these AChRs were characterized and cloned, and it reflects three sets of pharmacological and electrophysiological properties. AChRs of fetal muscle have the subunit composition $(\alpha1)_2\beta1\gamma\delta$, whereas AChRs of adult muscle have the subunit composition $(\alpha1)_2\beta1\epsilon\delta$ (Changeux 1990, 1991; Karlin, 1991). Neuronal AChRs that do not bind αBgt are formed from combinations of $\alpha2$, $\alpha3$, or $\alpha4$ subunits with $\beta2$, $\beta4$, and/or $\alpha5$ subunits (Sargent, 1993; Role, 1992). Neuronal AChRs that do bind αBgt are formed from $\alpha7$, $\alpha8$, and/or $\alpha9$ subunits, perhaps in combination with unknown subunits (Clarke, 1992; Elgoyhen et al., 1994). Thus far, $\alpha6$ and $\beta3$ subunits have not been reported to associate with the other AChR subunits to form functional AChRs (Deneris et al., 1991). In subsequent sections of this review, the three branches of the AChR gene family will be discussed in order.

2. MUSCLE NICOTINIC RECEPTORS

2.1. Introduction

More is known about the structure, function, synthesis, and developmental regulation of skeletal muscle AChRs than about any other ligand-gated ion channel. This is because muscle has provided a simple homogeneous system for electrophysiological, morphological, and developmental studies; because αBgt and related snake venom toxins serve as highly specific, high-affinity, competitive antagonists which are useful for locating, quantitating, and affinity-purifying these AChRs; and because fish electric organs have provided an abundant and homogeneous source of AChR protein for biochemical and structural studies (reviewed in Changeux 1990, 1991; Karlin 1991, 1993).

In order to understand neuronal AChRs, it is necessary to first appreciate what has been learned about muscle AChRs. This background will predict similarities and contrast differences in neuronal AChRs as they are discovered. Contemporary knowledge about muscle AChRs will be briefly reviewed starting with their basic functional properties, continuing with their three-dimensional structure, then considering the sequences of AChR subunits from their N-terminus to the C-terminus. Throughout, comparisons with neuronal AChRs will be noted.

2.1.1. A Bit of History, Myasthenia Gravis, and Other Pathological Considerations

In the early seventies, Kiefer *et al.* (1970) developed photoaffinity labeling reagents for muscle AChRs to provide a means for biochemically identifying AChRs. Much later, similar photoaffinity labeling reagents were extensively used by Changeux and coworkers to identify contact amino acids from three parts of the N-terminal extracellular domain, which contribute to the structure of the acetylcholine binding site (Galzi *et al.*, 1990). When it was found that snake venom toxins provided specific labeling reagents (Lee *et al.*, 1967), Lindstrom and Patrick (1974) succeeded in purifying milligram amounts of AChR protein from electric eels.

In order to prove that the electric organ protein that had been purified was an AChR, in those days before reconstitution, cloning, or expression techniques had been developed, we took an immunological approach. The plan was to immunize rabbits with the purified AChR and then test whether the antisera could block the electrophysiological function of AChRs in the eel electric organ single cell preparation. If antiserum to the purified protein blocked AChR function, we could conclude that we had purified at least part of the authentic AChR. Not only did the antisera block AChR function in the electric organ cells (Patrick *et al.*, 1973), but the rabbits also suffered a flaccid paralysis and died (Patrick and Lindstrom, 1973). Years before, Simpson (1960) had proposed that myasthenia gravis might be caused by autoantibodies to AChRs acting as curare-like competitive antagonists. We concluded that myasthenia gravis could be caused by an autoimmune response to AChRs. Autoantibodies to AChRs were detected in myasthenia gravis patients (Lindstrom *et al.*, 1976). This provided a useful diagnostic assay and focused therapy development toward specific immunosuppression. It turned out that the autoantibodies were primarily of specificities which did not competitively inhibit the acetylcholine binding site or allosterically block function. Instead, the majority of autoantibodies to AChRs were directed at a part of the extracellular surface of $\alpha 1$ subunits which we called the main immunogenic region (MIR) (Tzartos and Lindstrom, 1980; Tzartos *et al.*, 1982, 1991). These autoantibodies produce their pathological effects primarily by two mechanisms: 1) by reducing the amount of AChRs *via* cross-linking and facilitating endocytosis (termed antigenic modulation) and 2) by causing complement-dependent focal lysis which destroys the folded structure of the postsynaptic membrane and disrupts the efficient positioning of AChRs at the tips of folds in the postsynaptic membrane right across from active zones for transmitter release in the presynaptic membrane (reviewed in Lindstrom *et al.*, 1988).

Recently, Rogers and coworkers discovered that when they immunized rabbits with bacterially expressed peptides from glutamate receptors the rabbits developed seizures and died (Rogers *et al.*, 1994; Twyman *et al.*, 1995). These studies revealed that autoantibodies to GluR3 are involved in a severe form of childhood epilepsy called Rasmussen's encephalitis and act as receptor agonists. Thus there is now a known example of an autoantibody response to receptors in brain. Curiously, Lambert-Eaton myasthenic syndrome is a result of a paraneoplastic autoimmune response to small-cell carcinomas of the lung which results in antitumor autoantibodies to voltage-sensitive Ca^{2+} channels that are also involved in acetylcholine release at neuromuscular junctions (Vincent *et al.*, 1989; Takamori *et al.*, 1994). Recent studies have revealed autoantibodies to glutamate receptor subtypes in paraneoplastic cerebellar degeneration (Gahring *et al.*, 1995). The cause of the autoimmune response to AChR in myasthenia gravis remains unknown.

Our studies of myasthenia gravis and the antigenic structure of muscle AChR led to the development of a library of several hundred monoclonal antibodies (mAbs) for use as model autoantibodies and structural probes (Tzartos and Lindstrom, 1980; Tzartos *et al.*, 1981, 1983, 1986; Lindstrom *et al.*, 1981). The mAbs to the MIR provided our initial probe for neuronal AChRs and revealed structural homologies between muscle and neuronal AChRs (Swanson *et al.*, 1983; Jacob *et al.*, 1984, 1986; Smith *et al.*, 1985; Whiting and Lindstrom, 1986a,b). The discovery of the autoimmune basis for most myasthenia gravis cases helped us recognize a small fraction of myasthenia cases which were due to congenital defects in AChRs or other components of the neuromuscular junction (Engel, 1990, 1994). The first of these muscle AChR mutants to be sequenced causes a prolonged open channel time and has just been mapped to ε subunits (Ohno *et al.*, 1995). Curiously, the first mutation of a neuronal AChR has also been recently reported. This mutation, which truncates α4 subunits, causes a form of epilepsy known as benign neonatal familial convulsions (Steinlein *et al.*, 1994, 1995). Thus, from many perspectives, including the clinical, there are interesting parallels and contrasts between AChRs of muscles and nerves.

2.2. Function

AChRs at skeletal neuromuscular junctions play a classic postsynaptic role in neuromuscular transmission. The basic functional role of neuromuscular transmission is to sufficiently amplify the small current involved in the action potential propagated over great distances along the motor nerve axon to the nerve ending so that the relatively huge muscle fiber is depolarized sufficiently to fire an action potential which can then trigger contraction of the muscle fiber. There is typically a single synapse on each muscle fiber. It is critical that the fiber contract every time the nerve fires, so a safety factor of about 10 is built in to this synapse in terms of excess acetylcholine release and excess AChRs over what would be minimally necessary to trigger a response. In ganglia and some other synapses, neuronal α3 AChRs may play a similarly straightforward postsynaptic role (Jacob *et al.*, 1984; Vernallis *et al.*, 1993), though probably not with such a large safety factor. At many brain synapses the probability of transmitter release is much less than 1 (Südhof, 1995), which emphasizes the importance of presynaptic modulation. Many neuronal α3 and α4 AChRs are located presynaptically where they may affect the probability of successful transmission by modulating the release of other transmitters (Swanson *et al.*, 1987; Henley *et al.*, 1986a,b, 1988; Clarke *et al.*, 1986; Clarke, 1990;

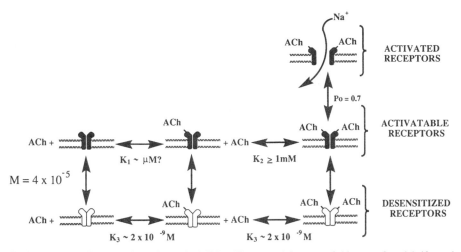

FIGURE 1. Model for activation of muscle AChRs. The general features of this sort of model, if not the particular binding affinities, open probability (P_o), isomerization constant for desensitization (M), or precise ion selectivities apply to neuronal AChRs as well. Although some subtypes (e.g., $\alpha4\beta2$ AChRs) exhibit higher affinity for acetylcholine, most neuronal subtypes exhibit inward rectification (i.e., their opening probability (P_o) decreases as the membrane depolarizes), and most neuronal subtypes which have been examined are more permeable to Ca^{2+} than are the relatively nonselective cation channels of muscle AChRs.

Whiting and Lindstrom, 1988; Grady *et al.*, 1994). At least some neuronal $\alpha7$ AChRs are located perisynaptically where their functional role is much less clear (Jacob and Berg, 1983; Vernallis *et al.*, 1993; Sargent and Wilson, 1995).

The functional properties of AChRs as determined by electrophysiological and biochemical studies are described in Fig. 1 (Changeux, 1990; Jackson, 1989; Sine, 1988; Sine and Taylor, 1980, 1982; Sine *et al.*, 1990). Muscle AChRs have two binding sites for acetylcholine, and when both have acetylcholine bound there is a high probability that the channel will flicker open for a millisecond or so at a time during a burst of activity before desensitizing to a conformation characterized by high affinity for agonists and a closed channel. In reality there may be several desensitized states with longer or shorter lifetimes. The desensitized states can also be reached from liganded states without passing through the open state. Although in the absence of acetylcholine virtually no AChRs are desensitized, in the continued presence of even low concentrations of agonists most or all of the AChRs will end up desensitized. The net effect of this is shown in Fig. 2. Continuous exposure of purified reconstituted *Torpedo* electric organ AChRs to agonist concentrations low enough to activate only a small fraction of the AChRs when applied acutely are sufficient, when applied chronically, to completely desensitize the AChRs and prevent an acute response.

The significance, if any, of desensitization in normal neuromuscular transmission is unknown, but desensitization is an intrinsic property of all the receptors in the super-family. In theory, it could permit the sensitivity of the synapse to be controlled by low levels of acetylcholine release, or it could provide a mechanism for "memory" of recent activation of the synapse. In reality, neuromuscular synapses are built with a big safety

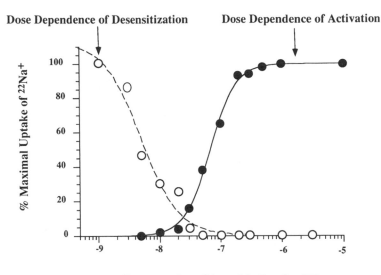

FIGURE 2. Dose dependence of AChR activation and desensitization. The data shown are replotted from Anholt *et al.* (1982). It shows the responses of AChRs purified from *Torpedo* electric organs and reconstituted into liposomes where their function is measured by agonist-stimulated entry of $^{22}Na^+$ into the liposomes. The activation curve used the indicated concentrations of acetylcholine for 10 sec. The desensitization curve used a 30-min preincubation in the indicated concentration of acetylcholine prior to a 10-sec exposure in the presence of $^{22}Na^+$ to a saturating concentration (10^{-4} M) of carbamylcholine to assay activity of the AChRs not desensitized during the preincubation with acetylcholine.

factor to ensure transmission every time. Desensitization only becomes evident under pharmacological manipulation by succinylcholine as a depolarizing blocker for surgical muscle relaxation, or when the doses of acetylcholine esterase inhibitors given to myasthenia gravis patients to compensate for the loss of functional AChRs are too high and cause muscle weakness similar to that which they were designed to prevent. In the central nervous system roles for desensitization of AChRs in normal physiological function are also unclear. Failure to desensitize can, under some circumstances, be damaging. Changeux and coworkers initially found that a single amino acid change in the second transmembrane domain which is thought to line the cation channel of α7 neuronal AChRs of chickens caused an array of pharmacological and electrophysiological changes in the properties of α7 homomers expressed in *Xenopus* oocytes. This could be explained best by assuming that these AChRs could be activated even when their acetylcholine binding site assumed a desensitized conformation (Revah *et al.*, 1991; Bertrand *et al.*, 1992, 1993a). Recently it was found that a similar mutation in AChRs of *C. elegans* produces channels which desensitize slowly, resulting in neuronal degeneration, perhaps as a result of excess influx of Cs^{2+} (Treinin and Chalfie, 1995).

The concept of desensitization also complicates thinking about the effects of chronic nicotine exposure in tobacco users. Nicotine is an agonist when applied acutely, but because it causes desensitization, it is a time-averaged antagonist when applied chronically. The typical serum concentration of nicotine in smokers has little effect on muscle AChRs but can cause extensive desensitization of neuronal $\alpha4\beta2$ AChRs which have high affinity for nicotine. The accumulation of desensitized neuronal AChRs can explain the phenomenon of tolerance among chronic tobacco users to higher doses of nicotine. The presence of multiple neuronal AChR subtypes with different affinities for nicotine, rates of desensitization, and histological localizations also complicates analysis of the behavioral effects of nicotine. What is most rewarding to a tobacco user? Is it the acute bolus dosages of nicotines activating AChRs which a smoker can so carefully modulate (by making the brain micromolar in nicotine within a minute of inhaling) or the chronic concentrations of serum nicotine persisting for hours in a chronic smoker (typically 0.2 μM) resulting in general AChR desensitization (Benowitz et al., 1990)? Is it overnight recovery of activatable AChRs that leads to withdrawal symptoms and the need of typical smokers to light up within half an hour of waking to relieve these withdrawal symptoms by once again desensitizing their AChRs?

Muscle AChR channels are usually thought of as relatively nonselective cation-specific channels whose conductance shows little voltage dependence. By contrast, most neuronal AChRs have been found to be more permeable to Ca^{2+} and to exhibit inward rectification, thus tending to remain closed at positive potentials (e.g., Lindstrom et al., 1995). Dani and coworkers (Costa et al., 1994) report the Ca^{2+}/Na^+ permeability ratio of muscle AChRs as 0.2 and that of $\alpha3\beta4$ AChRs as 1.1, but that of $\alpha7$ AChRs as 20 (Seguela et al., 1993), which is substantially higher than the value of 7 for NMDA receptors.

2.3.   Three-Dimensional Structure

More is known about the shape of muscle AChRs than about any other ligand-gated ion channel. The sequence homologies of the receptors in the superfamily to which AChRs belong suggest that all of these receptors share a basically similar shape. The shape of AChRs has been best characterized by electron microscopy (Mitra et al., 1989; Holtzman et al., 1982; Kubalek et al., 1987; Unwin et al., 1988; Unwin, 1993a,b, 1995). Diffraction analysis of two-dimensional crystalline arrays of Torpedo electric organ AChRs in fragments of their native membranes has provided resolution to 9 Å (Unwin, 1993a), which borders on the resolution necessary to resolve α-helices, but is not yet sufficient to resolve individual amino acids. In theory, this method can ultimately provide atomic resolution. Getting adequate crystals of Torpedo AChR has thus far frustrated X-ray crystallographic analysis of Torpedo AChRs, and getting adequate levels of expression of cloned neuronal AChRs has thus far frustrated more detailed analysis of two- or three-dimensional crystals of neuronal AChRs.

Images from diffraction analysis of electron micrographs of two-dimensional crystalline arrays of Torpedo AChRs have been interpreted in the context of much biochemical and molecular biological information about the AChR (Unwin, 1993a,b, 1995). Each Torpedo AChR is known to be composed of two $\alpha1$ subunits and one $\beta1$, one γ, and one δ subunit (Lindstrom et al., 1979; Raferty et al., 1980; Karlin, 1991, 1993). When the AChR is viewed from its top or bottom as it extends across the membrane, the five subunits are oriented around the central channel in a pentagonal array about 80 Å in diameter (Unwin, 1993a). Similar pentagonal arrays have been observed with other members of the superfamily, for example, purified 5-HT$_3$ receptors or cloned 5-HT$_3$ receptors expressed using the baculovirus system (Boess *et al.*, 1995; Green *et al.*, 1995). Coexpression of α1 subunits with either γ, δ, or ε subunits, but not β1 subunits, results in the formation of subunit dimers which exhibit the pharmacological properties of one or the other of the two acetylcholine binding sites on the AChR, suggesting that the acetylcholine binding sites are formed at the interfaces between α1 subunits and the γ, δ, or ε subunits (Blount and Merlie, 1989; Saedi *et al.*, 1991; Gu *et al.*, 1991a,b; Verrall and Hall, 1992; Yu and Hall, 1994). This idea is also supported by some affinity labeling experiments (Pederson and Cohen, 1990), especially by elegant affinity cross-linking and mutagenesis experiments which suggest that the anionic subsite which binds to the positively charged amine of agonists and antagonists is formed by the γ, δ, or ε subunit (Czajkowski *et al.*, 1993; Czajkowski and Karlin, 1995; Stauffer and Karlin, 1994). Mutagenesis experiments also indicate that the bivalent antagonist curare bridges the α1 and γ subunits (Fu and Sine, 1994). Collectively, these experiments argue strongly that the five subunits of *Torpedo* AChRs are oriented around the central channel in the order α1, γ or ε, α1, δ, β1 (Karlin, 1993). In neuronal AChRs as well, both α subunits and structural subunits contribute to the pharmacological properties of their acetylcholine binding sites. For example, AChRs expressed from α3 subunits in combination with β2 or β4 subunits differ in pharmacological properties (Luetje and Patrick, 1991). The idea that the channel is located symmetrically in the center of the AChR and bordered by equivalent parts of each subunit surrounding it is consistent with the observation that channel-blocking noncompetitive antagonists are able to photoaffinity-label homologous amino acids in the second transmembrane domain of each subunit (Hucho *et al.*, 1986; Revah *et al.*, 1990). Viewed from the side, the AChR is roughly cylindrical, about 120 Å long, with about 65 Å extending on the extracellular surface and about 15 Å extending on the cytoplasmic surface (Unwin, 1993a). The external mouth of the channel has a diameter of about 25 Å and is surrounded by walls of about this same thickness. The channel narrows abruptly near the level of the lipid bilayer. On the cytoplasmic surface the channel flares open again. The subunits are thought to be relatively rod-shaped and oriented around the central channel like barrel staves. The main immunogenic region is located at the tips of the extracellular surface of α1 subunits (Kubalek *et al.*, 1987; Beroukhim and Unwin, 1995). In this position it is well oriented to permit binding of autoantibodies which can fix complement, and it is ideally oriented to permit cross-linking of AChRs by autoantibodies that then aggregate and facilitate the endocytosis and lysosomal destruction of these AChRs (Conti-Tronconi *et al.*, 1981; Lindstrom *et al.*, 1988).

Elegant electron microscopy studies have revealed relative movements of *Torpedo* AChR subunits between resting, activated, and desensitized AChRs (Unwin, 1993a,b, 1995). Activated AChRs imaged by a rapid mixing and freezing procedure exhibit some difference from resting AChRs starting about one-third the way down from the extracellular surface, a region in which the acetylcholine binding site might be located (Unwin, 1995). The biggest effects are in the region of the narrowest part of the ion channel, where domains thought to correspond to the second transmembrane domain of each subunit are thought to form α-helices, each kinked in the center, which through small rotations open the channel to a diameter of 9–10 Å during channel activation (Unwin, 1995). Desensitized AChRs were imaged after prolonged exposure to agonist (Unwin et al., 1988). There was slight (10°) tangential tipping of two of the structural subunits observed in desensitized AChRs (the identities of these subunits as assigned in the microscopy studies is controversial; Karlin, 1993).

2.4. Amino Acid Sequences of Muscle and Neuronal AChRs

Partial N-terminal sequencing of *Torpedo* AChR subunits led to the recognition that the sequences of all the subunits were homologous (Raferty et al., 1980) and led to the initial cloning of cDNAs for these subunits (Noda et al., 1982). Low-stringency hybridization screening of cDNA libraries, first from muscle and then from a pheochromocytoma cell line, led to the initial cloning of a neuronal AChR subunit (Boulter et al., 1986). Thus, it seems appropriate to consider the general features of the amino acid sequences of muscle and neuronal AChRs together.

All AChR subunits and all subunits of receptors in the superfamily contain six structural features in order along their sequences: 1) a signal sequence which is removed after translocation of the mature N-terminus to the external surface of the endoplasmic reticulum, 2) a large extracellular N-terminal sequence containing a disulfide-linked loop and one or more N-glycosylation sites, 3) three closely spaced putative transmembrane sequences (M1–M3) located just after the large extracellular domain, 4) a large domain on the cytoplasmic surface containing consensus sites for phosphorylation, 5) a fourth putative transmembrane segment (M4), and 6) a short C-terminal extracellular tail.

Figure 3 compares the sequences of all of the known AChR subunits. Rat sequences are shown because more sequences are known from this species than any other, except in the case of α8, which is currently known only from chickens. The α subunits were the first to be characterized, initially by affinity labeling with the competitive antagonist MBTA (Karlin and Cowburn, 1973). They are the best characterized subunits and are associated with the acetylcholine binding site, although it is clear that the adjacent γ, δ, ε, β2, or β4 structural subunit also contributes contact amino acids to the binding site (Blount and Merlie, 1989; Luetje and Patrick, 1991). MBTA or bromoacetylcholine covalently reacts with α subunits only after a disulfide bond between the adjacent cysteine pair at α1 positions 192 and 193 is reduced (Kao et al., 1984). The presence of this cysteine pair has become the defining feature of α subunits, although in the case of neuronal AChR α5 subunits it is not clear that α5 plays the same role as other α subunits; instead α5 might act more as a structural subunit. The following discussion of amino acid sequence will apply especially to the α1 subunits as shown in Fig. 4. There is strong sequence identity between species for corresponding subtypes of AChR subunits. This is greatest for α1 subunits which have 80% sequence identity between *Torpedo* and human (Noda et al., 1983). The greatest diversity in sequence between types of subunits and between species is in the large cytoplasmic domains of these subunits. Sequences of AChR subunits will be considered from the N-terminus to the C-terminus.

All AChR subunits exhibit signal sequences at their N-termini that are cleaved during translation and serve to target the N-terminal domain on the mature subunit to the extracellular surface as it is synthesized in the endoplasmic reticulum. Synthesis and

Signal Peptide

α1	MELSTVLLLGLSSAGLVLG
α2	MTLSHSALQFWTHLYLWCLLLVPAVLTQQGSH
α3	MGVVLLPPPLSMLMLVMLLLPAASA
α4	MEIGGPGAPPPLLLLPLLLLLGTGLLPASSHIETR
α5	MVQLLAGRWRPTGARRGTAGGLPELSSA
α6	MLNWGRGDLRSGLCLWICGFLAFFKGSRG
α7	MCGGRGGIWLALAAALLHVSLQ
α8	MLTEKCLGFFYSGLCLWASLFLSFFKVSQQ
α9	MNRPHSCLSFCWMTFAASGIRAVETAN
β1	MRLGALLLLGVLGTPLAPGARG
β2	MLACMAGHSNSMALFSFSLLWLCSGVLGTD
β3	MTGFLRVFLVLSATLSGSWVTLTATAGLSSV
β4	MRGTPLLLVSLFSILLQDGDCRLAN
γ	MHGGQPQLLLLLATCLGAQSR
δ	MAGPVPTLGLLAALVVCGSWG
ε	MTMALLGTLLLLALFGRSQG

Large N-Terminal Extracellular Domain

α1	SEHETRLVAKLFEDYSSVVRPVEDHREIVQVTVGLQLIQLINVDEVNQIVTTNVRLKQQWVDYNLKWNP
α2	THAEDRLFKHLFGGYNRWARPVPNTSDVVIVRFGLSIAQLIDVDEKNQMMTTNVWLKQEWNDYKLRWDP
α3	SEAEHRLFQYLFEDYNIEERPVANVSHPVIIQFEVSMSQLVKVDEVNQIMETNLWLKQIMNDYKLKWKP
α4	AHAEERLLKRLFSGYNKWSRPVGNISDVVLVRFGLSIAQLIDVDEQNQMMTTNVWVKQEWHDYKLRWNP
α5	AKHEDSLFRDLFEDYERWVRPVEHLSDKIKIKFGLAISQLIDVDEKNQLMTTNVWLKQEWIDVKLRWNP
α6	CVSEEQLFHTLFAHYNRFIRPVENVSDPVTVHFELAITQLANVDEVNQIMETNLRHVWKDYRLCWPE
α7	GEFQRRLLYKELVKNTNPLERPVANDSQPLTVYFSLSLLQIMDVDEKNQVLTTNIWLQMSWTDHYLQWNM
α8	GESQRRLYRDLLRNYNRLERPVMNDSQPIVVELQLSLLQIIDVDEKNQVLIINAWLQMYWDIYLSWDQ
α9	GKYAQKLFSDLFEDYSSALRPVEDTDAVLNVTLQVTLSQIKDMDERNQILTAYLWIROTWHDAYLTWDR
β1	SEAEGQLIKKLFSNYDSSVRPAREVGDRVGVSIGLTLAQLISLNEKDEEMSTKVYLIDLEWITDYRLSWDP
β2	TEERLVEHLLDPSRYNKLIRPATNGSELVTVQLMVSLAQLISVHEREQIMTTNVWITQEWEDYRLTWKP
β3	AEHEDALLRHLFQGYQKWVRPVLNSSDIIKVYFGLKISQLVDVDEKNQLMTTNVWLKQEWTDQKLRWNP
β4	AEEKLMDDLLNKTRYNNLIRPATSSSQLISIRLELSLSQLISVNEREQIMTTSIWLKQEWIDYRLAWNS
γ	NQEE RLLADLM RNYDPHLRPAERDSDVVNVSLKLFTTNLISLNEREEALTTNVWIEMQWCDYRLRWDP
δ	LNEEQRLIQHLFEEKGYNKELRPVARKEDIVDVALSITLSNLISLKEVEETLTTNVWIDHAWIDSRLQWNA
ε	KNEELSLYHHLF DNYDFECRKPVRPEDTVTLKVLLSLTQLISVNEKEETLTTNVWIGIEWQDYRLNFSK

50

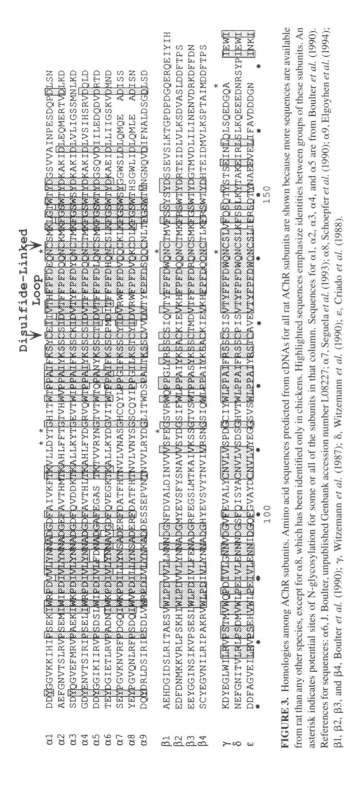

FIGURE 3. Homologies among AChR subunits. Amino acid sequences predicted from cDNAs for all rat AChR subunits are shown because more sequences are available from rat than any other species, except for α8, which has been identified only in chickens. Highlighted sequences emphasize identities between groups of these subunits. An asterisk indicates potential sites of N-glycosylation for some or all of the subunits in that column. Sequences for α1, α2, α3, α4, and α5 are from Boulter et al. (1990). References for sequences: α6, J. Boulter, unpublished Genbank accession number L08227; α7, Seguela et al. (1993); α8, Schoepfer et al. (1990); α9, Elgoyhen et al. (1994); β1, β2, β3, and β4, Boulter et al. (1990); γ, Witzemann et al. (1987); δ, Witzemann et al. (1990); ε, Criado et al. (1988).

ACh Binding Site Disulfide

M1

M2

α1	FMESGEWVMKEARGWKHWVFYSCCPTTPYLDITYHFV	MQRIPLFYVVNVIIPCLLFSFLTGLVFYLPTDS	GEKMTLSISVLLSLTVFLLVIVELIPS
α2	YWESGEWAIINATGTYNSKKYDCCAEIYPDVTYYFV	IRRLPLFYTINLIIPCLLISCLTVLVFYLPSEC	GEKITLCISVLLSLTVFLLLITEIIPS
α3	YWESGEWAIIKAPGYKHEIKYNCCEEIYQDITYSLY	IRRLPLFYTINLIIPCLLISFLTVLVFYLPSDC	GEKVTLCISVLLSLTVFLLLITEIIPS
α4	FWESGEWVIVDAVGTYNTRKYECCAEIYPDITYAFI	IRRLPLFYTLFLIIPCIGLSFLTVVVFYLPSEC	GEKISLCTSVLVSLTVFLLLITEIIPS
α5	FFDNGEWDISMSNKGNRTDSCCWY PYITYSFV	IRRLPMFYTINLIIPCLFISFLAVLVFYLPSDC	GEKVTLCISVLLSLTVFLLVITEIIPS
α6	FWENSEWEIVDASGYKHDIKYNCCEEIYIDIFYSFY	MRRRTLYYGLNLLIPCVLISALALLVFLLPADS	GEKISLGITVLLSLTVFMLLVAEIMPA
α7	YIPNGEWDLMGIPGKRNEKFYECCKE PYPDVTYTVT	MRRRTLYYGLNLLIPCVLISGLALLVFLLPADS	GEKISLGITVLLSLLVEMLLVAEIMPA
α8	YISNGEWDLVGVPGKRNELYYECCKE PYPDVTYTIT	LKRSSFYIVNLLIPCVMISFLAPLSFYLPAAS	GEKVSLGVTILLAMLVFQLMVAEIMPA
α9	FIEDVEWEVHGMPAVKNVISKCYCSE PYPDVBFTLL		
β1	EGTFIENGQWEIHKPSRLIQLFGDQRGGKEGHHEEVIFYLIIRRKPLFYLVNVTAPCLLITIAIFVYLPQDA	GEKMGISIFALLTVILLADKVPE	
β2	GEWDIALPGRRNENPDDSTYVD	ITYDFIIRRKPLFYTINLIIPCVLITSIAILVFYLPSDC	GEKMTLCISVLLALTVFLLLISKIVPP
β3	GEWDILNAKGMKGNRREGFYSYPF	VTYSFVLRRLPLFLIFLIIPCLGLSFITVLVFYLPSDE	GEKLSISTSVLVSLTVFLLVIEEIIPS
β4	GEWDIVALPGRRTVNPQDPSYVD	VTYDFIIKBRKLFYTINLIIPCVLITSIAILVFYLPSDC	GEKMTLCISVLLALTFLLLISKIVPP
γ	FIDPEAFTENGEWAIRHRPAKMLLDPVTPAEEAGHQKVVFYLL	IQRKPLFYVINIIVPCVLISSVAILIYFLPAKAGGQKCTVATNVLLAQTVFLFVAKKVPE	
δ	IIDPEGTENGEWEIVHRAAKVNVDPSVPMDSTNHQDTFYLI	IRKPLFYIINIIVPCVLISFMINLVFYLPGDC	GEKTSVAISVLLAQSVFLLLISKRLPA
ε	DLDTAAETENGEWAIDYCPGMIRHYEGGSTEDPGETDVIYTLI	IRRLFYVNIIVPCVLISGLVLLAYFLPAQAGGQKCTVSINVLLAQTVFLFLIAQKIPE	

200   250

M3 Large Cytoplasmic Domain

```
α1  TSSAVPLIGKYMLFTMVFVIASIIITVIVINTHHRSPST HIMPE WVRKVFIDTIPNIMFFSTMKRPSRDKQEKRIFTEDIDISGKPGPPPMGFH
α2  TSLVIPLIGEYLLFTMIFVTFSVITVFLVLNVHHRSPST HNMPN WVRVALEGRVFRWLM NRLLPPMELHGSPDLKLSPSYHWLETNMDAGEREETEEEE
α3  TSLVIPLIGEYLLFTMIFVTFSIVITVFLNVHHRTPTT HTMPT WVKAVFLNLLPRVMF TRTTSGEGDTPKTRFYGAELSNLNCFSRADSKSCKEGYPC
α4  TSLVIPLIGEYLLFTMIFVTFSIVITVFLNVHHRSERT HTMPA WVRRVFLDIVPRLLF MKRPSVVKDNCRRLIESMHKMANAPRFWPEPVGEPGILSDIC
α5  SSKVIPLIGEYLLFTMIFVTFLSHMVVFAINIHHRSSSI      WRKIFIHKLPKLLC MRSHADRYFTQREEAESGAGPKSRNT
α6  TSLVIPVGEYLLFTMIFVTFSIVVTFNIYHDPAT HIMPK WVKTMFLQVFSIIM MRRLDKTKEMDGVKDPKTHTKRPAKVFTHRKEPKLLKECR
α7  TSDSVPLIAQNFASEMIIVGLSVVVTVIVLRYHHDPDG GKMPK WTRIILLNWCAWFLR MKRPGEDKVRPACQHKPRCSLASVELSAGAGPPTSNGNLLY
α8  TSDSVPLIAQFFASIMVIVGLBVVVILQFHHDEQA GKMPR WVRVILLNWCAWFLR KKEGENIKPLSCKYSYPKHHPSLKNTEMNVLPGHQPSNGNM
α9  SENVRLIGKKYIATLALITAFALITIMIMVNIEFCGAEA RPVEH WAKVVILKYMSRILFVYDVGESCLCPRHSQEPEQVTKVYSKLPESNLKTSRNKDLSRKKE

β1  TSLAVFIIIKTMVTFMVLTHSVILSTVILNLHHRSPHT HQMPF WIRQIFIHKLPPYLG LKRPKPERDQLPEPHHSLSPRSGWSGWGRGTDEYFIRKPPSD
β2  TSLDVPLVGKYMLFTMVLVTFSIVTSVCVLNVHHRSPTT HTMAP WVKVVFLEKLPPLLF LQQPRHCARQRLRLRRQEREGEAVFREGPAADPCTCFV
β3  SSKVIPLIGEYLLFTMIFVTLSIIVTVFVINVHHRSSTY HPMAP WVKRLFLQRLPRWLC MKDPMDRFSFPDGKESDTAVRGKVSGKRKQTPASDGERVLVA
β4  TSLDIPLLGKYLLFTMVLVTFSIVTTVCVLNVHHRSPST HTMAS WVKECLHKLPTVF MKRPGLEVSLVRVPHPSQLHATADTAATSALGPTSPSNLYG

γ   TSQAVPLISKYFLTTLMVTILIWNSVWLNVSLRSPHT HSMAR GVRKVFLRLFPQIRMHVHPRAPAAVQDARLRLQNGSSGWP IMTREGDLCLPRSPLLF
δ   TSMAIPLVGKFLLTLGWVLVTMVVICWIVLNIHFRTPST HVLSE GVKKFFLETLPKLHM SRPEEDPGPRALIRRTSSLGY ISKAEYFSLKSRSDLM
ε   TSLSVPLLGRYLLMVATLIKMNCVILVNVSLRPTH HATSP RLRQILLEILRR GLSPPP EDPGAASPARRASSVGI LLRAELLILKKPESPLVE
                        300                                              350
```

FIGURE 3. (Continued).

```
α1  EEDENICVCAGLPDSSMGVLTGHGGLHLRAMEPETKTPSQA
α2  QDGTCGYCHHRRVKISHFSANLTRSSSESVNAVL
α3  NQGLSPAPTFCNFTDTAVETQPTCRSPPLEVPDLKTSEVEKASPCPSPGSCPPPKSSSGAPMLIKARSLSVQHVPSSQEAAEDGIRCRSRSIQYCVSQDGAASL
α4
α5
α6  HCHKSSEIAPGKRLSQQPAQWTENSEH
α7  IGFRGLEGMHCAPTPDSGVVCGRLACSPTHDEHLMHGAHPSDGD
α8  IYSHTMENPCCPQNNDLGSKSGKITCPLSEDNEHVQKKALMDTI
α9  VRKLLKNDLGYQGGIPQNTDSYC

β1  FLFPKLNRFQPESSAPDLRRFIDGPTRALGLPQELREVISSISYMARQLQEQEDHDALK
β2  NPASVQGLAGAFRAEPTAAGPGRSVGPCSCGLREAVDGVRFIADHMRSEDDDQSVR
β3  FLEKASESIRYISRHVKKEHFISQVV
β4  SSMYFVNPVPAAPKSAVSSHTAGLPRDARLRSSGRFREDLQEALEGVSFIAQHLESDDRDQSVI

γ   RQRQRNSLVQAVLEKLENGPPEMRQSQEFCGSLKQASPAIQACVDACNLMARARHQQSHFDSG
δ   EKQSERICLARRLTTARKPPASSEQVQQELFNEMKPAVDGANFIVNHMRDQNSYNEE
ε   EGQRHHGTWTAAALCQNLGAAAPEVRCCVDAVNFVAESTRDQEATGEE
         400                                    450
```

M4

```
α1   SPLIKHPEVKSAIEGVKYIAETMKSDQESNNAAEEWKYVAMVMDHILLGVFML
α2   SEILLSPQIQKALEGVHYIADRLRSEDADSSVKEDWKYVAMVVDRIFLWLFII
α3   SLSALSPEIKEAIQSVKYIAENMKAQNVAKEIQDDWKYVAMVIDRIFLWVFTL
α4   ADSKPTSSPTSLKARPSQLPVSDQASPCKCTCKEPSPVSPVTVLKAGGTKAPPQHLPLSALTRAVEGVQYIADHLKAEDTDFSVKEDWKYVAMVIDRIFLWMFII
α5                                                               LEAALDCIRYITRHVVKENDVREVVEDWKFIAQVLDRMFLWTFLL
α6                                                             PPDVEDVIDSVQFIAENMKSHNETKEVEDDWKYVAMVVDRVFLWVFII
α7                                                             PDLAKILEEVRYIANRNRCQDESEVICSEWKFAACVVDRLCLMAFSV
α8                                                             DVIVKILEEVQFIAMRFRKQDEGEEICSEWKFAAAVIDRLCLVAFTL
α9                                                             ARYEALAKNIFTAKCLKDHKATNSKGSEWKKVAKVIDRFFMWIFFA

β1                                                                     EDWQDFVAMVVDRLFLWTFIV
β2                                                                     EDWKYVAMVIDRLFLWMFVF
β3                                                                     QPWKFVAQVLDRIFLWLFLI
β4                                                                     EDWKFVAMVLDRLFLWVEVF

γ                                                                    NEEWLLVGRVLDRVCFLAMLS
δ                                                                    KDNWQVARTVDRLCLFVVTP
ε                                                                    LSDWVRMGKALDNVCFWAALV
```

500   550

FIGURE 3. (Continued).

α1 VCLISTLAVEAGRLIELHQQG
α2 VCFLSTHIGLPPPFLAGMI
α3 VCIISTAGLFLQPLMARDDT
α4 VCLLSTVGLFLPPWLAAC
α5 VSIISLGLFVPVIYKWANIIVPVHIGNTIK
α6 VCVFCTVGLFLQPLLGNTGAS
α7 FTIICTIGILMSAPNFVEAVSKDFA
α8 FAIICFFTILMSAPNFIEAVSKDFT
α9 MVFVMNVLITARAD

β1 FTSVGTLVIFLDATYHLPPPEPFP
β2 VCVFCTVGMFLQPLFQNYTATTFLHPDHSAPSSK
β3 ASVIGSILIFIPALKMWIHRFH
β4 VCIIGTMGIELPPLFQIHAPSKD

γ LFICSTAGIFFMAHYNQVPLEFFGDPRPYLPLPD
δ VMVVCTAWIFIQGVYNQPPPQPFFGDPFSYDEQDRRFI
ε LFSVGSTLIFIGGYFNQVPDLPYPCIQP
   •
   600

FIGURE 3. (continued).

conformational maturation of α1 subunits have been studied in greatest detail and provide a model for the behavior of the other subunits (Merlie *et al.*, 1983; Blount and Merlie, 1988, 1989, 1990, 1991a,b; Blount *et al.*, 1990). The N-terminal protein sequences of mature AChR subunits are known only from electric organ and muscle α1, β1, γ, and δ subunits (Raferty *et al.*, 1980; Conti-Tronconi *et al.*, 1982) and from neuronal α4, α7, and β2 subunits (Whiting *et al.*, 1987; 1991; Schoepfer *et al.*, 1988; Conti-Tronconi *et al.*, 1985).

Putative sites of N-glycosylation are found throughout the N-terminal half of the subunits that precedes the first putative transmembrane domain (M1) and which is thought to form a large extracellular domain. For example, neuronal α subunits, but not muscle α1 subunits, have a putative N-glycosylation site at position 24 of the consensus sequence shown in Fig. 3. All AChR subunits, except α7, α8, and α9, have a putative N-glycosylation site at position 141. In α1 subunits, at least, this site is known to be glycosylated (Gehle and Sumikawa, 1991).

A disulfide-linked loop is known to be present between cysteines 128 and 142 in α1 subunits (Kallaris *et al.*, 1989). Homologous cysteines are present in all other AChR subunits and in all other subunits of the superfamily (Cockroft *et al.*, 1992). Formation of this disulfide bond appears to be associated with the conformational maturation of α1 subunits, which precedes assembly with other subunits and is marked by the acquisition of the ability by α1 to bind with high affinity both αBgt and mAbs to the MIR (Blount and Merlie, 1990). Disruption of this loop by *in vitro* mutagenesis prevents binding of αBgt to α1 subunits and inhibits expression of assembled AChR on the cell surface (Sumikawa and Gehle, 1992).

2.4.1. The MIR

The main immunogenic region (MIR) of α1 subunits is a conformation-dependent epitope that is responsible for provoking more than half of the autoantibodies to muscle AChRs, which cause the weakness characteristics of myasthenia gravis (MG) and its animal model, experimental autoimmune myasthenia gravis (EAMG) (Lindstrom *et al.*, 1988; Tzartos *et al.*, 1991; Protti *et al.*, 1993). The MIR was discovered through the recognition that many of the mAbs made to electric organ AChRs were mutually competitive for binding to these AChRs (Tzartos and Lindstrom, 1980; Tzartos *et al.*, 1981) and that one of these mAbs could inhibit the binding of a large fraction of myasthenia gravis patient autoantibodies to human AChR (Tzartos *et al.*, 1982). Some of these mAbs could bind weakly to denatured α1 subunits. In this way their epitope could be localized through binding to peptide fragments of α1 or to synthetic α1 peptides, thus mapping some or all of the amino acids forming the MIR to within the sequence α1 66–76 as illustrated in Fig. 5 (Gullick and Lindstrom, 1983; Das and Lindstrom, 1989). The mAbs to the MIR usually cross-react well between species, but it was found that they did not bind to *Xenopus* neuromuscular junctions (Sargent *et al.*, 1984). There are two nonconservative substitutions at positions α1 68 and 71 in the *Xenopus* sequence (see Fig. 6). MIR mAbs did not bind to synthetic peptides with the *Xenopus* sequence (Das and Lindstrom, 1989). More important, MIR mAbs did not bind to intact *Torpedo* AChRs in which one or both amino acids α1 69 and 71 were changed to the *Xenopus* sequence, as

1-437

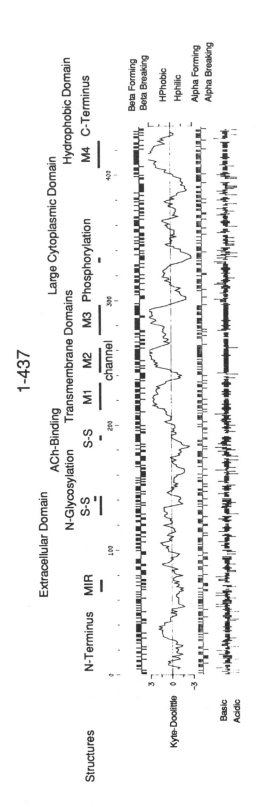

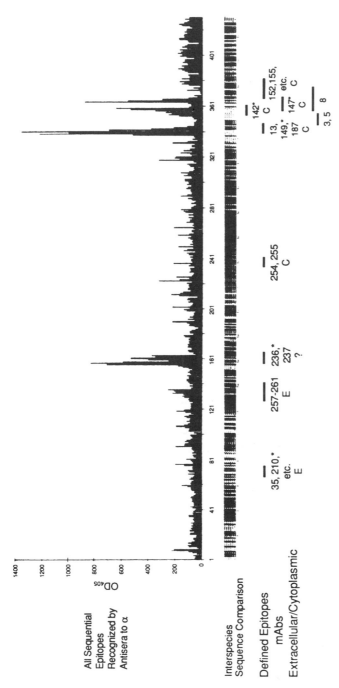

FIGURE 4. Structural features of *Torpedo* α1 subunits. This figure from Das and Lindstrom (1991) summarizes the antigenic structure of α1 subunits and compares this with other features of the sequence such as hydrophobicity, charge, and conservation of sequence of α1 subunits between species. The white gaps in the sequence indicate regions of unconserved sequence (see Luther *et al.*, 1989 for a legible sequence comparison.) The antigenic sequence of α1 subunits is typical of that of other AChR subunits to the extent that these are known (e.g., Gullick and Lindstrom, 1983). The most prominent epitopes on denatured α1 subunits occur at unconserved hydrophilic sequences, and most are in the large cytoplasmic domain between the third (M3) and fourth (M4) hydrophilic sequences. Data for binding of serum antibodies to SDS-denatured *Torpedo* α1 subunits shows the pattern of binding to overlapping eight amino acid peptides corresponding to the sequences 1–8, 2–9, 3–10, etc., for the entire α1 sequence synthesized on plastic pegs. The amount of antibody bound to each peptide was determined using peroxidase-coupled antibodies. Binding sites for the mAbs shown were determined by binding to synthetic peptides and/or native α1 peptide fragments.

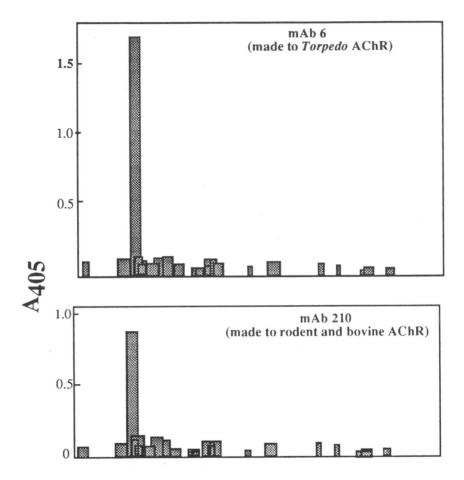

shown in Fig. 6 (Saedi *et al.*, 1990). Even MIR mAbs absolutely dependent on the native conformation of the α1 subunit did not bind when these substitutions were made, showing that all of the MIR mAbs shared overlapping epitopes. The MIR has been localized by electron crystallography to the extracellular tips of the α subunits, oriented so that antibody fragments bound to it angle away from the axis of the AChR (Beroukhim and Unwin, 1995). The endogenous function of the MIR, if any, is not known. The mAbs to this region do not block function but are especially effective at cross-linking AChRs on muscle surfaces, which increases the rate of AChR turnover. They are also effective at fixing complement, which results in focal lysis of the postsynaptic membrane (Lindstrom *et al.*, 1988; Tzartos *et al.*, 1991). Sequences homologous to the MIR are prominent in α5 and β3 subunits and binding of MIR mAbs has been detected to chicken α5 subunits (Conroy *et al.*, 1992) and to human α3 subunits (Wang and Lindstrom, unpublished). It is unknown if this cross-reaction with neuronal AChRs has any pathological significance. MIR mAbs and myasthenia gravis patient antibodies do not bind to human brain AChRs with high affinity for nicotine or for αBgt (Whiting *et al.*, 1987c). There are

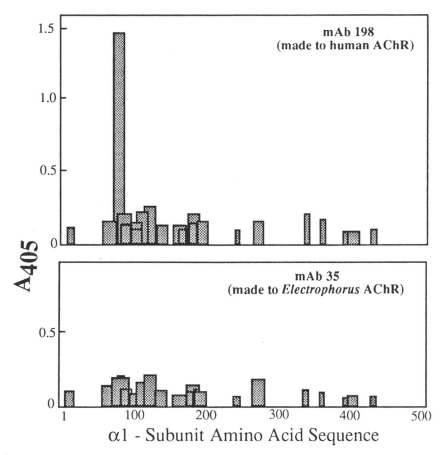

FIGURE 5. Binding of some MIR mAbs to synthetic peptides corresponding to *Torpedo* α1 sequences. Synthetic α1 peptides were covalently cross-linked to polylysine-coated microwells, and then binding of several MIR mAbs was determined by ELISA. Note that although several different MIR mAbs bind detectably to a synthetic α1 66–83 peptide, the absolutely conformation-dependent mAb does not. Adapted from Das and Lindstrom (1989).

numerous other B-lymphocyte epitopes on AChR subunits, as illustrated in Fig. 4 (Gullick and Lindstrom, 1983; Das and Lindstrom, 1991). There are also numerous T-lymphocyte epitopes on AChR subunits, but none predominates as does the MIR for antibodies on α1 subunits (Protti *et al.*, 1993).

2.4.2. Assembly of Subunits

Initial specific association of AChR subunits is thought to be mediated primarily by sequences in the N-terminal extracellular domain (Verrall and Hall, 1992; Krienkamp *et al.*, 1995). Some of the critical amino acids responsible for specific association are 106 and 115 (Gu *et al.*, 1991a,b) and 117 (Fu and Sine, 1994; Krienkamp *et al.*, 1995). Position

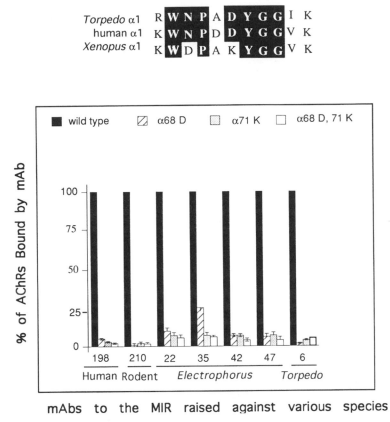

mAbs to the MIR raised against various species

FIGURE 6. Binding of some MIR mAbs to *Torpedo* AChRs with mutations in their α1 subunits at positions 68 and/or 71 to correspond to the sequence of *Xenopus* α1 subunits. The insert at the top of figure compares the sequences of α1 subunits in the region 66–76 from *Torpedo*, human, and *Xenopus*. Note that at positions 68 and 71 *Xenopus* has strongly nonconservative substitutions. Binding is determined by immune precipitation assays using AChRs labeled with ^{125}I-αBgt and mAb 142 to an epitope on the cytoplasmic surface of α1 subunits as a control for determining the 100% value. Note that substitution of the nonconserved *Xenopus* amino acid at position α68 or α71 or both prevents binding of all MIR mAbs, even mAb 35 whose binding is absolutely conformation-dependent and thus could not be localized using synthetic peptides in the experiment shown in Fig. 5. Adapted from Saedi *et al.* (1990).

117 is also thought to be adjacent or identical to amino acids forming the acetylcholine binding site, as are 145 and 150, which also appear to be involved in early steps of subunit association (Krienkamp *et al.*, 1995). Sequences in the large cytoplasmic loop appear to be involved in a late stage of subunit assembly (Yu and Hall, 1994). Evidence from several systems suggests that α1 subunits assemble first with γ, δ, or ε subunits to form complexes with intact acetylcholine binding sites at the subunit interfaces, followed by assembly with β subunits. Complete assembly proceeds more or less according to the mechanism depicted in Fig. 7 (Blount and Merlie, 1989, 1991b; Blount *et al.*, 1990; Saedi *et al.*, 1991; Gu *et al.*, 1991b). Such studies have not been conducted on neuronal AChRs,

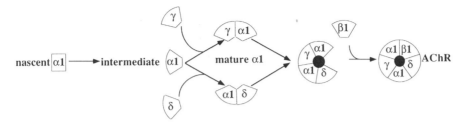

FIGURE 7. Major synthetic intermediates in assembly of muscle AChRs. α1 subunits are known to undergo a conformational maturation after they are synthesized and before association with other subunits, indicated by their acquisition of the ability to bind αBgt and MIR mAbs (Merlie and Lindstrom, 1983). Mature acetylcholine binding sites are formed at the interfaces between α1 subunits and γ and δ subunits (Blount and Merlie, 1989), whereas assembly of α and β subunits is not seen in the absence of γ or δ subunits (Saedi *et al.*, 1991). Adapted from Saedi *et al.* (1991).

but similar basic processes are likely to apply. Because of the circular arrangement of homologous subunits, specific structural features on the clockwise side of one subunit should interact with corresponding features on the counterclockwise side of the adjacent subunit, and these interactions should regulate which subunits can associate and in what order subunits can be organized around the central ion channel. Thus, for example, the plus face of the α1 subunit would interact with the minus face of the γ, δ, and ε subunits to form the acetylcholine binding sites, and the minus side of α1 subunits would interact with the plus sides of γ or β1 subunits (Krienkamp *et al.*, 1995). Muscle AChRs are expressed on the cell surface efficiently only when all subunits are present, though AChRs without γ or δ subunits can be expressed on the surface inefficiently because these subunits can poorly substitute for each other (Gu *et al.*, 1991b). Similarly, α4 subunits are expressed on the surface of *Xenopus* oocytes only when β2 subunit cDNAs are coexpressed (Wang and Lindstrom, unpublished).

2.4.3. The Acetylcholine Binding Site

The acetylcholine binding site appears to be formed from amino acids from three parts of the N-terminal extracellular domain of α subunits and also from two parts of the N-terminal domain of adjacent structural subunits, except in the case of α7, α8, and α9 subunits, which form functional homomers and thus must provide all of these contact amino acids. The initial affinity labeling experiments of Karlin and coworkers identified a disulfide-linked cysteine pair at α1 positions 192 and 193 near the acetylcholine binding site, and subsequently the most intense labeling by several reagents has been found in this vicinity (Kao *et al.*, 1984; Kao and Karlin, 1986). In subsequent affinity labeling experiments, lofotoxin labeled α1 tyrosine 190 (Abramson *et al.*, 1989), nicotine labeled α1 tyrosine 198 (Middleton and Cohen, 1991), and acetylcholine mustard labeled tyrosine 93 (Cohen *et al.*, 1991). The most extensive photoaffinity labeling and mutagenesis studies of α subunits of muscle and nerve have been done by Changeux and coworkers, who identified amino acids contributing to the acetylcholine binding site from three loops in the large N-terminal domain (using α1 numbering): loop A (tryptophan 86, tyrosine 93), loop B (tryptophan 149), and loop C (tyrosine 190, cysteines 192 and 193, and tyrosine 198) Galzi *et al.*, 1990, 1991; Galzi and Changeux, 1994; O'Leary and White, 1992; Tomaselli *et al.*, 1991). Amino acids from adjacent γ or δ subunits which contribute to the acetylcholine binding site have been identified at the homologous tryptophans γ55 and δ57 (Pederson and Cohen, 1990; Cohen *et al.*, 1992), the homologous tryptophan 54 of α7 (Corringer *et al.*, 1995), and aspartate δ180 and glutamate δ189, which also have homologues on α7 subunits (Czajkowski and Karlin, 1993, 1995; Karlin, 1993). The anionic subsite which interacts with the quaternary amine of acetylcholine contains both negatively charged amino acids and aromatic amino acids on structural subunits (Czajkowski and Karlin, 1993, 1995; Cohen *et al.*, 1992) and aromatic amino acids on α subunits (Cohen *et al.*, 1991; Galzi and Changeux, 1994). Neuronal AChRs which can function as homomers such as α7 must provide both the α subunit components of the acetylcholine binding site as well as the regions normally provided by structural subunits. Tryptophan 54 of α7 subunits is homologous to tryptophan 55 of δ or 57 of δ which are labeled by curare (Pederson and Cohen, 1990; Sine, 1993). Mutagenesis experiments on this amino acid suggest that it contributes to the binding site (Corringer *et al.*, 1995). The location of the acetylcholine binding site at the interface between subunits puts it at an ideal position to trigger cooperative changes in the orientation of subunits which may be involved in regulating opening of the cation channel and in a position which is very sensitive to allosterically medicated changes in the relative orientation of subunits (Bertrand and Changeux, 1995).

2.4.4. M1–M3 and the Cation Channel

The first of three closely spaced putative transmembrane domains, termed M1–M3, is located immediately C-terminal of amino acids in the vicinity of the α1 cysteine pair 192, 193 which contribute to the acetylcholine binding site (see Fig. 3). The putative transmembrane domains M1–M4 were initially identified by Kyte-Doolittle hydropathy plots and were generally assumed to be α-helical (Noda *et al.*, 1982). Actually mapping the transmembrane structure of the subunits has proven to be difficult and not without ambiguity (Criado *et al.*, 1985; Pederson *et al.*, 1990; Lei *et al.*, 1995) but is basically consistent with the idea of four transmembrane domains (Chavez and Hall, 1991, 1992; Anand *et al.*, 1993a). Electron diffraction studies revealed only one apparent α-helical transmembrane domain of the four domains expected, and it was suggested that all but M2 might exist in β- rather than α-helical conformations (Unwin, 1993a). The M1 domain may be in a uniquely good position to act as a linkage between ligand binding and channel gating because it extends between components of the acetylcholine binding site and M2, which is thought to line the cation channel. The M1 segment may also contribute to the channel lining (DiPaola *et al.*, 1990). During activation a conformation change takes place that involves movement around a proline-cysteine pair found in the middle of all M1 sequences (Lo *et al.*, 1991).

The M2 sequence is thought to line the cation channel for three reasons: 1) homologous amino acids in M2 from all subunits can be affinity-labeled by channel-blocking noncompetitive antagonists (Hucho *et al.*, 1986; Giraudat *et al.*, 1987); 2) *in vitro* mutagenesis studies reveal three critical sets of acidic amino acids in each M2 which are thought to form rings of negative charges lining either end of the channel that serve as selectivity filters (e.g., in the consensus sequence of Fig. 3 aspartate $\alpha 252$ and glutamate $\alpha 256$ on the cytoplasmic surface and glutamate $\alpha 277$ on the extracellular surface) (Imoto *et al.*, 1988; Konno *et al.*, 1991) and contribute to a generally stratified architecture of the channel (Bertrand *et al.*, 1993b); and 3) certain M2 amino acids were shown to be accessible from within the channel (Akabas *et al.*, 1994) and critical to its function (Lester, 1992; Cohen *et al.*, 1992; Bertrand *et al.*, 1993b). The M2 transmembrane domain appears to form an α-helix interrupted by a short extended segment with the gate near the cytoplasmic end relatively near the junction between M1 and M2 (Akabas *et al.*, 1994). The M2 sequences contain serine and threonine residues whose hydroxyl-containing side chains are thought to provide a hydrophilic environment for hydrated cations (Lester, 1992; Akabas *et al.*, 1994). The three acidic amino acids initially recognized by Imoto *et al.* (1988) to border M2 are replaced by neutral or positively charged amino acids in anion-selective receptors from the gene superfamily (Betz, 1990; Barnard, 1992). The $\alpha 7$ subunits have proven to be especially useful for channel mutagenesis studies because $\alpha 7$ forms functional homomers. Thus a single amino acid change in M2 results in a ring of five charged amino acids around the channel. Introduction of three amino acids from the M2 segments of anion-selective glycine or $GABA_A$ receptors into the $\alpha 7$ homomer was sufficient to convert $\alpha 7$ channels from cation-selective to anion selective (i.e., in the consensus sequence shown in Fig. 3, after α 7 glycine 255 insert a proline, replace $\alpha 7$ glutamate 256 with alanine, and replace $\alpha 7$ valine 268 with threonine) (Galzi *et al.*, 1992). The characteristic high Ca^{2+} permeability of $\alpha 7$ homomers can be eliminated without affecting other properties by a single mutation (in the consensus sequence of $\alpha 7$ glutamate 256 to alanine) (Bertrand *et al.*, 1993). Changing two rings of leucines at the extracellular end of $\alpha 7$M2 (in Fig. 3, $\alpha 7$ 273, 274) to threonine abolishes Ca^{2+} permeability but also increases the affinity for acetylcholine (Bertrand *et al.*, 1993). Mutation of a single amino acid in the middle of M2 has multiple effects perhaps associated with making what would otherwise be a desensitized state into a conducting state (Bertrand *et al.*, 1992; Revah *et al.*, 1991). Changing the $\alpha 7$ leucine 266 in Fig. 3 to a threonine resulted in an 80-pS conducting state in addition to the 45 pS wild-type conductance, loss of desensitization, and activation by antagonists such as curare. Clearly M2 is a critical region for channel structure and state transitions of AChRs.

The M3 domain is located immediately C-terminal of M2. Its sequence is rather well conserved, but not as well conserved as the sequence of M2, suggesting that it may function along with conserved parts of M1 and M2 in forming the core of the central transmembrane domain of AChRs.

2.4.5. The Large Cytoplasmic Domain

A large cytoplasmic domain is located between M3 and M4. Its cytoplasmic orientation has been shown by locating binding sites for sequence-specific mAbs to muscle AChR subunits to parts of this sequence (Ratnam *et al.*, 1986a,b), by the use of reporter epitopes in functionally expressed cloned muscle AChRs (Anand *et al.*, 1993a), by expression of inactive subunit fusion proteins (Chavez and Hall, 1992), and by using glycosylation consensus sequences as reporter domains (Chavez and Hall, 1991). Sequences in α1 and β1 corresponding to 564–580 in Fig. 3 have been shown to play a role in a late stage of AChR assembly (Yu and Hall, 1994).

The large cytoplasmic domain contains phosphorylation sites that might be involved in regulating AChR function, turnover, or location (Miles and Huganir, 1988; Huganir and Greengard, 1990; Wallace et al., 1991). The cytoplasmic surfaces of all muscle AChR subunits are thought to contain regions for interacting with a 43,000 molecular weight extrinsic membrane protein. This membrane protein is thought to be associated in a 1:1 ratio with AChRs and responsible for mediating their association with cytoskeletal elements that position AChRs at the tips of folds in the postsynaptic membrane (Froehner, 1993; Maimone and Merlie, 1993). It is currently unknown whether neuronal AChRs have similar associated proteins, but an associated protein has been found to copurify with glycine receptors and to be important for their localization (Kirsch et al., 1993).

The large cytoplasmic domain is the most variable region of AChR subunits in sequence between types of subunits and between species in the same type of subunits. Some of the differences in sequence between types of subunits may reflect their special functions, but the variability in sequence between species in particular may reflect the lack of strong functional constraints on parts of the structure of the large cytoplasmic domain. The most variable sequences between species in the cytoplasmic domain often provide prominent epitopes, as shown in Fig. 4 (Das and Lindstrom, 1991). The mAbs that efficiently recognize both native and denatured AChR subunits are usually directed at the large cytoplasmic domain (Das and Lindstrom, 1991; Ratnam et al., 1986a,b). This suggests that the conformation of the AChR in the vicinity of these epitopes is not rigidly constrained. Subunit-specific antibodies made to the large cytoplasmic domain of bacterially expressed subunit peptides often react well with native AChRs (Schoepfer et al., 1989, 1990).

2.4.6. M4 and the C-Terminus

The M4 segment is located at the C-terminal end of the large cytoplasmic domain. It is the most hydrophobic and least conserved of the four putative transmembrane domains and may be on the surface of the AChR where it could interact with the lipid annulus surrounding it. AChRs have been shown to retain function after replacing M4 with an extraneous transmembrane domain (Tobimatsu et al., 1987). However, recently it has been shown that mutation of α1 cysteine 592 in the consensus sequence of Fig. 3 altered AChR function, so some changes in M4, affecting perhaps the lipid interface, can alter AChR function (Lee et al., 1994).

All AChR subunits terminate shortly after M4. Several lines of evidence indicate that the C-terminal tails are expressed on the extracellular surface (McCrea et al., 1987; DiPaola et al., 1989; Dwyer, 1991; Chavez and Hall, 1992; Anand et al., 1993a).

2.5. Development of Muscle AChRs

AChRs of fetal muscle prior to innervation have the subunit stoichiometry $(\alpha 1)_2 \beta 1 \gamma \delta$, whereas at mature neuromuscular junctions a second subtype predominates in which γ has been replaced by ε (Witzemann *et al.*, 1990). The mature AChRs have a shorter channel open time and a much longer half-life for turnover (Changeux, 1990). Whether similar developmental subunit substitutions occur in neuronal AChRs is as yet unknown, but it seems likely that such changes will take place. Neuronal AChR subtypes appear early in development and the various subtypes are differentially regulated (Matter *et al.*, 1990; Hoover and Goldman, 1992; Hamassaki-Britto *et al.*, 1994a,b). As an aside, the γ subunits of *Torpedo* AChRs were initially named in order of molecular weight, but their localization in a mature synapse and the observation that they share epitopes with mammalian ε subunits (Nelson *et al.*, 1992) suggest that they are actually what are now termed ε subunits in adult muscle AChRs. The regulation of muscle AChR localization by factors like agrin is being studied in great detail (Froehner, 1993), and similar regulation will probably be discovered for neuronal AChRs. Transcriptional regulation of muscle AChR subunits has been studied in some detail (Laufer and Changeux, 1989; Changeux, 1991), and similar studies of neuronal AChRs are in progress (Bessis *et al.*, 1993; Hernandez *et al.*, 1995). These studies will not be reviewed in detail here.

3. NEURONAL AChRs THAT DO NOT BIND α-BUNGAROTOXIN

3.1. Initial History of Molecular Studies

Molecular studies of neuronal AChRs began in parallel using either cDNA probes (Boulter *et al.*, 1986; Goldman *et al.*, 1986) or MIR mAbs initially developed for muscle AChR $\alpha 1$ subunits (Swanson *et al.*, 1983; Jacob *et al.*, 1984, 1986; Smith *et al.*, 1985, 1986; Whiting and Lindstrom, 1986a,b; Henly *et al.*, 1986a,b; Stollberg *et al.*, 1986). cDNA studies of neuronal AChRs began when an $\alpha 1$ homologue that turned out to be $\alpha 3$ was discovered by low-stringency hybridization in PC12 cells (Boulter *et al.*, 1986). Then $\alpha 3$ cDNA was used in subsequent experiments that initially identified $\alpha 4$ and $\beta 2$ subunit cDNAs, then localized their mRNAs by *in situ* hybridization, and finally formed functional acetylcholine-gated cation channels by coexpression of $\alpha 3$ or $\alpha 4$ with $\beta 2$ subunits in *Xenopus* oocytes (Goldman *et al.*, 1986; Boulter *et al.*, 1987). Studies leading to the identification of neuronal AChR proteins began with the demonstration that mAb 35 and other mAbs to the MIR bound to specific regions of chicken brain (Swanson *et al.*, 1983). It was then shown that mAb 35 bound to the postsynaptic membrane at cholinergic synapses in chick ciliary ganglion neurons (Jacob *et al.*, 1984, 1986), that neuronal bungarotoxin caused downregulation of mAb 35–labeled AChRs (Smith *et al.*, 1985, 1986), that mAb 35 could be used to affinity-purify AChRs from chick brain with high affinity for nicotine (Whiting and Lindstrom, 1986a,b), and that antisera to these purified AChRs blocked the ligand-gated ion channel function of AChRs in chick ciliary ganglion neurons (Stollberg *et al.*, 1986). A library of mAbs was made to the AChRs purified from chick brain (Whiting *et al.*, 1987b). These provided subunit-specific mAbs such as mAb 270 to what are now known as $\beta 2$ subunits. Thus, there were multiple neuronal AChR subtypes differing in what are now called their α subunits, consisting of only two kinds of subunits and sharing a common $\beta 2$ structural subunit (Whiting *et al.*,

1987b). mAb 270 provided a probe that cross-reacted with rat brain AChRs and mAb 270 was used to histologically map β2 subunits in rat brain (Swanson *et al.*, 1987; Whiting *et al.*, 1987d). This localization was found to correspond to that of high-affinity nicotine binding sites in rat brain (Clarke *et al.*, 1985). Much later this histological localization of β2 subunits in rat brain was confirmed using an antiserum to a bacterially expressed fragment of β2 subunits (Hill *et al.*, 1993). mAb 270 was shown to bind to and downregulate functional AChRs in a rat pheochromocytoma cell line (Whiting *et al.*, 1987d). mAb 270 was also used to immunoaffinity-purify AChRs from rat brain, confirming that brain AChRs with high affinity for nicotine were composed to two subunits (Whiting and Lindstrom, 1987a). This also provided immunogen for preparation of another library of subunit-specific mAbs that could react with other mammalian AChRs, such as mAb 299 to what are now called α4 subunits or mAb 290 to what are now called β2 subunits (Whiting and Lindstrom, 1988).

3.1.1. Subunit Nomenclature

Differences in nomenclature initially confused the cDNA and protein studies. The cDNA studies by Heinemann and Patrick initially designated every homologous cDNA that contained homologues of the α1 cysteine pair 192, 193, as α2, α3, α4, α5, α6, *etc.*, in order of discovery, and termed subunit homologues that didn't contain this cysteine pair β2, β3, β4, *etc.*, in order of discovery (Deneris *et al.*, 1991). Ballivet and coworkers initially termed their binding subunit cDNAs the same way but termed their structural subunits non-α1 (for β2), non-α2 (for β3), *etc.* (Couturier *et al.*, 1990a). In our protein studies we followed the convention of the initial studies of purified electric organ and muscle AChR subunits, in which the subunits were named α, β, γ, δ, in order of increasing apparent molecular weight from 38,000 Da for α1, 50,000 Da for β1, 57,000 Da for γ, and 64,000 Da for δ. The problem was that when we did affinity-labeling experiments with ^{3}H-MBTA to identify the α homologue which contained a cysteine 192, 193 homologue near its acetylcholine binding site, we found that we labeled the 75,000-Da subunit rather than the 49,000-Da subunits in our purified AChRs (Whiting and Lindstrom, 1987b).

Nomenclature was resolved when we determined the N-terminal amino acid sequences of the two subunits in our purified AChRs and found that the affinity-labeled subunit of the mammalian brain high-affinity, nicotine-binding AChR was encoded by α4 cDNAs and that the structural subunit of this AChR was encoded by β2 cDNAs as shown in Fig. 8 (Whiting *et al.*, 1987a; Schoepfer *et al.*, 1988). Collectively, these studies demonstrated by using multiple subunit-specific mAbs and by N-terminal protein sequencing of the purified subunits that the mammalian brain AChR accounting for greater than 90% of the high-affinity nicotine binding was composed of α4 and β2 subunits (Whiting and Lindstrom, 1986b, 1987a, 1988; Whiting *et al.*, 1987a,b). Subsequently this subunit composition of the high-affinity nicotine-binding component of rat brain was also confirmed by affinity purification on a bromoacetylcholine affinity column and use of the same subunit-specific mAbs (Nakayama *et al.*, 1990, 1993). Much later this subunit composition was also confirmed using antipeptide sera to α4 and β2 subunits, and it was further demonstrated that the nicotine-induced increase in brain nicotine

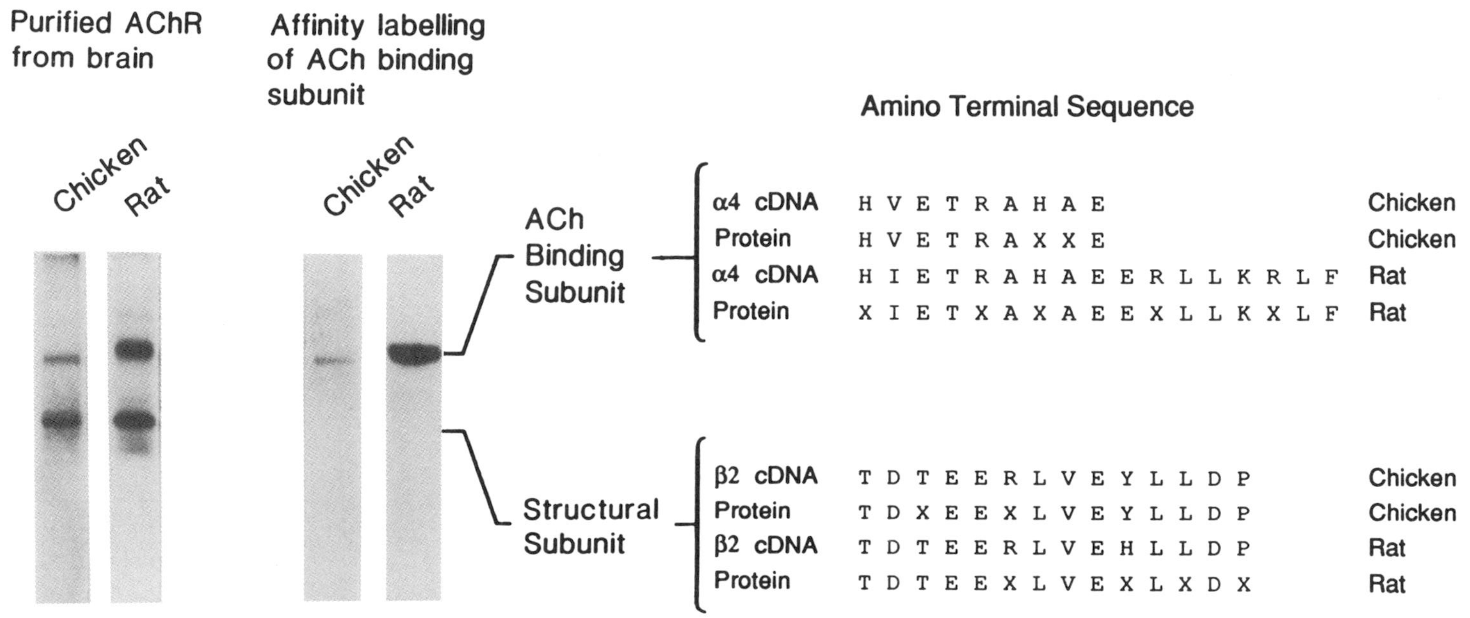

FIGURE 8. Subunit structure of α4β2 AChRs from brains of chickens and rats. The first panel shows the band pattern of silver-stained 49,000-Da and 75,000-Da subunits of immunoaffinity-purified AChRs resolved by electrophoresis on acrylamide gels in SDS. The second panel shows the autoradiograph of AChRs affinity-labeled with ³H-MBTA before electrophoresis to identify the α subunits. The third panel compares the N-terminal amino acid sequences of the purified subunits with the N-terminal sequences deduced from the corresponding cDNAs to show how the cDNAs encoding these subunits were identified. Data were taken from Whiting et al. (1987a,b,c,d, 1991a,b), Whiting and Lindstrom (1987 a,b), and Schoepfer et al. (1988), and Lindstrom et al. (1990). This figure was taken from Lindstrom et al. (1990).

binding sites resulted from an increase in α4β2 AChRs (Flores *et al.*, 1992). In summary, the current accepted nomenclature for subunits of neuronal AChRs which do not bind nicotine is α2, α3, α4, α5, α6 and β2, β3, β4, although functional AChRs with β3 or α6 subunits have not been reported.

3.2. Subunit Composition and Stoichiometry of AChR Subtypes

As detailed in the preceding section, in mammalian brain the predominant AChR subtype with high affinity for nicotine is composed of α4 and β2 subunits (Whiting and Lindstrom, 1988). In chick brain this subtype accounts for only about half of the high-affinity nicotine binding, and the other half is accounted for by β2 probably in combination with α2 subunits (Whiting *et al.*, 1987b). When α4 and β2 are coexpressed in *Xenopus* oocytes, the AChRs produced have the subunit stoichiometry $(\alpha 4)_2 (\beta 2)_3$ (Anand *et al.*, 1991; Cooper *et al.*, 1991). We demonstrated this by a method which can be applied to any cloned heteromeric receptor (Anand *et al.*, 1991). *Xenopus* oocytes injected with α4 and β2 cRNAs were incubated with ^{35}S-methionine for four or five days to ensure equilibrium labeling, then detergent extracts of oocytes were affinity-purified using a bromoacetylcholine affinity column and elution with carbamylcholine. To further ensure that only fully assembled functional AChRs were studied, the eluted AChRs were then sedimented on a sucrose gradient. The major ^{35}S-labeled peak cosedimented with native ^{3}H-nicotine–labeled brain α4β2 AChRs at about 10S, a size intermediate between monomers of *Torpedo* AChRs and the 13S dimers of *Torpedo* electric organ AChRs. These dimers are formed by a disulfide bond between δ subunits (and are unique to AChR from this species) (Karlin, 1991; Changeux, 1990). The ^{35}S-methionine–labeled AChRs from the peak fractions of the gradient were then concentrated by absorption on either a bromoacetylcholine affinity column or a mAb 270 affinity column. The AChRs were then eluted and denatured with SDS and the subunits resolved by electrophoresis. By determining the ratio of ^{35}S incorporated into each subunit and correcting for the known methionine content of each subunit, the ratio of the subunits in the assembled AChRs was determined. Control experiments with *Torpedo* AChRs gave the expected $(\alpha 1)_2 \beta 1 \gamma \delta$ stoichiometry.

Stoichiometry of other AChR subtypes is less well known. It is virtually certain from homology considerations that each AChR is formed from five subunits. The typical Hill coefficients of about 2 for dose-response curves for activation suggest that there are two acetylcholine binding sites. The binding sites are probably located at the interfaces between α subunits and structural subunits. It is known that the pharmacological properties of AChRs formed from α2, α3, or α4 subunits expressed in pairwise combination with β2 or β4 depend on both the α and β subunits (Luetje and Patrick, 1991; Papke, 1993). This implies an αβαββ organization of subunits around the channel, paralleling what is seen with the α1γα1δβ organization of muscle AChRs (Karlin, 1993). The α5 subunits may behave like β1 structural subunits. They are found in combination with α3, β2, and β4 subunits (Conroy *et al.*, 1992) but do not form functional AChRs when expressed in pairwise combination with β2 or β4 subunits (Deneris *et al.*, 1991). Their sequences in regions associated with forming acetylcholine binding sites are not well conserved (see Fig. 3). It may be that although α5 has the cysteine pair homologous with α1 192, 193, it actually serves a structural role comparable to that of β1 subunits in muscle and is similarly not associated with the subunit interfaces that form acetylcholine binding sites. The α5 subunits have been reported to be associated with a small fraction of α4β2 AChRs (Conroy et al., 1992) perhaps in an α4β2α4β2α5 organization.

Ciliary ganglion neurons express α3, α5, β2, β4, and α7 mRNAs in the ratio 3:1:1:1:7 and related neurons express this same set of subunits (Corriveau and Berg, 1993; Lukas et al., 1993). Ciliary ganglion neurons assemble several subtypes of AChRs from combinations of α3, α5, β2, and β4 subunits (α3, β4, and α5 are always coassembled and 20% of the total also have β2 associated) (Vernallis et al., 1993; Conroy and Berg, 1995), but in neither ganglia nor brain do immune precipitation experiments reveal coassembly of α7 or α8 subunits with α3, α4, α5, β2, or β4 subunits (Conroy et al., 1992; Whiting et al., 1987c). It may be that the three subtypes of AChR expressed in ciliary ganglia have the subunit arrangement α3β4α3β4α5 in the predominant synaptic form (i.e., two identical binding site interfaces), α3β4α3β2α5 in the minority synaptic form (i.e., two different binding site interfaces), and α7α7α7α7α7 in homomeric αBgt binding perisynaptic forms which are tenfold more abundant than the α3 AChRs. It has been speculated that the presence of β2 in some of the α3 AChRs might target them for presynaptic expression or make these AChRs subject to regulation by cyclic AMP (Conroy and Berg, 1995).

All of these experiments depend primarily on immunoisolation experiments using mAbs from libraries made in this laboratory (Tzartos and Lindstrom, 1980; Tzartos et al., 1981, 1983, 1986; Whiting and Lindstrom, 1987a, 1988; Whiting et al., 1987, 1991a). mAb 35 to the MIR on α1, which had initially been used to characterize autoantibody specificities in myasthenia gravis patients (Tzartos et al., 1982), was the first mAb found to cross-react with neuronal AChRs (Swanson et al., 1983). It has been the staple mAb for characterizing the α3 AChRs in chick ciliary ganglia (Jacob et al., 1984; Conroy and Berg, 1995). It was later reported to cross-react with chick α5 subunits (Conroy et al., 1992). In studies of human subunits, we have recently found that it binds to both α5 and α3 subunits expressed in Xenopus oocytes (Wang and Lindstrom, unpublished). It does not bind to α4, α7, α8, β2, or β4 subunits.

The presence of α3, β4, and α5 subunits in the same AChR molecule (Conroy et al., 1992; Conroy and Berg, 1995) is paralleled by the observation that these three subunits form a contiguous gene cluster on the same chromosome (Couturier et al., 1990a; Boulter et al., 1990). However, although γ and δ genes are also contiguous in the genome (Nef et al., 1984), in general, subunits of both muscle and neuronal AChRs which are known to associate in mature AChR proteins are encoded by genes scattered over several chromosomes (Anand and Lindstrom, 1992).

The surprising observation has been made that the neuronal AChR subunits α4, α5, α7, and β4 but not α2, α3, α8, or β3 are expressed in muscle during development, and α7 assembles without associating with α1 (Corriveau et al., 1995). It is unknown whether this reflects the need of muscle to express AChRs with properties different from those of fetal and adult muscle AChRs or whether this simply reflects a tolerable level of noise in transcriptional regulation.

3.3.   Summary of Properties of Some mAbs Useful in Studies of AChRs

It is important to provide a clear summary of the properties of the mAbs which have been used to study neuronal AChRs for several reasons: 1) the importance of mAbs in sorting out various AChR subtypes in immunoisolation experiments such as those described above, 2) their importance in immunohistochemically localizing AChR subunits (Swanson *et al.*, 1983, 1987; Jacob *et al.*, 1984, 1986; Sargent *et al.*, 1984, 1989; Watson *et al.*, 1988; Whiting *et al.*, 1991a,b; Keyser *et al.*, 1988, 1993; Britto *et al.*, 1992a,b, 1994; Hamassaki-Britto *et al.*, 1991, 1994a,b; Del Toro *et al.*, 1994), 3) the changes in AChR nomenclature after the initial descriptions of some of the libraries of mAbs, and 4) the still emerging discoveries of cross-reactions of mAbs with various AChR subunits (Conroy *et al.*, 1992; Wang and Lindstrom, unpublished). Table I summarizes the properties of some of the most commonly used mAbs in our libraries. Overall, we have made more than 300 mAbs, many of which were subunit-specific and species-specific for AChRs from electric organs, so many of these mAbs are not listed in the table. The mAbs which have been repeatedly used on neuronal AChRs or which were made to neuronal AChRs are selectively considered in this table.

3.4.   Pathological Significance

The major known pathological significance of neuronal nicotinic AChRs is that their interaction with nicotine is responsible for dependence on tobacco (Bock and Marsh, 1990). Tobacco use has been predicted to cause a quarter of a billion premature deaths worldwide by the turn of the century (Peto *et al.*, 1992). Upregulation of brain $\alpha 4 \beta 2$ AChRs by up to twofold is a characteristic result of chronic exposure to nicotine in rats (Schwartz and Kellar, 1983, 1985; Wonnacott, 1990; Flores *et al.*, 1992; Marks *et al.*, 1985, 1992, 1993). In human smokers, brain high-affinity binding sites for nicotine (presumably $\alpha 4 \beta 2$ AChRs) are similarly increased (Benwell *et al.*, 1988).

A substantial decrease in brain AChRs is associated with both Alzheimer's and Parkinson's diseases (Whitehouse *et al.*, 1988; Wells *et al.*, 1993; Lange *et al.*, 1993). This AChR loss may be secondary to cell death rather than a primary pathological mechanism as in myasthenia gravis. However the characteristic large losses in nicotine binding sites in these chronic neurodegenerative disease have provoked a new interest in developing agonists for $\alpha 4 \beta 2$ AChRs (Arneric *et al.*, 1994; Decker *et al.*, 1994) and $\alpha 7$ AChRs (Hunter *et al.*, 1994) as potential cognitive enhancers.

Nicotine effects in several diseases suggest that nicotinic AChRs may have some role to play in their pathology or therapy. Experiments strongly suggest that nicotine released from transdermal patches to patients receiving the dopamine blocker halo-peridol is effective in reducing the severity and frequency of tics in Tourette's syndrome (Silver *et al.*, 1995). It is unclear whether this reflects a role of activating or desensitizing AChRs in modulating dopamine release or in some consecutive postsynaptic link in the neural pathways involved. Epidemiological studies indicate that not smoking is by far the strongest environmental risk factor for Parkinson's disease (Baron, 1995). Why this should be so is unknown, but the knowledge that loss of dopamine due to degeneration of the substantia nigra in this disease and evidence that presynaptic AChRs can modulate the release of dopamine (Wonnacott *et al.*, 1990) suggest that self-medication with nicotine through smoking may play a role.

It is especially interesting that a missense mutation in M2 of α4 appears to be responsible for causing autosomal dominant nocturnal frontal lobe epilepsy (Steinlein *et al.*, 1994, 1995). This is the first genetically determined epilepsy which has been genetically mapped. Its chromosomal location is 20q13.3. Further studies of these patients who presumably lack functional α4β2 AChRs would prove very instructive for revealing the functional roles of α4β2 AChRs.

3.5. Upregulation of α4β2 AChRs by Chronic Exposure to Nicotine

Tobacco use or chronic exposure to nicotine causes an increase in brain α4β2 AChRs up to twofold (Schwartz and Kellar, 1985; Benwell *et al.*, 1988; Flores *et al.*, 1992). It has been suggested that this increase in AChRs is an adaptive response of neurons to maintain nicotinic transmission despite the accumulation of desensitized AChRs due to chronic exposure to agonist (Wonnacott, 1990). In fact we have found that agonist-induced upregulation of α4β2 AChRs is an intrinsic property of this protein which is exhibited when the cloned α4β2 AChRs are expressed in *Xenopus* oocytes or in a transfected fibroblast cell line called M10 (Peng *et al.*, 1994a). This has also been confirmed by others (Zhang *et al.*, 1994).

Upregulation in brain is not due to an increase in brain α4 or β2 mRNA (Marks *et al.*, 1992). Similarly, upregulation in M10 cells does not occur at the transcriptional level. Demonstration of AChR upregulation by nicotine in *Xenopus* oocytes injected with fixed amounts of α4 and β2 mRNA is the best proof that upregulation occurs posttranscriptionally (Peng *et al.*, 1994a).

Chronic exposure to nicotine or other agonists causes an increase in the number of α4β2 AChRs in the surface membrane as a result of a decrease in the rate of destruction of the AChRs, as shown in Figs. 9–11 (Peng *et al.*, 1994a). This pathological change is an interesting contrast with myasthenia gravis, where chronic exposure to autoantibodies causes a decrease in the amount of muscle AChRs by cross-linking the AChRs and increasing their rate of endocytosis (Lindstrom *et al.*, 1988). The mechanism by which agonists cause upregulation of expressed α4β2 AChRs is not triggered by ion flow through the AChR channel because the open channel–blocking, noncompetitive antagonist mecamylamine also causes upregulation by itself and synergistically with nicotine (Peng *et al.*, 1994a). This parallels the effect of mecamylamine *in vivo* (Collins *et al.*, 1994). The competitive antagonist curare does not cause upregulation and prevents the nicotine effect. Thus, nicotine and other agonists as well as mecamylamine seem to cause a change in the AChR conformation which causes it to be turned over more slowly.

The EC_{50} for nicotinic-induced upregulation of α4β2 AChRs (2×10^{-7} M, see Fig. 9) is pathologically significant because it is essentially identical to the mean steady-state serum concentration of nicotine in smokers (Benowitz *et al.*, 1990). The steady-state concentration of nicotine in smokers is interesting because it is near the EC_{50} for activation of α4β2 AChRs by nicotine when applied acutely (3.5×10^{-7} M), but much greater than the K_D for binding nicotine by desensitized α4β2 AChRs (3.9×10^{-9} M, see Table II) (Peng *et al.*, 1994a). Thus, at a steady-state nicotine concentration of 2×10^{-7} M

TABLE I

Properties of Some mAbs Used to Study AChRs[a]

mAb	Specificity			Isotype	Initial description	Immunogen	Comments
	Subunit	Epitope	Species				
6	α1	MIR	Torpedo, Electrophorus, weak on mammals	IgG1	Tzartos and Lindstrom, 1980	Torpedo AChR	Studied most until it was superseded by mAbs 35 and 210, which are more reactive with mammalian AChRs and thus better for both studies of EAMG and reaction with neuronal AChRs. Binds to chick brain AChRs competitively with mAb 35 (Swanson et al., 1983).
22	α1 but also?	MIR	Electrophorus, frog, goldfish	IgG2b	Tzartos et al., 1981	Electrophorus AChR	Used for histology in Rana ganglia, retina, and brain (Sargent et al., 1989). This was the best for frog neuronal AChRs of 42 mAbs to electric organ AChRs tested by Sargent et al. (1989), but 18 other MIR mAbs also labeled strongly. Binds an αBgt-labeled AChR from goldfish retina and optic tectum (Henly et al., 1986a,b). Binds to chick brain AChRs competitively with mAb 35 (Swanson et al., 1983).
35	α1; α5 chick, human; α3 human (Wang and Lindstrom, unpublished)	MIR	α1 all species tested but Xenopus	IgG1	Tzartos et al., 1981	Electrophorus AChR	Used histologically in chick (e.g., Swanson et al., 1983, Jacob et al., 1984, 1986). Used for immunoaffinity purification (Whiting and Lindstrom, 1986a). Used for AChR quantitation (Smith et al., 1985, 1986; Conroy and Berg, 1995). Used to measure specificities of autoantibodies in MG patients (Tzartos et al., 1982). Used to locate MIR by electron crystallography (Beroukhim and Unwin, 1995). Does not bind well to denatured AChR (Das and Lindstrom, 1989). Does not bind to α3, α4, α7, α8, β2 (Conroy et al., 1992). Passively transfers EAMG (Tzartos et al., 1987)

mAb	Specificity	Epitope	Species	Isotype	Reference	Immunogen	Description
47	α1 but also?	MIR	Electrophorus, frog, goldfish	IgG2a	Tzartos et al., 1981	Electrophorus AChR	Binds an αBgt-labeled AChR from goldfish retina and optic tectum (Henly et al., 1986a,b, 1988) Binds to Rana neuronal AChRs (Sargent et al., 1989). Binds to chick brain AChR competitively with mAb 35 (Swanson et al., 1983).
198	α1	MIR	Human, Torpedo, etc.	IgG2a	Tzartos et al., 1983	Purified human muscle AChR	Binds to native and denatured α1 with higher affinity for MIR synthetic peptide than mAb 210 (Das and Lindstrom, 1989). Binds to chick brain AChR competitively with mAb 35 (Swanson et al., 1983).
203	α1	MIR	Human, Torpedo, etc.	IgG2a	Tzartos et al., 1983	Purified human muscle AChR	Binds to native and denatured α1.
207	α1	MIR	Human	IgG2a	Tzartos et al., 1983	Purified human muscle AChR	Binds to native and denatured α1.
208	α1	MIR	Human, Torpedo, etc.	IgG2a	Tzartos et al., 1987	Purified mouse and bovine muscle AChR	Binds to native and denatured α1. Passively transfers EAMG (Tzartos et al., 1987).
210	α1; α5 chick, human; α3 human (Wang and Lindstrom, unpublished)	MIR	α1, all species tested but Xenopus	IgG1	Tzartos et al., 1987	Bovine and mouse AChR	Used histologically in chick (e.g., Keyser et al., 1988). Binds weakly to denatured α1 (Das and Lindstrom, 1989) and strongly to native α1 (Saedi et al., 1990). Does not bind to chick α3, α4, α7, α8, β2 (Conroy et al., 1992). Passively transfers EAMG (Tzartos et al., 1987).
61	α1	Within α1 (371–386) (Ratnam et al., 1986a)	Torpedo, Electrophorus, mouse, human	IgG2a	Tzartos et al., 1981	Denatured Electrophorus AChR	Binds to native and denatured α1. Very useful for studying AChR synthesis and assembly (Merlie and Lindstrom, 1983; Blount and Merlie, 1988, 1989, 1990, 1991a,b; Blount et al., 1990; Luther et al., 1989; Verrall and Hall, 1992; Gu et al., 1991a,b).

(continued)

TABLE I
(*Continued*)

| mAb | Subunit | Specificity | | Isotype | Initial description | Immunogen | Comments |
		Epitope	Species				
142	α1	α1 (359–365) (Das and Lindstrom, 1991), cytoplasmic surface (Ratnam et al., 1986b; Sargent et al., 1984)	Torpedo, Rana	IgG2a	Tzartos et al., 1986	Denatured Torpedo AChR	Binds to native and denatured AChR. Useful as a reporter epitope (Anand et al., 1993a). Useful histologically (Sargent et al., 1984).
147	α1	α1 (368–375) (Das and Lindstrom, 1991), cytoplasmic surface (Sargent et al., 1984)	Torpedo, Rana, Xenopus	IgG2a	Tzartos et al., 1986	Denatured Torpedo AChR	Binds to native and denatured AChR. Useful histologically (Sargent et al., 1984).
149	α1	α1 (341–347) (Das and Lindstrom, 1991) cytoplasmic surface (Sargent et al., 1984)	Torpedo, Rana, Xenopus	IgM	Tzartos et al., 1986	Denatured Torpedo AChR	Binds to native and denatured AChR. Useful histologically (Sargent et al., 1984).
236	α1	α1 (152–167) (Das and Lindstrom, 1991)	Torpedo	IgG2a	Criado et al., 1985	α1 synthetic peptide 152–167	Does not bind well to native AChR, but does bind well to denatured α1. Useful as a reporter epitope (Anand et al., 1993a).
321–326	α2	—	Chick	?	Not previously reported from studies by Schoepfer, Whiting, and Lindstrom	Bacterially expressed chick α2 large cytoplasmic domain fragment	Not well characterized. Does not bind detergent-solubilized nicotine-labeled chick brain AChRs well.

mAb	Subunit	Epitope	Species	Ig class	Reference	Antigen	Comments
313–315	α3	Within α3 (315–441)	Chick	IgG2a	Whiting et al., 1991a	Bacterially expressed large cytoplasmic domain fragment α3 (315–441) (Schoepfer et al., 1989)	Binds to native and denatured detergent-solubilized AChR. Used to histologically localize α3 AChRs (Whiting et al., 1991a,b; Britto et al., 1992a,b; Hamassaki-Britto et al., 1994a,b). Does not bind α1, α4, α5, α7, α8, β2 (Conroy et al., 1992).
316–317	α3	Within α3 (315–441)	Chick	?	Whiting et al., 1991a	Bacterially expressed large cytoplasmic domain fragment α3 (315–441) (Schoepfer et al., 1989)	Binds only to denatured α3.
284–285	α4	—	Chick	IgG2a	Whiting et al., 1987b	Purified chick brain AChR	Binds native and denatured chick α4.
286	α4	—	Chick, rat, human	IgM	Whiting et al., 1987b	Purified chick brain AChR	Binds native and denatured chick α4.
289	α4	Within α4 (330–511) (Whiting et al., 1991a)	Chick	IgM	Whiting et al., 1987b	Purified chick brain AChR	Binds native and denatured chick α4. Does not bind to α1, α3, α5, α6, α8, β2 (Conroy et al., 1992).
292	α4	—	Rat	IgG1	Whiting and Lindstrom, 1988	Purified rat brain AChR	Binds native and denatured α4.
293	α4	—	Chicken, rat, human	IgG2a	Whiting and Lindstrom, 1988	Purified rat brain AChR	Binds native and denatured α4.
299	α4	Extracellular (Peng et al., 1994a; Wang and Lindstrom, unpublished)	Chicken, rat, human	IgG1	Whiting and Lindstrom, 1988	Purified rat brain AChR	Binds native and denatured α4. Can be used to quantitate α4 AChRs on the cell surface (Peng et al., 1994a).
267	α5 and α2?	—	Chicken	IgG2a	Whiting et al., 1987b	Purified chick brain AChR	Binds to denatured α5 (Conroy et al., 1992) but not to native AChR (Whiting et al., 1987b) or α1, α3, α4, α7, α8, β2 (Conroy et al., 1992).

(continued)

TABLE I
(Continued)

| mAb | Specificity | | Species | Isotype | Initial description | Immunogen | Comments |
	Subunit	Epitope					
268	α5	α5 (91–100) (Wahlsten, Conti-Fine, Lindstrom, unpublished)	Chick, human	IgG1/2a	Whiting et al., 1987b	Purified chick brain AChR	Binds to denatured α5 (Conroy et al., 1992) but not to native AChR (Whiting et al., 1987b) or α1, α3, α4, α7, α8, β2 (Conroy et al., 1992).
306	α7	Within α7 (380–400) (McLane et al., 1992)	Chick, rat, human	IgG1	Schoepfer et al., 1990	Purified chicken and rat αBgt binding AChR	Binds to native and denatured chick α7 (Schoepfer et al., 1990) and to denatured rat and human α7 (Peng et al., 1994b; Del Toro et al., 1994) but not well to native mammalian α7. Useful for histology (Keyser et al., 1993; Britto et al., 1992a,b, 1994; Del Toro et al., 1994; Hamassaki-Britto et al., 1994b).
307	α7	Within α7 (380–400) (McLane et al., 1992)	Chick, rat, human	IgG1	Schoepfer et al., 1990	Purified chicken and rat αBgt binding AChR	Binds to native and denatured chick α7 (Schoepfer et al., 1990) and to denatured rat and human α7 (Peng et al., 1994b; Del Toro et al., 1994) but not well to native mammalian α7. Useful for histology (Keyser et al., 1993; Britto et al., 1992a,b, 1994; Del Toro et al., 1994; Hamassaki-Britto et al., 1994b).
318	α7	Within α7 (380–400) (McLane et al., 1992)	Chick	IgG	Schoepfer et al., 1990	Bacterially expressed α7 large cytoplasmic domain	Similar to mAb 306 except not reactive with mammalian. Useful for immunoisolation (Anand et al., 1993a,b).
319	α7	Within α7 (365–384) (McLane et al., 1992)	Chick, rat, human	IgG	Schoepfer et al., 1990	Bacterially expressed α7 large cytoplasmic domain	Similar to mAb 306. Does not bind to α8 (Conroy et al., 1992).
320	α7	Within α7 (380–400) (McLane et al., 1992)	Chick	IgG	Schoepfer et al., 1990	Bacterially expressed α7 large cytoplasmic domain	Binds native and denatured (Schoepfer et al., 1990).

No.	Subunit	Epitope/location	Species	Ig class	Reference	Antigen used	Properties
305	α8	Extracellular (Anand and Lindstrom, unpublished)	Chicken	IgG2c	Schoepfer et al., 1990	Purified chicken αBgt binding AChR	Has high affinity, but only for native α8. Used for immunoisolation (Schoepfer et al., 1990; Keyser et al., 1993; Anand et al., 1993b; Gerzanich et al., 1994) and for histological localization (Keyser et al., 1993; Britto et al., 1992a,b, 1994; Hamassaki-Britto et al., 1994a,b).
308	α8	Within α8 (323–342) (McLane et al., 1992)	Chicken	IgG2b	Schoepfer et al., 1990	Bacterially expressed large cytoplasmic domain fragment	Binds native and denatured α8. Used for histology like mAb 305. Does not bind α4, α7, β (Conroy et al., 1992).
309	α8	Within α8 (323–342) (McLane et al., 1992)	Chicken	IgG	Schoepfer et al., 1990	Bacterially expressed large cytoplasmic domain fragment	Binds native and denatured α8.
310	α8	Within α8 (353–372) (McLane et al., 1992)	Chicken	IgG	Schoepfer et al., 1990	Bacterially expressed large cytoplasmic domain fragment	Binds native and denatured α8.
311	α8	Within α8 (323–342) (McLane et al., 1992)	Chicken	IgG2b	Schoepfer et al., 1990	Bacterially expressed large cytoplasmic domain fragment	Binds native and denatured α8.
312	α8	—	Chicken	IgG2a	Schoepfer et al., 1990	Bacterially expressed large cytoplasmic domain fragment	Binds native and denatured α8.
73	β1	External surface	Bovine, rat, mouse	IgG1	Tzartos et al., 1986	Purified bovine muscle AChR	Binds to AChR in surface membrane of muscle, but doesn't cross-link them unless anti-antibody is added (Tzartos et al., 1985). Doesn't passively transfer EAMG (Tzartos et al., 1987). Used to study specificities of autoantibodies in MG patients (Tzartos et al., 1982).
111	β1	β1 (360–410) cytoplasmic surface (Sargent et al., 1984; Ratnam et al., 1986b	Torpedo, Rana, Xenopus, rat, mouse, human	IgG1	Tzartos et al., 1986	Purified denatured Torpedo AChR	Binds to native and denatured β1. Useful for histology (Sargent et al., 1984). Used to study assembly of subunits (Saedi et al., 1991).

(continued)

TABLE I
(Continued)

mAb	Specificity			Isotype	Initial description	Immunogen	Comments
	Subunit	Epitope	Species				
124	β1	Cytoplasmic surface (Sargent et al., 1984; Ratnam et al., 1986b)	Torpedo, Rana, Xenopus, bovine, human	IgG1	Tzartos et al., 1986	Purified denatured Torpedo AChR	Binds to native and denatured β1. Useful for histology (Sargent et al., 1984). Used to study assembly of AChR subunits (Yu and Hall, 1994; Gu et al., 1991a,b).
148	β1	β1 (360–410) cytoplasmic surface (Sargent et al., 1984; Ratnam et al., 1986b)	Torpedo, Rana, Xenopus, bovine, human	IgG2a	Tzartos et al., 1986	Purified denatured Torpedo AChR	Binds to native and denatured β1. Useful for histology (Sargent et al., 1984). Used to study assembly of α and β subunits (Smith et al., 1987; Conroy et al., 1990).
172	β1 and?	β1 C-terminus (Ratnam et al., 1986b)	Torpedo, goldfish	IgG1	Tzartos et al., 1986	Torpedo, AChR	Binds to a goldfish neuronal AChR that binds nicotine but not αBgt (Henley et al., 1988).
270	β2	Extracellular (Whiting et al., 1987d)	Chicken, mouse, rat	IgG2a	Whiting et al., 1987b	Chicken brain AChR purified using mAb 35	Has high affinity for native AChR but also works on denatured AChR. Used to immunoisolate and pharmacologically characterize AChRs (Whiting and Lindstrom, 1986b; Whiting et al., 1987d). Used to histologically localize AChRs (Swanson et al., 1987; Whiting et al., 1987d). Used to immunoaffinity purify AChRs (Whiting and Lindstrom, 1987a; Schoepfer et al., 1988). Does not bind α1, α4, α7, α8 (Conroy et al., 1992).
287	β2	Intracellular (Wang and Lindstrom, unpublished)	Chicken	IgM	Whiting et al., 1987b	Chicken brain AChR purified using mAB 35	Does not bind native AChR.
290	β2	Extracellular (Peng et al., 1994a)	Chicken, rat, bovine, human	IgG1	Whiting and Lindstrom, 1988	AChRs purified from rat brain using mAb 270	Binds only native not denatured β2. Useful for immunoisolating and surface labeling of β2-containing AChRs (Peng et al., 1994a) Useful for histology (Sargent et al., 1987).

mAb	Subunit	Epitope location	Species	Ig class	Reference	Antigen	Comments
291	β2	Intracellular (Wang and Lindstrom, unpublished)	Chicken, rat	IgG1	Whiting and Lindstrom, 1988	AChRs purified from rat brain using mAb 270	Binds only native, not denatured β2.
295	β2	Extracellular (Wang and Lindstrom, unpublished)	Chicken, rat, bovine, human	IgG2a	Whiting and Lindstrom, 1988	AChRs purified from rat brain using mAb 270	Binds only native, not denatured β2.
297	β2	Extracellular (Wang and Lindstrom, unpublished)	Chicken, rat, bovine, human	IgG2a	Whiting and Lindstrom, 1988	AChRs purified from rat brain using mAb 270	Binds only native, not denatured β2.
298	β2	Extracellular (Wang and Lindstrom, unpublished)	Chicken, rat, bovine, human	IgG2a	Whiting and Lindstrom, 1988	AChRs purified from rat brain using mAb 270	Binds only native, not denatured β2.
7	γ and δ	γ (374–384) (Nelson et al., 1992) cytoplasmic surface (Sargent et al., 1984)	Torpedo, Rana, Xenopus, canine	IgG1	Tzartos and Lindstrom, 1980	Torpedo AChR	Binds native and denatured AChR, weaker on mammalian. Can be used for mammalian histology (Nelson et al., 1992) as well as amphibian histology (Sargent et al., 1984).
66	γ	Extracellular ?	Bovine human (weakly)	IgG2a	Tzartos et al., 1986	Bovine AChR	Used to study specificities of autoantibodies in MG patient sera (Tzartos et al., 1982). Useful histologically (Tzartos et al., 1986).
145	γ	γ (390–405) (Nelson et al., 1992)	Torpedo	IgG2a	Tzartos et al., 1986	Torpedo AChR	Torpedo-specific mAb binds both native and denatured γ.
154	γ Torpedo, ε mammals (weakly) (Nelson et al., 1992)	γ/ε (364–374) (Nelson et al., 1992), cytoplasmic surface (Sargent et al., 1984)	Torpedo, Rana, Xenopus, canine	IgG1	Tzartos et al., 1986	Purified denatured Torpedo AChR	Binds native and denatured, strong on Torpedo but weak on mammals.

(continued)

TABLE I
(Continued)

	Specificity				Initial		
mAb	Subunit	Epitope	Species	Isotype	description	Immunogen	Comments
168	γ and ε	γ/ε (364–374) (Nelson et al., 1992), cytoplasmic surface (Sargent et al., 1984)	Torpedo, Rana, Xenopus, bovine	IgG1	Tzartos et al., 1986	Torpedo AChR	Binds native and denatured Torpedo AChR, much weaker on mammalian. Can be used for mammalian and amphibian histology (Sargent et al., 1984; Nelson et al., 1992).
137	δ	Within δ (300–410) (Ratnam et al., 1986b)	Torpedo	IgM	Tzartos et al., 1986	Torpedo AChR	Binds native and denatured δ and cross-reacts weakly with mammalian.
141	δ	Within δ (300–410) (Ratnam et al., 1986b), cytoplasmic surface (Sargent et al., 1994)	Torpedo, Rana, Xenopus	IgG2a	Tzartos et al., 1986	Torpedo AChR	Binds native and denatured (Tzartos et al., 1986). Can be used histologically on amphibians (Sargent et al., 1984)
166	δ	Within δ (300–410) (Ratnam et al., 1986b), cytoplasmic surface (Anderson et al., 1983)	Torpedo, Rana, Xenopus	IgG2a	Tzartos et al., 1986	Torpedo AChR	Binds native and denatured Tzartos et al., 1986. Can be used histologically on amphibians (Sargent et al., 1984).

[a]Many other mAbs have been made in this laboratory, and most are listed in the various articles included in the "initial description" part of the table. The mAbs listed in this table include virtually all of those now known to react with neuronal AChRs, but many subunit-specific mAbs which were raised against electric organ AChRs and which do not react well with mammalian muscle AChRs have been excluded. Not all of the properties of the mAbs listed in the table are known or listed. It is reasonable to infer, but not guaranteed, that similar mAbs will have similar properties, even if these properties have not all been explicitly tested for all of these mAbs. For example, all mAbs to the MIR on α1 compete for binding to muscle AChR α1 subunits, but it is clear that there are differences between mAbs to the MIR in species specificity and dependence of binding on the native conformation of the α1. Thus, although some mAbs to the MIR have been shown to bind to α3 or α5 subunits, or might in the future be found to bind to β3 subunits, it does not necessarily follow that all mAbs to the MIR will have those properties. In all studies the rat was the immunized species, except for mAbs 306 and 307 for which the mouse was used.

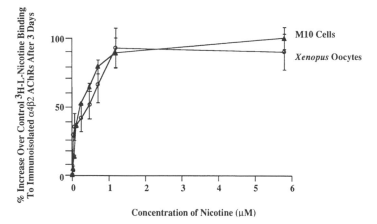

FIGURE 9. Dose dependence of nicotine-induced upregulation. M10 cells are mouse fibroblasts permanently transfected with chicken α4 and β2 subunits that exhibit the pharmacological properties of brain α4β2 AChRs and that function electrophysiologically (Whiting *et al.*, 1991b; Periera *et al.*, 1994). *Xenopus* oocytes were injected with cRNA for α4 and β2 RNAs. After three days in the indicated concentrations of nicotine, α4β2 AChRs were solubilized using Triton X-100, immunoisolated on microwells coated with mAb 290 to their β2 subunits and quantitated using 20 nM ^{3}H-nicotine. Data are reproduced from Peng *et al.* (1994a).

all α4β2 AChRs in a tobacco user should be desensitized and the net behavioral effect should reflect inhibition of these AChRs. Short-term desensitization is a reversible process. However, we found that after exposing oocytes expressing α4β2 AChRs to high concentrations of nicotine for three days, their AChR function was greatly reduced even though the total number of their AChRs was doubled (Peng *et al.*, 1994a). This reduction only recovered over a period of three days, which is about the rate expected for synthesis of new AChRs, as shown in Fig. 12. This suggests that long-term exposure to nicotine can result in a permanently inactivated AChR conformation. It is unknown whether this reflects the same nicotine-induced conformation that is turned over slowly. The loss of functional α4β2 AChRs on chronic exposure to nicotine due to both reversible and permanent desensitization provides an explanation for tolerance. Thus, chronic tobacco users can tolerate much higher nicotine concentrations than nonusers even though they may have twice as many α4β2 AChRs, because most of these excess AChRs can bind nicotine but don't work.

Upregulation by agonists is not a universal property of all AChR subtypes. AChRs in cultured muscle cells have been reported to be downregulated about 40% by chronic exposure to agonists (Noble *et al.*, 1978; Gardner and Fambrough, 1979; Appel *et al.*, 1981). However human muscle α1β1γδ AChRs expressed in the TE671 rhabdomyosarcoma cell line are upregulated by nicotine (Siegel and Lukas, 1988; Luther *et al.*, 1989). The α3β4α5 AChRs expressed in chick ciliary ganglion are reduced about 30% by chronic exposure of cultures to carbamylcholine (Smith *et al.*, 1986). Brain α-bungarotoxin binding sites (presumably α7AChRs; Anand *et al.*, 1993b) are upregulated by chronic nicotine treatment (Marks *et al.*, 1985), but to a lesser degree than α4β2 AChRs. The α7 AChRs in cultured rat hippocampal neurons treated with high concentrations of

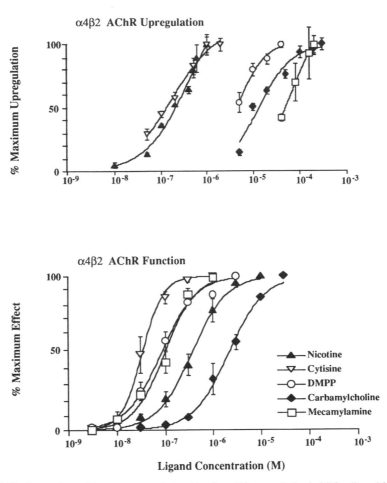

FIGURE 10. Comparison of dose-response relationships for α4β2 upregulation in M10 cells and function in *Xenopus* oocytes. Expression of α4β2 AChR expression in M10 cells grown to confluence was induced by 1 μM dexamethasone for three days, then the indicated concentrations of ligands were added for three days, and finally AChRs were quantitated in solid phase assays using ³H-nicotine to label AChRs immunoisolated through their β2 subunit as in Fig. 9. Voltage-clamp measurements on *Xenopus* oocytes normalized the maximum currents to saturating agonist concentrations. Blockage by mecamylamine was measured at 1 μM nicotine. Data are reproduced from Peng *et al.* (1994a).

nicotine (10 μM) for long periods (14 days) increase slightly (40%) (Barrantes *et al.*, 1995).

3.6. Epibatidine, a Potent Agonist for Neuronal AChRs

Good labeling reagents for AChRs can be hard to come by. The high affinity of αBgt for muscle AChRs and use of ¹²⁵I-αBgt as a labeling reagent for quantitating these

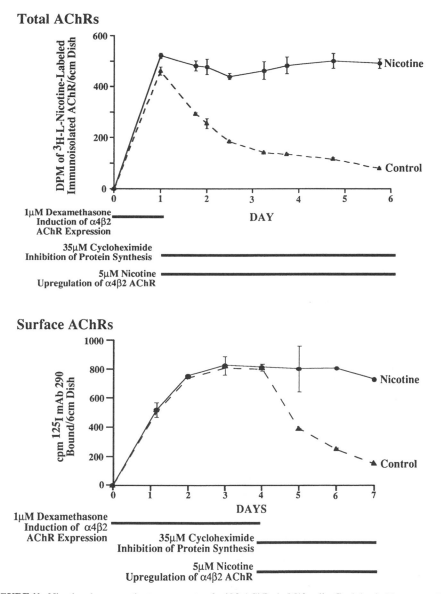

FIGURE 11. Nicotine decreases the turnover rate of $\alpha 4 \beta 2$ AChRs in M10 cells. Cycloheximide was used to inhibit the synthesis of new $\alpha 4 \beta 2$ AChRs in dexamethasone-induced M10 cells. In the experiment shown at the top, cycloheximide was added one day after dexamethasone induction. Then total $\alpha 4 \beta 2$ AChRs solubilized from the cells was measured by ^{3}H-nicotine binding to immunoisolated AChRs in microwells. In the experiment shown in the lower panel, dexamethasone induction of $\alpha 4 \beta 2$ AChR synthesis was allowed to maximize over four days before nicotine was added to one set of cultures. In this case only surface AChRs were measured by binding of ^{125}I-labeled mAb290 to $\beta 2$ subunits in intact cells. Data are from Peng *et al.* (1994a).

TABLE II.
Pharmacological Properties of α4β2 AChRs

| | Upregulation[a] $EC_{50}(\mu M)$ | Function[a] | | Equilibrium Binding $K_I(\mu M)$ |
		Activation $EC_{50}(\mu M)$	Blocking EC_{50} (μM)	
Agonists				
cytisine	0.17 ± 0.02	0.031 ± 0.003	—	0.00014 ± 0.00003^b
nicotine	0.21 ± 0.04	0.35 ± 0.02	—	0.0039 ± 0.00021^b
DMPP	3.0 ± 0.3	0.073 ± 0.01	—	0.0094 ± 0.0002
carbamylcholine	15.0 ± 4.0	2.5 ± 0.3	—	0.36 ± 0.013^b
Competitive antagonist				
curare	None	—	2.7 ± 0.5	25.0 ± 14.0^b
Noncompetitive antagonist				
mecamylamine	65.0 ± 6.0	—	0.37 ± 0.5	$>1000^b$

[a]Data from Fig. 10
[b]Data from Whiting et al., 1991b.

AChRs have been critical for biochemical studies. This value continues in studies of neuronal α7, α8, and α9 AChRs, which also bind αBgt. Tritiated nicotine has been useful for labeling α4β2 AChRs, though it falls an order of magnitude or two short of the affinity of αBgt for muscle AChRs. Labeling α3 AChRs has been more of a problem. Neuronal bungarotoxin is available in limited supply and it doesn't bind well to detergent-solubilized AChRs (Fiordalisi et al., 1991; Sorenson and Chiappinelli, 1992; Nooney et al., 1992). Iodine-125-labeled mAb35 has been used extensively to quantitate α3 AChRs (Smith et al., 1985; Conroy and Berg, 1995), but a labeled cholinergic ligand that could be used in competitive binding experiments would be more convenient.

Epibatidine, exo-2-(6-chloro-3-pyridyl)-7-azabicyclo[2.2.1]heptane, was initially discovered by John Daly in the skin of Ecuadoran poison frogs (Spande et al., 1992). The frogs apparently derive it from an unknown component of their diet in the wild (Daly, 1995). Epibatidine has now been synthesized and is commercially available (Huang and Shen, 1993; Fletcher et al., 1994). It was initially discovered to be 200-fold more potent than morphine in preventing nociception. Later it was discovered to be a potent nicotinic agonist (Badio and Daly, 1994) and to elicit a variety of nicotinic effects (Sullivan et al., 1994). Its antinociceptive effects, along with those already known for nicotine, suggest an important role for nicotinic AChRs in pain perception (Aceto et al., 1982; Yang et al., 1992). Epibatidine has been found to be a potent agonist on cultured rat hippocampal neurons (Alkondon and Albuquerque, 1995). It has also been used in brain membrane labeling studies and found to interact in several brain regions with sites of differing affinity (Houghtling et al., 1994). The greatest current value of epibatidine from many research perspectives is as a labeling reagent.

We have investigated the interaction of epibatidine with several types of cloned AChRs expressed in Xenopus oocytes (Gerzanich et al., 1995). Epibatidine is the most potent neuronal AChR agonist that we have studied and is an excellent labeling reagent,

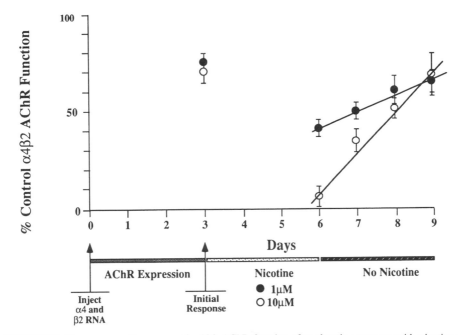

FIGURE 12. Time course of recovery of α4β2 AChR function after chronic treatment with nicotine of *Xenopus* oocytes. After measuring the initial response to 100μM acetylcholine, one group was incubated with 1μM nicotine, another with 10μM nicotine, and a third group was left untreated as a control. After three days the oocytes were washed and then electrophysiological responses to 100μM acetylcholine were measured at the indicated times over the next three days. Data are from Peng *et al.* (1994a).

especially for α3 AChRs that have otherwise been difficult to quantitate biochemically. The following brief review of our studies with epibatidine also shows some general pharmacological properties of various types of AChRs.

Figure 13 shows that epibatidine is more potent than nicotine as an agonist for α4β2 AChRs. The prolonged slow onset of responses to low concentrations of epibatidine probably reflects the low concentration used, whereas the prolonged decay probably reflects the high affinity with which epibatidine is bound. Figure 14 shows that epibatidine is much more potent than acetylcholine or nicotine as an agonist for human α3β2 AChRs or α3β4 AChRs. This figure also points out the important role of β2 and β4 in determining the affinity and efficacy of ligands. The roles of rat β2 versus β4 subunits in determining the properties of AChRs formed in association with α3 and α4 subunits have been studied in particular detail by Papke and coworkers (reviewed in Papke, 1993). There are generalities of pharmacological properties of AChR subtypes which usually extend across species, but this is not always the case. Figure 15 shows that chicken α3β2 AChRs, like human α3β2 AChRs, respond to nicotine as a partial agonist, but the shape of the time dependence of the responses to epibatidine is reversed between chickens and humans. Figure 16 shows the extremely high binding affinity of epibatidine in a direct binding experiment with human α3β2 AChRs. Epibatidine is still a more

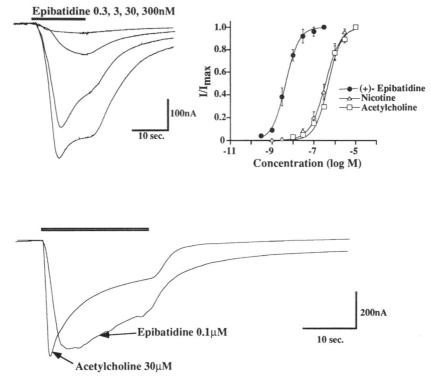

FIGURE 13. Activation by epibatidine of chicken α4β2 AChRs expressed in *Xenopus* oocytes. The top panel shows the responses to different concentrations of epibatidine and compares dose/response curves for epibatidine with those of nicotine and acetylcholine. The bottom panel compares the time courses of responses to saturating concentrations of epibatidine and nicotine. Data are from Gerzanich *et al.* (1995).

potent agonist than nicotine or acetylcholine on both α7 and electric organ AChRs, but much less potent than on α4 or α3 AChRs, as shown in Figure 17. Epibatidine is likely to become an important labeling reagent for several types of neuronal AChRs.

3.7.   Electrophysiological Properties

Much more is known about the properties of expressed cloned neuronal AChR subunit combinations than about the properties of different subtypes of neuronal AChRs expressed in intact neurons. This has been reviewed by Papke (1993) and will not be discussed in detail here. Albuquerque and coworkers, for example, have studied properties of AChRs expressed in rat hippocampal neurons which are thought to correspond to α4β2 AChRs, α3β4 AChRs, and α7 AChRs (Alkondon *et al.*, 1994). Changeux and coworkers have characterized several subtypes of AChRs expressed in the rat habenulo-interpeduncular system (Mulle *et al.*, 1991). Section 4.4. provides a few simple figures which contrast some basic properties of α4β2 AChRs and α7 AChRs with α1β1γδ AChRs, indicating that the neuronal AChRs exhibit rectification and higher calcium

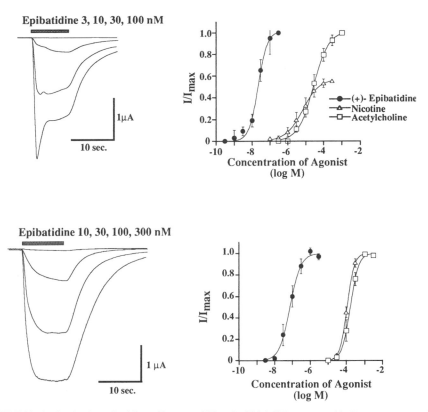

FIGURE 14. Activation by epibatidine of human α3β2 and α3β4 AChRs expressed in *Xenopus* oocytes. The top panel shows responses of α3β2 AChRs to various concentrations of epibatidine and compares dose/response curves for epibatidine, nicotine, and acetylcholine. The bottom panel makes a similar comparison for α4β4 AChRs. Data are from Gerzanich *et al.* (1995).

permeabilities than muscle AChRs. Influx of Ca^{2+} through both α3 AChRs and α7 AChRs is very likely to play important functional roles, extending the role of the neuronal AChRs from simply providing a mechanism for depolarizing the membrane to triggering an action potential (Mulle *et al.*, 1992; Rathouz and Berg, 1994; Vijayaraghavan *et al.*, 1992). Berg and coworkers have also suggested the intriguing hypothesis that a cAMP-dependent process can convert α3 AChRs from an inactivatable state to an activatable state (Vijayaraghavan *et al.*, 1990). If such complex modulation of synaptic function is revealed at ganglionic synapses, might not even more complex mechanisms be acting at cortical synapses? Neuronal AChRs in many cases are likely to play functional roles less straightforward than those at neuromuscular junctions.

3.8. Functional Roles

The α3 AChRs in ganglia would seem to play a straightforward postsynaptic role, yet these neurons have turned out to be quite complex. Single neurons express multiple

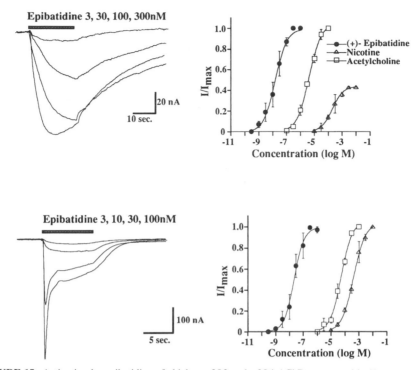

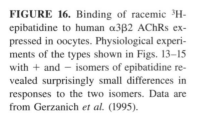

FIGURE 15. Activation by epibatidine of chicken α3β2 and α3β4 AChRs expressed in *Xenopus* oocytes. Note that the time courses of the responses are reversed from the human AChRs shown in Fig. 14, and that there are species-specific differences in EC_{50} values, but that in both chickens and humans α3β2 AChRs respond to nicotine as a partial agonist, whereas α3β4 AChRs respond to nicotine as a full agonist. Data are from Gerzanich *et al.* (1995).

FIGURE 16. Binding of racemic ³H-epibatidine to human α3β2 AChRs expressed in oocytes. Physiological experiments of the types shown in Figs. 13–15 with + and − isomers of epibatidine revealed surprisingly small differences in responses to the two isomers. Data are from Gerzanich *et al.* (1995).

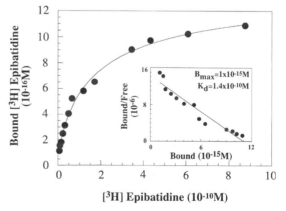

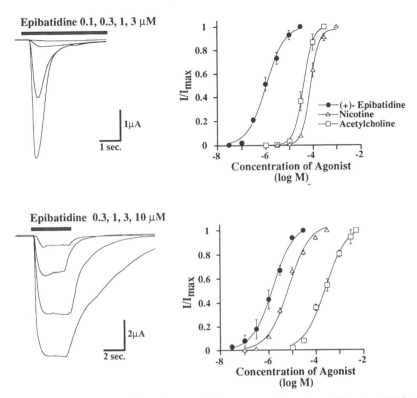

FIGURE 17. Activation by epibatidine of human $\alpha7$ homomers and muscle type *Torpedo* $\alpha1\beta1\gamma\delta$ AChRs expressed in oocytes. Data are from Gerzanich *et al.* (1995).

$\alpha3$ AChR subtypes postsynaptically as well as $\alpha7$ AChRs perisynaptically (Conroy and Berg, 1995).

Many, perhaps the majority, of neuronal AChRs in brain identified by binding of ^{3}H-nicotine or mAb 270 to $\beta2$ subunits seem to be localized presynaptically (Clarke *et al.*, 1986; Swanson *et al.*, 1987). AChRs on synaptosomes have been shown to mediate dopamine release (Grady *et al.*, 1992) and nicotine has been shown to modulate dopamine efflux from the striatum (Harsing *et al.*, 1992). The effects of AChRs on dopamine release may be especially important in addiction to nicotine (Clarke, 1990). AChRs have also been shown to modulate the release of several other transmitters (Wonnacott, 1990). Electrophysiological evidence has even been presented for AChRs on axons of GABA-ergic neurons which can modulate postsynaptic currents in response to nicotinic ligands (Léna *et al.*, 1993). Clarke (1995) has reviewed the histochemical localization of brain AChRs by *in situ* hybridization, mAbs, ^{3}H-nicotine, and ^{125}-I-αBgt and pointed out the difficulties in electrophysiologically identifying postsynaptic actions of acetylcholine in the brain. However, the death toll due to tobacco usage, which depends on nicotine to produce dependence, is strong testimony to the potent effects on nicotinic AChRs (Peto *et al.*, 1992).

It may be a mistake to think that all AChRs act postsynaptically to locally released acetylcholine as at neuromuscular junctions, or even presynaptically. Some AChRs may be involved in tonic responses to acetylcholine diffusing over relatively large areas. Cumulative Ca^{2+} influx through such AChRs could provide a mechanism whereby neuronal mechanisms could be subtly modulated without triggering endplate potentials. Low levels of ambient acetylcholine could also tune the sensitivity of AChRs by regulating the fraction that is desensitized. The nucleus basalis provides cholinergic innervation to the cortex and itself contains many high-affinity nicotine binding sites (presumably $\alpha4\beta2$ AChRs). These AChRs are severely depleted in Alzheimer's disease (Whitehouse et al., 1988). Learning and memory are impaired in Alzheimer's patients and in rats with nucleus basalis lesions, but grafting to cortex fibroblasts genetically modified to release acetylcholine caused lesioned rats to improve in a spatial navigation task (Winkler et al., 1995). It is unknown whether this effect is mediated through nicotinic AChRs, but this does provide a thought-provoking experiment. The nucleus basalis is rich in AChRs, but so is the thalamus (Swanson et al., 1987). Thus $\alpha4\beta2$ AChRs are in a position to influence a lot of neuronal traffic to the cortex, but their actual roles are unknown. Overall, the distribution of ^{3}H-nicotine binding sites and binding sites for mAb 270 to $\beta2$ subunits reveals that AChRs (predominantly $\alpha4\beta2$ AChRs) are present in small amounts in many areas of the brain, and high concentrations are found in a few areas such as superior colliculus, medial habenula, and interpeduncular nuclei (Clarke et al., 1986; Swanson et al., 1987).

Mice with their $\beta2$ subunits knocked out have been shown, as expected, to lack high-affinity binding sites for nicotine (Picciotto et al., 1995). In these mice nicotine no longer enhanced performance on a passive avoidance test, as expected, but, surprisingly, the mutant mice performed better than wild-type on this task. A possible explanation for this effect is that the net effect of injected nicotine is to block $\alpha4\beta2$ AChR function by causing desensitization, so that $\beta2$ knock-out mice with their $\alpha4\beta2$ AChRs nonfunctional behave like nicotine-treated mice. The minimal phenotypic effects of the $\beta2$ gene knock-out actually parallel the minimal phenotypic effects of the $\alpha4$ mutation in epilepsy patients (Steinlein et al., 1994, 1995). In both $\beta2$ knock-out mice and $\alpha4$ mutation patients compensatory mechanisms may also develop in the absence of these subunits to mask all the roles of AChRs of which they are part. For example, chick ciliary ganglia neurons normally express and coassemble $\alpha3$, $\beta2$, $\beta4$, and $\alpha5$ subunits (Conroy and Berg, 1995), yet it is clear that $\alpha3\beta4$ and $\alpha3\beta4\alpha5$ AChRs could function in the absence of $\beta2$ subunits. In neurons which expressed only $\alpha4$ and $\beta2$ subunits and in which $\beta4$ expression was not induced by the loss of $\beta2$ (as appears to be the case), loss of $\beta2$ would mean loss of all functional AChRs in these cells. If $\alpha4\beta2$ AChRs function primarily to modulate transmitter release, then their loss and the resulting loss of modulation would be no more traumatic than their total desensitization in a chronic smoker, but under certain conditions the system might not perform optimally, as revealed by the convulsions only at night in the $\alpha4$ mutation epilepsy patients.

Neuronal AChRs have been detected very early during development; for example, $\alpha3$ was detected in chick retina at embryonic day 4.5, and $\alpha8$ was detected a day later (Hamassaki-Britto et al., 1994a,b; Zoli et al., 1995). This is before choline acetyltransferase is detected and long before synaptogenesis takes place, suggesting that these AChRs may have a role in development. AChRs affect retinal ganglion process outgrowth (Lipton et al., 1988). It has been suggested that acetylcholine could act to stop process outgrowth in order to initiate synapse formation during development and synaptic plasticity (Lipton and Kater, 1989). The $\alpha 7$ AChRs have also been shown to mediate neurite retraction by a Ca^{2+}-dependent mechanism (Pugh and Berg, 1994; Quick, 1995).

Lung neuroendocrine cells and lung carcinoma cells have AChRs through which nicotine can act under some circumstances to promote proliferation (Maneckjie and Minna, 1990; Schuller, 1995). This may be of direct pathological significance in tobacco-related lung cancer. It also may provide an example of nicotinic AChRs turning up in unexpected tissues. Finally, this may be an example of AChR activation controlling cellular responses other than membrane depolarization or neurite outgrowth. Activation of neuronal AChRs is known to induce rapid gene transcription (Greenburg et al., 1986).

4. NEURONAL NICOTINE RECEPTORS THAT BIND α-BUNGAROTOXIN

4.1. Initial History of Molecular Studies and Subunit Nomenclature

Although neuronal binding sites for αBgt were discovered soon after this toxin was initially applied to studies of muscle AChRs (Green et al., 1973; Hunt and Schmidt, 1978), biochemical studies of these sites were deferred because αBgt did not appear to block the function of AChRs on neurons (Patrick and Stallcup, 1977; Carbonetto et al., 1978). The functional activity of these $\alpha 7$ AChRs has been discovered only recently because of their very rapid desensitization and because large, slowly desensitizing responses from $\alpha 3$ AChRs in many of the cells examined obscured the activity of the $\alpha 7$ AChRs (Couturier et al., 1990a,b; Alkondon and Albuquerque, 1991, 1993; Vijaya-raghavan et al., 1992).

αBgt binding proteins affinity-purified from chick brain yielded a partial N-terminal amino acid sequence (Conti-Tronconi et al., 1985), which we used to design an oligonucleotide probe. This probe was used to identify a cDNA for a subunit we initially termed αBgt binding protein $\alpha 1$ (Schoepfer et al., 1990). This cDNA was used to identify a closely related sequence from chick brain which we termed αBgt binding protein $\alpha 2$. We raised subunit-specific mAbs to bacterially expressed peptides corresponding to putative large cytoplasmic domains of these subunits and used these to prove that these cDNAs encoded subunits of chick brain αBgt binding proteins. Ballivet and coworkers then reported a chick cDNA identical to αBgt binding protein $\alpha 1$ which they called $\alpha 7$ and demonstrated that when expressed in Xenopus oocytes it formed functional homomers which were blocked by αBgt (Couturier et al., 1990a,b). For euphony we then called αBgt binding protein $\alpha 1$ "$\alpha 7$" and called αBgt binding protein $\alpha 2$ "$\alpha 8$". The $\alpha 7$ protein has subsequently been cloned from rats (Seguela et al., 1993) and humans (Peng et al., 1994b; Doncettestamm et al., 1993; Chini et al., 1994). The $\alpha 8$ protein has not yet been found in other species, but while looking for it in rats, α9 was discovered (Elgoyhen *et al.*, 1994).

4.2.   Subunit Composition and Stoichiometry of AChR Subtypes

Most chick brain AChRs that could bind αBgt (75%) were found by immune precipitation to contain α7 subunits, and the remainder were found to contain both α7 and α8 subunits (Schoepfer *et al.*, 1990). In contrast, in chick retina the majority of AChRs that can bind αBgt (69%) contain α8 subunits, and both α7α8 AChRs (17%) and α7 AChRs comprise minority populations (Keyser *et al.*, 1993). The complete subunit composition of these AChRs is not known. Lack of immune precipitation by mAbs to other known subunits indicates that they do not contain these subunits. Purified preparations typically contain multiple bands on acrylamide gel electrophoresis which could reflect unknown subunits, contamination, aggregates, or proteolytic degradation. It may be that these subunits naturally occur as homomers. The α7 subunit is expressed in oocytes as a homomer as efficiently as α4β2 AChRs or α1β1γδ AChRs, in the sense that half of the total AChRs are found on the cell surface (Lindstrom *et al.*, 1995). However, α8 is expressed in oocytes as a homomer much less efficiently, suggesting that it may usually be associated with a structural subunit (Gerzanich *et al.*, 1994).

Homology of α7 AChRs with other AChRs suggests that they should assemble into pentamers. Clever physiological experiments also support this concept (Palma *et al.*, 1995). Methyllyccanitine (MLA) is an insecticide produced by delphiniums and is an extremely potent and selective antagonist of α7 AChRs (Wonnacott *et al.*, 1993). Analysis of kinetics of inhibition of α7 AChRs by MLA and recovery from this blockage indicate that there are five binding sites for MLA in the AChR, but that binding of one MLA molecule is sufficient to prevent function (Palma *et al.*, 1995). The Hill coefficient for agonists on α7 and α AChRs is in the same 1.5–2 range as that for $(α1)_2β1γδ$ AChRs and $(α4)_2(β2)_3$ AChRs, which suggests that binding of agonists to two of the binding sites of α7 AChRs is necessary for activation, as is the case for the heteromeric AChRs (Gerzanich *et al.*, 1994).

4.3.   Pharmacological Properties

Despite the similarities in their N-terminal extracellular domain amino acid sequence, α7 and α8 AChRs differ in their pharmacological properties (Keyser *et al.*, 1993; Anand *et al.*, 1993b; Gerzanich *et al.*, 1994). The α7 AChRs have much lower affinity for αBgt (K_D = 2 nM) than does muscle AChR, but α7 AChRs have much higher affinity for αBgt than do α8 AChRs (K_D = 20 nM) (Keyser *et al.*, 1993; Anand *et al.*, 1993). However, α7 AChRs have much lower affinity for small cholinergic ligands than do α8 AChRs (Anand *et al.*, 1993). This is also reflected by the properties of α7 and α8 homomers. This is illustrated in the case of agonists by Fig. 18. Note that for chick α7 homomers DMPP is a very low-efficacy agonist, but for human α7 homomers DMPP is the most potent full agonist of those shown in Fig. 19. The contrasting affinities for antagonists of α7 and α8 homomers are shown in Fig. 20. Note that for α8 the classic

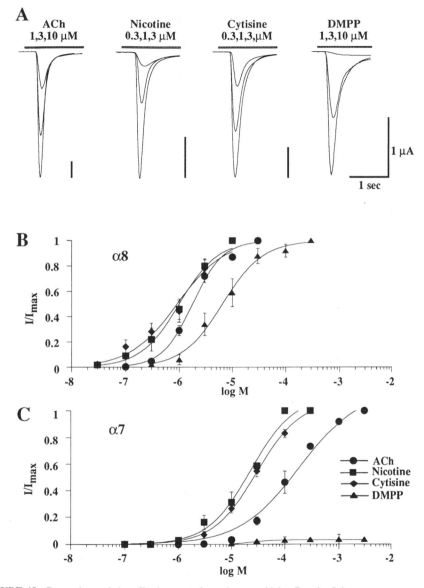

FIGURE 18. Comparison of the effectiveness of agonists on chick $\alpha7$ and $\alpha8$ homomers expressed in *Xenopus* oocytes. Data are from Gerzanich *et al.* (1994).

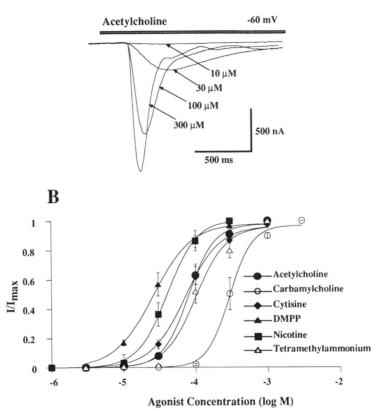

FIGURE 19. Effectiveness of agonists on human α7 homomers expressed in *Xenopus* oocytes. Data are from Peng *et al.* (1994b).

muscarinic antagonist atropine and the classic glycinergic antagonist strychnine are as potent as the classic nicotinic antagonist curare.

4.4. Channel Properties

The virtually identical sequences in the M1–M3 transmembrane domain regions of α7 and α8 subunits suggest that they should have virtually identical channel properties, and this is the case (Gerzanich *et al.*, 1994). Both α7 and α8 homomers exhibit the strong inward rectification that is also observed for some other neuronal AChRs, but not muscle AChRs, as shown in Fig. 21. Both α7 and α8 homomers exhibit very rapid desensitization, which contrasts the slower rate of desensitization of α4β2 AChRs or α1β1γδ AChRs, as shown in Fig. 22. The most striking feature of α7 and α8 AChRs is their high permeability for Ca^{2+} (Fig. 23). Both the rapid desensitization and the inward rectifica-

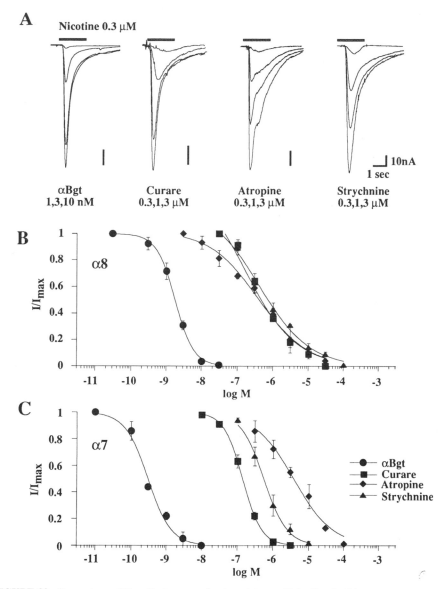

FIGURE 20. Comparison of the effectiveness of antagonists on chick α7 and α8 homomers expressed in *Xenopus* oocytes. Data are from Gerzanich *et al.* (1994).

tion combine to minimize sustained influx through these AChRs. These self-limiting features are especially interesting when considering the possible functional roles of such AChRs, especially those at extrasynaptic locations where they would presumably be exposed to minimal amounts of acetylcholine. Ca^{2+} influx through α7 and α8 homomers can trigger Ca^{2+}-sensitive Cl^- channels in *Xenopus* oocytes (Gerzanich *et al.*, 1994). It is

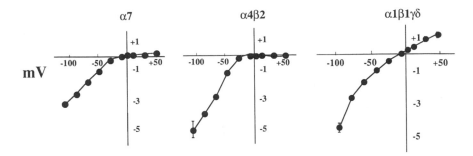

Normalized Current (I/I-50mV)

FIGURE 21. Comparison of rectification properties of chicken $\alpha 7$ homomers, chicken $\alpha 4 \beta 2$ AChRs, and *Torpedo* $\alpha 1 \beta 1 \gamma \delta$ AChRs expressed in *Xenopus* oocytes. Data are from Lindstrom *et al.* (1995).

tempting to think that this might model the role of Ca^{2+} entering through such AChRs as a second messager to activate ion channels and other cellular mechanisms.

It is interesting to compare and contrast $\alpha 7$ AChRs and NMDA receptors. Both are found in high concentration in the hippocampus (Del Toro *et al.*, 1994). Seguela *et al.* (1993) point out that the Ca^{2+} conductance of $\alpha 7$ AChRs is at least as great as that of NMDA receptors. The two receptors have opposite rectification properties (Gerzanich *et al.*, 1994). Depolarization of the membrane shuts off $\alpha 7$ AChRs and turns on NMDA receptors. While $\alpha 7$ AChRs desensitize very rapidly, NMDA receptors do not. It is reported that both can be involved in long-term potentiation in the hippocampus (Hunter *et al.*, 1994). Albuquerque and coworkers are studying in detail the electrophysiological properties of $\alpha 7$ AChRs in the hippocampus (Alkondon *et al.*, 1994, 1995). It will be intriguing to learn in the future, from a combination of high-resolution electrophysiology and histological localization, what exactly the functional roles of $\alpha 7$ and related subunits are in the hippocampus and elsewhere.

4.5. Functional Roles

The $\alpha 7$ AChRs have been histologically localized by binding of labeled αBgt (Clarke *et al.*, 1985; Jacob and Berg, 1983), mAbs (Britto *et al.*, 1992a,b; Keyser *et al.*, 1993; Del Toro *et al.*, 1994), and *in situ* hybridization (Seguela *et al.*, 1993). The perisynaptic localization of $\alpha 7$ on ganglionic neurons which express $\alpha 3$ AChRs at synapses is especially interesting (Sargent and Wilson, 1995), suggesting, perhaps, that the $\alpha 3$ AChRs are involved in synaptic responses to acetylcholine, whereas the $\alpha 7$ AChRs may respond to acetylcholine leaking from the synapse in the course of a trophic role. The $\alpha 7$ AChRs are histologically prominent in the hippocampus (Del Toro *et al.*, 1994; Barrantes *et al.*, 1995) and corresponding electrophysiological responses have been detected on virtually all hippocampal neurons (Alkondon *et al.*, 1994). Their functional role is not yet clear. The $\alpha 7$ AChRs on cultured neurons can regulate neurite outgrowth via Ca^{2+} influx, suggesting that these AChRs could have a role in synapse

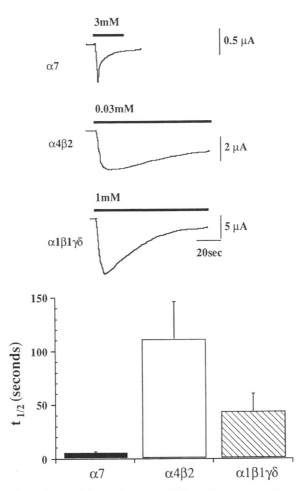

FIGURE 22. Comparison of rates of desensitization of chicken α7 homomers, chicken α4β2 AChRs, and *Torpedo* α1β1γδ AChRs expressed in *Xenopus* oocytes. α7 homomers desensitize most rapidly of the three, but this rate may be underestimated due to the time necessary to perfuse the occytes. Top: Typical responses to saturating concentrations of ACh. Bottom: Average halftimes for desensitization at −90 mV. Data are from Lindstrom *et al.* (1995).

formation during development or plasticity as well as more conventional signaling roles (Pugh and Berg, 1994; Quick, 1995).

The α8 AChRs have been immunohistologically localized in chick brain and retina (Britto *et al.*, 1992a,b; Keyser *et al.*, 1993; Hamassaki-Britto *et al.*, 1994a,b). No specific electrophysiological studies have been performed. Although α8 has thus far been found only in chicks, it is not yet clear whether there is a mammalian homologue. The physiological roles of α8 would be expected to be similar to those of α7.

Cochlear hair cells provide an example of a novel synaptic functional role for αBgt-binding AChRs, which may reflect the sorts of roles they exhibit elsewhere in the

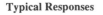

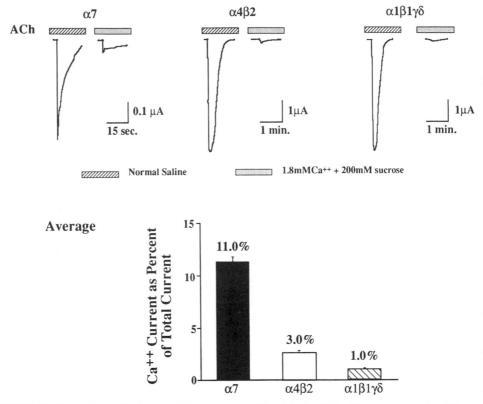

FIGURE 23. Comparison of the fraction of the current carried by Ca^{2+} for chick $\alpha 7$ homomers, chick $\alpha 4\beta 2$ AChRs, and *Torpedo* $\alpha 1\beta 1\gamma \delta$ AChRs expressed in *Xenopus* oocytes. All other cations were replaced by 200 mM sucrose leaving only 1.8 mM $CaCl_2$ in the bathing solution. Data are from Lindstrom *et al.* (1995).

nervous system. Fuchs and coworkers found that chick hair cells exhibited a response to acetylcholine from efferent endings of brain stem neurons which could be blocked by αBgt (Fuchs and Murroro, 1992a,b). Acetylcholine provoked Ca^{2+} influx through these AChRs, which triggered a prolonged inhibitory hyperpolarizing response due to activation of Ca^{2+}-dependent K^+ channels. These AChRs could be blocked by αBgt, curare, atropine, and strychnine, like $\alpha 7$ homomers, but αBgt had very low affinity and nicotine behaved as an antagonist. Rat $\alpha 9$ homomers were found to have these properties, and $\alpha 9$ was found in rat hair cells by *in situ* hybridization (Elgoyhen *et al.*, 1994). We have immunoisolated $\alpha 7$ protein and cloned $\alpha 7$ cDNAs from chick cochlea (Anand, Fuchs, and Lindstrom, unpublished). Thus, like ciliary ganglion neurons, cochlear hair cells may turn out to express several AChR subtypes, again challenging us to understand the roles of multiple neuronal AChR subtypes expressed not only throughout the nervous system, but also within single neurons.

ACKNOWLEDGMENTS. Research in the laboratory is supported by grants from the NINCDS (NS11323-22), the Muscular Dystrophy Association, the Smokeless Tobacco Research Council, Inc., and the Council for Tobacco Research–USA, Inc. Rene Anand and Gregg Wells provided valuable comments on the manuscript.

5. REFERENCES

Abramson, S., Culver, Y., and Taylor, P., 1989, An analog of lophotoxin reacts covalently with Tyr190 in the α subunit of the nicotinic acetylcholine receptor, *J. Biol. Chem.* **264**:1266–1267.

Aceto, M., Awaya, H., Martin, B., and May, E., 1983, Antinociceptive action of nicotine and its methiodide derivatives in mice and rats, *Br. J. Pharmacol.* **79**:869–876.

Akabas, M., Kaufmann, C., Archdeacon, P., and Karlin, A., 1994, Identification of acetylcholine receptor channel-lining residues in the entire M2 segment of the α subunit, *Neuron* **13**:919–927.

Alkondon, M., and Albuquerque, E., 1991, Initial characterization of the nicotinic acetylcholine receptors in rat hippocampal neurons, *J. Recept. Res.* **11**:1101–1201.

Alkondon, M., and Albuquerque, E., 1993, Diversity of nicotinic acetylcholine receptors in rat hippocampal neurons I: Pharmacological and functional evidence for distinct structural subtypes, *J. Pharmacol. Exp. Ther.* **265**:1455–1473.

Alkondon, M., and Albuquerque, E., 1995, Diversity of nicotinic acetylcholine receptors in rat hippocampal neurons III: Agonist actions of the novel alkaloid epibatidine and analysis of type II current, *J. Pharmacol. Exp. Ther.* **274**:771–782.

Alkondon, M., Reinhardt, S., Lobron, C., Hermsen, B., Maelicke, A., and Albuquerque, E., 1994, Diversity of nicotinic acetylcholine receptors in rat hippocampal neurons. II. The rundown and inward rectification of agonist-elicited whole cell currents and identification of receptor subunits by in situ hybridization, *J. Pharmacol. Exp. Ther.* **271**:494–506.

Anand, R., and Lindstrom, J., 1992, Chromosomal localization of seven neuronal nicotinic receptor subunit genes in humans, *Genomics* **13**:962–967.

Anand, R., Conroy, W. G., Schoepfer, R., Whiting, P., and Lindstrom, J., 1991, Chicken neuronal nicotinic acetylcholine receptors expressed in Xenopus oocytes have a pentameric quaternary structure, *J. Biol. Chem.* **266**:11192–11198.

Anand, R., Bason, L., Saedi, M., Gerzanich, V., Peng, X., and Lindstrom, J., 1993a, Reporter epitopes: A novel approach to examine transmembrane topology of integral membrane proteins applied to the α1 subunit of the nicotinic acetylcholine receptor, *Biochemistry* **32**:9975–9984.

Anand, R., Peng, X., Ballesta, J., and Lindstrom, J., 1993b, Pharmacological characterization of α bungarotoxin sensitive AChRs immunoisolated from chick retina: Contrasting properties of α7 and α8 subunit-containing subtypes, *Mol. Pharmacol.* **44**:1046–1050.

Anand, R., Peng, X., and Lindstrom, J., 1993c, Homomeric and native α7 acetylcholine receptors exhibit remarkably similar but non-identical pharmacological properties, suggesting that the native receptor is a heteromeric protein complex, *FEBS Lett.* **327**:241–246.

Anderson, D., Blobel, G., Tzartos, S., Bullick, W., and Lindstrom, J., 1983, Transmembrane orientation of an early biosynthetic form of acetylcholine receptor delta subunit determined by proteolytic dissection in conjunction with monoclonal antibodies, *J. Neurosci.* **3**:1773–1784.

Anholt, R., Fredkin, D., Deerinck, T., Ellisman, M., Montal, M., and Lindstrom, J., 1982, Incorporation of acetylcholine receptors into liposomes: Vesicle structure and acetylcholine receptor, *J. Biol. Chem.* **25**:7122–7134.

Appel, S. H., Blosser, J. C., McMaraman, J. L., Ashizawa, T., and Elias, S. B., 1981, The effects of carbamylcholine, calcium and cyclic nucleotides on acetylcholine receptor synthesis in cultured myotubes, *Ann. N.Y. Acad. Sci.* **377**:189–197.

Arneric, S., Sullivan, J., Briggs, C., Donnelly-Roberts, D., Anderson, D., Roszkiewicz, J., Hughes, M., Cadman, E., Adams, P., Garvey, D., Wasicak, J., and Williams, M., 1994, (S)-3-methyl-5-(1-methyl-2-

pyrrolidinyl) isoxazole (ABT418): A novel cholinergic ligand with cognition-enhancing and anxiolytic activities: 1. in vitro characterization. *J. Pharmacol. Exp. Ther.* **270**:310–318.

Badio, B., and Daly, J., 1994, Epibatidine, a potent analgesic and nicotinic agonist, *Mol. Pharmacol.* **45:** 563–569.

Barnard, E., 1992, Receptor classes and the transmitter-gated ion channels, *Trends Biol. Sci.* **17**:368–374.

Baron, J., 1995, The epidemiology of cigarette smoking and Parkinson's disease, in: *Effects of Nicotine on Biological Systems II* (P. Clarke *et al.*, eds.), Birkhauser, Basel, pp. 313–319.

Barrantes, G., Rodger, A., Lindstrom, J., and Wonnacott, S., 1995, α Bungarotoxin binding sites in rat hippocampal and cortical cultures: Initial characterization, co-localization with α 7 subunits and up-regulation by chronic nicotine treatment, *Brain Res.* **672**:228–236.

Benowitz, N., Porchet, H., and Jacob, P., 1990, Pharmacokinetics, metabolism, and pharmacodynamics of nicotine, in: *Nicotine Psychopharmacology* (S. Wonnocott *et al.*, eds.), Oxford Science Publications, Oxford, pp. 112–157.

Benwell, M., Balfour, D., and Anderson, J., 1988, Evidence that tobacco smoking increases the density of (-)-[^{3}H]nicotine binding sites in human brain, *J. Neurochem.* **50**:1243–1247.

Beroukhim, R., and Unwin, N., 1995, Three dimensional location of the main immunogenic region of the acetylcholine receptor, *Neuron* **15**:323–331.

Bertrand, D., and Changeux, J. P., 1995, Nicotinic receptor: An allosteric protein specialized for intercellular communication, *Neurosciences* **7**:75–90.

Bertrand, D., Devillers-Thiery, A., Revah, F., Galzi, J. L., Hussy, N., Mulle, C., Bertrand, S., Ballivet, M., and Changeux, J. P., 1992, Unconventional pharmacology of a neuronal nicotinic receptor mutated in the channel domain, *Proc. Natl. Acad. Sci. USA* **89**:1261–1265.

Bertrand, D., Galzi, J. L., Devillers-Thiery, A., Bertrand, S., and Changeux, J. P., 1993a, Mutations at two distinct sites within the channel domain M2 alter calcium permeability of neuronal α 7 nicotinic receptor, *Proc. Natl. Acad. Sci. USA* **90**:6971–6975.

Bertrand, D., Galzi, U. L., Devillers-Thiery, A., Bertrand, S., and Changeux, V. P., 1993b, Stratification of the channel domain in neurotransmitter receptors, *Curr. Opin. Cell Biol.* **5**:688–693.

Bessis, A., Savatier, N., Devillers-Thiery, A., Benjamin, A., and Changeux, V. P., 1993, Negative regulatory elements upstream of a novel exon of the neuronal nicotinic acetylcholine receptor α2 subunit gene, *Nucleic Acids Res.* **21**:2185–2192.

Betz, H., 1990, Ligand-gated ion channels in the brain: The amino acid receptor superfamily, *Neuron* **5:** 383–392.

Blount, P., and Merlie, J., 1988, Native folding of an acetylcholine receptor α subunit expressed in the absence of other receptor subunits, *J. Biol. Chem.* **262**:4367–4376.

Blount, P., and Merlie, J. P., 1989, Molecular basis of the two nonequivalent ligand binding sites of the muscle nicotinic acetylcholine receptor, *Neuron* **3**:349–357.

Blount, P., and Merlie, J. P., 1990, Mutational analysis of muscle nicotinic acetylcholine receptor subunit assembly, *J. Cell Biol.* **111**:2612–2622.

Blount, P., and Merlie, J. P., 1991a, BIP associates with newly synthesized subunits of the mouse muscle nicotinic receptor, *J. Cell Biol.* **113**:1125–1132.

Blount, P., and Merlie, J. P., 1991b, Characterization of an adult muscle acetylcholine receptor subunit by expression in fibroblasts, *J. Biol. Chem.* **266**:14692–14696.

Blount, P., Smith, M., and Merlie, J., 1990, Assembly intermediates of the mouse muscle nicotinic acetylcholine receptor in stably transfected fibroblasts, *J. Cell Biol.* **111**:2601–2611.

Bock, G., and Marsh, J. (eds.), 1990, *The Biology of Nicotine Dependence, Ciba Foundation Symposium 152*, John Wiley and Sons, New York.

Boess, F., Beroukhim, R., and Martin, I., 1995, Ultrastructure of the 5-hydroxytryptamine$_3$ receptor, *J. Neurochem.* **64**:1401–1405.

Boulter, J., Evans, K., Goldman, D., Martin, G., Treco, D., Heinemann, S., and Patrick, J., 1986, Isolation of a cDNA clone coding for a possible neural nicotinic acetylcholine receptor α subunit, *Nature* **319**:368–374.

Boulter, J., Connolly, J., Deneris, E., Goldman, D., Heinemann, S., and Patrick, J., 1987, Functional expression of two neuronal nicotinic acetylcholine receptors from cDNA clones identifies a gene family, *Proc. Natl. Acad. Sci. USA* **84**:7763–7767.

Boulter, J., O'Shea-Greenfield, A., Duvoisin, R., Connolly, J., Wada, E., Jensen, A., Gardner, P., Ballivet, M., Deneris, E., McKinnon, D., Heinemann, S., and Patrick, J., 1990, α3, α5, and β4: Three members of the rat neuronal nicotinic acetylcholine receptor-related gene family form a gene cluster, *J. Biol. Chem.* **265:**4472–4482.

Brake, A., Wagenbach, M., and Julius, D., 1994, New structural motif for ligand-gated ion channels defined by an ionotropic ATP receptor, *Nature* **371:**519–523.

Britto, L. R., Hamassaki-Britto, D. E., Ferro, E. S., Keyser, K. T., Karten, H. J., and Lindstrom, J. M., 1992a, Neurons of the chick brain and retina expressing both α bungarotoxin-sensitive and α bungarotoxin-insensitive nicotinic acetylcholine receptors: An immunohistochemical analysis, *Brain Res.* **590:** 193–200.

Britto, L., Keyser, K., Lindstrom, J., and Karten, H., 1992b, Immunohistochemical localization of nicotinic acetylcholine receptor subunits in the mesencephalon and diencephalon of the chick (Gallus gallus), *J. Comp. Neurol.* **317:**325–340.

Britto, L. R., Torrao, A. S., Hamassaki-Britto, D. E., Mpodozis, J., Keyser, K. T., Lindstrom, J. M., and Karten, H. J., 1994, Effects of retinal lesions upon the distribution of nicotinic acetylcholine receptor subunits in the chick visual system, *J. Comp. Neurol.* **350:**473–484.

Carbonetto, S., Fambrough, D., and Muller, K., 1978, Nonequivalence of α bungarotoxin receptors and acetylcholine receptors in chick sympathetic neurons, *Proc. Natl. Acad. Sci. USA* **75:**1016–1020.

Changeux, J. P., 1990, Functional architecture and dynamics of the nicotinic acetylcholine receptor: An allosteric ligand-gated ion channel, in: *1988–1989 Fidia Research Foundation: Neuroscience Award Lectures*, Vol. 4, pp. 21–168.

Changeux, J. P., 1991, Compartmentalized transcription of acetylcholine receptor genes during motor endplate epigenesis, *New Biol.* **3:**413–429.

Chavez, R., and Hall, Z., 1991, The transmembrane topology of the amino terminus of the α subunit of the nicotinic acetylcholine receptor, *J. Biol. Chem.* **266:**15532–15538.

Chavez, R., and Hall, Z., 1992, Expression of fusion proteins of the nicotinic acetylcholine receptor from mammalian muscle identifies the membrane-spanning regions in the α and δ subunits, *J. Cell Biol.* **116:**385–393.

Chini, B., Raimond, E., Elgoyhen, E., Moralli, D., Bolzaretti, M., and Heinemann, S., 1994, Molecular cloning and chromosomal localization of the human α 7 nicotinic receptor subunit gene (CHRNA 7), *Genomics* **19:**379–381.

Clarke, P., 1990, Mesolimbic dopamine activation—the key to nicotine reinforcement? in: *The Biology of Nicotine Dependence, Ciba Foundation Symposium 152m* (G. Bock and J. Marsh, eds.), John Wiley and Sons, Chichester, pp. 153–162.

Clarke, P. B. S., 1992, The fall and rise of neuronal α bungarotoxin binding proteins, *Trends Pharmacol.* **13:**407–413.

Clarke, P., 1995, Nicotinic receptors and cholinergic transmission in the central nervous system, *Ann. N.Y. Acad. Sci,* **757:**73–83.

Clarke, P., Schwartz, R., Paul, S., Pert, C., and Pert, A., 1985, Nicotinic binding in rat brain: Autoradiographic comparison of [³H]acetylcholine, [³H]nicotine, and [¹²⁵I] α bungarotoxin, *J. Neurosci.* **5:**1307–1315.

Clarke, P., Hamill, G., Nadi, N., Jacobwitz, D., and Pert, A., 1986, ³H-Nicotine and ¹²⁵I-α bungarotoxin-labeled nicotinic receptors in the interpenduncular nucleus of rats. II. Effects of habenular deafferentation, *J. Comp. Neurol.* **251:**407–413.

Cockcroft, V., Osguthorpe, D., Barnard, E., Friday, A., and Lunt, G., 1992, Ligand-gated ion channels—homology and diversity, *Mol. Neurobiol.* **4:**129–169.

Cohen, B., Labarca, C., Davidson, N., and Lester, H., 1992, Mutations in M2 alter the selectivity of the mouse nicotinic acetylcholine receptor for organic and alkali metal cations, *J. Gen. Physiol.* **100:**373–400.

Cohen, J., Sharp, S., and Lu, W., 1991, Structure of the agonist binding site of the nicotinic acetylcholine receptor, *J. Biol. Chem.* **266:**23354–23364.

Collins, A., Luo, Y., Selvaag, S., and Marks, M., 1994, Sensitivity to nicotine and brain nicotinic receptors are altered by chronic nicotine and mecamylamine infusion, *J. Pharmacol. Exp. Ther.* **271:**125–133.

Conroy, W. G., and Berg, D. K., 1995, Neurons can maintain multiple classes of nicotinic acetylcholine receptors distinguished by different subunit compositions, *J. Biol. Chem.* **270:**4424–4431.

Conroy, W., Saedi, M., and Lindstrom, J., 1990, TE671 cells express an abundance of a partially mature acetylcholine receptor α subunit which has characteristics of an assembly intermediate, *J. Biol. Chem.* **265:**21642–21651.

Conroy, W., Vernallis, A., and Berg, D., 1992, The α 5 gene product assembles with multiple acetylcholine receptor subunits to form distinctive receptor subtypes in brain, *Neuron* **9:**1–20.

Conti-Tronconi, B., Gotti, G., Hunkapiller, M., and Raferty, M., 1982, Mammalian muscle acetylcholine receptor: A supramolecular structure formed by four related proteins, *Science* **218:**1227–1229.

Conti-Tronconi, B., Tzartos, S., and Lindstrom, J., 1981, Monoclonal antibodies as probes of acetylcholine receptor structure. II: Binding to native receptor, *Biochemistry* **20:**2181–2191.

Conti-Tronconi, B., Dunn, S., Barnard, E., Dolly, J., Lai, F., Ray, N., and Raferty, M., 1985, Brain and muscle nicotinic acetylcholine receptors are different but homologous proteins, *Proc. Natl. Acad. Sci. USA* **82:**5208–5212.

Cooper, E., Couturier, S., and Ballivet, M., 1991, Pentameric structure and subunit stoichiometry of a neuronal nicotinic acetylcholine receptor, *Nature* **350:**235–238.

Corringer, P. V., Galzi, V. L., Eisele, J. L., Bertrand, S., Changeux, V. P., and Bertrand, D., 1995, Identification of a new component of the agonist binding site of the nicotinic α7 homoligomeric receptor, *J. Biol. Chem.* **270:**11749–11752.

Corriveau, R., and Berg, D., 1993, Coexpression of multiple acetylcholine receptor genes in neurons: Quantification of transcripts during development, *J. Neurosci.* **13:**2662–2671.

Corriveau, R., Romano, S., Conroy, W., Olivia, L., and Berg, D., 1995, Expression of neuronal acetylcholine receptor genes in vertebrate skeletal muscle during development, *J. Neurosci.* **15:**1372–1383.

Costa, A., Patrick, J., and Dani, J., 1994, Improved technique for studying ion channels expression in *Xenopus* oocytes, including fast perfusion, *Biophys. J.* **67:**1–7.

Couturier, S., Erkman, L., Valera, S., Rungger, D., Bertrand, S., Boulter, J., Ballivet, M., and Bertrand, D., 1990a, α5, α3, and non α3. Three clustered avian genes encoding neuronal nicotinic acetylcholine receptor related subunits, *J. Biol. Chem.* **265:**17560–17567.

Couturier, S., Bertrand, D., Matter, J., Hernandez, M., Bertrand, S., Millar, N., Valera, S., Barkas, T., and Ballivet, M., 1990b, A neuronal nicotinic acetylcholine receptor subunit (α 7) is developmentally regulated and forms a homomeric channel blocked by α bungarotoxin, *Neuron* **5:**847–856.

Criado, M., Hochschwender, S., Sarin, V., Fox, J. L., and Lindstrom, J., 1985, Evidence for unpredicted transmembrane domains in acetylcholine receptor subunits, *Proc. Natl. Acad. Sci. USA* **82:**2004–2008.

Criado, M., Witzemann, V., Koenen, M., and Sakmann, B., 1988, Nucleotide sequence of rat muscle acetylcholine receptor epsilon subunit, *Nucleic Acids Res.* **16:**10920.

Czajkowski, C., and Karlin, A., 1995, Structure of the nicotinic receptor acetylcholine binding site, *J. Biol. Chem.* **270:**3160–3164.

Czajkowski, C., Kaufmann, C., and Karlin, A., 1993, Negatively charged amino acid residues in the nicotinic receptor δ subunit that contribute to the binding of acetylcholine, *Proc. Natl. Acad. Sci. USA* **90:**6285–6289.

Daly, J., 1995, The chemistry of poisons in amphibian skin, *Proc. Natl. Acad. Sci. USA* **92:**9–13.

Das, M., and Lindstrom, J., 1989, The main immunogenic region of the nicotinic acetylcholine receptor: Interaction of monoclonal antibodies with synthetic peptides, *Biochem. Biophys. Res. Commun.* **165:**865–871.

Das, M., and Lindstrom, J., 1991, Epitope mapping of antibodies to acetylcholine receptor, *Biochemistry* **30:**2470–2477.

Decker, M., Brioni, J., Sullivan, J., Buckley, M., Rodek, R., Rasziewicz, V., Kang, C., Kim, D., Giardina, W., Wasicak, J., Garvey, D., Williams, M., and Arneric, S., 1994, (S)-3-methyl-5-(1-methyl-2-pyrrolidinyl)isoxazole (ABT418): A novel cholinergic ligand with cognition-enhancing and anxiolytic activities: II. in vivo characterization, *J. Pharmacol. Exp. Ther.* **270:**319–328.

Del Toro, E., Juiz, J., Peng, X., Lindstrom, J., and Criado, M., 1994, Immunocytochemical localization of the α 7 subunit of the nicotinic acetylcholine receptor in the rat central nervous system, *J. Comp. Neurol.* **349:**325–342.

Deneris, E., Connolly, J., Rogers, S., and Duvoisin, R., 1991, Pharmacological and functional diversity of neuronal nicotinic acetylcholine receptors, *Trends Pharmacol. Sci.* **12:**34–40.

DiPaola, M., Czajkowski, C., and Karlin, A., 1989, The sideness of the COOH terminus of the acetylcholine δ subunit, *J. Biol. Chem.* **264**:15457–15463.

DiPaola, M., Kao, P., and Karlin, A., 1990, Mapping the subunit site photolabeled by the noncompetitive inhibitor ^{3}H-quinacrine azide in the active state of the nicotinic acetylcholine receptor, *J. Biol. Chem.* **265**:11017–11029.

Doncellestamm, L., Monteggia, L., Donnelly-Roberts, D., Wong, M., Lee, J., Tian, J., and Giordano, T., 1993, Cloning and sequence of the human α 7 nicotinic acetylcholine receptor, *Drug Dev. Res.* **30**:252–256.

Dwyer, B., 1991, Topological dispositions of lysine α 380 and lysine γ 486 in the acetylcholine receptor from *Torpedo californica*, *Biochemistry* **30**:4105–4112.

Eisile, J. L., Bertrand, S., Galzi, J. L., Devillers-Thiery, A., Changeux, J. P., and Bertrand, D., 1993, Chimaeric nicotinic-serotonergic receptor combines distinct ligand binding and channel specificities, *Nature* **366**: 479–483.

Elgoyhen, A., Johnson, D., Boulter, J., Vetter, D., and Heinemann, S, 1994, α9: An acetylcholine receptor with novel pharmacological properties expressed in rat cochlear hair cells, *Cell* **79**:705–715.

Engel, A., 1990, Congenital disorder of neuromuscular transmission, *Seminars Neurol.* **10**:12–26.

Engel, A., 1994, Myasthenic syndromes, in: *Myology*, 2nd edition, Vol. 2, (A. Engel and C. Franzini-Armstrong, eds.), McGraw-Hill, New York, pp. 1798–1835.

Fiodalisi, J., Fetter, C., Ten Harmsel, A., Gigowski, R., Chiappinelli, V., and Grant, G., 1991, Synthesis and expression in *Escherichia coli* of a gene for κ-bungarotoxin, *Biochemistry* **30**:10337–10343.

Fletcher, S., Baker, R., Chambers, M., Herbert, R., Hobbs, S., Thomas, S., Verrier, H., Watt, A., and Ball, R., 1994, Total synthesis and determination of the absolute configuration of epibatidine, *J. Org. Chem.* **59**: 1771–1778.

Flores, C., Rogers, S., Pabreza, L., Wolfe, B., and Kellar, K., 1992, A subtype of nicotinic cholinergic receptor in rat brain is composed of α4 and β2 subunits and is upregulated by chronic nicotine treatment, *Mol. Pharmacol.* **41**:31–37.

Froehner, S., 1991, The submembrane machinery for nicotinic acetylcholine receptor clustering, *J. Cell Biol.* **114**:1–7.

Froehner, S., 1993, Regulation of ion channel distribution at synapses, *Annu. Rev. Neurosci.* **16**:347–368.

Fu, D., and Sine, S., 1994, Competitive antagonists bridge α-γ subunit interface of the acetylcholine receptor through quaternary aromatic interactions, *J. Biol. Chem.* **269**:26152–26157.

Fuchs, P., and Murrow, B., 1992a, Cholinergic inhibition of short (outer) hair cells of the chick's cochlea, *J. Neurosci.* **12**:800–809.

Fuchs, P., and Murrow, B., 1992b, A novel cholinergic receptor mediates inhibition of chick cochlear hair cells. *Proc. R. Soc. London Ser. B* **248**:35–40.

Gahring, L., Twyman, R., Greenlee, J., and Rogers, S., 1995, Autoantibodies to neuronal glutamate receptors in patients with paraneoplastic neurodegenerative syndrome enhance receptor activation, *Mol. Med.* **1**: 245–253.

Galzi, J. L., and Changeux, J. P., 1994, *Curr. Opin. Struct. Bio.*. **4**:554–565.

Galzi, J. L., Revah, F., Black, D., Goeldner, M., Hirth, C., and Changeux, J. P., 1990, Identification of a novel amino acid α tyrosine 93 within the cholinergic ligand-binding sites of the acetylcholine receptor by photoaffinity labeling, *J. Biol. Chem.* **265**:10430–10437.

Galzi, J. L., Bertrand, D., Devillers-Thiery, A., Revah, F., Bertrand, S., and Changeux, J. P., 1991, Functional significance of aromatic amino acids from three peptide loops of the α7 neuronal nicotinic receptor site investigated by site directed mutagenesis, *FEBS Lett.* **294**:198–202.

Galzi, J. L., Devillers-Thiery, A., Hussy, N., Bertrand, S., Changeux, J. P., and Bertrand, D., 1992, Mutations in the channel domain of a neuronal nicotinic receptor convert ion selectivity from cationic to anionic, *Nature* **359**:500–505.

Gardner, J. M., and Fambrough, D. M., 1979, Acetylcholine receptor degradation measured by density labeling: Effects of cholinergic ligands and evidence against recycling, *Cell* **16**:661–674.

Gehle, V., and Sumikawa, K., 1991, Site directed mutagenesis of the conserved N-glycosylation site on the nicotinic acetylcholine receptor subunits, *Mol. Brain Res.* **11**:17–25.

Gerzanich, V., Anand, R., and Lindstrom, J., 1994, Homomers of α 8 subunits nicotinic receptors functionally expressed in *Xenopus* oocytes exhibit similar channel but contrasting binding site properties compared to α 7 homomers, *Mol. Pharmacol.* **45**:212–220.

Gerzanich, V., Peng, X., Wang, F., Wells, G., Anand, R., Fletcher, S., and Lindstrom, J., 1995, Comparative pharmacology of epibatidine a potent agonist for neuronal nicotinic acetylcholine receptors, *Mol. Pharmacol.* **48:**774–782.

Giraudat, J., Dennis, M., Heidmann, T., Hanmont, P. Y., Lederer, F., and Changeux, J. P., 1987, Structure of the high-affinity binding site for noncompetitive blockers of the acetylcholine receptor: [^{3}H] Chlorpromazine labels homologous residues in the β and δ chains, *Biochemistry* **26:**2410–2418.

Goldman, D., Simmons, D., Swanson, L., Patrick, J., and Heinemann, S., 1986, Mapping of brain areas expressing RNA homologous to two different acetylcholine receptor α subunit cDNAs, *Proc. Natl. Acad. Sci. USA* **83:**4076–4080.

Grady, S., Marks, M., Wonnacott, S., and Collins, A., 1992, Characterization of nicotinic receptor-mediated ^{3}H-dopamine release from synaptosomes prepared from mouse striatum, *J. Neurochem.* **59:**848–856.

Grady, S., Marks, M., and Collins, A., 1994, Desensitization of nicotine-stimulated ^{3}H-dopamine release from mouse striatal synaptosomes, *J. Neurochem.* **62:**1390–1398.

Green, L., Sytkowski, A., Vogel, A., and Nirenberg, M., 1973, α Bungarotoxin used as a probe for acetylcholine receptors of cultured neurons, *Nature* **243:**163–166.

Green, T., Stouffer, K., and Lummis, S., 1995, Expression of recombinant homo-oligomeric 5-hydroxytryptamine$_3$ receptors provides new insights into their maturation and structure, *J. Biol. Chem.* **270:**6056–6061.

Greenburg, M., Ziff, E., and Greene, L., 1986, Stimulation of neuronal acetylcholine receptors induces rapid gene transcription, *Science* **234:**80–83.

Gu, Y., Camacho, P., Gardner, P., and Hall, Z., 1991a, Identification of two amino acid residues in the ε subunit that promote mammalian muscle acetylcholine receptor assembly in COS cells, *Neuron* **6:**879–887.

Gu, Y., Forsayeth, J., Verrall, S., Yu, X., and Hall, Z., 1991b, Assembly of the mammalian muscle acetylcholine receptor in transfected COS cells, *J. Cell Biol.* **114:**799–807.

Gullick, W., and Lindstrom, J., 1983, Mapping the binding of monoclonal antibodies to the acetylcholine receptor from *Torpedo californica*, *Biochemistry* **22:**3312–3320.

Gullick, W., Tzartos, S., and Lindstrom, J., 1981, Monoclonal antibodies as probes of acetylcholine receptor structure. I. Peptide mapping, *Biochemistry* **20:**2173–2180.

Hamassaki-Britto, D., Brzozowska-Prechtl, A., Karten, H., Lindstrom, J., and Keyser, K., 1991, GABA-like immunoreactive cells containing nicotinic acetylcholine receptors in the chick retina, *J. Comp. Neurol.* **313:**394–408.

Hamassaki-Britto, D., Brzozowska-Prechtl, A., Karten, H., and Lindstrom, J., 1994a, Bipolar cells of the chick retina containing α bungarotoxin-sensitive nicotinic acetylcholine receptors, *Vis. Neurosci.* **11:**63–70.

Hamassaki-Britto, D., Gardino, P. F., Hokoc, J. N., Keyser, K. T., Karten, H. J., Lindstrom, J. M., and Britto, L. R., 1994b, Differential development of α-bungarotoxin-sensitive and α-bungarotoxin-insensitive nicotinic acetylcholine receptors in the chick retina, *J. Comp. Neurol.* **347:**161–170.

Harsing, L., Sershen, H., and Lajtha, A., 1992, Dopamine efflux from striatum after chronic nicotine: Evidence for autoreceptor desensitization, *J. Neurochem.* **59:**48–54.

Henley, J., Lindstrom, J., and Oswald, R., 1986a, Acetylcholine receptor synthesis in retina and transport to the optic tectum in goldfish, *Science* **232:**1627–1629.

Henley, J., Mynlieff, M., Lindstrom, J., and Oswald, R., 1986b, Interaction of monoclonal antibodies to electroplaque acetylcholine receptors with the α bungarotoxin binding site of goldfish brain, *Brain Res.* **364:**405–408.

Henley, J. M., Lindstrom, J. M., and Oswald, R. E., 1988, Interaction of monoclonal antibodies with α-bungarotoxin and (−) nicotine binding sites in goldfish brain, *J. Biol. Chem.* **263:**9686–9691.

Hernandez, M. C., Erkman, L., Matter-Sadzinski, L., Roztocil, T., Ballivet, M., and Matter, J. M., 1995, Characterization of the nicotinic acetylcholine receptor β3 gene, *J. Biol. Chem.* **270:**3224–3233.

Hill, J., Zoli, M., Bourgeois, J. P., and Changeux, J. P., 1993, Immunocytochemical localization of a neuronal nicotinic receptor: The β2 subunit, *J. Neurosci.* **13:**1551–1568.

Holtzman, E., Wise, D., Wall, J., and Karlin, A., 1982, Electron microscopy of complexes of isolated acetylcholine receptor, biotinyl-toxin and avidin, *Proc. Natl. Acad. Sci. USA* **79:**310–314.

Hoover, F., and Goldman, D., 1992, Temporarily correlated expression of nAChR genes during development of the mammalian retina, *Exp. Eye Res.* **54:**561–570.

Houghtling, R., Davila-Garcia, M., Hurt, S., and Kellar, K., 1994, [³H] Epibatidine binding to nicotinic receptors in brain, *Med. Chem. Res.* **4**:538–546.

Huang, D., and Shen, T., 1993, A versatile total synthesis of epibatidine and analogs, *Tetrahedron Lett.* **34**:3251–3254.

Hucho, F., Oberthur, W., and Lottspeich, F., 1986, The ion channel of the nicotinic acetylcholine receptor is formed by homologous helices of the receptor subunits, *FEBS Lett.* **205**:137–142.

Huganir, R., and Greengard, P., 1990, Regulation of neurotransmitter receptor desensitization by protein phosphorylation, *Neuron* **5**:555–567.

Hunt, S., and Schmidt, J., 1978, Some observations on the binding patterns of α bungarotoxin in the central nervous system of the rat, *Brain Res.* **157**:213–232.

Hunter, B., deFiebre, C., Papke, R., Kem, W., and Meyer, E., 1994, A novel nicotinic agonist facilitates induction of long-term potentiation in the rat hippocampus, *Neurosci. Lett.* **168**:130–134.

Imoto, K., Busch, C., Sakmann, B., Mishina, M., Konno, T., Nakai, J., Bujo, H., Mori, Y., Fukuda, K., and Numa, S., 1988, Rings of negatively charged amino acids determine the acetylcholine receptor channel conductance, *Nature* **335**:645–648.

Jackson, M., 1989, Perfection of a synaptic receptor: Kinetics and energetics of the acetylcholine receptor, *Proc. Natl. Acad. Sci. USA* **86**:2199–2203.

Jacob, M., and Berg, D., 1983, The ultrastructural localization of α bungarotoxin binding sites in relation to synapses on chick ciliary ganglion neurons, *J. Neurosci.* **3**:260–271.

Jacob, M., Berg, D., and Lindstrom, J., 1984, A shared antigenic determinant between the *Electrophorus* acetylcholine receptor and a synaptic component on chick ciliary ganglion neurons, *Proc. Natl. Acad. Sci. USA* **81**:3223–3227.

Jacob, M., Lindstrom, J., and Berg, D., 1986, Surface and intracellular distribution of a putative neuronal nicotinic acetylcholine receptor, *J. Cell Biol.* **103**:205–214.

Kao, P., and Karlin, A., 1986, Acetylcholine receptor binding site contains a disulfide crosslink between adjacent half-cystinyl residues, *J. Biol. Chem.* **261**:8085–8088.

Kao, P., Dwork, A., Kaldany, R., Silver, M., Wideman, J., Stein, S., and Karlin, A., 1984, Identification of the α subunit half cysteine specifically labeled by an affinity reagent for the acetylcholine receptor binding site, *J. Biol. Chem.* **259**:11662–11665.

Karlin, A., 1991, Exploration of the nicotinic acetylcholine receptor, *Harvey Lectures Series* **85**:71–107.

Karlin, A., 1993, Structure of nicotinic acetylcholine receptors, *Curr. Opin. Neurobiol.* **3**:299–309.

Karlin, A., and Cowburn, D., 1973, The affinity-labeling of partially purified acetylcholine receptor from electric tissue of *Electrophorus*, *Proc. Natl. Acad. Sci. USA* **70**:3636–3640.

Kellaries, K., Ware, D., Smith, S., and Kyte, J., 1989, Assessment of the number of free cysteines and isolation and identification of cysteine-containing peptides from acetylcholine receptor, *Biochemistry* **28**:3469–3482.

Keyser, K., Hughes, T., Whiting, P., Lindstrom, J., and Karten, H., 1988, Cholinoceptive neurons in the retina of the chick: An immunohistochemical study of the nicotinic acetylcholine receptors, *Vis. Neurosci.* **1**:349–366.

Keyser, K., Britto, L., Schoepfer, R., Whiting, P., Cooper, J., Conroy, W., Karten, H., Lindstrom, J., 1993, Three subtypes of α-bungarotoxin-sensitive nicotinic acetylcholine receptors are expressed in chick retina, *J. Neurosci.* **13**:442–454.

Kiefer, H., Lindstrom, J., Lennox, E., and Singer, S., 1970, Photo-affinity labeling of specific acetylcholine binding sites on membranes, *Proc. Natl. Acad. Sci. USA* **67**:1688–1694.

Kirsch, V., Walters, I., Triller, A., and Betz, H., 1993, Gepherin antisense oligonucleotides prevent glycine receptor clustering in spinal neurons, *Nature* **366**:745–748.

Konno, T., Busch, C., Von Kitzing, E., Imoto, K., Wang, F., Nakai, J., Mishina, M., Numa, S., and Sakmann, B., 1991, Rings of anionic amino acids as structural determinants of ion selectivity in the acetylcholine receptor channel, *Proc. R. Soc. London Ser. B* **244**:69–79.

Krienkamp, H. J., Maeda, R., Sine, S., and Taylor, P., 1995, Intersubunit contacts governing assembly of the mammalian nicotinic acetylcholine receptor, *Neuron* **14**:635–644.

Kubalek, E., Ralston, S., Lindstrom, J., and Unwin, N., 1987, Location of subunits within the acetylcholine receptor: Analysis of tubular crystals from *Torpedo marmorata*, *J. Cell Biol.* **105**:9–18.

Lange, K., Wells, F., Jenner, P., and Marsden, P., 1993, Altered muscarinic and nicotinic receptor densities in cortical and subcortical regions in Parkinson's disease, *J. Neurochem.* **60:**197–203.

Langosch, D., Thomas, L., and Betz, H., 1988, Conserved quaternary structure of ligand-gated ion channels: The postsynaptic glycine receptor is a pentamer, *Proc. Natl. Acad. Sci. USA* **85:**7394–7398.

Laufer, R., and Changeux, J. P., 1989, Activity-dependent regulation of gene expression in muscle and neuronal cells, *Mol. Neurobiol.* **3:**1–53.

Lee, C., Tseng, L., and Chiu, T., 1967, Influence of denervation on localization of neurotoxins from clapid venoms in rat diaphragm, *Nature* **215:**1177–1178.

Lee, Y., Li, L., Lasalde, J., Rojas, L., McNamee, M., Ortiz-Miranda, S., and Pappone, P., 1994, Mutations in the M4 domain of *Torpedo californica* acetylcholine receptor dramatically alter ion channel function, *Biophys. J.* **66:**646–653.

Lei, S., Okita, D., and Conti-Fine, B., 1995, Binding of monoclonal antibodies against the carboxyl terminal segment of the nicotinic receptor δ subunit suggests an unusual transmembrane disposition of this sequence region, *Biochemistry* **34:**6675–6688.

Léna, C., Changeux, J. P., and Mulle, C., 1993, Evidence for "preterminal" nicotinic receptors on GABAergic axons in the rat interpeduncular nucleus, *J. Neurosci.* **13:**2680–2688.

Lester, H., 1992, The permeation pathway of neurotransmitter-gated ion channels, *Annu. Rev. Biophys. Biomol. Struct.* **21:**267–292.

Lindstrom, J., 1995, Nicotinic acetylcholine receptors, in: *CRC Handbook of Receptors and Channels, Ligand and Voltage-Gated Ion Channels* (A. North, ed.), CRC Press, Boca Raton, pp. 153–175.

Lindstrom, J., and Patrick, J., 1974, Purification of the acetylcholine receptor by affinity chromatography, in: *Synaptic Transmission and Neuronal Interaction* (M. V. L. Bennet, ed.), Raven Press, New York, pp. 191–216.

Lindstrom, J., Seybold, M., Lennon, V., Whittingham, S., and Duane, D., 1976, Antibody to acetylcholine receptor in myasthenia gravis: Prevalence, clinical correlates, and diagnostic value, *Neurology* **26:**1054–1059.

Lindstrom, J., Merlie, J., and Yogeeswaran, 1979, Biochemical properties of acetylcholine receptor subunits from *Torpedo californica*, *Biochemistry* **18:**4465–4470.

Lindstrom, J., Einarson, B., and Tzartos, S., 1981, Production and assay of antibodies to acetylcholine receptors, *Methods Enzymol.* **74:**432–460.

Lindstrom, J., Shelton, G. D., and Fuji, Y., 1988, Myasthenia gravis, *Adv. Immunol.* **42:**233–284.

Lindstrom, J., Schoepfer, R., Conroy, W. G., and Whiting, P., 1990, Structural and functional heterogeneity of nicotinic receptors, in: *The Biology of Nicotine Dependence, Ciba Foundation Symposium 152* (G. Bock and J. Marsh, eds.), John Wiley and Sons, New York, pp. 43–61.

Lindstrom, J., Anand, R., Peng, X., Gerzanich, V., Wang, F., and Li, Y., 1995, Neuronal nicotinic receptor subtypes, *Ann. N.Y. Acad. Sci.* **757:**100–116.

Lipton, S., and Kater, S., 1989, Neurotransmitter regulation of neuronal outgrowth, plasticity, and survival, *Trends Neurosci.* **12:**265–270.

Lipton, S., Aizenman, E., and Loring, R., 1987, Neural nicotinic acetylcholine responses in solitary mammalian retinal ganglion cells, *Pflügers Arch.* **410:**37–43.

Lipton, S., Frosch, M., Phillips, M., Tauck, D., and Aizenman, E., 1988, Nicotinic antagonists enhance process outgrowth by rat retinal ganglion cells in culture, *Science* **239:**1293–1296.

Lo, D., Pinkham, J., and Stevens, C., 1991, Role of a key cysteine residue in the gating of the acetylcholine receptor, *Neuron* **6:**31–40.

Luetje, C., and Patrick, J., 1991, Both α and β subunits contribute to the agonist sensitivity of neuronal nicotinic acetylcholine receptors, *J. Neurosci.* **11:**837–845.

Lukas, R., Norman, S., and Lucero, L., 1993, Characterization of nicotinic acetylcholine receptors expressed by cells of the SH-SY5Y human neuroblastoma clonal line, *Mol. Cell. Neurosci.* **4:**1–12.

Luther, M., Schoepfer, R., Whiting, P., Blatt, Y., Montal, M. S., Montal, M., and Lindstrom, J., 1989, Muscle acetylcholine receptor is expressed in the human cerebellar medulloblastoma cell line TE671, *J. Neurosci* **9:**1082–1096.

Maimone, M., and Merlie, J., 1993, Interaction of the 43kd postsynaptic protein with all subunits of the muscle nicotinic acetylcholine receptor, *Neuron* **11:**53–66.

Maneckjie, R., and Minna, J., 1990, Opioid and nicotine receptors affect growth regulation of human lung cancer cell lines, *Proc. Natl. Acad. Sci. USA* **87:**3294–3298.

Marks, M., Stitzel, J., and Collins, A., 1985, Time course study of the effects of chronic nicotine infusion on drug response and brain receptor, *J. Pharmacol. Exp. Ther.* **235**:619–628.

Marks, M., Pauly, J., Gross, D., Deneris, E., Hermans-Borgmeyer, I., Heinemann, S., and Collins, A., 1992, Nicotine binding and nicotinic receptor subunit RNA after chronic nicotine treatment, *J. Neurosci.* **12**:2765–2784.

Marks, M., Grady, S., and Collins, A., 1993, Downregulation of nicotinic receptor function after chronic nicotine infusion. *J. Pharmacol. Exp. Ther.* **266**:1268–1275.

Matter, J., Matter-Sadzinski, L., and Ballivet, M., 1990, Expression of neuronal nicotinic acetylcholine receptor genes in the developing chick visual system, *EMBO J.* **9**:1021–1026.

McCrea, P. D., Popot, J. L., and Engelman, D. M., 1987, Transmembrane topography of the nicotinic acetylcholine receptor δ subunit, *EMBO J.* **6**:3619–3626.

McGehee, D., and Role, L., 1995, Physiological diversity of nicotinic acetylcholine receptors expressed by vertebrate neurons, *Annu. Rev. Physiol.* **57**:521–546.

McLane, K., Wu, X., Lindstrom, J., and Conti-Tronconi, B., 1992, Epitope mapping of polyclonal and monoclonal antibodies against two α bungarotoxin binding subunits from neuronal nicotinic receptors, *J. Neuroimmunol.* **38**:115–128.

Merlie, J. P., and Lindstrom, J., 1983, Assembly in vivo of mouse muscle acetylcholine receptor: Identification of an α subunit species which may be an assembly intermediate, *Cell* **34**:747–757.

Middleton, R., and Cohen, J., 1991, Mapping of the acetylcholine binding site of the nicotinic acetylcholine receptor: ^{3}H-nicotine as an agonist photoaffinity label, *Biochemistry* **30**:6987–6997.

Miles, K., and Huganir, R., 1988, Regulation of nicotinic acetylcholine receptors by protein phosphorylation, *Mol. Neurobiol.* **2**:91–124.

Mitra, A., McCarthy, M., and Stroud, R., 1989, Three-dimensional structure of the nicotinic acetylcholine receptor and location of the major associated 43kD cytoskeletal protein, determined at 22 Å by low-dose electron microscopy and x-ray diffraction of 12.5 Å, *J. Cell Biol.* **109**:755–774.

Mulle, C., Vidal, C., Benoit, P., and Changeux, J. P., 1991, Existence of different subtypes of nicotinic acetylcholine receptors in the rat habenulo-interpeduncular system, *J. Neurosci.* **11**:2588–2597.

Mulle, C., Choquet, D., Korn, H., and Changeux, J. P., 1992, Calcium influx through nicotinic receptor in rat central neurons and its relevance to cellular regulation, *Neuron* **8**:135–143.

Nakayama, H., Shirase, M., Nakashima, T., Kurogochi, Y., and Lindstrom, J. M., 1990, Affinity purification of nicotinic acetylcholine receptor from rat brain, *Mol. Brain Res.* **7**:221–226.

Nakayama, H., Okuda, H., and Nakashima, T., 1993, Phosphorylation of rat brain nicotinic acetylcholine receptor by cAMP-dependent protein kinase in vitro, *Mol. Brain Res.* **20**:171–177.

Nef, P., Mauron, A., Stalder, R., Alliod, C., and Ballivet, M., 1984, Structure, linkage, and sequence of the two genes encoding the δ and γ subunits of the nicotinic acetylcholine receptor, *Proc. Natl. Acad. Sci. USA* **81**:7975–7979.

Nelson, S., Shelton, G., Lei, S., Lindstrom, J., and Conti-Tronconi, B., 1992, Epitope mapping of monoclonal antibodies to *Torpedo* acetylcholine receptor γ subunits, which specifically recognize the ε subunit of mammalian muscle acetylcholine receptor, *J. Neuroimmunol.* **36**:13–27.

Nobel, M., Brown, T., and Peakcock, J., 1978, Regulation of acetylcholine receptor levels by a cholinergic agonist in mouse muscle cell cultures, *Proc. Natl. Acad. Sci.* **75**:3488–3492.

Noda, M., Takahashi, H., Tanabe, T., Toyosato, M., Furutani, Y., Hirose, T., Asai, M., Inayama, S., Miyata, T., and Numa, S., 1982, Primary structure of α-subunit precursor of *Torpedo californica* acetylcholine receptor deduced from cDNA sequence, *Nature* **299**:793–797.

Noda, M., Furutani, Y., Takahashi, H., Toyosato, M., Tanabe, T., Shimizu, S., Kikyotani, S., Kayano, T., Hirose, T., Inayama, S., and Numa, S., 1983, Cloning and sequence analysis of calf cDNA and human genomic DNA encoding α subunit precursor of muscle acetylcholine receptor, *Nature* **305**:818–823.

Nooney, J., Lambert, J., and Chiappinelli, V., 1992, The interaction of κ-bungarotoxin with the nicotinic receptor of bovine chromaffin cells, *Brain Res.* **573**:77–82.

Ohno, K., Hutchinson, D., Milone, M., Brengman, J., Bouzat, C., Sine, S., and Engel, A., 1995, Congenital myasthenia syndrome caused by prolonged acetylcholine receptor channel openings due to a mutation in the M2 domain of the ε subunit, *Proc. Natl. Acad. Sci. USA* **92**:758–762.

O'Leary, M., and White, M., 1992, Mutational analysis of ligand-induced activation of the *Torpedo* acetylcholine receptor, *J. Biol. Chem.* **267**:8360–8365.

Palma, E., Bertrand, S., Binzoni, T., and Bertrand, D., 1995, Homomeric neuronal nicotinic α7 receptors present five putative high affinity binding sites for the toxin MLA, in press.

Papke, R., 1993, The kinetic properties of neuronal nicotinic receptor: Genetic basis of functional diversity, *Prog. Neurobiol.* **41:**509–531.

Patrick, J., and Lindstrom, J., 1973, Autoimmune response to acetylcholine receptor, *Science* **180:**871–872.

Patrick, J., and Stallcup, W., 1977, Immunological distinction between acetylcholine receptor and the α bungarotoxin binding component on sympathetic neurons, *Proc. Natl. Acad. Sci. USA* **74:**4689–4692.

Patrick, J., Lindstrom, J., Culp, B., and McMillan, J., 1973, Studies on purified eel acetylcholine receptor and anti-acetylcholine receptor antibody, *Proc. Natl. Acad. Sci. USA* **70:**3334–3338.

Pederson, S., and Cohen, J., 1990, D-Tubucurarine binding sites are located at α-γ and α-δ subunit interfaces of the nicotinic acetylcholine receptor, *Proc. Natl. Acad. Sci. USA* **87:**2785–2789.

Pederson, S., Bridgman, P., Sharp, S., Cohen, J., 1990, Identification of a cytoplasmic region of the *Torpedo* nicotinic acetylcholine receptor α subunit by epitope mapping, *J. Biol. Chem.* **265:**569–581.

Peng, X., Anand, R., Whiting, P., and Lindstrom, J., 1994a, Nicotine-induced upregulation of neuronal nicotinic receptors results from a decrease in the rate of turnover, *Mol. Pharmacol.* **46:**523–530.

Peng, X., Katz, M., Gerzanich, V., Anand, R., and Lindstrom, J., 1994b, Human α7 acetylcholine receptor: Cloning of the α7 subunit from the SH-SY5Y cell line and determination of pharmacological properties of native receptors and functional α7 homomers expressed in *Xenopus* oocytes, *Mol. Pharmacol.* **45:** 546–554.

Pereira, E., Alkondon, M., Reinhardt-Maelicke, S., Maelicke, A., Peng, X., Lindstrom, J., Whiting, P., and Albuquerque, E., 1994, Physostigmine and galanthamine reveal the presence of the novel binding site on the α4 β2 subtype of neuronal nicotinic acetylcholine receptor stably expressed in fibroblast cells, *J. Pharmacol. Exp. Ther.* **270:**768–778.

Peto, R., Lopez, A., Boreham, J., Thun, M., and Heath, C., 1992, Mortality from tobacco in developed countries: Indirect estimation from national vital statistics, *Lancet* **339:**1268–1278.

Picciotto, M., Zoll, M., Léna, C., Bessis, A., Lallemand, Y., LeNovére, N., Vincent, P., Pich, M., Brúlet, P., and Changeux, J. P., 1995, Abnormal avoidance learning in mice lacking functional high affinity nicotine receptor in the brain, *Nature* **374:**65–67.

Protti, M., Manfredi, A., Horton, R., Bellone, M., and Conti-Tronconi, B., 1993, Myasthenia gravis: Recognition of a human autoantigen at the molecular level, *Immunol. Today* **14:**363–368.

Pugh, P., and Berg, D., 1994, Neuronal acetylcholine receptors that bind α bungarotoxin mediate neurite retraction in a calcium-dependent manner, *J. Neurosci.* **14:**889–896.

Quick, M., 1995, Growth related role for the nicotinic α bungarotoxin receptor, in: *Effects of Nicotine on Biological Systems II* (P. Clarke *et al.*, eds.), Birkhäuser, Basel, pp. 145–150.

Raferty, M., Hunkapillar, M., Strader, C., and Hood, L., 1980, Acetylcholine receptor: Complex of homologous subunits, *Science* **208:**1454–1457.

Rathouz, M., and Berg, D., 1994, Synaptic-type acetylcholine receptors raise intracellular calcium levels by two mechanisms, *J. Neurosci.* **14:**6935–6945.

Ratnam, M., Le Nguyen, D., Rivier, J., Sargent, P. B., and Lindstrom, J., 1986a, Transmembrane topography of nicotinic acetylcholine receptor: Immunochemical tests contradict theoretical prediction based on hydrophobicity profiles, *Biochemistry* **25:**2633–2643.

Ratnam, M., Sargent, P., Sarin, V., Fox, J., Nguyen, D., Rivier, J., Criado, M., and Lindstrom, J., 1986b, Location of antigenic determinants on primary sequences of subunits of nicotinic acetylcholine receptor by peptide mapping, *Biochemistry* **25:**2621–2632.

Revah, F., Galzi, J. L., Giraudat, J., Haumont, P. Y., Lederer, F., and Changeux, J. P., 1990, The noncompetitive blocker ^{3}H-chlorpromazine labels three amino acids of the acetylcholine receptor γ subunit implications for the α helical organization of region MII and for the structure of the ion channel, *Proc. Natl. Acad. Sci. USA* **87:**4675–4679.

Revah, F., Bertrand, D., Galzi, J. L., Devillers-Thiery, A, Mulle, C., Hussy, N., Bertrand, S., Ballivet, M., and Changeux, J. P., 1991, Mutations in the channel domain alter desensitization of a neuronal nicotinic receptor, *Nature* **353:**846–849.

Rogers, S., Andrews, J., Gahring, L., Whisemand, T., Caulay, K., Crain, B., Hughes, T., Heinemann, S., and McNamara, J., 1994, Autoantibodies to glutamate receptor GluR3 in Rasmussen's encephalitis, *Science* **265:**648–651.

Role, L., 1992, Diversity in primary structure and function of neuronal nicotinic acetylcholine receptor channels, *Curr. Opin. Neurobiol.* **2**:254–262.

Saedi, M. S., Anand, R., Conroy, W. G., and Lindstrom, J., 1990, Determination of amino acids critical to the main immunogenic region of intact acetylcholine receptors by in vitro mutagenesis, *FEBS Lett.* **267**:55–59.

Saedi, M., Conroy, W. G., and Lindstrom, J., 1991, Assembly of *Torpedo* acetylcholine receptor in *Xenopus* oocytes, *J. Cell Biol.* **112**:1007–1015.

Sargent, P., 1993, The diversity of neuronal nicotinic acetylcholine receptors, *Annu. Rev. Neurosci.* **16**: 403–443.

Sargent, P., and Wilson, H., 1995, Distribution of nicotinic acetylcholine receptor subunit immunoreactivities on the surface of chick ciliary ganglion neurons, in: *Effects of Nicotine on Biological Systems II* (P. Clarke *et al.*, eds.), Birkhäuser, Basel, pp. 355–361.

Sargent, P., Hedges, B., Tsavaler, L., Clemmons, L., Tzartos, S., and Lindstrom, J., 1984, The structure and transmembrane nature of the acetylcholine receptor in amphibian skeletal muscles revealed by crossreacting monoclonal antibodies, *J. Cell Biol.* **98**:609–618.

Sargent, P., Pike, S., Nadel, D., and Lindstrom, J., 1989, Nicotinic acetylcholine receptor-like molecules in the retina, retinotectal pathway, and optic tectum of the frog, *J. Neurosci.* **9**:565–573.

Schoepfer, R., Whiting, P., Esch, F., Blacher, R., Shimasaki, S., and Lindstrom, J., 1988, cDNA clones coding for the structural subunit of a chicken brain nicotinic acetylcholine receptor, *Neuron* **1**:241–248.

Schoepfer, R., Halvorsen, S., Conroy, W. G., Whiting, P., and Lindstrom, J., 1989, Antisera against an α-3 fusion protein bind to ganglionic but not to brain nicotinic acetylcholine receptors, *FEBS Lett.* **257**: 393–399.

Schoepfer, R., Conroy, W. G., Whiting, P., Gore, M., and Lindstrom, J., 1990, Brain α-bungarotoxin binding protein cDNAs and mAbs reveal subtypes of this branch of the ligand-gated ion channel gene superfamily, *Neuron* **5**:35–48.

Schuller, H., 1995, Mechanisms of nicotine stimulated cell proliferation in normal and neoplastic neuroendocrine lung cells, in: *Effects of Nicotine on Biological Systems II* (P. Clarke *et al.*, eds.), Birkhäuser, Basel, pp. 151–158.

Schwartz, R., and Kellar, K., 1983, Nicotinic cholinergic receptor binding sites in the brain: Regulation in vivo, *Science* **220**:214–216.

Schwartz, R., and Kellar, K., 1985, In vivo regulation of [^{3}H] acetylcholine recognition sites in brain by nicotinic cholinergic drugs, *J. Neurochem.* **45**:427–433.

Seeburg, P., 1993, The molecular biology of mammalian glutamate receptor channels, *Trends Neurosci.* **16**:359–364.

Seguela, P., Wadiche, J., Dinelly-Miller, K., Dani, J., and Patrick, J., 1993, Molecular cloning, functional properties, and distribution of rat brain α7: A nicotinic cation channel highly permeable to calcium, *J. Neurosci.* **13**:596–604.

Siegel, H., and Lukas, R., 1988, Nicotinic agonists regulate α bungarotoxin binding sites of TE671 human medulloblastoma cells, *J. Neurochem.* **50**:1272–1278.

Silver, A., Shytle, R., Philipp, M., and Sanberg, P., 1995, Transdermal nicotine in Tourette's syndrome, in: *Effects of Nicotine on Biological Systems II* (P. Clarke *et al.*, eds.), Birkhäuser, Basel, pp. 293–299.

Simpson, J., 1960, Myasthenia gravis: A new hypothesis, *Scot. Med. J.* **5**:419–436.

Sine, S., 1988, Functional properties of human skeletal muscle acetylcholine receptors expressed by the TE671 cell line, *J. Biol. Chem.* **263**:18052–18062.

Sine, S., 1993, Molecular dissection of subunit interfaces in the acetylcholine receptor: Identification of residues that determine curare selectivity, *Proc. Natl. Acad. Sci. USA* **90**:9436–9440.

Sine, S., and Taylor, P., 1980, The relationship between agonist occupation and the permeability response of the cholinergic receptor revealed by bound cobra α-toxin, *J. Biol. Chem.* **255**:10144–10156.

Sine, S., and Taylor, P., 1982, Local anesthetics and histrionicotoxin are allosteric inhibitors of the acetylcholine receptor, *J. Biol. Chem.* **257**:8106–8114.

Sine, S., Claudio, T., and Sigworth, F., 1990, Activation of *Torpedo* acetylcholine receptors expressed in mouse fibroblasts, *J. Gen. Physiol.* **96**:395–437.

Smith, M., Stollberg, J., Lindstrom, J., and Berg, D. K., 1985, Characterization of a component in chick ciliary ganglia that cross-reacts with monoclonal antibodies to muscle and electric organ acetylcholine receptor, *J. Neurosci.* **5**:2726–2731.

Smith, M., Margiotta, J., Franco, A., Lindstrom, J., and Berg, D., 1986, Cholinergic modulation of an acetylcholine receptor-like antigen on the surface of chick ciliary ganglion neurons in cell culture, *J. Neurosci.* **6**:946–953.

Smith, M., Lindstrom, J., and Merlie, J. P., 1987, Formation of the α-bungarotoxin binding site and assembly of the nicotinic acetylcholine receptor subunits occur in the endoplasmic reticulum, *J. Biol. Chem.* **262**:4367–4376.

Sorenson, E., and Chiappinelli, V., 1992, Localization of ^{3}H-nicotine, ^{125}I-κappa-bungarotoxin, and ^{125}I-α-bungarotoxin binding to nicotinic sites in the chicken forebrain and midgrain, *J. Comp. Neurol.* **323**:1–12.

Spande, T., Carroffo, M., Edwards, M., Yeh, H., Panel, L., and Daly, J., 1992, Epibatidine: A novel (chloropyridyl) azabicyclo-heptane with potent analgesic activity from Ecuadoran poison frog, *J. Am. Chem. Soc.* **114**:3475–3478.

Stauffer, D., and Karlin, A., 1994, Electrostatic potential of the acetylcholine binding sites in the nicotinic receptor probed by reactions of binding site cysteines with charged methanethiosulfonates, *Biochemistry* **33**:6840–6849.

Steinlein, O., Mulley, J., Propping, P., Wallace, R., Phillips, H., Sutherland, G., Schafer, J., and Berkovic, S., 1995, A missense mutation in the neuronal nicotinic acetylcholine receptor α4 subunit is associated with autosomal dominant nocturnal frontal lobe epilepsy, *Nature Genetics* **11**:201–203.

Steinlein, O., Smigrodzki, R., Lindstrom, J., Anand, R., Kohler, M., Tocharoentanophol, C., and Vogel, F., 1994, Refinement of the localization of the gene for neuronal nicotinic acetylcholine receptor α4 subunit (CHRNA4) to human chromosome 20q 13.2–q13.3, *Genomics* **22**:493–495.

Stollberg, J., Whiting, P. J., Lindstrom, J., and Berg, D. K., 1986, Functional blockade of neuronal acetylcholine receptors by antisera to a putative receptor from brain, *Brain Res.* **378**:179–182.

Sudhof, T. 1995, The synaptic vesicle cycle: A cascade of protein–protein interactions, *Nature* **375**:645–653.

Sullivan, J., Decker, M., Brioni, J., Donnelly, Roberts, D., Anderson, D., Bannon, A., Kang, C., Adems, P., Piattoni-Kaplan, M., Buckley, M., Gopalakrishnan, M., Williams, M., and Arneric, S., 1994, (±) Epibatidine elicits a diversity of in vitro and in vivo effects mediated by nicotinic acetylcholine receptor, *J. Pharmacol. Exp. Ther.* **271**:624–663.

Sumikawa, K., and Gehle, V., 1992, Assembly of mutant subunits of the nicotinic acetylcholine receptor lacking the conserved disulfide loop structure, *J. Biol. Chem.* **267**:6286–6290.

Swanson, L., Lindstrom, J., Tzartos, S., Schmued, L., O'Leary, D., and Cowan, W., 1983, Immunohistochemical localization of monoclonal antibodies to the nicotinic acetylcholine receptor in the midbrain of the chick, *Proc. Natl. Acad. Sci. USA* **80**:4532–4536.

Swanson, L., Simmons, D., Whiting, P., and Lindstrom, J., 1987, Immunohistochemical localization of neuronal nicotinic receptors in the rodent central nervous system, *J. Neurosci.* **7**:3334–3342.

Takamori, M., Hamada, T., Komai, K., Takakashi, M., and Yoshida, A., 1994, Synaptotagmin can cause an immune-mediated model of Lambert-Eaton myasthenic syndrome in rats, *Ann. Neurol.* **35**:74–80.

Tobimatsu, T., Fujita, Y., Fukuda, K., Tanaka, K., Mori, Y., Konno, T., Mishina, M., and Numa, S., 1987, Effects of substitution of putative transmembrane segments on nicotinic acetylcholine receptor function, *FEBS Lett.* **222**:56–62.

Tomaselli, G., McLaughlin, J., Jurman, M., Hawrot, E., and Yellen, G., 1991, Mutations affecting agonist sensitivity of the nicotinic acetylcholine receptor, *Biophys. J.* **60**:721–727.

Treinin, M., and Chalfie, M., 1995, A mutated acetylcholine receptor subunit causes neuronal degeneration in *C. elegans*, *Neuron* **14**:871–877.

Twyman, R., Gahring, L., Spiess, J., and Rogers, S., 1995, Glutamate receptor antibodies activate a subset of receptors and reveal an agonist binding site, *Neuron* **14**:755–762.

Tzartos, S., and Lindstrom, J., 1980, Monoclonal antibodies used to probe acetylcholine receptor structure: Localization of the main immunogenic region and detection of similarities between subunits, *Proc. Natl. Acad. Sci. USA* **77**:755–759.

Tzartos, S., Rand, D., Einarson, B., and Lindstrom, J., 1981, Mapping of surface structures on *Electrophorus* acetylcholine receptor using monoclonal antibodies, *J. Biol. Chem.* **256**:8635–8645.

Tzartos, S., Seybold, M., and Lindstrom, J., 1982, Specificity of antibodies to acetylcholine receptors in sera from myasthenia gravis patients measured by monoclonal antibodies, *Proc. Natl. Acad. Sci. USA* **79**:188–192.

Tzartos, S., Hochschwender, S., Langeberg, L., and Lindstrom, J., 1983, Demonstration of a main immunogenic region on acetylcholine receptors from human muscle using monoclonal antibodies to human receptor, *FEBS Lett.* **158:**116–118.

Tzartos, S. J., Sophianos, D., and Efthimiadis, A., 1985, Role of the main immunogenic region of acetylcholine receptor in myasthenia gravis. An Fab monoclonal antibody protects against antigenic modulation by human sera, *J. Immunol.* **134:**2343–2349.

Tzartos, S., Langeberg, L., Hochschwender, S., Swanson, L. W., and Lindstrom, J., 1986, Characteristics of monoclonal antibodies to denatured *Torpedo* and to native calf acetylcholine receptors: Species, subunit and region specificity, *J. Neuroimmunol.* **10:**235–253.

Tzartos, S., Hochschwender, S., Vasquez, P., and Lindstrom, J., 1987, Passive transfer of experimental autoimmune myasthenia gravis by monoclonal antibodies to the main immunogenic region of the acetylcholine receptor, *J. Neuroimmunol.* **15:**185–194.

Tzartos, S., Barkas, T., Cung, M., Kordossi, A., Loutrari, H., Marraud, M., Papadouli, I., Sakarellos, C., Sophianos, D., and Tsikaris, V., 1991, The main immunogenic region of the acetylcholine receptor, structure and role in myasthenia gravis, *Autoimmunity* **8:**259–270.

Unwin, N., 1993a, Nicotinic acetylcholine receptor at 9Å resolution, *J. Mol. Biol.* **229:**1101–1124.

Unwin, N., 1993b, Neurotransmitter action: Opening of ligand-gated ion channels, *Cell* **10:**31–41.

Unwin, N., 1995, Acetylcholine receptor channel imaged in the open state, *Nature* **373:**37–43.

Unwin, N., Toyoshima, C., and Kubalek, E., 1988, Arrangement of the acetylcholine receptor subunits in the resting and desensitized states determined by cryoelectron microscopy of crystallized *Torpedo* postsynaptic membranes, *J. Cell Biol.* **107:**1123–1138.

Valera, S., Hussy, N., Evans, R., Adami, N., North, R., Surprenant, A., and Buell, G., 1994, A new class of ligand-gated ion channel defined by P_{2x} receptor for extracellular ATP, *Nature* **371:**516–519.

Vernallis, A., Conroy, W., and Berg, D., 1993, Neurons assemble acetylcholine receptors with as many as three kinds of subunits while maintaining subunit segregation among receptor subtypes, *Neuron* **10:**451–464.

Verrall, S., and Hall, Z., 1992, The N-terminal domains of acetylcholine receptor subunits contain recognition signals for the initial steps of receptor assembly, *Cell* **68:**23–31.

Vijayaraghavan, S., Schmid, H., Halvorsen, S., and Berg, D., 1990, Cyclic AMP-dependent phosphorylation of a neuronal acetylcholine receptor α type subunit, *J. Neurosci.* **10:**3255–3262.

Vijayaraghavan, S., Rathouz, M., Pugh, P., and Berg, D., 1992, Nicotinic receptors that bind α bungarotoxin on neurons raise intracellular free Ca^{++}, *Neuron* **8:**353–362.

Vincent, A., Lang, B., Newsom-Davis, J., 1989, Autoimmunity to the voltage-gated calcium channel underlies the Lambert-Eaton myasthenic syndrome, a paraneoplastic disorder, *Trends Neurosci.* **12:**496–502.

Wallace, B., Qu, Z., and Huganir, R., 1991, Agrin induces phosphorylation of the nicotinic acetylcholine receptor, *Neuron* **6:**869–878.

Watson, J., Adkins-Regan, E., Whiting, P., Lindstrom, J., and Podleski, T., 1988, Autoradiographic localization of nicotinic acetylcholine receptors in the brain of the zebra finch (*Poephila guttata*), *J. Comp. Neurol.* **274:**255–264.

Whitehouse, P., Martino, A., Marcus, K., Zweig, R., Singer, H., Price, D., and Kellar, K., 1988, Reduction in acetylcholine and nicotine binding in several degenerative diseases, *Arch. Neurol.* **45:**722–724.

Whiting, P., and Lindstrom, J., 1986a, Purification and characterization of a nicotinic acetylcholine receptor from chick brain, *Biochemistry* **25:**2082–2093.

Whiting, P., and Lindstrom, J., 1986b, Pharmacological properties of immunoisolated neuronal nicotinic receptors, *J. Neurosci.* **6:**3061–3069.

Whiting, P. J., and Lindstrom, J., 1987a, Purification and characterization of a nicotinic acetylcholine receptor from rat brain, *Proc. Natl. Acad. Sci. USA* **84:**595–599.

Whiting, P., and Lindstrom, J., 1987b, Affinity labeling of neuronal acetylcholine receptors localizes the neurotransmitter binding site to the β subunit, *FEBS Lett.* **213:**55–60.

Whiting, P., and Lindstrom, J., 1988, Characterization of bovine and human neuronal nicotinic acetylcholine receptors using monoclonal antibodies, *J. Neurosci.* **8:**3395–3404.

Whiting, P., Esch, F., Shimasaki, S., and Lindstrom, J., 1987a, Neuronal nicotinic acetylcholine receptor β-subunit is coded for by the cDNA clone α4, *FEBS Lett.* **219:**459–463.

Whiting, P., Liu, R., Morley, B., and Lindstrom, J., 1987b, Structurally different neuronal nicotinic acetylcholine receptor subtypes purified and characterized using monoclonal antibodies, *J. Neurosci.* **7:**4005–4016.

Whiting, P., Cooper, J., and Lindstrom, J., 1987c, Antibodies in sera from patients with myasthenia gravis do not bind to acetylcholine receptors from human brain, *J. Neuroimmunol.* **16:**205–213.

Whiting, P., Schoepfer, R., Swanson, L., Simmons, D., and Lindstrom, J., 1987d, Functional acetylcholine receptor in PC12 cells reacts with a monoclonal antibody to brain nicotinic receptors, *Nature* **327:** 515–518.

Whiting, P., Schoepfer, R., Conroy, W., Gore, M., Keyser, K., Shimasaki, S., Esch, F., and Lindstrom, J., 1991a, Differential expression of nicotinic acetylcholine receptor subtypes in brain and retina, *Mol. Brain Res.* **10:**61–70.

Whiting, P., Schoepfer, R., Lindstrom, J., and Priestly, T., 1991b, Structural and pharmacological characterization of the major brain nicotinic acetylcholine receptor subtype stably expressed in mouse fibroblasts, *Mol. Pharmacol.* **40:**463–472.

Winkler, J., Suhr, S., Gage, F., Thal, L., and Fisher, L., 1995, Essential role of neocortical acetylcholine in spatial memory, *Nature* **375:**484–487.

Witzemann, V., Barg, B., Nishikawa, Y., Sakmann, B., and Numa, S., 1987, Differential regulation of muscle acetylcholine receptor γ and ε subunit mRNAs, *FEBS Lett.* **223:**104–112.

Witzemann, V., Stein, E., Barg, B., Konno, T., Koenen, M., Kues, W., Criado, M., Hofmann, M., and Sakmmann, B., 1990, Primary structure and functional expression of α-, β-, γ-, δ-, and ε-subunits of the acetylcholine receptor from rat muscle, *Eur. J. Biochem.* **194:**437–448.

Wonnacott, S., 1990, The paradox of nicotinic acetylcholine receptor upregulation by nicotine, *Trends Pharmacol. Sci.* **11:**216–219.

Wonnacott, S., Drasdo, A., Sanderson, E., and Rowell, P. 1990, Presynaptic nicotinic receptors and modulation of transmitter release, in: *The Biology of Nicotine Dependence, Ciba Foundation Symposium* (G. Bock and J. Marsh, eds.), John Wiley and Sons, Chichester, pp. 87–105.

Wonnacott, S., Albuquerque, E., and Bertrand, D., 1993, Methylcaconitine: A new probe that discriminates between nicotinic receptor subclasses, *Methods Neurosci.* **12:**263–275.

Yang, C., Wu, W., and Zbuzek, V., 1992, Antinociceptive effect of chronic nicotine and nociceptive effect of its withdrawal measured by hot-plate and tail-flick in rats, *Psychopharmacology* **106:**417–420.

Yu, X., and Hall, Z., 1994, A sequence in the main cytoplasmic loop of the α subunit is required for assembly of mouse muscle nicotinic acetylcholine receptor, *Neuron* **13:**247–255.

Zhang, X., Gong, Z., Helstrom-Lindahl, E., and Nordberg, A., 1994, Regulation of $\alpha 4\beta 2$ nicotinic acetylcholine receptors in M10 cells following treatment with nicotinic agents, *Neuroreport* **6:**313–317.

Zoli, M., Lenovere, N., Hill, J., and Changeux, J. P., 1995, Developmental regulation of nicotinic ACh receptor mRNAs in the central and peripheral nervous systems, *J. Neurosci.* **15:**1912–1939.

INDEX

451